Applied Probability and Statistics

BAILEY · The Elements of Stochastic Processes with Applications to the Natural Sciences

BAILEY · Mathematics, Statistics and Systems for Health

BARTHOLOMEW · Stochastic Models for Social Processes, *Second Edition*

BECK and ARNOLD · Parameter Estimation in Engineering and Science

BENNETT and FRANKLIN · Statistical Analysis in Chemistry and the Chemical Industry

BHAT · Elements of Applied Stochastic Processes

BLOOMFIELD · Fourier Analysis of Time Series: An Introduction

BOX · R. A. Fisher, The Life of a Scientist

BOX and DRAPER · Evolutionary Operation: A Statistical Method for Process Improvement

BOX, HUNTER, and HUNTER · Statistics for Experimenters: An Introduction to Design, Data Analysis, and Model Building

BROWN and HOLLANDER · Statistics: A Biomedical Introduction

BROWNLEE · Statistical Theory and Methodology in Science and Engineering, *Second Edition*

BURY · Statistical Models in Applied Science

CHAMBERS · Computational Methods for Data Analysis

CHATTERJEE and PRICE · Regression Analysis by Example

CHERNOFF and MOSES · Elementary Decision Theory

CHOW · Analysis and Control of Dynamic Economic Systems

CLELLAND, deCANI, BROWN, BURSK, and MURRAY · Basic Statistics with Business Applications, *Second Edition*

COCHRAN · Sampling Techniques, *Third Edition*

COCHRAN and COX · Experimental Designs, *Second Edition*

COX · Planning of Experiments

COX and MILLER · The Theory of Stochastic Processes, *Second Edition*

DANIEL · Application of Statistics to Industrial Experimentation

DANIEL · Biostatistics: A Foundation for Analysis in the Health Sciences, *Second Edition*

DANIEL and WOOD · Fitting Equations to Data

DAVID · Order Statistics

DEMING · Sample Design in Business Research

DODGE and ROMIG · Sampling Inspection Tables, *Second Edition*

DRAPER and SMITH · Applied Regression Analysis

DUNN · Basic Statistics: A Primer for the Biomedical Sciences, *Second Edition*

DUNN and CLARK · Applied Statistics: Analysis of Variance and Regression

ELANDT-JOHNSON · Probability Models and Statistical Methods in Genetics

FLEISS · Statistical Methods for Rates and Proportions

GALAMBOS · The Asymptotic Theory of Extreme Order Statistics

GIBBONS, OLKIN, and SOBEL · Selecting and Ordering Popula- tions: A New Statistical Methodolo

GNANADESIKAN · Methods for St tivariate Observations

GOLDBERGER · Econometric Theo

GOLDSTEIN and DILLON · Discret

Introduction to
Statistical Time Series

A WILEY PUBLICATION IN APPLIED STATISTICS

INTRODUCTION TO STATISTICAL TIME SERIES

WAYNE A. FULLER

Iowa State University

JOHN WILEY & SONS

New York • Chichester • Brisbane • Toronto

To Evelyn

Library of Congress Cataloging in Publication Data

Fuller, Wayne A

Introduction to statistical time series.

(Wiley series in probability and mathematical statistics)

Bibliography: p.

Includes index.

1. Time-series analysis. 2. Regression analysis.

I. Title.

QA280.F84 519.5 76-6954

ISBN 0-471-28715-6

Printed in the United States of America

10 9 8 7 6 5 4

Preface

This textbook was developed from a course in time series given at Iowa State University. The classes were composed primarily of graduate students in economics and statistics. Prerequisites for the course were an introductory graduate course in the theory of statistics and a course in linear regression analysis. Since the students entering the course had varied backgrounds, chapters containing elementary results in Fourier analysis and large sample statistics, as well as a section on difference equations, were included in the presentation.

The theorem-proof format was followed because it offered a convenient method of organizing the material. No attempt was made to present the most general results available. Instead, the objective was to give results with practical content whose proofs were generally consistent with the prerequisites. Since many of the statistics students had completed advanced courses, a few theorems were presented at a level of mathematical sophistication beyond the prerequisites. Homework requiring application of the statistical methods was an integral part of the course.

By emphasizing the relationship of the techniques to regression analysis and using data sets of moderate size, most of the homework problems can be worked with any of a number of statistical packages. One such package is SAS (Statistical Analysis System, available through the Institute of Statistics, North Carolina State University). SAS contains a segment for periodogram computations that is particularly suited to this text. The system also contains a segment for regression with time series errors compatible with the presentation in Chapter 9. Another package is available from International Mathematical and Statistical Library, Inc.; this package has a chapter on time series programs.

There is some flexibility in the order in which the material can be covered. For example, the major portions of Chapters 1, 2, 5, 6, 8, and 9 can be treated in that order with little difficulty. Portions of the later chapters deal with spectral matters, but these are not central to the development of those chapters. The discussion of multivariate time series is positioned in separate sections so that it may be introduced at any point.

I thank A. R. Gallant for the proofs of several theorems and for the repair of others; J. J. Goebel for a careful reading of the manuscript that led to numerous substantive improvements and the removal of uncounted mistakes; and D. A. Dickey, M. Hidiroglou, R. J. Klemm, and G. H. K. Wang for computing examples and for proofreading. G. E. Battese, R. L. Carter, K. R. Crouse, J. D. Cryer, D. P. Hasza, J. D. Jobson, B. Macpherson, J. Mellon, D. A. Pierce and K. N. Wolter also read portions of the manuscript. I also thank my colleagues, R. Groeneveld, D. Isaacson, and O. Kempthorne, for useful comments and discussions. I am indebted to a seminar conducted by Marc Nerlove at Stanford University for the organization of some of the material on Fourier analysis and spectral theory. A portion of the research was supported by joint statistical agreements with the U. S. Bureau of the Census.

I thank Margaret Nichols for the repeated typings required to bring the manuscript to final form and Avonelle Jacobson for transforming much of the original illegible draft into typescript.

WAYNE A. FULLER

Ames, Iowa
February 1976

Contents

CHAPTER 1

Introduction

The analysis of time series applies to many fields. In economics the recorded history of the economy is often in the form of time series. Economic behavior is quantified in such time series as the consumer price index, unemployment, gross national product, population, and production. The natural sciences also furnish many examples of time series. The water level in a lake, the air temperature, the yields of corn in Iowa, and the size of natural populations are all collected over time. Growth models that arise naturally in biology and in the social sciences represent an entire field in themselves.

The mathematical theory of time series has been an area of considerable activity in recent years. Applications in the physical sciences such as the development of designs for airplanes and rockets, the improvement of radar and other electronic devices, and the investigation of certain production and chemical processes have resulted in considerable interest in time series analysis. This recent work should not disguise the fact that the analysis of time series is one of the oldest activities of scientific man.

A successful application of statistical methods to the real world requires a melding of statistical theory and knowledge of the material under study. We shall confine ourselves to the statistical treatment of moderately well-behaved time series, but we shall illustrate some techniques with real data.

1.1. PROBABILITY SPACES

When investigating outcomes of a game, an experiment, or some natural phenomenon, it is useful to have a representation for all possible outcomes. The individual outcomes denoted by ω are called *elementary events*. The set of all possible elementary events is called the *sure* event and is denoted by Ω. An example is the tossing of a die where we could take $\Omega=\{$one spot shows, two spots show,..., six spots show$\}$ or, more simply, $\Omega=\{1,2,3,4,5,6\}$.

Let A be a subset of Ω and let $\mathcal{Q}$ be a collection of such subsets. If we observe the outcome ω and ω is in A, we say that A has occurred. Intuitively, it is possible to specify $P(A)$, the probability that (or expected long-run frequency with which) A will occur. It is reasonable to require that the function $P(A)$ satisfy:

AXIOM 1. $P(A) \geqslant 0$ for every A in $\mathcal{Q}$.
AXIOM 2. $P(\Omega) = 1$.
AXIOM 3. If $A_1, A_2, \ldots$ is a countable sequence from $\mathcal{Q}$ and $A_i \cap A_j$ is the null set for all $i \neq j$, then $P(\cup_{i=1}^{\infty} A_i) = \Sigma_{i=1}^{\infty} P(A_i)$.

Using our die tossing example, if the die is fair we would take $P(A) = \frac{1}{6}$[the number of elementary events ω in A]. Thus $P(\{1,3,5\}) = \frac{1}{6} \times 3 = \frac{1}{2}$. It may be verified that Axioms 1 to 3 are satisfied for $\mathcal{Q}$ equal to $\mathcal{P}(\Omega)$ where $\mathcal{P}(\Omega)$ is the collection of all possible subsets of Ω.

Unfortunately, for technical mathematical reasons, it is not always possible to define $P(A)$ for all A in $\mathcal{P}(\Omega)$ and also to satisfy Axiom 3. To eliminate this difficulty, the class of subsets $\mathcal{Q}$ of Ω on which P is defined is restricted. The collection $\mathcal{Q}$ is required to satisfy:

1. If A is in $\mathcal{Q}$, then the complement A^c is also in $\mathcal{Q}$.
2. If $A_1, A_2, \ldots$ is a countable sequence from $\mathcal{Q}$, then $\cup_{i=1}^{\infty} A_i$ is in $\mathcal{Q}$.
3. The null set is in $\mathcal{Q}$.

A nonempty collection $\mathcal{Q}$ of subsets of Ω that satisfies conditions 1 to 3, is said to be a *sigma-algebra* or *sigma-field.*

We are now in a position to give a formal definition of a probability space. A *probability space*, represented by $(\Omega, \mathcal{Q}, P)$, is the sure event Ω together with a sigma-algebra $\mathcal{Q}$ of subsets of Ω and a function $P(A)$ defined on $\mathcal{Q}$ that satisfies Axioms 1 to 3.

For our purposes it is unnecessary to consider the subject of probability spaces in detail. In simple situations such as tossing a die Ω is easy to enumerate, and P satisfies Axioms 1 to 3 for $\mathcal{Q}$ equal to the set of all subsets of Ω.

Although it is conceptually possible to enumerate all possible outcomes of an experiment, it may be a practical impossibility to do so and, for most purposes, it is unnecessary to do so. It is usually enough to record the outcome by some function that assumes values on the real line. That is, we assign to each outcome ω a real number $X(\omega)$ and, if ω is observed, we record $X(\omega)$. In our die tossing example we could take $X(\omega) = 1$ if the player wins and -1 if the house wins.

Formally, a *random variable* X is a real valued function defined on Ω such that the set $\{\omega\colon X(\omega) \leqslant x\}$ is a member of $\mathcal{Q}$ for every real number x. The function $F_X(x) = P(\{\omega\colon X(\omega) \leqslant x\})$ is called the *distribution function* of the random variable X.

The reader who wishes to explore further the subjects of probability spaces, random variables, and distribution functions for stochastic processes may read Tucker (1967, pp. 1–33). The preceding brief introduction will suffice for our purposes.

1.2. TIME SERIES

Let $(\Omega, \mathcal{Q}, P)$ be a probability space and let T be an index set. A real valued *time series* (or *stochastic process*) is a real valued function $X(t,\omega)$ defined on $T \times \Omega$ such that for each fixed t, $X(t,\omega)$ is a random variable on $(\Omega, \mathcal{Q}, P)$. The function $X(t,\omega)$ is often written $X_t(\omega)$ or X_t, and a time series can be considered as a collection $\{X_t\colon t \in T\}$ of random variables.

For fixed ω, $X(t,\omega)$ is a real valued function of t. This function of t is called a *realization* or a *sample function*. If we look at a plot of some recorded time series such as gross national product, it is important to realize that conceptually we are looking at a plot of $X(t,\omega)$ with ω fixed. The collection of all possible realizations is called the *ensemble* of functions or the ensemble of realizations.

If the index set contains exactly one element, the stochastic process is a single random variable and we have defined the distribution function of the process. For stochastic processes with more than one random variable we need to consider the *joint distribution function*. The joint distribution function of a finite set of random variables $\{X_{t_1}, X_{t_2}, \ldots, X_{t_n}\}$ from the collection $\{X_t\colon t \in T\}$ is defined by

$$F_{X_{t_1}, X_{t_2}, \ldots, X_{t_n}}(x_{t_1}, x_{t_2}, \ldots, x_{t_n}) = P\{\omega\colon X(t_1,\omega) \leqslant x_{t_1}, \ldots, X(t_n,\omega) \leqslant x_{t_n}\}. \tag{1.2.1}$$

A time series is called *strictly stationary* if

$$F_{X_{t_1}, X_{t_2}, \ldots, X_{t_n}}(x_{t_1}, x_{t_2}, \ldots, x_{t_n}) = F_{X_{t_1+h}, X_{t_2+h}, \ldots, X_{t_n+h}}(x_{t_1}, x_{t_2}, \ldots, x_{t_n}),$$

where the equality must hold for all possible (nonempty finite distinct) sets of indices $t_1, t_2, \ldots, t_n$ and $t_1+h, t_2+h, \ldots, t_n+h$ in the index set and all $(x_{t_1}, x_{t_2}, \ldots, x_{t_n})$ in the range of the random variable X_t. Note that the indices $t_1, t_2, \ldots, t_n$ are not necessarily consecutive. If a time series is strictly stationary we see that the distribution function of the random variable is the same at every point in the index set. Furthermore, the joint distribution depends only on the distance between the elements in the index set, and not on their actual values. Naturally this does not mean a particular realization will appear the same as another realization.

If $\{X_t: t \in T\}$ is a strictly stationary time series with $E\{|X_t|\} < \infty$, then the expected value of X_t is a constant for all t, since the distribution function is the same for all t. Likewise, if $E\{X_t^2\} < \infty$, then the variance of X_t is a constant for all t.

A time series is defined completely in a probabilistic sense if one knows the cumulative distribution (1.2.1) for any finite set of random variables $(X_{t_1}, X_{t_2}, \ldots, X_{t_n})$. However, in most applications, the form of the distribution function is not known. A great deal can be accomplished, however, by dealing only with the first two moments of the time series. In line with this approach we define a time series to be *weakly stationary* if:

1. The expected value of X_t is a constant for all t.
2. The covariance matrix of $(X_{t_1}, X_{t_2}, \ldots, X_{t_n})$ is the same as the covariance matrix of $(X_{t_1+h}, X_{t_2+h}, \ldots, X_{t_n+h})$ for all nonempty finite sets of indices $(t_1, t_2, \ldots, t_n)$ and all h such that $t_1, t_2, \ldots, t_n$, $t_1+h, t_2+h, \ldots, t_n+h$ are contained in the index set.

As before, $t_1, t_2, \ldots, t_n$ are not necessarily consecutive members of the index set. Also, since the expected value of X_t is a constant, it may conveniently be taken as 0. The covariance matrix, by definition, is a function only of the distance between observations. That is, the covariance of X_{t+h} and X_t depends only on the distance, h, and we may write

$$\text{Cov}\{X_t, X_{t+h}\} = E\{X_t X_{t+h}\} = \gamma(h),$$

where $E\{X_t\}$ has been taken to be zero. The function $\gamma(h)$ is called the *autocovariance* of X_t. When there is no danger of confusion, we shall abbreviate the expression to *covariance*.

The terms *stationary in the wide sense, covariance stationary, second order stationary*, and *stationary* are also used to describe a weakly stationary time series. It follows from the definitions that a strictly stationary process with the first two moments finite is also weakly stationary. However, a strictly stationary time series may not possess finite moments and hence may not be covariance stationary.

Many time series as they occur in practice are not stationary. For example, the economies of many countries are developing or growing. Therefore, the typical economic indicators will be showing a "trend" through time. This trend may be either in the mean, the variance, or both. Such nonstationary time series are sometimes called evolutionary. A good portion of the practical analysis of time series is connected with the transformation of an evolving time series into a stationary time series. In later sections we shall consider several of the procedures used in this connection. The reader will find himself familiar with many of these techniques, since they are closely related to least squares and regression.

1.3. EXAMPLES OF STOCHASTIC PROCESSES

Example 1. Let the index set be $T=\{1,2\}$ and let the space of outcomes be the possible outcomes associated with tossing two dice, one at "time" $t=1$ and one at time $t=2$. Then,

$$\Omega=\{1,2,3,4,5,6\}\times\{1,2,3,4,5,6\}.$$

Define

$$X(t,\omega)=t+[\text{value on die } t]^2.$$

Therefore, for a particular ω, say $\omega_3=(1,3)$, the realization or sample function would be $(1+1,2+9)=(2,11)$. In this case, both Ω and T are finite, and we are able to determine that there are 36 possible realizations.

Example 2. One of the most important time series is a sequence of uncorrelated random variables, say $\{e_t: t\in(0,\pm1,\pm2,\ldots)\}$, each with zero mean and finite variance, $\sigma^2>0$. We say that e_t is a sequence of uncorrelated $(0,\sigma^2)$ random variables.

This time series is sometimes called *white noise*. Note that the index set is the set of all integers. The set Ω is determined by the range of the random variables. Let us assume that the e_t are normally distributed and therefore have range $(-\infty,\infty)$. Then $\omega\in\Omega$ is a real valued infinite dimensional vector with an element associated with each integer. The covariance function of $\{e_t\}$ is given by

$$\begin{aligned}\gamma_e(h)&=\sigma^2, && h=0\\ &=0, && \text{otherwise.}\end{aligned}$$

Because of the importance of this time series, we shall reserve the symbol e_t for a sequence of uncorrelated random variables with positive finite variance. On occasion we shall further assume that the variables are independent and perhaps of a specified distribution. These additional assumptions will be stated when used.

In a similar manner the most commonly used index set will be the set of all integers. If the index set is not stated the reader may assume that it is the set of all integers.

Example 3. Consider the time series

$$X_t=\hat{\beta}_1+\hat{\beta}_2 t, \qquad t\in[0,1],$$

where $\hat{\boldsymbol{\beta}}=(\hat{\beta}_1,\hat{\beta}_2)'$ is distributed as a bivariate normal random variable

with mean $\boldsymbol{\beta}=(\beta_1,\beta_2)'$ and covariance matrix

$$\begin{pmatrix} \sigma_{11} & \sigma_{12} \\ \sigma_{12} & \sigma_{22} \end{pmatrix}.$$

Any realization of this process yields a function continuous on the interval [0, 1], and the process therefore is called *continuous*. Such a process might represent the outcome of an experiment to estimate the linear response to an input variable measured on the interval [0, 1]. The $\hat{\beta}$'s are then the estimated regression coefficients. The set Ω is the two dimensional space of real numbers. Each experiment conducted to obtain a set of estimates of $\hat{\beta}_1$ and $\hat{\beta}_2$ constitutes a realization of the process.

The mean function of X_t is given by

$$E\{X_t\}=E\{\hat{\beta}_1+\hat{\beta}_2 t\}=\beta_1+\beta_2 t$$

and the covariance function by

$$\begin{aligned} &E\{[X_t-E\{X_t\}][X_{t+h}-E\{X_{t+h}\}]\} \\ &\quad=E\{[(\hat{\beta}_1-\beta_1)+(\hat{\beta}_2-\beta_2)t][(\hat{\beta}_1-\beta_1)+(\hat{\beta}_2-\beta_2)(t+h)]\} \\ &\quad=\sigma_{11}+\sigma_{12}(t+h)+\sigma_{12}t+\sigma_{22}t(t+h), \qquad t,t+h\in[0,1]. \end{aligned}$$

It is clear that this process is not stationary.

Example 4. For the reader familiar only with the study of multivariate statistics, the idea of a random variable that is both continuous and random in time may require a moment of reflection. On the other hand, such processes occur in nature. For example, the water level of a lake or river can be plotted as a continuous function of time. Furthermore, such a plot might appear so smooth as to support the idea of a derivative or instantaneous rate of change in level.

Consider such a process, $\{X_t : t\in(-\infty,\infty)\}$, and let the covariance of the process be given by

$$\gamma(h)=E\{(X_t-\mu)(X_{t+h}-\mu)\}=Ae^{-\alpha h^2}, \qquad \alpha>0.$$

Thus, for example, if time is measured in hours and the river level at 7:00 A.M. is reported daily, then the covariance between daily reports is $Ae^{-576\alpha}$. Likewise, the covariance between the change from 7:00 A.M. to 8:00 A.M. and the change from 8:00 A.M. to 9:00 A.M. is given by

$$E\{(X_{t+1}-X_t)(X_{t+2}-X_{t+1})\}=-A[1-2e^{-\alpha}+e^{-4\alpha}].$$

Note that the variance of the change from t to $t+h$ is

$$\operatorname{Var}\{X_{t+h}-X_t\}=2A[1-e^{-\alpha h^2}]$$

and

$$\lim_{h\to 0}\operatorname{Var}\{X_{t+h}-X_t\}=0.$$

A process with this property is called *mean square continuous*.

We might define the rate of change per unit of time by

$$R_t(h)=\frac{X_{t+h}-X_t}{h}.$$

For a fixed $h>0$, $R_t(h)$ is a well-defined random variable and

$$\operatorname{Var}\{R_t(h)\}=\frac{2A[1-e^{-\alpha h^2}]}{h^2}.$$

Furthermore, by L'Hospital's rule,

$$\begin{aligned}\lim_{h\to 0}\operatorname{Var}\{R_t(h)\} &= \lim_{h\to 0}\frac{2A[2\alpha h e^{-\alpha h^2}]}{2h}\\ &=2A\alpha.\end{aligned}$$

Stochastic processes for which this limit exists are called *mean square differentiable*.

1.4. PROPERTIES OF THE AUTOCOVARIANCE AND AUTOCORRELATION FUNCTIONS

To compare the basic properties of time series, it is often useful to have a function that is not influenced by the units of measurement. To this end, we define the *autocorrelation function* of a stationary time series by

$$\rho(h)=\frac{\gamma(h)}{\gamma(0)}.$$

Thus the autocorrelation function is the autocovariance function normalized to be one at $h=0$. As with the autocovariance function, we shall abbreviate the expression to correlation function when no confusion will result.

The autocovariance and autocorrelation functions of stationary time series possess several distinguishing characteristics. Recall that a function $f(x)$ defined for $x \in \chi$ is said to be *positive semidefinite* if it satisfies

$$\sum_{j=1}^{n} \sum_{k=1}^{n} a_j a_k f(t_j - t_k) \geqslant 0 \tag{1.4.1}$$

for any set of real numbers $(a_1, a_2, \ldots, a_n)$ and for any $(t_1, t_2, \ldots, t_n)$ such that $t_i - t_j$ is in χ for all (i,j). We then have:

Theorem 1.4.1. The covariance function of the stationary time series $\{X_t : t \in T\}$ is positive semidefinite.

Proof. Without loss of generality, let $E\{X_t\}=0$. Let $(t_1, t_2, \ldots, t_n) \in T$, let $(a_1, a_2, \ldots, a_n)$ be any set of real numbers, and let $\gamma(t_j - t_k)$ be the covariance between X_{t_j} and X_{t_k}. We know that the variance of a random variable, when it is defined, is nonnegative. Therefore,

$$0 \leqslant \operatorname{Var}\left\{ \sum_{j=1}^{n} a_j X_{t_j} \right\} = E\left\{ \sum_{j=1}^{n} \sum_{k=1}^{n} a_j a_k X_{t_j} X_{t_k} \right\}$$

$$= \sum_{j=1}^{n} \sum_{k=1}^{n} a_j a_k \gamma(t_j - t_k). \qquad \blacktriangle$$

If we set $n=2$ in (1.4.1), we have

$$0 \leqslant a_1^2 \gamma(0) + a_2^2 \gamma(0) + 2a_1 a_2 \gamma(t_1 - t_2),$$

which implies

$$\tfrac{1}{2}(a_1^2 + a_2^2) \geqslant -a_1 a_2 \frac{\gamma(t_1 - t_2)}{\gamma(0)} .$$

For $t_1 - t_2 = h$ we set $-a_1 = a_2 = 1$ and then set $a_1 = a_2 = 1$ to obtain the well-known property of correlations:

$$|\rho(h)| \leqslant 1. \tag{1.4.2}$$

The concepts of *even* and *odd* functions (about zero) will prove useful in our study. Cos t is an even function. The use of this description is apparent when one notes the symmetry of cos t on the interval $[-\pi, \pi]$. Similarly, sin t is an *odd function*. In general, an even function, $f(t)$, defined on a domain T, is a function that satisfies $f(t) = f(-t)$ for all t and $-t$ in T. An odd function, $g(t)$, is a function satisfying $g(t) = -g(-t)$ for all t and

$-t$ in T. Many simple properties of even and odd functions follow immediately. For example, if $f(t)$ is an even function and $g(t)$ is an odd function, where both are integrable on the interval $[-A,A]$, then, for $0 \leq b \leq A$,

$$\int_{-b}^{b} g(t)\,dt = 0,$$

$$\int_{-b}^{b} f(t)\,dt = 2\int_{0}^{b} f(t)\,dt.$$

As an exercise, the reader may verify that the product of two even functions is even, the product of two odd functions is even, the product of an odd and an even function is odd, the sum of even functions is even, and the sum of odd functions is odd.

Theorem 1.4.2. The covariance function of a real valued stationary time series is an even function of h. That is, $\gamma(h) = \gamma(-h)$.

Proof. We assume, without loss of generality, that $E\{X_t\} = 0$. By stationarity,

$$E\{X_t X_{t+h}\} = \gamma(h)$$

for all t and $t+h$ contained in the index set. Therefore, if we set $t_0 = t_1 - h$,

$$\gamma(h) = E\{X_{t_0} X_{t_0+h}\} = E\{X_{t_1-h} X_{t_1}\} = \gamma(-h). \qquad \blacktriangle$$

Given this theorem, we shall often evaluate $\gamma(h)$ for real valued time series for nonnegative h only. Should we fail to so specify, the reader may always safely substitute $|h|$ for h in a covariance function.

In the study of statistical distribution functions the characteristic function of a distribution function is defined by

$$\varphi(h) = \int e^{\iota xh}\, dG(x),$$

where the integral is a Lebesgue-Stieltjes integral, $G(x)$ is the distribution function, and $e^{\iota xh}$ is the complex exponential defined by $e^{\iota xh} = \cos xh + \iota \sin xh$. It is readily established that the function $\varphi(h)$ satisfies[1]

1. $\varphi(0) = 1$.
2. $|\varphi(h)| \leq 1$ for all $h \in (-\infty, \infty)$.
3. $\varphi(h)$ is uniformly continuous on $(-\infty, \infty)$.

[1] See, for example, Tucker (1967, p. 42 ff.) or Gnedenko (1967, p. 266 ff.)

It can be shown that a continuous function $\rho(h)$ with $\rho(0)=1$ is a characteristic function if and only if it is positive semidefinite.[2] It follows that:

Theorem 1.4.3. If the function $\rho(h)$ is the correlation function of a mean square continuous stationary time series with index set $T=(-\infty,\infty)$, then it is representable in the form

$$\rho(h)=\int_{-\infty}^{\infty} e^{ixh}\,dG(x),$$

where $G(x)$ is a distribution function, and the integral is a Lebesgue-Stieltjes integral.

For the index set $T=\{0,\pm 1,\pm 2,\dots\}$, the corresponding theorem is:

Theorem 1.4.4. If the function $\rho(h)$ is the correlation function of a stationary time series with index set $T=\{0,\pm 1,\pm 2,\dots\}$, then it is representable in the form

$$\rho(h)=\int_{-\pi}^{\pi} e^{ixh}\,dG(x),$$

where $G(x)$ is a distribution function.

This theorem will be discussed in Chapter 3.

1.5. COMPLEX VALUED TIME SERIES

Occasionally it is advantageous, from a theoretical point of view, to consider complex valued time series. Letting X_t and Y_t be two real valued time series, we define the complex valued time series, Z_t, by

$$Z_t=X_t+iY_t. \tag{1.5.1}$$

The expected value of Z_t is given by

$$E\{Z_t\}=E\{X_t\}+iE\{Y_t\} \tag{1.5.2}$$

and we note that

$$E^*\{Z_t\}=E\{Z_t^*\}=E\{X_t\}-iE\{Y_t\}, \tag{1.5.3}$$

where the symbol "*" is used to denote the complex conjugate.

[2] See Gnedenko (1967, p. 290 and p. 387) for relevant theorems.

The covariance of Z_t and Z_{t+h} is defined as

$$\begin{aligned}\operatorname{Cov}\{Z_t, Z_{t+h}\} &= E\{(Z_t - E\{Z_t\})(Z^*_{t+h} - E\{Z^*_{t+h}\})\}\\ &= \operatorname{Cov}\{X_t, X_{t+h}\} + i\operatorname{Cov}\{Y_t, X_{t+h}\}\\ &\quad - i\operatorname{Cov}\{X_t, Y_{t+h}\} + \operatorname{Cov}\{Y_t, Y_{t+h}\}. \qquad (1.5.4)\end{aligned}$$

Note that the variance of a complex valued process is always real and nonnegative, since it is the sum of the variances of two real valued random variables.

The definitions of stationarity for complex time series are completely analogous to those for real time series. Thus, a complex time series, Z_t, is weakly stationary if the expected value of Z_t is a constant for all t and the covariance matrix of $(Z_{t_1}, Z_{t_2}, \ldots, Z_{t_n})$ is the same as the covariance matrix of $(Z_{t_1+h}, Z_{t_2+h}, \ldots, Z_{t_n+h})$, where all indices are contained in the index set.

From (1.5.4), the autocovariance of a stationary complex time series Z_t with zero mean is given by

$$\begin{aligned}\gamma_Z(h) &= [E\{X_t X_{t+h}\} + E\{Y_t Y_{t+h}\}] + i[E\{Y_t X_{t+h}\} - E\{X_t Y_{t+h}\}]\\ &\overset{\text{(say)}}{=} g_1(h) + ig_2(h). \qquad (1.5.5)\end{aligned}$$

We see that $g_1(h)$ is a symmetric or even function of h, and $g_2(h)$ is an odd function of h. By the definition of the complex conjugate, we have

$$\gamma^*_Z(h) = \gamma_Z(-h). \qquad (1.5.6)$$

Therefore, the autocovariance function of a complex time series is *skew symmetric*, where (1.5.6) is the definition of a skew symmetric function.

A complex valued function, $\gamma(\cdot)$, defined on the integers, is positive semidefinite if

$$\sum_{j=1}^{n} \sum_{k=1}^{n} v_j v^*_k \gamma(t_j - t_k) \geqslant 0 \qquad (1.5.7)$$

for any set of n complex numbers $(v_1, v_2, \ldots, v_n)$ and any integers $(t_1, t_2, \ldots, t_n)$. Thus, as in the real valued case, we have:

Theorem 1.5.1. The covariance function of a stationary complex valued time series is positive semidefinite.

It follows from Theorem 1.5.1 that the correlation inequality holds for

complex random variables; that is,

$$\rho_Z(h)\rho_Z^*(h)=\frac{\gamma_Z(h)\gamma_Z^*(h)}{[\gamma_Z(0)]^2}\leqslant 1.$$

In the sequel, if we use the simple term "time series," the reader may assume that we are speaking of a real valued time series. All complex valued time series will be identified as such.

1.6. PERIODIC FUNCTIONS AND PERIODIC TIME SERIES

Periodic functions play an important role in the analysis of empirical time series. We define a function $f(t)$ with domain T to be *periodic* if there exists an $H>0$ such that for all t, $t+H\in T$,

$$f(t+H)=f(t),$$

where H is the *period* of the function. That is, the function $f(t)$ takes on all of its possible values in an interval of length H. For any positive integer k, kH is also a period of the function. While the examples of perfectly periodic functions are rare, there are situations where observed time series may be decomposed into the sum of two time series, one of which is periodic or nearly periodic. Even casual observation of many economic time series will disclose seasonal variation wherein peaks and troughs occur at approximately the same month each year. Seasonal variation is apparent in many natural time series, such as the water level in a lake, daily temperature, wind speeds and velocity, and the levels of the tides. Many of these time series also display regular daily variation that has the appearance of rough "cycles."

The trigonometric functions have traditionally been used to approximate periodic behavior. A function obeying the sine-cosine type periodicity is completely specified by three parameters. Thus, we write

$$f(t)=A\sin(\lambda t+\varphi), \tag{1.6.1}$$

where the *amplitude* is the absolute value of A, λ is the *frequency*, and φ is the *phase angle*. The frequency is the number of times the function repeats itself in a period of length 2π. The phase angle is a "shift parameter" in that it determines the points ($-\lambda^{-1}\varphi$ plus integer multiples of π) where the function is zero. A parameterization that is more useful in the estimation of such a function can be constructed by using the trigonometric identity

$$\sin(\lambda t+\varphi)=\sin\lambda t\cos\varphi+\cos\lambda t\sin\varphi.$$

Thus (1.6.1) becomes

$$f(t)=B_1\cos\lambda t+B_2\sin\lambda t, \tag{1.6.2}$$

where $B_1=A\sin\varphi$ and $B_2=A\cos\varphi$.

Let us consider a simple type of time series whose realizations will display perfect periodicity. Define $\{X_t\colon t\in(0,\pm1,\pm2,\ldots)\}$ by

$$X_t=e_1\cos\lambda t+e_2\sin\lambda t, \tag{1.6.3}$$

where e_1 and e_2 are independent drawings from a normal $(0,1)$ population. Note that the realization is completely determined by the two random variables e_1 and e_2. The amplitude and phase angle vary from realization to realization, but the period is the same for every realization.

The stochastic properties of this time series are easily derived:

$$\begin{aligned}E\{X_t\}&=E\{e_1\cos\lambda t+e_2\sin\lambda t\}\\&=0;\end{aligned}$$

$$\begin{aligned}\gamma(h)&=E\{(e_1\cos\lambda t+e_2\sin\lambda t)(e_1\cos\lambda(t+h)+e_2\sin\lambda(t+h))\}\\&=E\{e_1^2\cos\lambda t\cos\lambda(t+h)+e_2^2\sin\lambda t\sin\lambda(t+h)\\&\qquad+e_1e_2\cos\lambda t\sin\lambda(t+h)+e_2e_1\sin\lambda t\cos\lambda(t+h)\}\\&=\cos\lambda t\cos\lambda(t+h)+\sin\lambda t\sin\lambda(t+h)\\&=\cos\lambda h.\end{aligned} \tag{1.6.4}$$

We see that the process is stationary. This example also serves to emphasize the fact that the covariance function is obtained by averaging the product X_tX_{t+h} over realizations.

It is clear that any time series defined by the finite sum

$$X_t=\sum_{j=1}^{M}e_{1j}\cos\lambda_j t+\sum_{j=1}^{M}e_{2j}\sin\lambda_j t, \tag{1.6.5}$$

where e_{1j} and e_{2j} are independent drawings from a normal $(0,\sigma_j^2)$ population and $\lambda_j\neq\lambda_k$ for $k\neq j$, will display periodic behavior. The representation (1.6.5) is a useful approximation for portions of some empirical time series, and we shall see that ideas associated with this representation are important to the theoretical study of time series as well.

1.7. VECTOR VALUED TIME SERIES

Most of the concepts associated with univariate time series have immediate generalizations to vector valued time series. We now introduce representations for multivariate processes.

The k-dimensional time series $\{\mathbf{X}_t: t=0, \pm 1, \pm 2, \ldots\}$ is defined by

$$\mathbf{X}_t = \begin{bmatrix} X_{1t} \\ X_{2t} \\ \vdots \\ X_{kt} \end{bmatrix}, \tag{1.7.1}$$

where $\{X_{it}: t=0, \pm 1, \pm 2, \ldots\}$, $i=1,2,\ldots,k$, are scalar time series. The expected value of $\mathbf{X}_t$ is

$$E\{\mathbf{X}_t\} = \begin{bmatrix} E\{X_{1t}\} \\ E\{X_{2t}\} \\ \vdots \\ E\{X_{kt}\} \end{bmatrix}.$$

Assuming the mean is zero, we define the covariance matrix of $\mathbf{X}_t$ and $\mathbf{X}_{t+h}$ by

$$E\{\mathbf{X}_t \mathbf{X}'_{t+h}\} = \begin{bmatrix} E\{X_{1t}X_{1,t+h}\} & E\{X_{1t}X_{2,t+h}\} & \cdots & E\{X_{1t}X_{k,t+h}\} \\ E\{X_{2t}X_{1,t+h}\} & E\{X_{2t}X_{2,t+h}\} & \cdots & E\{X_{2t}X_{k,t+h}\} \\ \vdots & \vdots & & \vdots \\ E\{X_{kt}X_{1,t+h}\} & E\{X_{kt}X_{2,t+h}\} & \cdots & E\{X_{kt}X_{k,t+h}\} \end{bmatrix}. \tag{1.7.2}$$

As with scalar time series, we define $\mathbf{X}_t$ to be covariance stationary if:

1. The expected value of $\mathbf{X}_t$ is a constant function of time.
2. The covariance matrix of $\mathbf{X}_t$ and $\mathbf{X}_{t+h}$ is the same as the covariance matrix of $\mathbf{X}_j$ and $\mathbf{X}_{j+h}$ for all t, $t+h$, j, $j+h$ in the index set.

If a vector time series is stationary, then every component scalar time series is stationary. However, a vector of scalar stationary time series is not necessarily a vector stationary time series.

The second stationarity condition means that we can express the covariance matrix as a function of h only, and, assuming $E\{\mathbf{X}_t\}=0$, we

write

$$\boldsymbol{\Gamma}(h) = E\{\mathbf{X}_t \mathbf{X}'_{t+h}\}$$

for stationary time series. Note that the diagonal elements of this matrix are the autocovariances of the X_{jt}. The off-diagonal elements are the *cross covariances* of X_{it} and X_{jt}. The element $E\{X_{it}X_{j,t+h}\}$ is not necessarily equal to $E\{X_{i,t+h}X_{jt}\}$ and, hence, $\boldsymbol{\Gamma}(h)$ is not necessarily equal to $\boldsymbol{\Gamma}(-h)$. For example, let

$$\begin{aligned} X_{1t} &= e_t \\ X_{2t} &= e_t + \beta e_{t-1}. \end{aligned} \tag{1.7.3}$$

Then

$$\begin{aligned} \gamma_{12}(1) &= E\{X_{1t}X_{2,t+1}\} = \beta\sigma^2 \\ \gamma_{12}(-1) &= E\{X_{1,t+1}X_{2t}\} = 0. \end{aligned}$$

However,

$$\begin{aligned} \gamma_{21}(1) &= E\{X_{2t}X_{1,t+1}\} = 0 = \gamma_{12}(-1) \\ \gamma_{21}(-1) &= E\{X_{2t}X_{1,t-1}\} = \beta\sigma^2 = \gamma_{12}(1) \end{aligned}$$

and it is clear that

$$\boldsymbol{\Gamma}(h) = \boldsymbol{\Gamma}'(-h). \tag{1.7.4}$$

It is easy to verify that (1.7.4) holds for all vector stationary time series, and we state the result as a lemma.

Lemma 1.7.1. The autocovariance matrix of a vector stationary time series satisfies $\boldsymbol{\Gamma}(h) = \boldsymbol{\Gamma}'(-h)$.

The positive semidefinite property of the scalar autocovariance function is maintained essentially unchanged for vector processes.

Lemma 1.7.2. The covariance function of a vector stationary time series $\{\mathbf{X}_t : t \in T\}$ is a positive semidefinite function in that

$$\sum_{j=1}^{n} \sum_{i=1}^{n} \mathbf{a}'_j \boldsymbol{\Gamma}(t_j - t_i)\mathbf{a}_i \geqslant 0$$

for any set of real vectors $\{\mathbf{a}_1, \mathbf{a}_2, \ldots, \mathbf{a}_n\}$ and any set of indices $\{t_1, t_2, \ldots, t_n\} \in T$.

Proof. The result can be obtained by evaluating the variance of $\sum_{j=1}^{n} \mathbf{a}_j'\mathbf{X}_{t_j}$. ▲

We define the *correlation matrix* of $\mathbf{X}_t$ and $\mathbf{X}_{t+h}$ by

$$\mathbf{P}(h) = \mathbf{D}_0^{-1}\boldsymbol{\Gamma}(h)\mathbf{D}_0^{-1}, \tag{1.7.5}$$

where $\mathbf{D}_0$ is a diagonal matrix with the square root of the variances of the X_{jt} as diagonal elements; that is,

$$\mathbf{D}_0^2 = \operatorname{diag}\{\gamma_{11}(0), \gamma_{22}(0), \ldots, \gamma_{kk}(0)\}.$$

The ijth element of $\mathbf{P}(h)$, $\rho_{ij}(h)$, is called the *cross correlation* of X_{it} and X_{jt}.

For the time series of (1.7.3) we have

$$\mathbf{P}(h) = \begin{cases} \begin{bmatrix} 1 & (1+\beta^2)^{-1/2} \\ (1+\beta^2)^{-1/2} & 1 \end{bmatrix}, & h = 0 \\ \begin{bmatrix} 0 & (1+\beta^2)^{-1/2}\beta \\ 0 & (1+\beta^2)^{-1}\beta \end{bmatrix}, & h = 1 \\ \begin{pmatrix} 0 & 0 \\ 0 & 0 \end{pmatrix}, & |h| > 1. \end{cases}$$

REFERENCES

Chung (1968), Gnedenko (1967), Loève (1963), Rao (1965a), Tucker (1967), Yaglom (1962).

EXERCISES

1. Determine the mean and covariance function for Example 1 of Section 3.
2. Discuss the stationarity of the following time series:

(a) $\{X_t: t \in (0, \pm 1, \pm 2, \ldots)\}$ = Value of a randomly chosen observation from a normal distribution with mean $\frac{1}{2}$ and variance $\frac{1}{4}$ } t odd

= 1 if toss of a true coin results in a head
= 0 if toss of a true coin results in a tail } t even;

(b) $\{X_t: t \in (0, \pm 1, \pm 2, \ldots)\}$ is a time series of independent identically distributed random variables whose distribution function is that of Student's t-distribution with one degree of freedom;

(c) $X_t = e_0, \quad t=0$
$\quad = \rho X_{t-1} + e_t, \quad t = 1,2,\ldots,$
where $|\rho|<1$ and the e_t are independent identically distributed $(0,1)$ random variables.

3. Which of the following processes is covariance stationary for $T = \{t: t \in (0,1,2,\ldots)\}$, where the e_t are independent identically distributed $(0,1)$ random variables and a_1, a_2 are fixed real numbers?

(a) $e_1 + e_2 \cos t$.
(b) $e_1 + e_2 \cos t + e_3 \sin t$.
(c) $a_1 + e_1 \cos t$.
(d) $a_1 + e_1 \cos t + e_2 \sin t$.
(e) $e_t + a_1 \cos t$.
(f) $a_1 + e_1 a_2^t + e_2, \quad 0 < a_2 < 1$.

4. (a) Prove the following theorem.

Theorem. If X_t and Y_t are independent covariance stationary time series, then $aX_t + bY_t$, where a and b are real numbers, is a covariance stationary time series.

(b) Let $Z_t = X_t + Y_t$. Give an example to show that Z_t covariance stationary does not imply X_t is covariance stationary.

5. Give a covariance stationary time series such that $\rho(h) \neq 0$ for all h.

6. Which of the following functions is the covariance function of a stationary time series? Explain why or why not.

(a) $g(h) = 1 + |h|, \quad h = 0, \pm 1, \pm 2, \ldots;$
(b) $g(h) = 1, \quad h = 0$
$\quad = -\frac{1}{2}, \quad h = \pm 1$
$\quad = 0, \quad$ otherwise;
(c) $g(h) = 1 + \frac{1}{4}\sin 4h, \quad h = 0, \pm 1, \pm 2, \ldots;$
(d) $g(h) = 1 + \frac{1}{4}\cos 4h, \quad h = 0, \pm 1, \pm 2, \ldots;$
(e) $g(h) = 1, \quad h = 0, \pm 1$
$\quad = 0, \quad$ otherwise.

7. Let e_t be a time series of normal independent $(0, \sigma^2)$ random variables. Define

$$
\begin{aligned}
Y_t &= e_t, && \text{if } t \text{ is odd} \\
&= -e_t, && \text{if } t \text{ is even.}
\end{aligned}
$$

Is the complex valued time series

$$Z_t = e_t + \iota Y_t$$

stationary?

CHAPTER 2

Moving Average and Autoregressive Processes

In any study, one of the first undertakings is to describe the item under investigation. Typically there is no unique description and, as the discipline develops, alternatives appear that prove more useful in some applications.

The joint distribution function (1.2.1) gives a complete description of the distributional properties of a time series. The covariance function provides a less complete description, but one from which useful conclusions can be drawn. As we have noted, the correlation function has functional properties analogous to those of the characteristic function of a statistical distribution function. Therefore, the distribution function associated with the correlation function provides an alternative representation for a time series. This representation will be introduced in Chapter 3 and discussed in Chapter 4.

Another method of describing a time series is to express the time series as a function of more elementary time series. A sequence of independent identically distributed random variables is a very simple type of time series. In applications it is often possible to express the observed time series as a relatively simple function of this elementary time series. In particular, many time series have been represented as linear combinations of uncorrelated $(0, \sigma^2)$ or of independent $(0, \sigma^2)$ random variables. We now consider such representations.

2.1. MOVING AVERAGE PROCESSES

We shall call the time series $\{X_t: t \in (0, \pm 1, \pm 2, \ldots)\}$, defined by

$$X_t = \sum_{j=-M}^{M} \alpha_j e_{t-j}, \tag{2.1.1}$$

where M is a nonnegative integer, α_i are real numbers, $\alpha_{-M} \neq 0$, and the e_t are uncorrelated $(0, \sigma^2)$ random variables, a *finite moving average time series* or a *finite moving average process*.

If there exists no finite M such that $\alpha_j = 0$ for all j with absolute value greater than M, then $\{X_t\}$ is an *infinite moving average process* and is given the representation

$$X_t = \sum_{j=-\infty}^{\infty} \alpha_j e_{t-j}. \tag{2.1.2}$$

The time series defined by

$$X_t = \sum_{j=0}^{M} \alpha_j e_{t-j}, \tag{2.1.3}$$

where $\alpha_0 \neq 0$ and $\alpha_M \neq 0$, is called a *one-sided moving average* of *order M*. Note that the number of terms in the sum on the right-hand side of (2.1.3) is $M+1$. In particular, (2.1.3) is a *left-moving average*, since nonzero weights are applied only to e_t and to e's with indexes less than t. We shall most frequently use the one-sided representation of moving average time series. Note that there would be no loss of generality in considering only the one-sided representation. If in the representation (2.1.1) we define the random variable ϵ_t by $\epsilon_t = e_{t+M}$, we have, for (2.1.1),

$$\begin{aligned} X_t &= \sum_{j=-M}^{M} \alpha_j e_{t-j} = \sum_{s=0}^{2M} \alpha_{s-M} \epsilon_{t-s} \\ &= \sum_{s=0}^{2M} \beta_s \epsilon_{t-s}, \end{aligned} \tag{2.1.4}$$

where $\beta_s = \alpha_{s-M}$.

Likewise we can, without loss of generality, take β_0 in (2.1.4) to be one. If β_0 is not one we simply define

$$u_{t-s} = \beta_0 \epsilon_{t-s}$$

$$\delta_s = \beta_s \beta_0^{-1}, \qquad s = 0, 1, 2, \ldots, 2M,$$

and obtain the representation

$$X_t = \sum_{s=0}^{2M} \delta_s u_{t-s}, \tag{2.1.5}$$

where $\delta_0 = 1$ and the u_t are uncorrelated $(0, \beta_0^2\sigma^2)$ random variables.

The covariance function of an Mth order moving average is distinctive in that it is zero for all h satisfying $|h| > M$. For the time series (2.1.3) we have

$$E\{X_t\} = 0,$$

$$\gamma_X(h) = E\{X_t X_{t+h}\} = \sum_{i=0}^{M-|h|} \alpha_i \alpha_{i+|h|}\sigma^2, \qquad 0 \leq |h| \leq M$$

$$= 0, \qquad |h| > M. \tag{2.1.6}$$

As an example, consider the simple time series

$$X_t = \sum_{i=0}^{4} e_{t-i}.$$

Clearly,

$$E\{X_t\} = \sum_{i=0}^{4} E\{e_{t-i}\} = 0, \quad \text{for all } t,$$

and

$$E\{X_t X_{t+h}\} = E\left\{\left(\sum_{i=0}^{4} e_{t-i}\right)\left(\sum_{i=0}^{4} e_{t+h-i}\right)\right\}$$

$$= \begin{cases} (5-|h|)\sigma^2, & |h| \leq 4 \\ 0, & \text{otherwise.} \end{cases}$$

By (2.1.6) only the first autocorrelation of a first order moving average is nonzero, while all autocorrelations of order greater than two for a second order moving average are zero. For a qth order moving average, $\rho(q) \neq 0$, and all higher order autocorrelations are zero. There are further restrictions on the correlation function of moving average time series. Consider the first order moving average

$$X_t = e_t + \alpha e_{t-1}. \tag{2.1.7}$$

Then

$$\rho(1) = \frac{\alpha}{1+\alpha^2}. \tag{2.1.8}$$

Using elementary calculus, it is easy to prove that $\rho(1)$ achieves a maximum of 0.5 for $\alpha=1$ and a minimum of -0.5 for $\alpha=-1$.

Thus, the first autocorrelation of a first order moving average process will always fall in the interval $[-0.5, 0.5]$. For any ρ in $(0, 0.5)$ [or in $(-0.5, 0)$] there are two α values yielding such a ρ, one in the interval $(0, 1)$ [or in $(-1, 0)$] and one in the interval $(1, \infty)$ [or $(-\infty, -1)$].

In Figure 2.1.1 we display 100 observations from a realization of the time series

$$X_t = e_t + 0.7e_{t-1},$$

where the e_t are normal independent $(0, 1)$ random variables. The positive correlation between adjacent observations is clear to the eye and is emphasized by the plot of X_t against X_{t-1} of Figure 2.1.2. The correlation between X_t and X_{t-1} is $(1.49)^{-1}(0.7) \doteq 0.4698$, and this is the slope of the line plotted on the figure. On the other hand, since the e_t are independent, X_t is independent of X_{t-2} and this is clear from the plot of X_t against X_{t-2} in Figure 2.1.3.

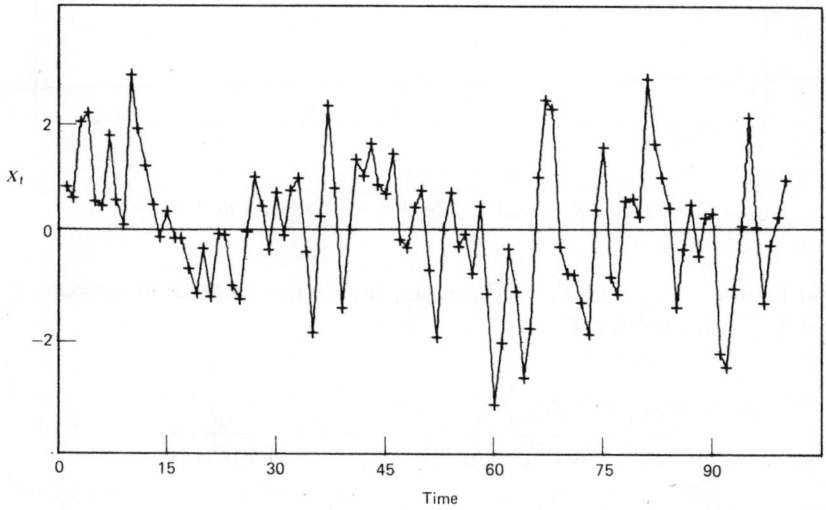

Figure 2.1.1. One hundred observations from the time series $X_t = e_t + 0.7e_{t-1}$.

While the simple correlation between X_t and X_{t-2} is zero, the *partial correlation* between X_t and X_{t-2} adjusted for X_{t-1} is not zero. The correlation between X_t and X_{t-1} is $\rho(1) = (1+\alpha^2)^{-1}\alpha$, which is the same as

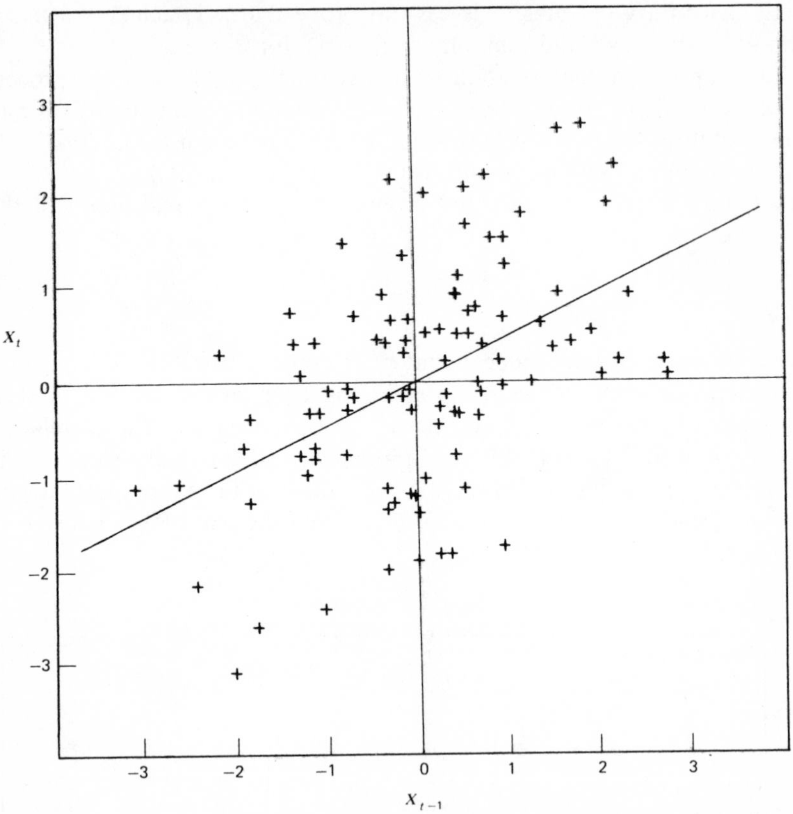

Figure 2.1.2. Plot of X_t against X_{t-1} for 100 observations from $X_t = e_t + 0.7e_{t-1}$.

that between X_{t-1} and X_{t-2}. Therefore, the partial correlation between X_t and X_{t-2} adjusted for X_{t-1} is

$$\frac{\operatorname{Cov}\{X_t - cX_{t-1}, X_{t-2} - cX_{t-1}\}}{\operatorname{Var}\{X_t - cX_{t-1}\}} = -\frac{\alpha^2}{1+\alpha^2+\alpha^4}$$

where $c = (1+\alpha^2)^{-1}\alpha$.

In Figure 2.1.4, $X_t - (0.7)(1.49)^{-1}X_{t-1}$ is plotted against $X_{t-2} - (0.7)(1.49)^{-1}X_{t-1}$ for the time series of Figure 2.1.1. For this time series the theoretical partial autocorrelation is -0.283, and this is the slope of the line plotted in Figure 2.1.4.

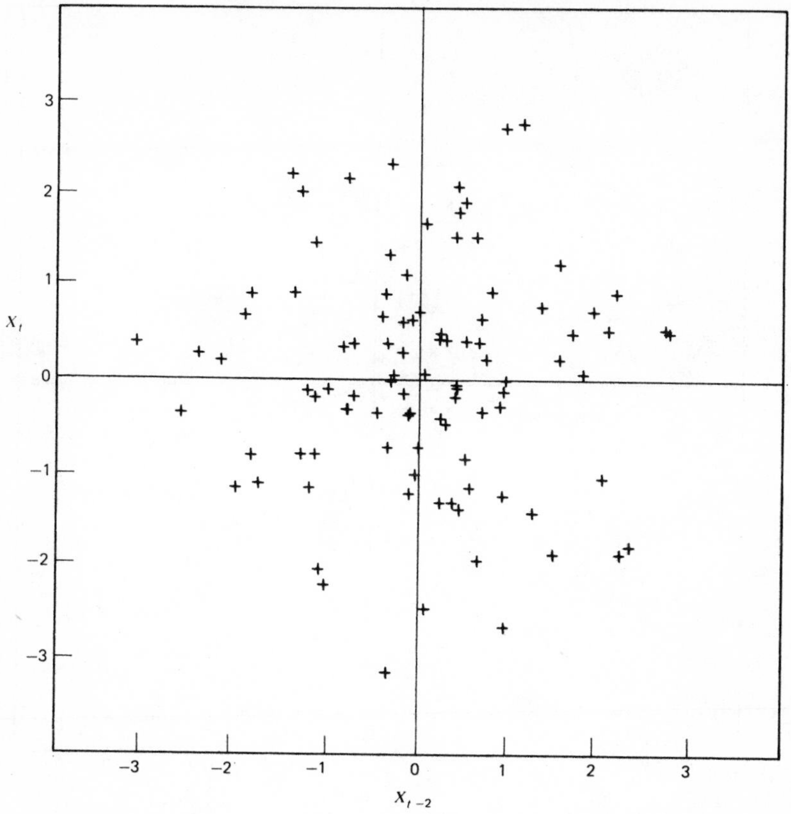

Figure 2.1.3. Plot of X_t against X_{t-2} for 100 observations from $X_t = e_t + 0.7e_{t-1}$.

One hundred observations from the second order moving average

$$X_t = e_t - 1.40e_{t-1} + 0.48e_{t-2}$$

are displayed in Figure 2.1.5. The correlation function of this process is

$$\begin{aligned} \rho(h) &\doteq -0.65, & h &= 1, \\ &\doteq \;\; 0.15, & h &= 2, \\ &= \;\; 0\,, & h &\geqslant 3. \end{aligned}$$

Once again the nature of the correlation can be observed from the plot.

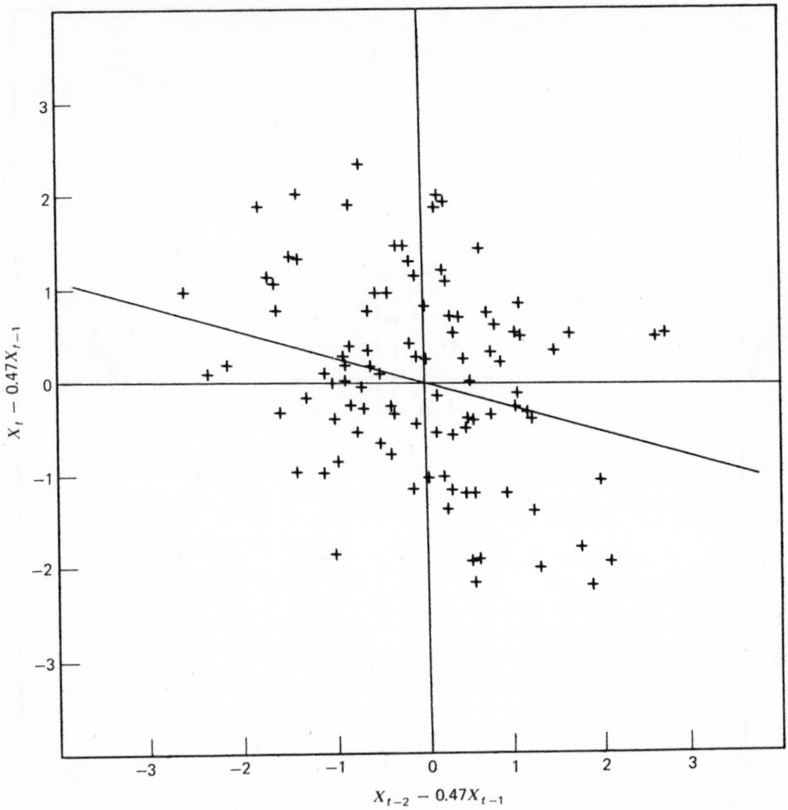

Figure 2.1.4. Plot of $X_t - 0.47X_{t-1}$ against $X_{t-2} - 0.47X_{t-1}$ for 100 observations from $X_t = e_t + 0.7e_{t-1}$.

Adjacent observations are often on opposite sides of the mean, indicating that the first order autocorrelation is negative. This correlation is also apparent in Figure 2.1.6, where X_t is plotted against X_{t-1}.

At the beginning of this section we defined a moving average time series to be a linear combination of a sequence of uncorrelated random variables. It is also common to speak of forming a moving average of an arbitrary time series, X_t. Thus, Y_t defined by

$$Y_t = \sum_{i=-M}^{M} \alpha_i X_{t+i}$$

is a moving average of the time series X_t. The reader may prove:

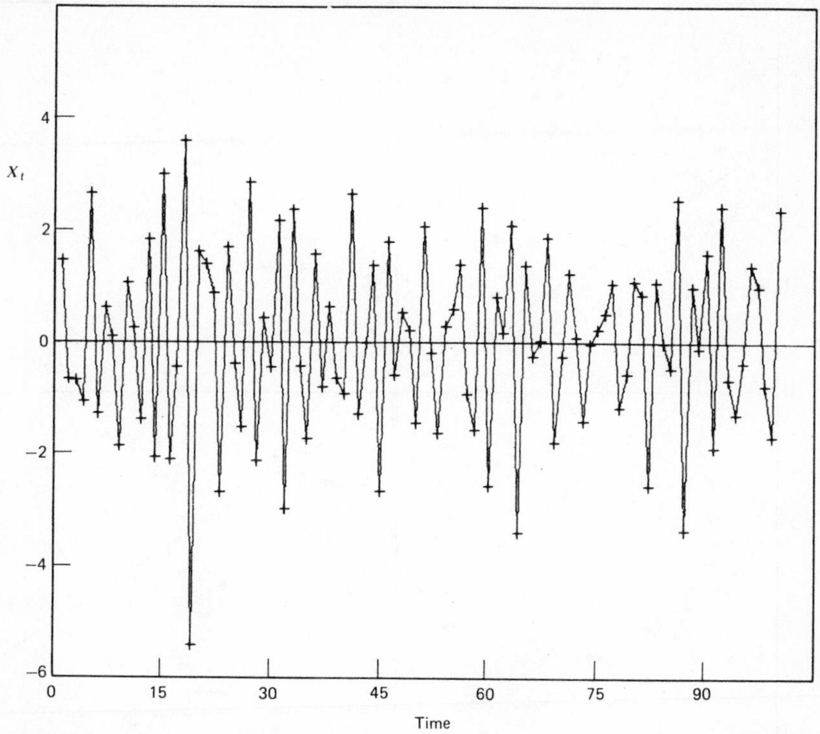

Figure 2.1.5. One hundred observations from the time series $X_t = e_t - 1.40e_{t-1} + 0.48e_{t-2}$.

Theorem 2.1.1. If $\{X_t : t \in (0, \pm 1, \pm 2, \ldots)\}$ is a stationary time series with mean zero and covariance function $\gamma_X(h)$, then

$$Y_t = \sum_{i=-M}^{M} \alpha_i X_{t+i},$$

where M is finite and the α_i are real numbers, is a stationary time series with mean zero and covariance function

$$\gamma_Y(h) = \sum_{i=-M}^{M} \sum_{j=-M}^{M} \alpha_i \alpha_j \gamma_X(h+j-i).$$

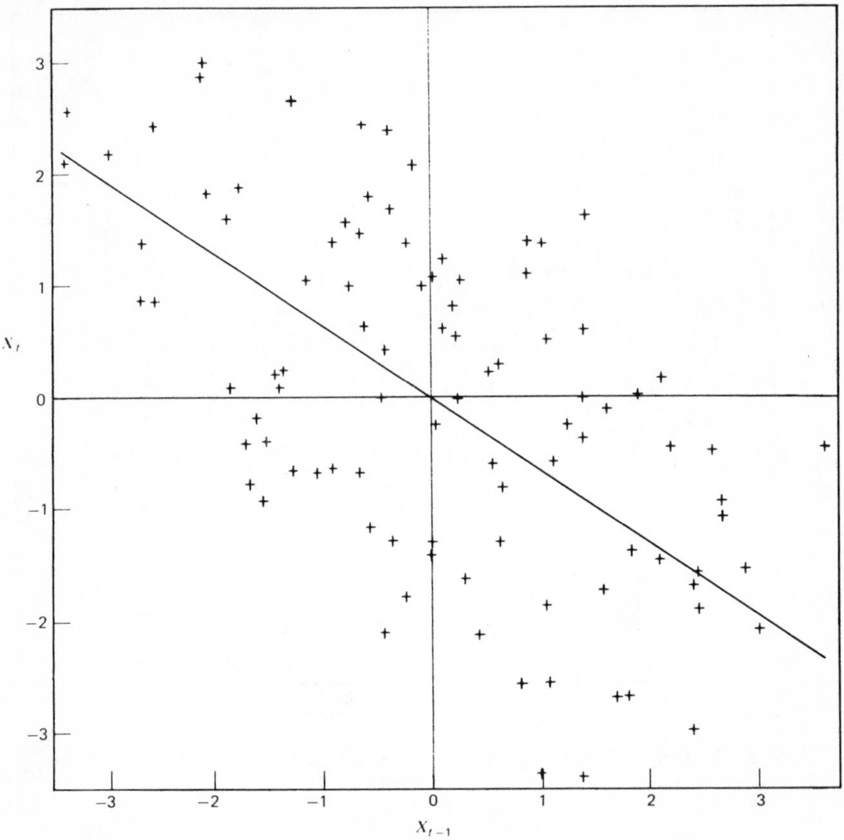

Figure 2.1.6. Plot of X_t against X_{t-1} for 100 observations from $X_t = e_t - 1.40e_{t-1} + 0.48e_{t-2}$.

2.2. ABSOLUTELY SUMMABLE SEQUENCES AND INFINITE MOVING AVERAGES

Before investigating the properties of infinite moving average time series, we review some results on the convergence of partial sums of infinite sequences of real numbers.

Recall that an *infinite sequence* is a function whose domain is the set $T_1 = (1, 2, \ldots)$ of natural numbers. A *doubly infinite sequence* is a function whose domain is the set $T_2 = (0, \pm 1, \pm 2, \ldots)$ of all integers. When no confusion will result, we shall use the word sequence for a function whose domain is either T_1 or T_2. We denote the infinite sequence by $\{a_j\}_{j=1}^{\infty}$ and

the doubly infinite sequence by $\{a_j\}_{j=-\infty}^{\infty}$. When the domain is clear, the notation will be abbreviated to $\{a_j\}$ or perhaps to a_j.

The *infinite series* created from the sequence $\{a_j\}_{j=1}^{\infty}$ is the sequence of partial sums $\{s_j\}_{j=1}^{\infty}$ defined by

$$s_j = \sum_{i=1}^{j} a_i, \qquad j = 1, 2, \ldots.$$

The symbols $\sum_{j=1}^{\infty} a_j$ and $\Sigma\{a_j\}$ are often used to represent both the infinite sequence $\{s_j\}_{j=1}^{\infty}$ generated by the sequence $\{a_j\}$ and the limit

$$\lim_{n\to\infty} \sum_{j=1}^{n} a_j = \lim_{n\to\infty} s_n$$

when the limit is defined. If the sequence $\{s_j\}$ has a finite limit we say that the series $\Sigma\{a_j\}$ is *convergent*. If the limit is not defined the series is *divergent*.

If the limit

$$\lim_{n\to\infty} \sum_{j=1}^{n} |a_j|$$

is finite, the series $\Sigma\{a_j\}$ is said to be *absolutely convergent*, and we write $\sum_{j=1}^{\infty} |a_j| < \infty$. We also describe this situation by saying the sequence $\{a_j\}$ is *absolutely summable*. For a doubly infinite sequence, if the limit

$$\lim_{n\to\infty} \sum_{j=-n}^{n} |a_j|$$

exists and is finite, then the series is absolutely convergent, and we write

$$\sum_{j=-\infty}^{\infty} |a_j| < \infty,$$

and say that the sequence $\{a_j\}$ is absolutely summable.

The following elementary properties of absolutely summable sequences will be useful in our investigation of moving average time series.

1. If $\{a_j\}$ is absolutely summable, then

$$\sum_{j=-\infty}^{\infty} a_j^2 < \infty.$$

The reader can verify that the converse is not true by considering $\Sigma_{j=1}^{\infty} j^{-2}$ and $\Sigma_{j=1}^{\infty} j^{-1}$.

2. Given two absolutely summable sequences $\{a_j\}$ and $\{b_j\}$, then the sequences $\{a_j+b_j\}$ and $\{a_j b_j\}$ are absolutely summable:

$$\sum_{j=-\infty}^{\infty} |a_j+b_j| \leqslant \sum_{j=-\infty}^{\infty} (|a_j|+|b_j|) = \sum_{j=-\infty}^{\infty} |a_j| + \sum_{j=-\infty}^{\infty} |b_j| < \infty;$$

$$\sum_{j=-\infty}^{\infty} |a_j b_j| = \sum_{j=-\infty}^{\infty} (|a_j||b_j|) \leqslant \sum_{j=-\infty}^{\infty} \frac{1}{2}(|a_j|+|b_j|)^2 < \infty.$$

Of course, $\Sigma_{j=-\infty}^{\infty}(|a_j|+|b_j|)^2 < \infty$, since $\{(|a_j|+|b_j|)^2\}$ is the square of an absolutely summable sequence.

3. The *convolution* of two absolutely summable sequences $\{a_j\}$ and $\{b_j\}$, defined by

$$c_j = \sum_{k=-\infty}^{\infty} a_k b_{j-k},$$

is absolutely summable:

$$\sum_{j=-\infty}^{\infty} |c_j| \leqslant \sum_{j=-\infty}^{\infty} \sum_{k=-\infty}^{\infty} |a_k||b_{j-k}| \leqslant \sum_{k=-\infty}^{\infty} |a_k| \sum_{s=-\infty}^{\infty} |b_s| < \infty.$$

This result generalizes and, for example, if we define

$$d_j = \sum_{k=-\infty}^{\infty} a_k \sum_{m=-\infty}^{\infty} b_m c_{j-k-m},$$

where $\{a_j\}$, $\{b_j\}$, and $\{c_j\}$ are absolutely summable, then

$$\sum_{j=-\infty}^{\infty} |d_j| < \infty.$$

The infinite moving average time series

$$X_t = \sum_{j=-\infty}^{\infty} \alpha_j e_{t-j}, \qquad t=0, \pm 1, \pm 2, \ldots,$$

was introduced in equation (2.1.2) of Section 2.1. In investigating the behavior of infinite moving averages we will repeatedly interchange the summation and expectation operations. As justification for this procedure, we give the following theorems. The theorems may be skipped by those not familiar with real analysis.

Theorem 2.2.1. Let the sequence of real numbers $\{a_j\}$ and the sequence of random variables $\{Z_t\}$ satisfy

$$\sum_{j=-\infty}^{\infty} |a_j| < \infty$$

and

$$E\{Z_t^2\} \leqslant K, \qquad t = 0, \pm 1, \pm 2, \ldots,$$

for some finite K. Then there exists a sequence of random variables $\{X_t\}$ such that for $t = 0, \pm 1, \pm 2, \ldots$,

$$E\left\{\left|X_t - \sum_{j=-n}^{n} a_j Z_{t-j}\right|^2\right\} \to 0 \quad \text{as} \quad n \to \infty,$$

$$\sum_{j=-n}^{n} a_j Z_{t-j}$$

converges in probability to X_t, and

$$E\{X_t^2\} < \infty.$$

Moreover, the sequence of random variables $\{X_t\}$ is uniquely determined except on an event of probability zero. If, in addition, $\{Z_t\}$ is a sequence of independent random variables, then

$$\sum_{j=-n}^{n} a_j Z_{t-j} \to X_t \quad \text{almost surely.}$$

Proof. Since $(|Z_t| - |Z_j|)^2 \geqslant 0$, then $E\{|Z_t||Z_j|\} \leqslant K$ for all t, j. Let $\epsilon > 0$ be given and choose N such that $[\sum_{j=N}^{\infty}(|a_j| + |a_{-j}|)]^2 K < \epsilon$. Fix t and let

$n>m>N$. Then,

$$
\begin{aligned}
E\left\{\left|\sum_{j=-n}^{n} a_j Z_{t-j}-\sum_{j=-m}^{m} a_j Z_{t-j}\right|^2\right\}
&=E\left\{\left|\sum_{j=-n}^{-m-1} a_j Z_{t-j}+\sum_{s=m+1}^{n} a_s Z_{t-s}\right|^2\right\}\\
&\leqslant E\left\{\left|\sum_{j=-n}^{-m-1} a_j Z_{t-j}\right|^2\right\}+2E\left\{\left|\sum_{j=-n}^{-m-1} a_j Z_{t-j}\sum_{s=m+1}^{n} a_s Z_{t-s}\right|\right\}\\
&\quad+E\left\{\left|\sum_{s=m+1}^{n} a_s Z_{t-s}\right|^2\right\}\\
&\leqslant K\left(\sum_{j=-n}^{-m-1}|a_j|\right)^2+2K\left(\sum_{j=-n}^{-m-1}\sum_{s=m+1}^{n}|a_j||a_s|\right)+K\left(\sum_{s=m+1}^{n}|a_s|\right)^2\\
&\leqslant\left[\sum_{j=N}^{\infty}(|a_j|+|a_{-j}|)\right]^2 K\\
&<\epsilon.
\end{aligned}
$$

Thus, the sequence $\{\sum_{j=-n}^{n} a_j Z_{t-j}\}$ is Cauchy in squared mean, and there is a random variable X_t such that $E\{X_t^2\}<\infty$ and

$$
E\left\{\left|X_t-\sum_{j=-n}^{n} a_j Z_{t-j}\right|^2\right\}\to 0 \quad \text{as} \quad n\to\infty
$$

(Tucker, 1967, pp. 105, 106). Convergence in squared mean implies convergence in probability; therefore,

$$
\sum_{j=-n}^{n} a_j Z_{t-j} \xrightarrow{P} X_t.
$$

(See definitions 5.1.3 and 5.1.8.) By letting the index t assume the values $t=0,\pm 1,\pm 2,\ldots,$ we generate the sequence $\{X_t\}_{t=-\infty}^{\infty}$ of random variables.

In order to show that $\{X_t\}$ is determined uniquely except on an event of probability zero, consider a sequence of random variables $\{Y_t\}$ with the

properties of $\{X_t\}$. Let $\epsilon>0$ and fix t. Then,

$$E\{(X_t-Y_t)^2\}=E\left\{\left(X_t-\sum_{j=-n}^{n}a_jZ_{t-j}+\sum_{j=-n}^{n}a_jZ_{t-j}-Y_t\right)^2\right\}$$

$$\leqslant\left\{\left[E\left|X_t-\sum_{j=-n}^{n}a_jZ_{t-j}\right|^2\right]^{1/2}+\left[E\left|\sum_{j=-n}^{n}a_jZ_{t-j}-Y_t\right|^2\right]^{1/2}\right\}^2$$

by the Minkowski inequality (Royden, 1963, p. 95), and

$$E\{(X_t-Y_t)^2\}<4\epsilon$$

for n sufficiently large. Thus, $E\{(X_t-Y_t)^2\}=0$, since $\epsilon>0$ was arbitrary. Then $X_t=X(t,\omega)=Y(t,\omega)=Y_t$ except for $\omega\in G_t$ with $P(G_t)=0$. Since $P(\cup_{t=-\infty}^{\infty}G_t)=0$, we have that $X_t=Y_t$ for all t and for all ω not in G, where $P(G)=0$.

Finally, we show that if $\{Z_t\}$ is a sequence of independent random variables then the sum converges almost surely. Let $\{Z_t\}$ be a sequence of independent random variables with $E\{Z_t^2\}<K$ for all t. Then,

$$\sum_{j=-n}^{n}|E\{a_jZ_{t-j}\}|\leqslant\sum_{j=-n}^{n}|a_j|E\{|Z_{t-j}|\}$$

$$\leqslant K^{1/2}\sum_{j=-\infty}^{\infty}|a_j|<\infty$$

and

$$\sum_{j=-n}^{n}\operatorname{Var}\{a_jZ_{t-j}\}\leqslant\sum_{j=-n}^{n}E\{a_jZ_{t-j}\}^2$$

$$\leqslant K\sum_{j=-\infty}^{\infty}a_j^2<\infty.$$

Therefore, $\Sigma_{j=-n}^{n}a_jZ_{t-j}$ converges almost surely to a random variable Y_t, and $Y_t=X_t$ except for $\omega\in G_t$ where $P(G_t)=0$. See Tucker (1967, pp. 102, 112). ▲

In the sequel, when we write

$$\sum_{j=-\infty}^{\infty}a_jZ_{t-j}$$

we are referring to the random variable X_t given by the previous theorem. Thus,

$$X_t = \sum_{j=-\infty}^{\infty} a_j Z_{t-j}$$

should be taken to mean

$$E\left\{\left|\sum_{j=-n}^{n} a_j Z_{t-j} - X_t\right|^2\right\} \to 0 \quad \text{as} \quad n \to \infty.$$

Convergence in mean square (squared mean) is discussed further in Section 5.1.

Theorem 2.2.2. Let the sequences of real numbers $\{a_j\}$, $\{b_j\}$ be absolutely summable. Let $\{Z_t\}$ be a sequence of random variables such that $E\{Z_t^2\} \leqslant K$ for all $t = 0, \pm 1, \pm 2, \ldots$ and for some finite K, and define

$$X_t = \sum_{j=-\infty}^{\infty} a_j Z_{t-j}$$

and

$$Y_t = \sum_{k=-\infty}^{\infty} b_k Z_{t-k}.$$

Then

$$E\{X_t\} = \lim_{n\to\infty} \sum_{j=-n}^{n} a_j E\{Z_{t-j}\}$$

and

$$E\{X_t Y_t\} = \lim_{n\to\infty} \sum_{j=-n}^{n} \sum_{k=-n}^{n} a_j b_k E\{Z_{t-j} Z_{t-k}\}.$$

Proof. By Theorem 2.2.1 we can construct the sequences of random variables $\{X_t^\dagger\}$ and $\{Y_t^\dagger\}$ such that $X_t^\dagger = \sum_{j=-\infty}^{\infty} |a_j||Z_{t-j}|$ and $Y_t^\dagger = \sum_{j=-\infty}^{\infty} |b_j||Z_{t-j}|$ for $t = 0, \pm 1, \pm 2, \ldots$. Then $E\{(X_t^\dagger)^2\} < \infty$ for all t, and it follows that

$$E\{|X_t^\dagger|\} < E\{1 + (X_t^\dagger)^2\} < \infty.$$

We also have $\sum_{j=-n}^{n} a_j Z_{t-j} \xrightarrow{P} X_t$ as $n \to \infty$ and

$$\left| \sum_{j=-n}^{n} a_j Z_{t-j} \right| \leq \sum_{j=-n}^{n} |a_j| |Z_{t-j}| \leq X_t^{\dagger}.$$

Applying the dominated convergence theorem, which holds for convergence in probability as well as almost surely (Royden, 1963, p. 78), we have

$$E\{X_t\} = \lim_{n\to\infty} E\left\{ \sum_{j=-n}^{n} a_j Z_{t-j} \right\}.$$

Since $(|x|-|y|)^2 \geq 0$, we have $|xy| \leq \frac{1}{2}(x^2+y^2)$; therefore,

$$\begin{aligned}
\left| \left(\sum_{j=-n}^{n} a_j Z_{t-j} \right)\left(\sum_{k=-n}^{n} b_k Z_{t-k} \right) \right| & \\
&\leq \left(\sum_{j=-n}^{n} |a_j Z_{t-j}| \right)\left(\sum_{k=-n}^{n} |b_k Z_{t-k}| \right) \\
&\leq \frac{1}{2}\left(\sum_{j=-n}^{n} |a_j Z_{t-j}| \right)^2 + \frac{1}{2}\left(\sum_{k=-n}^{n} |b_k Z_{t-k}| \right)^2 \\
&\leq \frac{1}{2}(X_t^{\dagger})^2 + \frac{1}{2}(Y_t^{\dagger})^2.
\end{aligned}$$

Applying the dominated convergence theorem,

$$\begin{aligned}
E\{X_t Y_t\} &= \lim_{n\to\infty} E\left\{ \left(\sum_{j=-n}^{n} a_j Z_{t-j} \right)\left(\sum_{k=-n}^{n} b_k Z_{t-k} \right) \right\} \\
&= \lim_{n\to\infty} \sum_{j=-n}^{n} \sum_{k=-n}^{n} a_j b_k E\{Z_{t-j} Z_{t-k}\}.
\end{aligned}$$

▲

Corollary 2.2.2.1. Let the sequences of real numbers $\{a_j\}$, $\{b_j\}$ be absolutely summable. Let $\{e_t\}$ be a sequence of uncorrelated $(0, \sigma^2)$ random variables and define for $t = 0, \pm 1, \pm 2, \ldots$,

$$X_t = \sum_{j=-\infty}^{\infty} a_j e_{t-j}$$

and

$$Y_t = \sum_{k=-\infty}^{\infty} b_k e_{t-k}.$$

Then,

$$E\{X_t\} = 0$$

and

$$E\{X_t Y_t\} = \sigma^2 \sum_{j=-\infty}^{\infty} a_j b_j.$$

Proof. The sequence $\{e_t\}$ satisfies the hypotheses of the previous theorem with $K = \sigma^2$. Since $E\{e_t\} = 0$ and $E\{e_t e_{t'}\} = 0$ for $t \neq t'$, we have the results. ▲

Corollary 2.2.2.2. Let $\{X_t\}$ be given by

$$X_t = \sum_{j=-\infty}^{\infty} a_j e_{t-j}$$

and define

$$Y_t = \sum_{j=-\infty}^{\infty} b_j X_{t-j},$$

where $\{e_t\}$ is a sequence of uncorrelated $(0, \sigma^2)$ random variables and $\{a_j\}$ and $\{b_j\}$ are absolutely summable sequences of real numbers. Then $\{Y_t\}$ is a stationary moving average time series and

$$\begin{aligned} \gamma_Y(h) &= \sum_{j=-\infty}^{\infty} \sum_{k=-\infty}^{\infty} b_j b_k \gamma_X(k-j+h) \\ &= \sum_{j=-\infty}^{\infty} \alpha_j \alpha_{j-h} \sigma^2, \end{aligned} \tag{2.2.2}$$

where

$$\alpha_j = \sum_{k=-\infty}^{\infty} b_k a_{j-k}.$$

Proof. By definition,

$$\begin{aligned} Y_t &= \sum_{j=-\infty}^{\infty} b_j X_{t-j} \\ &= \sum_{j=-\infty}^{\infty} b_j \sum_{k=-\infty}^{\infty} a_k e_{t-j-k}. \end{aligned}$$

Since $\{a_j\}$ and $\{b_j\}$ are absolutely summable, we may interchange the order of summation (Fubini's Theorem). Setting $m=j+k$, we have

$$\begin{aligned} Y_t &= \sum_{m=-\infty}^{\infty} \sum_{j=-\infty}^{\infty} b_j a_{m-j} e_{t-m} \\ &= \sum_{m=-\infty}^{\infty} \alpha_m e_{t-m}, \end{aligned}$$

where

$$\alpha_m = \sum_{j=-\infty}^{\infty} b_j a_{m-j}.$$

By result 3 (p. 28) on absolutely summable sequences, $\{\alpha_m\}$ is absolutely summable. Therefore, using Theorem 2.2.2,

$$\begin{aligned} \gamma_Y(h) &= \sum_{m=-\infty}^{\infty} \alpha_m \alpha_{m-h} \sigma^2 \\ &= \sum_{m=-\infty}^{\infty} \sum_{j=-\infty}^{\infty} b_j a_{m-j} \sum_{k=-\infty}^{\infty} b_k a_{m-k-h} \sigma^2 \\ &= \sum_{j=-\infty}^{\infty} \sum_{k=-\infty}^{\infty} b_j b_k \sum_{m=-\infty}^{\infty} a_{m-j} a_{m-k-h} \sigma^2 \\ &= \sum_{j=-\infty}^{\infty} \sum_{k=-\infty}^{\infty} b_j b_k \gamma_X(k-j+h). \end{aligned}$$ ▲

We may paraphrase this corollary by saying that an infinite moving average (where the coefficients are absolutely summable) of an infinite moving average stationary time series is itself an infinite moving average stationary time series.

2.3. AN INTRODUCTION TO AUTOREGRESSIVE TIME SERIES

Many time series encountered in practice are well approximated by the representation

$$\sum_{i=0}^{p} \alpha_i X_{t-i} = e_t, \qquad t = 0, \pm 1, \pm 2, \ldots, \tag{2.3.1}$$

where $\alpha_0 \neq 0$, $\alpha_p \neq 0$, and the e_t are uncorrelated $(0, \sigma^2)$ random variables. The sequence $\{X_t\}$ is called a pth *order autoregressive* time series. The defining equation (2.3.1) is sometimes called a *stochastic difference equation*.

To study such processes, let us consider the first order autoregressive time series, which we express in its most common form,

$$X_t = \rho X_{t-1} + e_t. \tag{2.3.2}$$

By repeated substitution of

$$X_{t-i} = \rho X_{t-i-1} + e_{t-i}$$

for $i = 1, 2, \ldots, N$ into (2.3.2), we obtain

$$X_t = \rho^N X_{t-N} + \sum_{i=0}^{N-1} \rho^i e_{t-i}. \tag{2.3.3}$$

Under the assumptions that $|\rho| < 1$ and $E\{X_t^2\} < K < \infty$, we have

$$\lim_{N \to \infty} E\left\{\left(X_t - \sum_{i=0}^{N} \rho^i e_{t-i}\right)^2\right\} = 0. \tag{2.3.4}$$

Thus, if e_t is defined for $t \in \{0, \pm 1, \pm 2, \ldots\}$ and X_t satisfies the difference equation (2.3.2) with $|\rho| < 1$, then we may express X_t as an infinite moving average of the e_t:

$$X_t = \sum_{i=0}^{\infty} \rho^i e_{t-i}.$$

It follows that $E\{X_t\}=0$ for all t, and

$$\gamma(h)=E\{X_tX_{t+h}\}=E\left\{\left(\sum_{i=0}^{\infty}\rho^ie_{t-i}\right)\left(\sum_{j=0}^{\infty}\rho^je_{t+h-j}\right)\right\}$$

$$=\sigma^2\sum_{i=0}^{\infty}\rho^i\rho^{i+h}$$

$$=\frac{\rho^h}{1-\rho^2}\sigma^2, \tag{2.3.5}$$

for $h=0,1,\ldots$. The covariance function for $h\neq 0$ is also rapidly obtained by making use of a form of (2.3.3),

$$X_t=\rho^hX_{t-h}+\sum_{j=0}^{h-1}\rho^je_{t-j}, \qquad h=1,2,\ldots. \tag{2.3.6}$$

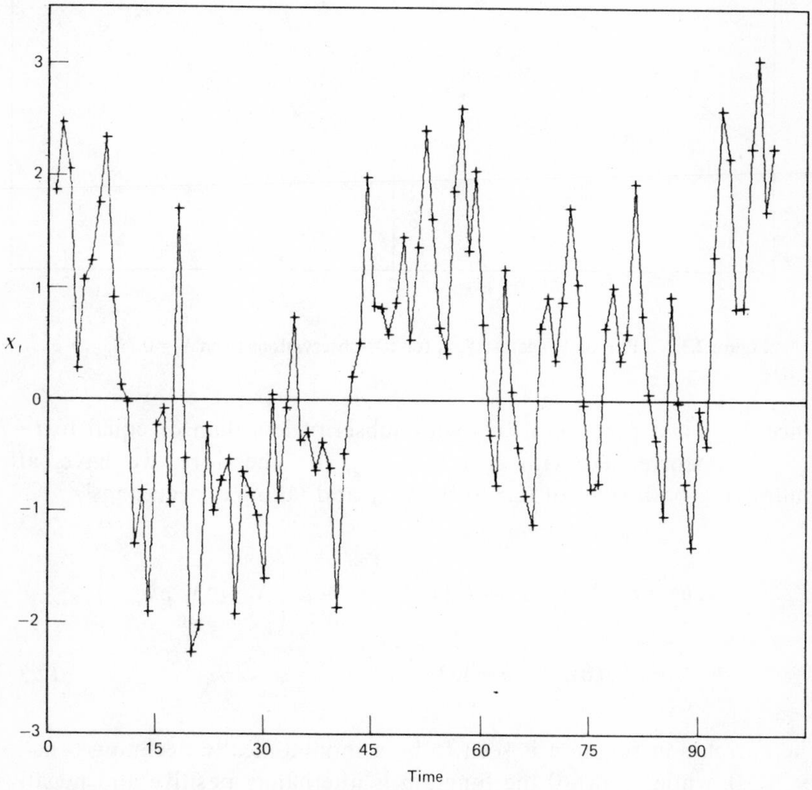

Figure 2.3.1. One hundred observations from the time series $X_t=0.7X_{t-1}+e_t$.

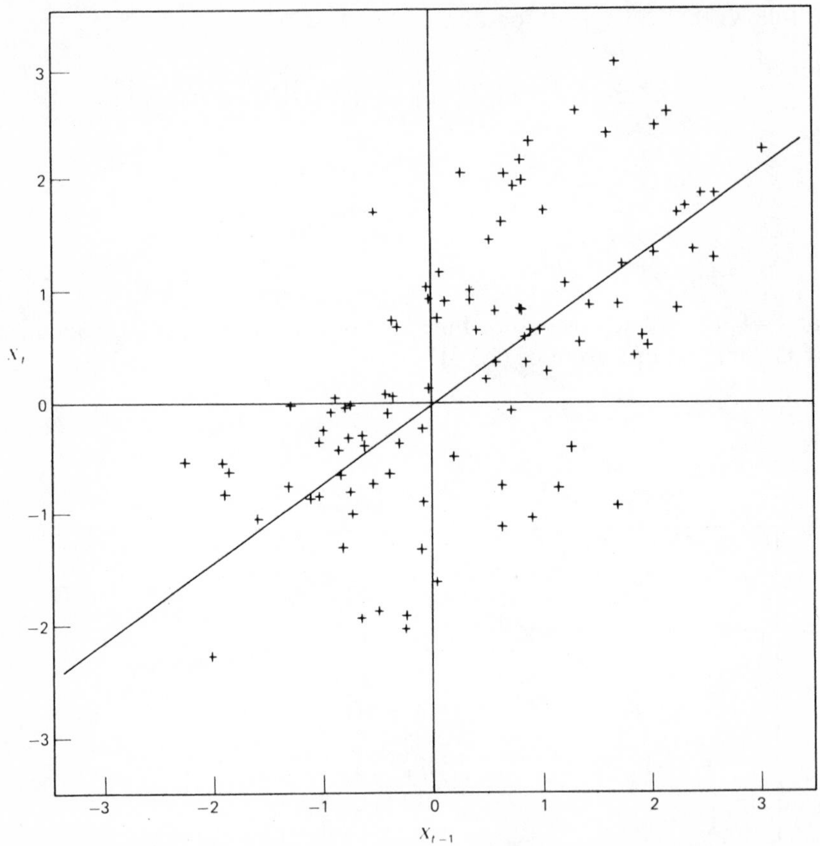

Figure 2.3.2. Plot of X_t against X_{t-1} for 100 observations from $X_t = 0.7X_{t-1} + e_t$.

Since X_{t-h} is a function of e's with subscript less than or equal to $t-h$, X_{t-h} is uncorrelated with $e_{t-h+1}, e_{t-h+2}, \ldots$. Therefore, we have, after multiplying both sides of (2.3.6) by X_{t-h} and taking expectations,

$$\gamma(h) = E\{X_t X_{t-h}\} = E\{\rho^h X_{t-h}^2\} + E\left\{X_{t-h} \sum_{j=0}^{h-1} \rho^j e_{t-j}\right\}$$

$$= \rho^h \gamma(0), \qquad h = 1, 2, \ldots. \tag{2.3.7}$$

The correlation function is seen to be a monotonically declining function for $\rho > 0$, while for $\rho < 0$ the function is alternately positive and negative, the absolute value declining at a geometric rate.

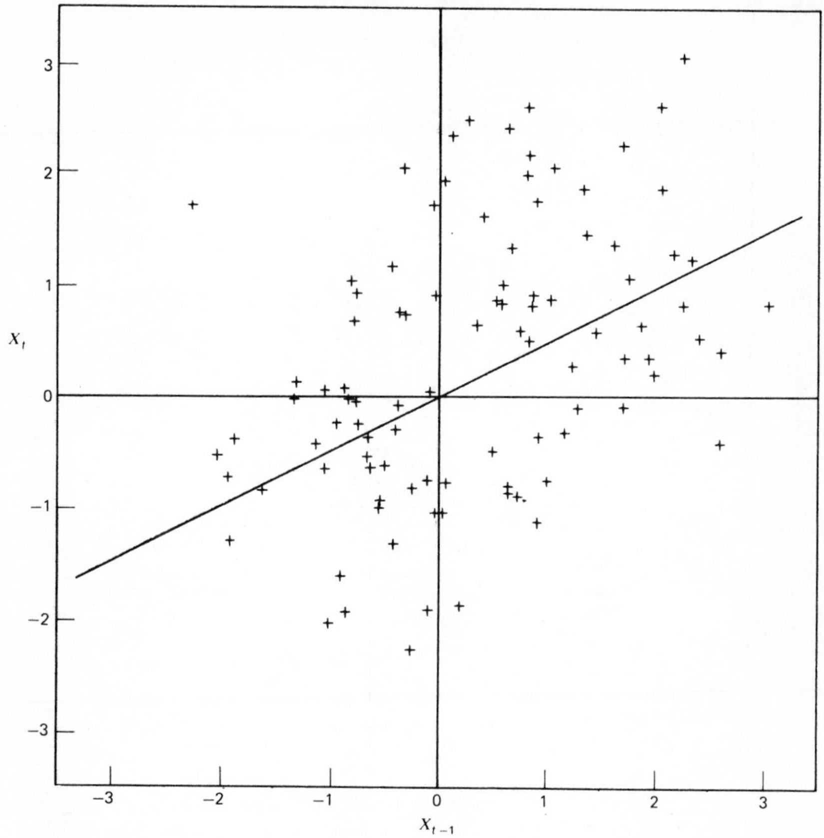

Figure 2.3.3. Plot of X_t against X_{t-2} for 100 observations from $X_t = 0.7X_{t-1} + e_t$.

Figure 2.3.1 displays 100 observations from the first order process

$$X_t = 0.7X_{t-1} + e_t.$$

The current observation, X_t, is plotted against the preceding observation, X_{t-1}, in Figure 2.3.2 and against X_{t-2} in Figure 2.3.3. The correlation between X_t and X_{t-2} is $(0.7)^2 = 0.49$, but the partial correlation between X_t and X_{t-2} adjusted for X_{t-1} is zero, since

$$\text{Cov}\{X_t - \rho X_{t-1}, X_{t-2} - \rho X_{t-1}\} = (\rho^2 - \rho^2 - \rho^2 + \rho^2)\gamma(0) = 0.$$

The zero partial correlation is illustrated by Figure 2.3.4, which contains

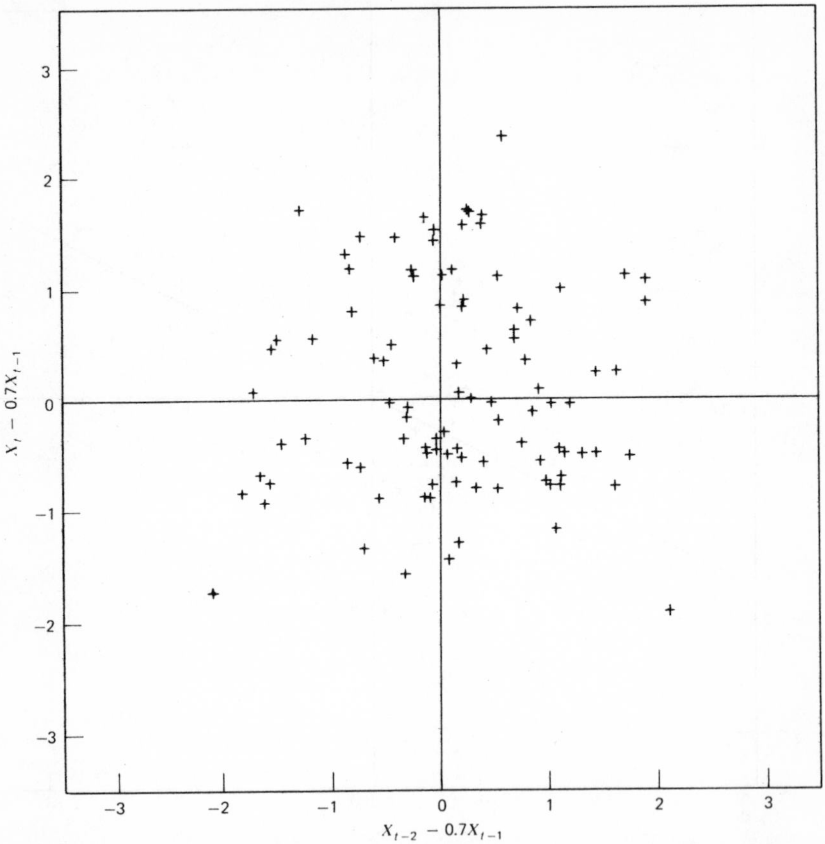

Figure 2.3.4. Plot of $X_t - 0.7X_{t-1}$ against $X_{t-2} - 0.7X_{t-1}$ for 100 observations from $X_t = 1.40X_{t-1} - 0.48X_{t-2} + e_t$.

a plot of $X_t - 0.7X_{t-1}$ against $X_{t-2} - 0.7X_{t-1}$. In fact, for any $h \in (2,3,\ldots)$, the partial correlation between X_t and X_{t-h} adjusted for X_{t-1} is zero. This follows, since

$$\operatorname{Cov}\{X_t - \rho X_{t-1}, X_{t-h} - \rho^{h-1}X_{t-1}\}$$
$$= (\rho^h - \rho^{h-1}\rho - \rho\rho^{h-1} + \rho^h)\gamma(0) = 0, \qquad h = 2,3,\ldots.$$

This important property of the first order autoregressive time series can be characterized by saying that all the useful (correlational) information about X_t in the entire realization previous to X_t is contained in X_{t-1}. Compare this result to the partial correlational properties of the moving average process discussed in Section 2.1.

Linear difference equations and their solutions play an important role in the study of autoregressive time series as well as in other areas of time series analysis. Therefore we digress to present the needed elementary results before studying higher order autoregressive processes. The initial development follows Goldberg (1958).

2.4. DIFFERENCE EQUATIONS

Given a sequence $\{y_t\}_{t=0}^{\infty}$, the *first difference* is defined by

$$\Delta y_t = y_t - y_{t-1}, \qquad t = 1, 2, \ldots, \tag{2.4.1}$$

and the nth *difference* is defined by

$$\Delta^n y_t = \Delta^{n-1} y_t - \Delta^{n-1} y_{t-1} = \sum_{r=0}^{n} (-1)^r \binom{n}{r} y_{t-r}, \qquad t = n, n+1, \ldots,$$

where

$$\binom{n}{r} = \frac{n!}{r!(n-r)!}$$

are the binomial coefficients. The equation

$$y_t + a_1 y_{t-1} + a_2 y_{t-2} + \cdots + a_n y_{t-n} = r_t, \qquad t = n, n+1, \ldots, \tag{2.4.2}$$

where the a_i are real constants, $a_n \neq 0$, and r_t is a real function of t is a *linear difference equation of order n* with constant coefficients. The values $y_{t-1}, y_{t-2}, \ldots, y_{t-n}$ are sometimes called *lagged* values of y_t. Equation (2.4.2) could be expressed in terms of y_{t-n} and the differences of y_t; for example,

$$\Delta^n y_t + b_1 \Delta^{n-1} y_{t-1} + b_2 \Delta^{n-2} y_{t-2} + \cdots + b_{n-1} \Delta y_{t-n+1} + b_n y_{t-n} = r_t, \tag{2.4.3}$$

where the b_i's are linear functions of the a_i's.

A third representation[1] of (2.4.2) is possible. Let the symbol $\mathcal{B}$ denote the operation of replacing y_t by y_{t-1}; that is,

$$\mathcal{B} y_t = y_{t-1}.$$

[1] The reader may wonder at the benefits associated with several alternative notations. All are heavily used in the literature and all have advantages for certain operations.

Using the *backward shift operator* $\mathcal{B}$ we have

$$\Delta y_t = y_t - \mathcal{B} y_t = [1 - \mathcal{B}] y_t.$$

As with the difference operator, we denote repeated application of the operator with the appropriate exponent; for example,

$$\mathcal{B}^3 y_t = \mathcal{B}\,\mathcal{B}\,\mathcal{B}\, y_t = y_{t-3}.$$

Therefore, we can use the operator symbol to write (2.4.2) as

$$(1 + a_1 \mathcal{B} + a_2 \mathcal{B}^2 + \cdots + a_n \mathcal{B}^n) y_t = r_t, \qquad t = n, n+1, \ldots.$$

Associated with (2.4.2) is the *reduced* or *homogeneous* difference equation

$$y_t + a_1 y_{t-1} + a_2 y_{t-2} + \cdots + a_n y_{t-n} = 0. \tag{2.4.4}$$

From the linearity of (2.4.2) and (2.4.4) we have the following properties.

1. If $y^{(1)}$ and $y^{(2)}$ are solutions of (2.4.4), then $b_1 y^{(1)} + b_2 y^{(2)}$ is a solution of (2.4.4) where b_1 and b_2 are arbitrary constants.
2. If Y is a solution of (2.4.4) and $y^\dagger$ a solution of (2.4.2), then $Y + y^\dagger$ is a solution of (2.4.2).

$y^\dagger$ is called a *particular solution* and $Y + y^\dagger$ a *general solution* of the difference equation (2.4.2).

Equations (2.4.2) and (2.4.4) specify y_t as a function of the preceding n values of y. By using an inductive proof it is possible to demonstrate that the linear difference equation of order n has one and only one solution for which values at n consecutive t values are arbitrarily prescribed. This important result means that when we obtain n linearly independent solutions to the difference equation, we can construct the general solution.

The first order homogeneous difference equation

$$y_t = a y_{t-1}, \qquad t = 1, 2, \ldots, \tag{2.4.5}$$

has the unique solution

$$y_t = a^t y_0, \tag{2.4.6}$$

where y_0 is the value of y at $t = 0$. Note that if $|a| < 1$, the solution tends to zero as t increases. Conversely, if $|a| > 1$, the absolute value of the solution

sequence increases without bound. If a is negative the solution oscillates with alternately positive and negative values. The first order nonhomogeneous difference equation

$$y_t = a y_{t-1} + b \tag{2.4.7}$$

has the unique solution

$$\begin{aligned} y_t &= y_0 a^t + b\left(\frac{1-a^t}{1-a}\right), && a \neq 1 \\ &= y_0 + bt, && a = 1. \end{aligned} \tag{2.4.8}$$

If the constant b in the difference equation (2.4.7) is replaced by a function of time, for example,

$$y_t = a y_{t-1} + b_t, \qquad t = 1, 2, \ldots, \tag{2.4.9}$$

then the solution is given by

$$y_t = y_0 a^t + \sum_{j=0}^{t-1} a^j b_{t-j}, \qquad t = 1, 2, \ldots, \tag{2.4.10}$$

where y_0 is the value of y_t at time 0.

The solutions (2.4.6), (2.4.8), and (2.4.10) are readily verified by differencing. The fact that the linear function is reduced by differencing to the constant function is of particular interest. The generalization of this result is important in the analysis of nonstationary time series.

Theorem 2.4.1. Let y_t be a polynomial of degree n whose domain is the integers. Then the first difference Δy_t is expressible as a polynomial of degree $n-1$ in t.

Proof. Since

$$y_t = \sum_{p=0}^{n} \beta_p t^p, \qquad \beta_n \neq 0,$$

and

$$y_{t-1} = \sum_{p=0}^{n} \beta_p (t-1)^p,$$

we have

$$\begin{aligned}
\Delta y_t &= \sum_{p=0}^{n} \beta_p \left[t^p - (t-1)^p \right] \\
&= \sum_{p=0}^{n} \beta_p \left[t^p - \sum_{s=0}^{p} \binom{p}{s} (-1)^s t^{p-s} \right] \\
&= \sum_{p=1}^{n} \beta_p \left[- \sum_{s=1}^{p} \binom{p}{s} (-1)^s t^{p-s} \right] \\
&\overset{\text{(say)}}{=} \sum_{p=0}^{n-1} \gamma_p t^p,
\end{aligned}$$

where $\gamma_{n-1} = n\beta_n$. ▲

It follows that by taking repeated differences, one can reduce a polynomial to the zero function. We state the result as a corollary.

Corollary 2.4.1. Let y_t be a polynomial of degree n whose domain is the integers. Then the nth difference $\Delta^n y_t$ is a constant function, and the $(n+1)$st difference $\Delta^{n+1} y_t$ is the zero function.

In investigating difference equations of higher order the nature of the solution is a function of the *auxiliary equation*, sometimes called the *characteristic equation*. For equation (2.4.2) the auxiliary equation is

$$m^n + a_1 m^{n-1} + \cdots + a_n = 0. \tag{2.4.11}$$

This polynomial equation will have n (not necessarily distinct) roots, and the behavior of the solution is intimately related to these roots.

Consider the second order linear homogeneous equation

$$y_t + a_1 y_{t-1} + a_2 y_{t-2} = 0, \tag{2.4.12}$$

where $a_2 \neq 0$. The two roots of the auxiliary equation

$$m^2 + a_1 m + a_2 = 0 \tag{2.4.13}$$

will fall into one of three categories.

1. Real and distinct.
2. Real and equal.
3. A complex conjugate pair.

The general solution of the homogeneous difference equation (2.4.12) for the three categories is:

1. $y_t = b_1 m_1^t + b_2 m_2^t$, (2.4.14)

where b_1 and b_2 are constants determined by the initial conditions, and m_1 and m_2 are the roots of (2.4.13).

2. $y_t = (b_1 + b_2 t) m^t$, (2.4.15)

where m is the repeated root of (2.4.13).

3. $y_t = b_1^* m_1^t + b_1 m_2^t$, (2.4.16)

where b_1^* is the complex conjugate of b_1 and $m_2 = m_1^*$ is the complex conjugate of m_1.

If the complex roots are expressed as

$$r(\cos\theta \pm i \sin\theta),$$

where

$$r = a_2^{1/2} = |m_1|,$$

$$\cos\theta = -\frac{a_1}{2r},$$

and

$$\sin\theta = \frac{|a_1^2 - 4a_2|^{1/2}}{2r},$$

then the solution can be expressed as

$$\begin{aligned} y_t &= r^t (d_1 \cos t\theta + d_2 \sin t\theta) \\ &= g_1 r^t \cos(t\theta + g_2), \end{aligned} \tag{2.4.17}$$

where d_1, d_2, g_1, and g_2 are real constants.

Substitution of (2.4.14) into (2.4.12) immediately establishes that (2.4.14) is a solution and, in fact, that m_1^t and m_2^t are independent solutions when $m_1 \neq m_2$. For a repeated root, substitution of (2.4.15) into (2.4.12) gives

$$(b_1 + b_2 t) m^{t-2} (m^2 + a_1 m + a_2) - b_2 m^{t-2} (a_1 m + 2a_2) = 0.$$

The first term is zero because m is a root of (2.4.13). For m a repeated root, $ma_1 + 2a_2 = 0$, proving that (2.4.15) is the general solution.

As with unequal real roots, substitution of (2.4.16) into (2.4.12) demonstrates that (2.4.16) is the solution in the case of complex roots.

The solution for a linear homogeneous difference equation of order n can be obtained using the roots of the auxiliary equation. The solution is a sum of n terms where:

1. For every real and distinct root, m, a term of the form bm^t is included.
2. For every real root of order p (a root repeated p times), a term of the form

$$(b_1 + b_2 t + b_3 t^2 + \cdots + b_p t^{p-1}) m^t$$

is included.
3. For each pair of unrepeated complex conjugate roots, a term of the form

$$\alpha r^t \cos(t\theta + \beta)$$

is included, where r and θ are defined following (2.4.16).
4. For a pair of complex conjugate roots repeated p times, a term of the form

$$r^t\left[\alpha_1 \cos(t\theta + \beta_1) + \alpha_2 t \cos(t\theta + \beta_2) + \cdots + \alpha_p t^{p-1} \cos(t\theta + \beta_p)\right]$$

is included.

Systems of difference equations in more than one function of time arise naturally in many models of the physical and economic sciences. To extend our discussion to such models, consider the system of difference equations in the two functions y_{1t} and y_{2t}:

$$y_{1t} = a_{11} y_{1,t-1} + a_{12} y_{2,t-1} + b_{1t},$$
$$y_{2t} = a_{21} y_{1,t-1} + a_{22} y_{2,t-1} + b_{2t}, \qquad (2.4.18)$$

where $a_{11}, a_{12}, a_{21}, a_{22}$ are real constants, b_{1t} and b_{2t} are real functions of time, and $t = 1, 2, \ldots$. The system may be expressed in matrix notation as

$$\mathbf{y}_t = \mathbf{A}\mathbf{y}_{t-1} + \mathbf{b}_t, \qquad (2.4.19)$$

where

$$\mathbf{y}_t' = (y_{1t}, y_{2t}),$$

$$\mathbf{A} = \begin{pmatrix} a_{11} & a_{12} \\ a_{21} & a_{22} \end{pmatrix},$$

and $\mathbf{b}_t' = (b_{1t}, b_{2t})$.

Proceeding from the solution (2.4.6) of the first order homogeneous scalar difference equation, we might postulate that the homogeneous equation

$$\mathbf{y}_t = \mathbf{A}\mathbf{y}_{t-1} \tag{2.4.20}$$

would have the solution

$$\mathbf{y}_t = \mathbf{A}^t\mathbf{y}_0, \tag{2.4.21}$$

and that the equation (2.4.19) would have the solution

$$\mathbf{y}_t = \mathbf{A}^t\mathbf{y}_0 + \sum_{j=0}^{t-1} \mathbf{A}^j\mathbf{b}_{t-j}, \tag{2.4.22}$$

where we understand that $\mathbf{A}^0 = \mathbf{I}$. Direct substitution of (2.4.21) into (2.4.20) and (2.4.22) into (2.4.19) verifies that these are the solutions of the matrix analogs of the first order scalar equations.

In the scalar case the behavior of the solution depends in a critical manner on the magnitude of the coefficient, a, for the first order equation and on the magnitude of the roots of the characteristic equation for the higher order equations. If the roots are all less than one in absolute value the effect of the initial conditions is transient, the initial conditions being multiplied by the product of powers of the roots and polynomials in t.

In the vector case the quantities analogous to the roots of the characteristic equation are the roots of the determinantal equation

$$|\mathbf{A} - m\mathbf{I}| = 0.$$

For the moment, we assume that the roots of the determinantal equation are distinct. Let the two roots of the matrix $\mathbf{A}$ of (2.4.19) be m_1 and m_2 and define the vectors $\mathbf{q}_{.1}$ and $\mathbf{q}_{.2}$ by the equations

$$(\mathbf{A} - m_1\mathbf{I})\mathbf{q}_{.1} = 0,$$

$$(\mathbf{A} - m_2\mathbf{I})\mathbf{q}_{.2} = 0,$$

$$\mathbf{q}'_{.1}\mathbf{q}_{.1} = 1,$$

$$\mathbf{q}'_{.2}\mathbf{q}_{.2} = 1.$$

By construction, the matrix $\mathbf{Q} = (\mathbf{q}_{.1}, \mathbf{q}_{.2})$ is such that

$$\mathbf{AQ} = \mathbf{QM}, \tag{2.4.23}$$

where $\mathbf{M}=\operatorname{diag}(m_1, m_2)$. Since the roots are distinct, the matrix $\mathbf{Q}$ is nonsingular.[2] Hence,

$$\mathbf{Q}^{-1}\mathbf{A}\mathbf{Q}=\mathbf{M}. \tag{2.4.24}$$

Define $\mathbf{z}_t'=(z_{1t}, z_{2t})$ and $\mathbf{c}_t'=(c_{1t}, c_{2t})$ by

$$\mathbf{z}_t=\mathbf{Q}^{-1}\mathbf{y}_t,$$
$$\mathbf{c}_t=\mathbf{Q}^{-1}\mathbf{b}_t.$$

Then, multiplying (2.4.19) by $\mathbf{Q}^{-1}$, we have

$$\mathbf{z}_t=\mathbf{Q}^{-1}\mathbf{A}\mathbf{Q}\mathbf{z}_{t-1}+\mathbf{c}_t$$

or

$$\begin{pmatrix} z_{1t} \\ z_{2t} \end{pmatrix}=\begin{pmatrix} m_1 & 0 \\ 0 & m_2 \end{pmatrix}\begin{pmatrix} z_{1,t-1} \\ z_{2,t-1} \end{pmatrix}+\begin{pmatrix} c_{1t} \\ c_{2t} \end{pmatrix}.$$

The transformation $\mathbf{Q}^{-1}$ reduces the system of difference equations to two simple first order difference equations with solutions

$$z_{1t}=\sum_{j=0}^{t} m_1^j c_{1,t-j},$$

$$z_{2t}=\sum_{j=0}^{t} m_2^j c_{2,t-j}, \tag{2.4.25}$$

where we have set $y_{10}=b_{10}$ and $y_{20}=b_{20}$ for notational convenience. Therefore, the effect of the initial conditions is transient if $|m_1|<1$ and $|m_2|<1$. The solutions (2.4.25) may be expressed in terms of the original variables as

$$\mathbf{Q}^{-1}\mathbf{y}_t=\sum_{j=0}^{t} \mathbf{M}^j\mathbf{Q}^{-1}\mathbf{b}_{t-j}$$

or

$$\mathbf{y}_t=\sum_{j=0}^{t} \mathbf{Q}\mathbf{M}^j\mathbf{Q}^{-1}\mathbf{b}_{t-j}$$

$$=\sum_{j=0}^{t} \mathbf{A}^j\mathbf{b}_{t-j}, \tag{2.4.26}$$

[2] See Bellman (1960, p. 184).

where the last representation follows from $\mathbf{A}^2 = \mathbf{QMQ}^{-1}\mathbf{QMQ}^{-1} = \mathbf{QM}^2\mathbf{Q}^{-1}$, etc.

If the $(n \times n)$ matrix $\mathbf{A}$ has multiple roots it cannot always be reduced to a diagonal by a transformation matrix $\mathbf{Q}$. However, $\mathbf{A}$ can be reduced to a matrix whose form is slightly more complicated.

Theorem 2.4.2. Let $\mathbf{\Lambda}_i$ be the $k_i \times k_i$ matrix

$$\mathbf{\Lambda}_i = \begin{bmatrix} m_i & 1 & 0 & \cdots & 0 & 0 \\ 0 & m_i & 1 & \cdots & 0 & 0 \\ 0 & 0 & m_i & \cdots & 0 & 0 \\ \vdots & \vdots & \vdots & & \vdots & \vdots \\ 0 & 0 & 0 & \cdots & 0 & m_i \end{bmatrix},$$

where $\mathbf{\Lambda}_i = m_i$ when $k_i = 1$. Then there exists a matrix $\mathbf{Q}$ such that

$$\mathbf{Q}^{-1}\mathbf{AQ} = \begin{bmatrix} \mathbf{\Lambda}_1 & \mathbf{0} & \cdots & \mathbf{0} \\ \mathbf{0} & \mathbf{\Lambda}_2 & \cdots & \mathbf{0} \\ \vdots & \vdots & & \vdots \\ \mathbf{0} & \mathbf{0} & \cdots & \mathbf{\Lambda}_r \end{bmatrix} = \mathbf{\Lambda}, \tag{2.4.27}$$

where $\sum_{i=1}^{r} k_i = n$ and the m_i, $i = 1, 2, \ldots, r$, are the characteristic roots of $\mathbf{A}$. If m_i is a repeated root it may appear in more than one block of $\mathbf{\Lambda}$, but the total number of times it appears is equal to its multiplicity.

Proof. See Finkbeiner (1960) or Miller (1963). ▲

The representation $\mathbf{\Lambda}$ is called the *Jordan canonical form*. The powers of the matrix $\mathbf{\Lambda}_i$ are given by

$$\mathbf{\Lambda}_i^j = \begin{bmatrix} m_i^j & \binom{j}{1} m_i^{j-1} & \binom{j}{2} m_i^{j-2} & \cdots & \binom{j}{k_i - 1} m_i^{j-k_i+1} \\ 0 & m_i^j & \binom{j}{1} m_i^{j-1} & \cdots & \binom{j}{k_i - 2} m_i^{j-k_i+2} \\ 0 & 0 & m_i^j & \cdots & \binom{j}{k_i - 3} m_i^{j-k_i+3} \\ \vdots & \vdots & \vdots & & \vdots \\ 0 & 0 & 0 & \cdots & m_i^j \end{bmatrix}. \tag{2.4.28}$$

Therefore, the solution of the first order vector difference equation with initial conditions $\mathbf{y}_0 = \mathbf{b}_0$ is given by

$$\mathbf{y}_t = \sum_{j=0}^{t} \mathbf{Q}\boldsymbol{\Lambda}^j\mathbf{Q}^{-1}\mathbf{b}_{t-j} \tag{2.4.29}$$

and the effect of the initial conditions goes to zero as $t \to \infty$ if all of the roots of $|\mathbf{A} - m\mathbf{I}| = 0$ are less than one in absolute value.

To further illustrate the solution, consider a system of dimension two with repeated root that has been reduced to the Jordan canonical form:

$$y_{1t} = my_{1,t-1} + y_{2,t-1}$$
$$y_{2t} = my_{2,t-1}.$$

It follows that

$$y_{2t} = y_{20}m^t$$

and y_{1t} is given as the solution of the nonhomogeneous equation,

$$y_{1t} = my_{1,t-1} + y_{20}m^{t-1}.$$

Whence

$$y_{1t} = y_{10}m^t + y_{20}tm^{t-1}.$$

This approach may be extended to treat higher order vector difference equations. For example, consider the second order vector difference equation

$$\mathbf{y}_t = \mathbf{A}_1\mathbf{y}_{t-1} + \mathbf{A}_2\mathbf{y}_{t-2}, \tag{2.4.30}$$

where $\mathbf{A}_1$ and $\mathbf{A}_2$ are $k \times k$ matrices and $\mathbf{y}_t' = (y_{1t}, y_{2t}, \ldots, y_{kt})$. The system of equations may also be written as

$$\mathbf{x}_t = \mathbf{A}\mathbf{x}_{t-1}, \tag{2.4.31}$$

where $\mathbf{x}_t' = (\mathbf{y}_t', \mathbf{y}_{t-1}')$ and

$$\mathbf{A} = \begin{pmatrix} \mathbf{A}_1 & \mathbf{A}_2 \\ \mathbf{I} & \mathbf{0} \end{pmatrix}. \tag{2.4.32}$$

Thus we have converted the original second order vector difference equation of dimension k to a first order difference equation of dimension $2k$.

The properties of the solution will depend on the nature of the Jordan canonical form of $\mathbf{A}$. In particular the effect of the initial conditions will be transient if all of the roots of $|\mathbf{A}-m\mathbf{I}|=0$ are less than one in absolute value. This condition is equivalent to the condition that the roots of

$$|\mathbf{I}m^2-\mathbf{A}_1m-\mathbf{A}_2|=0$$

be less than one in absolute value. To prove this equivalence, form the matrix product

$$\begin{pmatrix} \mathbf{I} & \mathbf{A}_2m^{-1} \\ \mathbf{0} & \mathbf{I} \end{pmatrix}\begin{pmatrix} \mathbf{A}_1-m\mathbf{I} & \mathbf{A}_2 \\ \mathbf{I} & -m\mathbf{I} \end{pmatrix}\begin{pmatrix} \mathbf{I} & \mathbf{0} \\ m^{-1}\mathbf{I} & \mathbf{I} \end{pmatrix}$$
$$=\begin{pmatrix} \mathbf{A}_1-m\mathbf{I}+\mathbf{A}_2m^{-1} & \mathbf{0} \\ \mathbf{0} & -m\mathbf{I} \end{pmatrix}$$

and take the determinant of both sides.

Let the nth order vector difference equation be given by

$$\mathbf{y}_t+\mathbf{A}_1\mathbf{y}_{t-1}+\mathbf{A}_2\mathbf{y}_{t-2}+\cdots+\mathbf{A}_n\mathbf{y}_{t-n}=\mathbf{r}_t, \qquad t=1,2,\ldots, \qquad (2.4.33)$$

where $\mathbf{y}_t$ is a k-vector and $\mathbf{r}_t$ is a real vector valued function of time. The associated $\mathbf{A}$-matrix is

$$\mathbf{A}=\begin{bmatrix} -\mathbf{A}_1 & -\mathbf{A}_2 & \cdots & -\mathbf{A}_{n-1} & -\mathbf{A}_n \\ \mathbf{I} & \mathbf{0} & \cdots & \mathbf{0} & \mathbf{0} \\ \vdots & \vdots & & \vdots & \vdots \\ \mathbf{0} & \mathbf{0} & \cdots & \mathbf{I} & \mathbf{0} \end{bmatrix}.$$

Then the solution is

$$\mathbf{y}_t=\sum_{j=0}^{t-1}\mathbf{W}_j\mathbf{r}_{t-j}+\boldsymbol{\tau}_t,$$

where the $\mathbf{W}_j$ are $k\times k$ matrices defined by

$$\begin{aligned}
&\mathbf{W}_0=\mathbf{I}\\
&\mathbf{W}_1+\mathbf{A}_1=\mathbf{0}\\
&\mathbf{W}_2+\mathbf{A}_1\mathbf{W}_1+\mathbf{A}_2=\mathbf{0}\\
&\vdots\\
&\mathbf{W}_j+\mathbf{A}_1\mathbf{W}_{j-1}+\cdots+\mathbf{A}_n\mathbf{W}_{j-n}=\mathbf{0}, \quad j=n,n+1,\ldots,
\end{aligned}$$

τ_t is the vector composed of the first k entries in the nk vector $\mathbf{x}_t = \mathbf{A}^t\mathbf{x}_0$, and $\mathbf{x}_0' = (\mathbf{y}_0', \mathbf{y}_{-1}', \ldots, \mathbf{y}_{1-n}')$. It follows that $\mathbf{W}_j$ is the upper left block of the matrix $\mathbf{A}^j$.

2.5. THE SECOND ORDER AUTOREGRESSIVE TIME SERIES

We consider the stationary second order autoregressive time series defined by the stochastic difference equation

$$X_t + \alpha_1 X_{t-1} + \alpha_2 X_{t-2} = e_t, \qquad t \in (0, \pm 1, \pm 2, \ldots), \tag{2.5.1}$$

where the e_t are uncorrelated $(0, \sigma^2)$ random variables, $\alpha_2 \neq 0$, and the roots of $m^2 + \alpha_1 m + \alpha_2 = 0$ are less than one in absolute value. Recalling that we were able to express the first order autoregressive process as a weighted average of past values of e_t, we postulate that X_t can be expressed as

$$X_t = \sum_{j=0}^{\infty} w_j e_{t-j}, \tag{2.5.2}$$

where the w_j are functions of α_1 and α_2. We display this representation as

$$X_t = w_0 e_t + w_1 e_{t-1} + w_2 e_{t-2} + w_3 e_{t-3} + \cdots$$

$$X_{t-1} = w_0 e_{t-1} + w_1 e_{t-2} + w_2 e_{t-3} + \cdots$$

$$X_{t-2} = w_0 e_{t-2} + w_1 e_{t-3} + \cdots .$$

If the X_t are to satisfy the difference equation (2.5.1), then we must have

$$\begin{aligned} w_0 &= 1 \\ w_1 + \alpha_1 w_0 &= 0 \\ w_j + \alpha_1 w_{j-1} + \alpha_2 w_{j-2} &= 0, \qquad j = 2, 3, \ldots . \end{aligned} \tag{2.5.3}$$

Thus the w_j are given by the general solution to the second order linear homogeneous difference equation subject to the initial conditions specified by the equations in w_0 and w_1. Therefore, if the roots are distinct, we have

$$w_j = (m_1 - m_2)^{-1} m_1^{j+1} + (m_2 - m_1)^{-1} m_2^{j+1}, \qquad j = 0, 1, 2, \ldots,$$

and, if the roots are equal,

$$w_j = (1 + j) m^j, \qquad j = 0, 1, 2, \ldots .$$

It follows that the covariance function for the second order autoregressive time series is given by

$$\gamma(h)=\sigma^2 \sum_{j=h}^{\infty} w_j w_{j-h}, \qquad h=0,1,2,\ldots . \tag{2.5.4}$$

We shall investigate the covariance function in a slightly different manner. Multiply (2.5.1) by $X_{t-h}(h \geqslant 0)$ to obtain

$$X_t X_{t-h} + \alpha_1 X_{t-1} X_{t-h} + \alpha_2 X_{t-2} X_{t-h} = e_t X_{t-h}. \tag{2.5.5}$$

Now we have demonstrated that X_t is expressible as a weighted average of e_t and *previous* e's. Thus, taking the expectation of both sides of (2.5.5), we have

$$\begin{aligned} \gamma(h)+\alpha_1\gamma(h-1)+\alpha_2\gamma(h-2)&=0, \qquad h>0 \\ &=\sigma^2, \qquad h=0. \end{aligned} \tag{2.5.6}$$

That is, the covariance function, for $h>0$, satisfies the homogeneous difference equation associated with the stochastic difference equation defining the time series.

The equations (2.5.6) are called the *Yule-Walker equations*. With the aid of these equations we can obtain a number of alternative expressions for the autocovariances, autocorrelations, and coefficients of the original representation. Using the two equations associated with $h=1$ and $h=2$, we may solve for α_1 and α_2 as functions of the autocovariances:

$$\alpha_1=\frac{\gamma(1)\gamma(2)-\gamma(0)\gamma(1)}{\gamma^2(0)-\gamma^2(1)}, \qquad \alpha_2=\frac{\gamma^2(1)-\gamma(0)\gamma(2)}{\gamma^2(0)-\gamma^2(1)}. \tag{2.5.7}$$

If we use the three equations (2.5.6) associated with $h=0,1,2$ we can obtain expressions for the autocovariances in terms of the coefficients:

$$\begin{aligned} \gamma(0)&=\frac{(1+\alpha_2)\sigma^2}{(1-\alpha_2)\left[(1+\alpha_2)^2-\alpha_1^2\right]} \\ \gamma(1)&=\frac{-\alpha_1\sigma^2}{(1-\alpha_2)\left[(1+\alpha_2)^2-\alpha_1^2\right]} \\ \gamma(2)&=\frac{\left[\alpha_1^2-\alpha_2(1+\alpha_2)\right]\sigma^2}{(1-\alpha_2)\left[(1+\alpha_2)^2-\alpha_1^2\right]}. \end{aligned} \tag{2.5.8}$$

Using $\gamma(0)$ and $\gamma(1)$ of (2.5.8) as the initial conditions together with the general solution of the homogeneous equation, we can obtain the expression for the covariance function.

However, it is somewhat simpler to derive the correlation function first. To obtain $\rho(1)$ as a function of the coefficients, we divide equation (2.5.6) for $h=1$ by $\gamma(0)$ giving

$$\rho(1)+\alpha_1+\alpha_2\rho(1)=0$$

and

$$\rho(1)=-\frac{\alpha_1}{1+\alpha_2}=\frac{m_1+m_2}{1+m_1m_2}. \tag{2.5.9}$$

Using (2.5.9) and $\rho(0)=1$ as initial conditions, we have, for unequal roots (real or complex),

$$\rho(h)=\left[(m_1-m_2)(1+m_1m_2)\right]^{-1}\left[m_1^{h+1}(1-m_2^2)-m_2^{h+1}(1-m_1^2)\right],$$
$$h=0,1,2,\ldots. \tag{2.5.10}$$

If the roots are complex the autocorrelation function may also be expressed as

$$\rho(h)=\frac{r^h\sin(h\theta+\delta)}{\sin\delta}, \tag{2.5.11}$$

where

$$r=\alpha_2^{1/2}=(m_1m_2)^{1/2}$$

$$\cos\theta=-\frac{\alpha_1}{2\alpha_2^{1/2}}=\frac{m_1+m_2}{2(m_1m_2)^{1/2}}$$

$$\tan\delta=\frac{1+\alpha_2}{1-\alpha_2}\tan\theta.$$

For roots real and equal the autocorrelation function is given by

$$\rho(h)=\left[1+h\left(\frac{1-m^2}{1+m^2}\right)\right]m^h, \qquad h=0,1,2,\ldots. \tag{2.5.12}$$

If we substitute $\alpha_1=-(m_1+m_2)$ and $\alpha_2=m_1m_2$ into (2.5.8) we can express the variance as a function of the roots:

$$\gamma(0)=\frac{1+m_1m_2}{(1-m_1m_2)(1-m_1^2)(1-m_2^2)}\sigma^2. \tag{2.5.13}$$

Equation (2.5.13) together with (2.5.10) and (2.5.12) may be used to express the covariance function in terms of the roots.

Figure 2.5.1 contains a plot of the correlation function, sometimes called the *correlogram*, for the time series

$$X_t = 1.40X_{t-1} - 0.48X_{t-2} + e_t.$$

The roots of the auxiliary equation are 0.8 and 0.6. The correlogram has much the same appearance as that of a first order autoregressive process with large ρ. The empirical correlograms of many economic time series have this general appearance, the first few correlations being quite large.

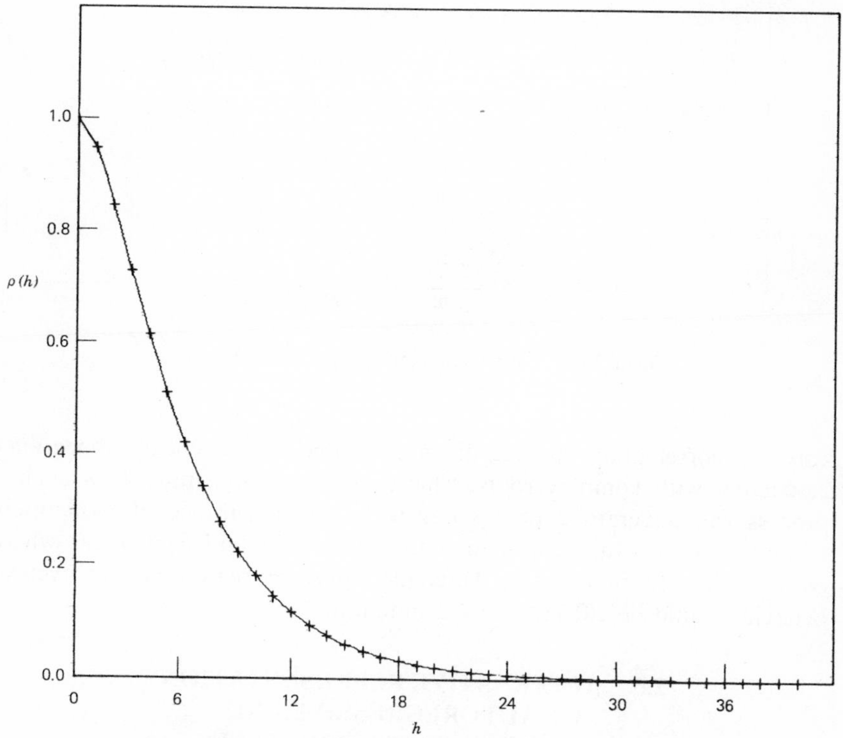

Figure 2.5.1. Correlogram for $x_t = 1.40X_{t-1} - 0.48X_{t-2} + e_t$.

The correlogram in Figure 2.5.2 is that associated with the time series

$$X_t = X_{t-1} - 0.89X_{t-2} + e_t.$$

The roots of the auxiliary equation are the complex pair $0.5 \pm 0.8i$. In this

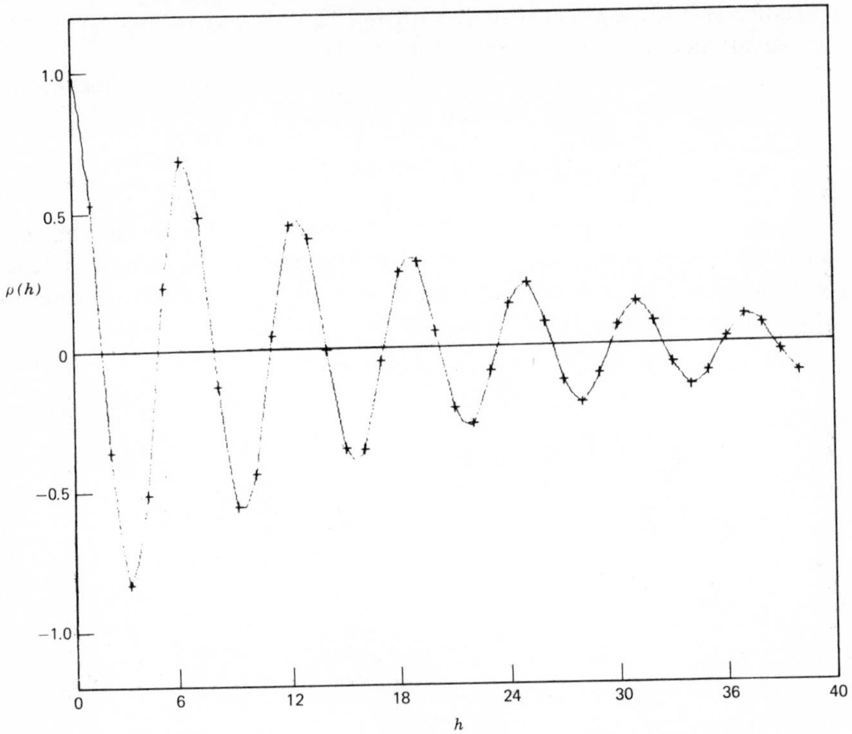

Figure 2.5.2. Correlogram for $X_t = X_{t-1} - 0.89X_{t-2} + e_t$.

case the correlogram has the distinctive "declining cyclical" appearance associated with complex roots. This correlogram illustrates how such a process can generate a time series with the appearance of moderately regular "cycles." In the present case $\cos\theta \doteq 0.53$ and $\theta \doteq 0.322\pi$, where $\cos\theta$ is defined in (2.5.11). Thus the apparent period of the cyclical behavior would be $2(0.322)^{-1} \doteq 6.2$ time units.

2.6. ALTERNATIVE REPRESENTATIONS OF AUTOREGRESSIVE AND MOVING AVERAGE PROCESSES

We have demonstrated that the stationary second order autoregressive time series can also be represented as an infinite moving average time series. This result generalizes immediately to:

Theorem 2.6.1. Let the roots of the polynomial

$$m^p + a_1 m^{p-1} + a_2 m^{p-2} + \cdots + a_p = 0$$

be less than one in absolute value, where $a_p \neq 0$, and let the weights $\{w_j\}_{j=0}^{\infty}$ be defined by the solution of the homogeneous difference equation

$$w_j + a_1 w_{j-1} + a_2 w_{j-2} + \cdots + a_p w_{j-p} = 0, \qquad j = p, p+1, \ldots,$$

subject to the boundary conditions

$$\begin{aligned} w_0 &= 1 \\ w_1 &= -a_1 w_0 \\ w_2 &= -a_1 w_1 - a_2 w_0 \\ w_3 &= -a_1 w_2 - a_2 w_1 - a_3 w_0 \\ &\vdots \\ w_{p-1} &= -a_1 w_{p-2} - a_2 w_{p-3} - a_3 w_{p-4} - \cdots - a_{p-1} w_0. \end{aligned}$$

Let $\{e_t\}$ be a sequence of uncorrelated $(0, \sigma^2)$ random variables. Then the mean square limit

$$X_t = \sum_{j=0}^{\infty} w_j e_{t-j}$$

is a stationary moving average process. Moreover, X_t satisfies the stochastic difference equation

$$X_t + a_1 X_{t-1} + a_2 X_{t-2} + \cdots + a_p X_{t-p} = e_t$$

for almost every realization of $\{e_t\}$.

Proof. Let $m_s (s = 1, 2, \ldots, d)$ be the real roots of the characteristic polynomial and let $m_s = r_s e^{i\theta_s}, m_{s+1} = r_s e^{-i\theta_s} (s = d+1, d+3, \ldots, p-1)$ be the conjugate pairs of complex roots. The weights w_j are of the form

$$w_j = \sum_{s=1}^{p} c_s w_j^{(s)},$$

where the c_s are real coefficients and the form of $w_j^{(s)}$ is

$$w_j^{(s)} = (j)^l (m_s)^j, \qquad s = 1, 2, \ldots, d,$$

$$\left[w_j^{(s)}, w_j^{(s+1)} \right] = \left[(j)^l (r_s)^j \cos j\theta_s, (j)^l (r_s)^j \sin j\theta_s \right]$$

$$s = d+1, d+3, \ldots, p-1.$$

[The term $(j)^l$ accounts for possible multiplicity of roots, where $0 \leqslant l < p$.]

Define λ to be such that $1>\lambda>\max_s\{|m_s|\}$. Then, by Exercise 2.24, $\sum_{j=1}^{\infty}|w_j|<\infty$. The existence and stationarity of the time series $\{X_t\}_{t=-\infty}^{\infty}$ follow from Theorem 2.2.1 and Corollary 2.2.2.2. Therefore,

$$E\{|X_t+a_1X_{t-1}+a_2X_{t-2}+\cdots+a_pX_{t-p}-e_t|^2\}$$

$$=\lim_{M\to\infty} E\left\{\left|\sum_{k=0}^{p-1} e_{t-k}\left[\sum_{j=0}^{k} a_j w_{k-j}\right]\right.\right.$$

$$\left.\left.+\sum_{j=p}^{M}(w_j+a_1w_{j-1}+\cdots+a_pw_{j-p})e_{t-j}-e_t\right|^2\right\}=0,$$

where $a_0=1$, and X_t satisfies the difference equation except for realizations of $\{e_j\}$ in a set A_t, where $P\{A_t\}=0$. Now $P\{\cup_{t=-\infty}^{\infty}A_t\}=0$, which completes the proof. ▲

Corollary 2.6.1.1. Let X_t be stationary and satisfy

$$X_t+a_1X_{t-1}+\cdots+a_pX_{t-p}=e_t,$$

where the e_t are uncorrelated $(0,\sigma^2)$ random variables and the roots of the characteristic polynomial

$$m^p+a_1m^{p-1}+\cdots+a_p=0$$

are less than one in absolute value. Then,

$$\gamma(h)+a_1\gamma(h-1)+\cdots+a_p\gamma(h-p)=0, \qquad h=1,2,\ldots.$$

Proof. Reserved for the reader. See equation (2.5.6). ▲

Whenever discussing autoregressive time series, we have assumed that the roots of the auxiliary equation were less than one in absolute value. This is because we have explicitly or implicitly visualized the time series as being created in a forward manner. If a time series is created in a forward manner the effect of the initial conditions will go to zero only if the roots are less than one in absolute value. For example, if we define

$$Y_t=\begin{cases} a_0e_0, & t=0\\ \rho Y_{t-1}+e_t, & t=1,2,\ldots, \end{cases} \tag{2.6.1}$$

where it is understood that Y_t is formed by adding e_t to ρY_{t-1}, then $\{Y_t$:

$t=0,1,2,\ldots\}$ is stationary for $|\rho|<1$ and $a_0=(1-\rho^2)^{-1/2}$. However, for $|\rho|\geqslant 1$, the time series formed by adding e_t to ρY_{t-1} is clearly nonstationary for all a_0.

On the other hand, there is a stationary time series that satisfies the difference equation

$$Y_t=\rho Y_{t-1}+e_t, \qquad t=0,\pm 1,\pm 2,\ldots, \tag{2.6.2}$$

for $|\rho|>1$. To see this, let us consider the stationary time series

$$Y_r=0.8\,Y_{r-1}+\epsilon_r \tag{2.6.3}$$

$$=\sum_{j=0}^{\infty}(0.8)^j\epsilon_{r-j}, \qquad r=0,\pm 1,\pm 2,\ldots, \tag{2.6.4}$$

where ϵ_r are uncorrelated $(0,\sigma^2)$ random variables. If we change the direction in which we count on the integers so that $-r=t$ and $r-1=-t+1$ and divide (2.6.3) by 0.8, we have

$$Y_{t+1}=1.25\,Y_t-1.25\epsilon_t \tag{2.6.5}$$

and

$$Y_t=\sum_{j=0}^{\infty}(0.8)^j\epsilon_{t+j}. \tag{2.6.6}$$

By setting $e_{t+1}=-1.25\epsilon_t$, equation (2.6.5) can be written in the form (2.6.2) with $\rho=1.25$.

The covariance function for the time series (2.6.6) is the same as the covariance function of the time series (2.6.4). Thus, if a time series $\{Y_t$: $t\in(0,\pm 1,\pm 2,\ldots)\}$ has an autocorrelation function of the form $\rho^{|h|}$, where $0<|\rho|<1$, it can be written as a forward moving average of uncorrelated random variables or as a backward moving average of uncorrelated random variables. Likewise,

$$Y_t-\rho Y_{t-1}=e_t \tag{2.6.7}$$

defines a sequence of uncorrelated random variables, as does

$$Y_t-\rho^{-1}Y_{t-1}=Z_t. \tag{2.6.8}$$

The representations (2.6.4) and (2.6.6) make clear why the only stationary time series with unit root is the trivial time series that is constant (with probability one) for all t.

While both the e_t of (2.6.7) and the Z_t of (2.6.8) are uncorrelated random variables, the variance of Z_t is larger than the variance of e_t by a factor of ρ^{-2}. This explains why the representation (2.6.7) is the one that appears in the applications of stationary autoregressive processes. That is, one typically chooses the ρ in an equation such as (2.6.7) to minimize the variance of e_t. It is worth mentioning that nonstationary autoregressive representations with roots greater than or equal to one in absolute value have appeared in practice. That the stationary autoregressive time series can be given either a forward or a backward representation also finds some use, and we state the result before proceeding.

Corollary 2.6.1.2. Let the covariance function of the time series $\{X_t: t\in(0,\pm 1,\pm 2,\ldots)\}$ with zero mean satisfy the difference equation

$$\sum_{j=0}^{p} \alpha_j \gamma_X(h-j)=0, \qquad h=1,2,\ldots,$$

where $\alpha_0=1$ and the roots of the characteristic equation

$$\sum_{j=0}^{p} \alpha_j m^{p-j}=0$$

are less than one in absolute value. Then X_t satisfies the stochastic difference equation

$$\sum_{j=0}^{p} \alpha_j X_{t-j}=e_t,$$

where $\{e_t\}$ is a sequence of uncorrelated $(0,\sigma^2)$ random variables, and also satisfies the stochastic difference equation

$$\sum_{j=0}^{p} \alpha_j X_{t+j}=v_t,$$

where $\{v_t\}$ is a sequence of uncorrelated $(0,\sigma^2)$ random variables.

Having demonstrated that a stationary finite autoregressive time series can be given an infinite moving average representation, we now obtain an alternative representation of the finite moving average time series. Since the finite moving average time series can be viewed as a difference equation in e_t, we have a result parallel to that of Theorem 2.6.1.

Theorem 2.6.2. Let the time series $\{X_t: t\in(0,\pm 1,\ldots)\}$ be defined by

$$X_t=e_t+b_1e_{t-1}+b_2e_{t-2}+\cdots+b_qe_{t-q}, \qquad t=0,\pm 1,\pm 2,\ldots,$$

where $b_q \neq 0$, the roots of the characteristic equation

$$m^q + b_1 m^{q-1} + b_2 m^{q-2} + \cdots + b_q = 0$$

are less than one in absolute value and $\{e_t\}$ is a sequence of uncorrelated $(0,\sigma^2)$ random variables. Then X_t can be expressed as an infinite autoregressive process

$$\sum_{j=0}^{\infty} c_j X_{t-j} = e_t,$$

where the coefficients c_j satisfy the homogeneous difference equation

$$c_j + b_1 c_{j-1} + b_2 c_{j-2} + \cdots + b_q c_{j-q} = 0, \qquad j = q, q+1, \ldots,$$

and the initial conditions

$$\begin{aligned}
c_0 &= 1 \\
c_1 &= -b_1 \\
c_2 &= -b_1 c_1 - b_2 \\
&\vdots \\
c_{q-1} &= -b_1 c_{q-2} - b_2 c_{q-3} - \cdots - b_{q-1}.
\end{aligned}$$

Proof. Reserved for the reader. ▲

In discussing finite moving averages we placed no restrictions on the coefficients and, for example, the time series

$$Y_t = e_t + e_{t-1}$$

is clearly stationary. The root of the auxiliary equation for this difference equation is negative one and, therefore, the condition of Theorem 2.6.2 is not met. An attempt to express e_t as an autoregressive process using that theorem will fail because the remainder associated with an autoregressive representation of order n will be $\pm e_{t-n-1}$.

Time series satisfying the conditions of Theorem 2.6.2 are sometimes called *invertible* moving averages. From the example we see that not all moving average processes are invertible.

In our earlier discussion of the moving average time series we demonstrated that we could always assign the value one to the coefficient of e_t.

We were also able to obtain all autocorrelation functions of the first order moving average type for a α restricted to the range $[-1,1]$. We are now in a position to generalize this result.

Theorem 2.6.3. Given a time series X_t with zero mean and autocorrelation function

$$\rho_X(h)=\begin{cases}1, & h=0\\ \rho(1), & h=1\\ 0, & h>1,\end{cases}$$

where $|\rho(1)|<0.5$, there exists an α, $|\alpha|<1$, and a sequence of uncorrelated random variables $\{e_t\}$ such that X_t is defined by

$$X_t=e_t+\alpha e_{t-1}\ .$$

Proof. The equation in α, $\rho(1)=(1+\alpha^2)^{-1}\alpha$, has one root less than one in absolute value and one greater than one in absolute value. The root of smaller absolute value is chosen, and we define e_t by

$$e_t=\sum_{j=0}^{\infty}(-\alpha)^j X_{t-j}\ .$$

By Theorem 2.2.1 this random variable is well defined as a limit in squared mean and, by Theorem 2.2.2,

$$\begin{aligned}\gamma_e(0)=E\{e_t^2\}&=\sum_{j=0}^{\infty}(-\alpha)^{2j}\gamma_X(0)-2\alpha\sum_{j=0}^{\infty}(-\alpha)^{2j}\gamma_X(1)\\ &=\left[\frac{1}{1-\alpha^2}-\frac{2\alpha}{1-\alpha^2}\,\frac{\alpha}{1+\alpha^2}\right]\gamma_X(0)\\ &=\frac{\gamma_X(0)}{1+\alpha^2},\\ \gamma_e(h)&=\sum_{j=0}^{\infty}(-\alpha)^{h+2j}\gamma_X(0)+\sum_{j=0}^{\infty}(1+\alpha^2)(-\alpha)^{h-1+2j}\gamma_X(1)\\ &=\left[\frac{(-\alpha)^h}{1-\alpha^2}+\frac{(1+\alpha^2)(-\alpha)^{h-1}}{1-\alpha^2}\,\frac{\alpha}{1+\alpha^2}\right]\gamma_X(0)\\ &=0,\qquad h>0.\end{aligned}$$

▲

The dualities between autoregressive and moving average representations established in Theorems 2.6.1 and 2.6.2 can also be described using formal operations with the backward shift operator. We recall that the first order autoregressive time series

$$Y_t - \rho Y_{t-1} = [1 - \rho \mathcal{B}] Y_t = e_t, \qquad |\rho| < 1,$$

can be written as

$$Y_t = \sum_{j=0}^{\infty} \rho^j e_{t-j} = \left[\sum_{j=0}^{\infty} (\mathcal{B} \rho)^j \right] e_t$$

$$= \frac{1}{1 - \rho \mathcal{B}} e_t,$$

where it is understood that $\mathcal{B}^0$ is the identity operator, and we have written

$$[1 - \rho \mathcal{B}]^{-1} = \sum_{j=0}^{\infty} (\mathcal{B} \rho)^j$$

in analogy to the expansion that holds for a real number $|\rho \mathcal{B}| < 1$.

The fact that the operator can be formally manipulated as if it were a real number with absolute value 1 furnishes a useful way of obtaining the alternative expressions for moving average and autoregressive processes whose characteristic equations have roots less than one in absolute value. Thus the second order autoregressive time series

$$X_t + a_1 X_{t-1} + a_2 X_{t-2} = e_t$$

can also be written as

$$(1 + a_1 \mathcal{B} + a_2 \mathcal{B}^2) X_t = e_t,$$

or

$$(1 - m_1 \mathcal{B})(1 - m_2 \mathcal{B}) X_t = e_t,$$

or

$$X_t = [(1 - m_1 \mathcal{B})(1 - m_2 \mathcal{B})]^{-1} e_t,$$

where m_1 and m_2 are the roots of the characteristic equation

$$m^2 + a_1 m + a_2 = 0.$$

Since the roots of the quadratic in $\mathcal{B}$,

$$1+a_1\mathcal{B}+a_2\mathcal{B}^2=0,$$

are the reciprocals of the roots of the characteristic equation, the restriction on the roots is often stated as the requirement that the roots of

$$1+a_1\mathcal{B}+a_2\mathcal{B}^2=0$$

be greater than one in absolute value.

Using the backward shift operator, the invertible second order moving average

$$Y_t=e_t+b_1e_{t-1}+b_2e_{t-2}$$

can be given the representation

$$Y_t=(1+b_1\mathcal{B}+b_2\mathcal{B}^2)e_t$$

or

$$Y_t=(1-g_1\mathcal{B})(1-g_2\mathcal{B})e_t$$

where g_1 and g_2, the roots of

$$g^2+b_1g+b_2=0,$$

are less than one in absolute value. The autoregressive representation of the moving average process is then given by

$$e_t=[(1-g_1\mathcal{B})(1-g_2\mathcal{B})]^{-1}Y_t.$$

The backward shift representation of the moving average process is useful in illustrating the fact that any finite moving average process whose characteristic equation has some roots greater than one and some less than one can be given a representation whose characteristic equation has all roots less than one in absolute value.

Theorem 2.6.4. Let $\{X_t\colon t\in(0,\pm1,\pm2,\ldots)\}$ have the representation

$$X_t=\prod_{i=1}^{p}(1-m_i\mathcal{B})e_t,$$

where the e_t are uncorrelated $(0,\sigma^2)$ random variables and none of the m_i,

$i=1,2,\ldots,p$, are of unit absolute value. Let $m_1, m_2, \ldots, m_L$, $0<L\leqslant p$, be greater than one in absolute value. Then X_t also has the representation

$$X_t=\begin{cases}\prod_{i=1}^{L}(1-m_i^{-1}\mathcal{B})\prod_{i=L+1}^{p}(1-m_i\mathcal{B})\epsilon_t, & 0<L<p\\ \prod_{i=1}^{p}(1-m_i^{-1}\mathcal{B})\epsilon_t, & L=p,\end{cases}$$

where the ϵ_t are uncorrelated $(0,\sigma^2[\Pi_{i=1}^{L} m_i]^2)$.

Proof. We define the polynomial in $\mathcal{B}$, $\mathcal{Q}(\mathcal{B})$, by

$$\mathcal{Q}(\mathcal{B})=\begin{cases}\prod_{i=1}^{L}(1-m_i^{-1}\mathcal{B})\prod_{i=L+1}^{p}(1-m_i\mathcal{B}), & 0<L<p\\ \prod_{i=1}^{p}(1-m_i^{-1}\mathcal{B}), & L=p.\end{cases}$$

The roots of $\mathcal{Q}(\mathcal{B})$ are greater than one in absolute value and the roots of $m^p\mathcal{Q}(m^{-1})$ are less than one in absolute value. Therefore the time series

$$\epsilon_t=\sum_{j=0}^{\infty} w_j X_{t-j},$$

where the w_j satisfy the difference equation

$$\mathcal{Q}(\mathcal{B})w_j=0, \qquad j=p,p+1,\ldots,$$

and the initial conditions outlined in Theorem 2.6.2, is well defined. Let m_1 be a real root greater than one in absolute value and define Z_t by

$$Z_t=\prod_{i=L+1}^{p}(1-m_i\mathcal{B})e_t.$$

Then the two time series $Z_t-m_1Z_{t-1}$ and $m_1(Z_t-m_1^{-1}Z_{t-1})$ have the same covariance function. Likewise if m_1 and $m_2=m_1^*$ are a complex conjugate pair of roots with absolute value greater than one, $Z_t-(m_1+m_2)Z_{t-1}+|m_1|^2Z_{t-2}$ has the same covariance function as $|m_1|^2[Z_t-(m_1^{-1}+m_2^{-1})Z_{t-1}+|m_1|^{-2}Z_{t-2}]$.

By repeated application of these arguments, we conclude that the original time series

$$X_t=\prod_{i=1}^{p}(1-m_i\mathcal{B})e_t$$

and the time series

$$Y_t = \left(\prod_{i=1}^{L} m_i \right) \mathcal{Q}(\mathcal{B}) e_t$$

have the same covariance function. Therefore,

$$\epsilon_t = \mathcal{Q}^{-1}(\mathcal{B}) X_t$$

and

$$\left(\prod_{i=1}^{L} m_i \right) e_t = \mathcal{Q}^{-1}(\mathcal{B}) Y_t$$

have the same covariance function. ▲

As an example of this theorem, the moving average

$$\begin{aligned} X_t &= e_t - 3.2e_{t-1} - 3.2e_{t-2} \\ &= (1 - 4.0\mathcal{B})(1 + 0.8\mathcal{B})e_t, \end{aligned}$$

where the e_t are uncorrelated $(0, \sigma^2)$, also has the representation

$$\begin{aligned} X_t &= \epsilon_t + 0.55\epsilon_{t-1} - 0.20\epsilon_{t-2} \\ &= (1 - 0.25\mathcal{B})(1 + 0.8\mathcal{B})\epsilon_t, \end{aligned}$$

where the ϵ_t are uncorrelated $(0, 16\sigma^2)$. The reader may check that $\gamma_X(0)$, $\gamma_X(1)$, and $\gamma_X(2)$ are $21.48\sigma^2$, $7.04\sigma^2$ and $-3.20\sigma^2$, respectively, for this moving average.

2.7. AUTOREGRESSIVE MOVING AVERAGE TIME SERIES

Having considered autoregressive and moving average processes, it is natural to investigate time series defined by the combination of low order autoregressive and moving average components. The sequence $\{X_t: t \in (0, \pm 1, \pm 2, \ldots)\}$, defined by

$$X_t + a_1 X_{t-1} + \cdots + a_p X_{t-p} = e_t + b_1 e_{t-1} + \cdots + b_q e_{t-q}, \tag{2.7.1}$$

where $a_p \neq 0$, $b_q \neq 0$, and $\{e_t\}$ is a sequence of uncorrelated $(0, \sigma^2)$ random variables, is called an *autoregressive moving average time series of order*

(p,q), which we shall abbreviate[3] to autoregressive moving average (p,q). The autoregressive moving average (1,1),

$$X_t - \theta_1 X_{t-1} = e_t + b_1 e_{t-1}, \tag{2.7.2}$$

where $|\theta_1| < 1$, has furnished a useful approximation for some time series encountered in practice. Note that we have defined the coefficient on X_{t-1} as $-\theta_1$ to simplify the representations to follow. If we let

$$u_t = e_t + b_1 e_{t-1}$$

and write

$$X_t = \theta_1 X_{t-1} + u_t,$$

we can express X_t of (2.7.2) as an infinite moving average of the u_t,

$$X_t = \sum_{j=0}^{\infty} \theta_1^j u_{t-j},$$

and hence of the e_t,

$$\begin{aligned} X_t &= \sum_{j=0}^{\infty} \theta_1^j e_{t-j} + b_1 \sum_{j=0}^{\infty} \theta_1^j e_{t-j-1} \\ &= e_t + (\theta_1 + b_1) \sum_{j=1}^{\infty} \theta_1^{j-1} e_{t-j}, \end{aligned} \tag{2.7.3}$$

where the random variables are defined by Theorem 2.2.1.

If $|b_1| < 1$, we can also express e_t as an infinite moving average of the u_t,

$$e_t = \sum_{j=0}^{\infty} (-b_1)^j u_{t-j},$$

and hence of the X_t,

$$e_t = X_t - (\theta_1 + b_1) \sum_{j=1}^{\infty} (-b_1)^{j-1} X_{t-j}. \tag{2.7.4}$$

If we let Z_t denote the first order autoregressive process

$$Z_t = \sum_{j=0}^{\infty} \theta_1^j e_{t-j},$$

[3] Box and Jenkins (1970) suggest the further abbreviated notation ARMA(p,q).

then

$$X_t = Z_t + b_1 Z_{t-1}. \tag{2.7.5}$$

That is, the time series can be expressed as a first order moving average of a first order autoregressive time series. Using this representation we express the autocovariance function of X_t in terms of that of Z_t,

$$\gamma_X(h) = (1+b_1^2)\gamma_Z(h) + b_1\gamma_Z(h-1) + b_1\gamma_Z(h+1),$$

to obtain

$$\gamma_X(h) = \left[\frac{1+b_1^2}{1-\theta_1^2}\theta_1^{|h|} + \frac{b_1}{1-\theta_1^2}\left(\theta_1^{|h-1|} + \theta_1^{|h+1|}\right)\right]\sigma^2. \tag{2.7.6}$$

Hence,

$$\begin{aligned} \gamma_X(h) &= \frac{1+b_1^2+2b_1\theta_1}{1-\theta_1^2}\sigma^2, & h&=0 \\ &= \frac{(1+b_1\theta_1)(\theta_1+b_1)}{1-\theta_1^2}\theta_1^{h-1}\sigma^2, & h&=1,2,\ldots, \end{aligned} \tag{2.7.7}$$

and

$$\begin{aligned} \rho_X(h) &= 1, & h&=0 \\ &= \frac{(1+b_1\theta_1)(\theta_1+b_1)}{1+b_1^2+2b_1\theta_1}\theta_1^{h-1}, & h&=1,2,\ldots . \end{aligned}$$

Thus, for $h \geqslant 1$, the autocorrelation function of the autoregressive moving average (1, 1) has the same appearance as that of a first order autoregressive time series in that it is declining at a geometric rate where the rate is θ_1.

From equation (2.7.7) we see that $\gamma_X(h)$ is zero for $h \geqslant 1$ if $b_1 = -\theta_1$. The autoregressive moving average (1, 1) process reduces to a sequence of uncorrelated random variables in such a case, which is also clear from the representation in (2.7.3).

These results generalize to higher order processes in the obvious manner. We state the generalizations of (2.7.3) and (2.7.4).

Theorem 2.7.1. Let the stationary time series X_t satisfy

$$\sum_{j=0}^{p} a_j X_{t-j} = \sum_{i=0}^{q} b_i e_{t-i},$$

where $a_0 = b_0 = 1$, the roots of

$$m^p + a_1 m^{p-1} + \cdots + a_p = 0$$

are less than one in absolute value, and $\{e_t\}$ is a sequence of uncorrelated $(0, \sigma^2)$ random variables. Then X_t has the representation

$$X_t = \sum_{j=0}^{\infty} v_j e_{t-j},$$

where

$$\begin{aligned}
v_0 &= b_0 = 1 \\
v_1 &= b_1 - a_1 \\
v_2 &= b_2 - a_2 - a_1 v_1 \\
&\vdots \\
v_j &= b_j - \sum_{i=1}^{\min(j,p)} a_i v_{j-i}, \qquad j \leqslant q \\
v_j &= - \sum_{i=1}^{\min(j,p)} a_i v_{j-i}, \qquad j > q.
\end{aligned}$$

Proof. Writing

$$\sum_{j=0}^{p} a_j X_{t-j} = u_t,$$

we obtain

$$\begin{aligned}
X_t &= \sum_{j=0}^{\infty} w_j u_{t-j} \\
&= \sum_{j=0}^{\infty} w_j \left(\sum_{i=0}^{q} b_i e_{t-j-i} \right) \\
&\overset{\text{(say)}}{=} \sum_{r=0}^{\infty} v_r e_{t-r},
\end{aligned}$$

where $u_t = \sum_{i=0}^{q} b_i e_{t-i}$ and the w_j are defined in Theorem 2.6.1. Since the w_j are absolutely summable, this representation is well defined as a limit in

mean square. We then write

$$\sum_{i=0}^{q} b_i e_{t-i} = \sum_{j=0}^{p} a_j X_{t-j} = \sum_{j=0}^{p} a_j \left(\sum_{r=0}^{\infty} v_r e_{t-j-r} \right)$$

and obtain the result by equating coefficients of e_{t-i}, $i=0,1,\ldots$. ▲

Theorem 2.7.2. Let the stationary time series X_t satisfy

$$\sum_{j=0}^{p} a_j X_{t-j} = \sum_{i=0}^{q} b_i e_{t-i},$$

where $a_0 = b_0 = 1$, the roots of

$$m^p + a_1 m^{p-1} + \cdots + a_p = 0$$

and of

$$m^q + b_1 m^{q-1} + \cdots + b_q = 0$$

are less than one in absolute value, and $\{e_t\}$ is a sequence of uncorrelated $(0, \sigma^2)$ random variables. Then X_t has the representation

$$e_t = \sum_{j=0}^{\infty} d_j X_{t-j},$$

where

$$d_0 = 1$$

$$d_1 = -b_1 + a_1$$

$$d_2 = -b_1 d_1 - b_2 + a_2$$

$$\vdots$$

$$d_j = -\sum_{i=1}^{\min(j,q)} b_i d_{j-i} + a_j, \qquad j \leq p$$

$$d_j = -\sum_{i=1}^{\min(j,q)} b_i d_{j-i}, \qquad j > p.$$

Proof. Reserved for the reader. ▲

2.8. VECTOR PROCESSES

From an operational standpoint the generalizations of moving average and autoregressive representations to the vector case are obtained by

substituting matrix and vector expressions for the scalar expressions of the preceding sections.

Consider a first order moving average process of dimension two:

$$X_{1t} = b_{11}e_{1,t-1} + b_{12}e_{2,t-1} + e_{1t}$$
$$X_{2t} = b_{21}e_{1,t-1} + b_{22}e_{2,t-1} + e_{2t}. \tag{2.8.1}$$

The system may be written in matrix notation as

$$\mathbf{X}_t = \mathbf{B}\mathbf{e}_{t-1} + \mathbf{e}_t,$$

where $\{\mathbf{e}_t\}$ is a sequence of uncorrelated $(\mathbf{0}, \mathbf{\Sigma})$ vector random variables,

$$\mathbf{\Sigma} = E\{\mathbf{e}_t\mathbf{e}_t'\} = \begin{pmatrix} \sigma_{11} & \sigma_{12} \\ \sigma_{21} & \sigma_{22} \end{pmatrix},$$

and we assume $\mathbf{B}$ contains at least one nonzero element. Then the autovariance function of $\mathbf{X}_t$ is

$$\mathbf{\Gamma}(h) = E\{\mathbf{X}_t\mathbf{X}_{t+h}'\} = E\{(\mathbf{B}\mathbf{e}_{t-1} + \mathbf{e}_t)(\mathbf{e}_{t+h-1}'\mathbf{B}' + \mathbf{e}_{t+h}')\}$$

and it follows that

$$\mathbf{\Gamma}(h) = \begin{cases} \mathbf{B\Sigma B}' + \mathbf{\Sigma}, & h = 0 \\ \mathbf{\Sigma B}', & h = 1 \\ \mathbf{B\Sigma}, & h = -1 \\ \mathbf{0}, & \text{otherwise.} \end{cases}$$

Note that $\mathbf{\Gamma}(h) = \mathbf{\Gamma}'(-h)$, as we would expect from Lemma 1.7.1.

The qth order moving average time series of dimension k is defined by

$$\mathbf{X}_t = \sum_{j=0}^{q} \mathbf{B}_j\mathbf{e}_{t-j},$$

where the $\mathbf{B}_j$ are $k \times k$ matrices with $\mathbf{B}_0 = \mathbf{I}$ and at least one element of $\mathbf{B}_q$ not zero, and the $\mathbf{e}_t$ are uncorrelated $(\mathbf{0}, \mathbf{\Sigma})$ random variables. Then,

$$\mathbf{\Gamma}(h) = \begin{cases} \sum_{j=0}^{q-h} \mathbf{B}_j\mathbf{\Sigma}\mathbf{B}_{j+h}', & 0 \leqslant h \leqslant q \\ \sum_{j=0}^{q+h} \mathbf{B}_{j-h}\mathbf{\Sigma}\mathbf{B}_j', & -q \leqslant h \leqslant 0 \\ \mathbf{0}, & |h| > q. \end{cases} \tag{2.8.2}$$

As with the scalar process, the autocovariance matrix is zero for $|h|>q$.

In investigating the properties of infinite moving average and autoregressive processes in the scalar case, we made extensive use of absolutely summable sequences. We now extend this concept to matrix sequences.

Definition 2.8.1. Let $\{\mathbf{G}_j\}_{j=1}^{\infty}$ be a sequence of $k\times r$ matrices where the elements of $\mathbf{G}_j$ are $g_{im}(j)$, $i=1,2,\ldots,k$, $m=1,2,\ldots,r$. If each of the kr sequences $\{g_{im}(j)\}_{j=1}^{\infty}$ is absolutely summable, we say that the sequence $\{\mathbf{G}_j\}_{j=1}^{\infty}$ is absolutely summable.

The infinite moving average

$$\mathbf{X}_t=\sum_{j=0}^{\infty}\mathbf{B}_j\mathbf{e}_{t-j},$$

where the $\mathbf{e}_t$ are uncorrelated $(\mathbf{0},\mathbf{\Sigma})$ random variables, will be well defined as a limit in mean square if $\{\mathbf{B}_j\}$ is absolutely summable. Thus, for example, if $\mathbf{X}_t$ is a vector stationary time series of dimension k with zero mean and absolutely summable covariance function and $\{\mathbf{W}_j\}_{j=-\infty}^{\infty}$ is an absolutely summable sequence of $k\times k$ matrices, then

$$\mathbf{Y}_t=\sum_{j=-\infty}^{\infty}\mathbf{W}_j\mathbf{X}_{t-j}$$

is well defined as a limit in mean square and

$$\mathbf{\Gamma}_Y(h)=\sum_{j=-\infty}^{\infty}\sum_{s=-\infty}^{\infty}\mathbf{W}_j\mathbf{\Gamma}_X(h-s+j)\mathbf{W}_s'.$$

The vector autoregressive process of order p and dimension k is defined by

$$\sum_{j=0}^{p}\mathbf{A}_j\mathbf{X}_{t-j}=\mathbf{e}_t, \tag{2.8.3}$$

where the $\mathbf{e}_t$ are uncorrelated k-dimensional $(\mathbf{0},\mathbf{\Sigma})$ random variables and the $\mathbf{A}_j$ are $k\times k$ matrices with $\mathbf{A}_0=\mathbf{I}$ and $\mathbf{A}_p\neq\mathbf{0}$. The process may be expressed as an infinte one-sided moving average of the $\mathbf{e}_t$ if the roots of the characteristic equation are less than one in absolute value.

Theorem 2.8.1. Let the stationary time series $\{\mathbf{X}_t\colon t\in(0,\pm1,\pm2,\ldots)\}$ satisfy (2.8.3) and let the roots of the determinantal equation

$$|\mathbf{A}_0m^p+\mathbf{A}_1m^{p-1}+\cdots+\mathbf{A}_p|=0$$

be less than one in absolute value. Then $\mathbf{X}_t$ can be expressed as an infinite moving average of the $\mathbf{e}_t$, say

$$\mathbf{X}_t = \sum_{j=0}^{\infty} \mathbf{W}_j \mathbf{e}_{t-j},$$

where the matrices $\mathbf{W}_j$ satisfy

$$\mathbf{W}_0 = \mathbf{I}$$

$$\mathbf{W}_1 = -\mathbf{A}_1$$

$$\mathbf{W}_2 = -\mathbf{A}_1\mathbf{W}_1 - \mathbf{A}_2$$

$$\vdots$$

$$\mathbf{W}_s = -\sum_{j=1}^{p} \mathbf{A}_j \mathbf{W}_{s-j}, \qquad s = p, p+1, \ldots .$$

Proof. Essentially the same argument as used at the beginning of Section 2.5 may be used to demonstrate that $\Sigma_{j=0}^{\infty} \mathbf{W}_j \mathbf{e}_{t-j}$ is the solution of the difference equation. Since the roots of the determinantal equation are less than one in absolute value, the sequence of matrices $\{\mathbf{W}_j\}$ is absolutely summable and the random variables are well defined as limits in mean square. ▲

Given Theorem 2.8.1, we can construct the multivariate Yule-Walker equations for the autocovariances of the process. Multiplying (2.8.3) on the right by $\mathbf{X}'_{t-s}$, $s \geq 0$ and taking expectations, we obtain

$$\sum_{j=0}^{p} \mathbf{A}_j \boldsymbol{\Gamma}(j) = \boldsymbol{\Sigma}, \qquad s = 0,$$

$$\sum_{j=0}^{p} \mathbf{A}_j \boldsymbol{\Gamma}(j-s) = \mathbf{0}, \qquad s = 1, 2, \ldots, \tag{2.8.4}$$

where we have used

$$E\{\mathbf{e}_t \mathbf{X}'_{t-s}\} = \boldsymbol{\Sigma}, \qquad s = 0$$

$$= \mathbf{0}, \qquad \text{otherwise.}$$

These equations are of the same form as those given in equation (2.5.6) and Corollary 2.6.1.1, but they contain a larger number of elements.

Theorem 2.8.2. Let the time series $\mathbf{X}_t$ be defined by

$$\mathbf{X}_t = \sum_{j=0}^{q} \mathbf{B}_j \mathbf{e}_{t-j},$$

where the $\mathbf{e}_t$ are uncorrelated k-dimensional vector random variables with zero mean and covariance matrix $\mathbf{\Sigma}$, the $\mathbf{B}_j$ are $k \times k$ matrices with $\mathbf{B}_0 = \mathbf{I}$, $\mathbf{B}_q \neq \mathbf{0}$, and the kq roots of the determinantal equation

$$|\mathbf{B}_0 m^q + \mathbf{B}_1 m^{q-1} + \cdots + \mathbf{B}_q| = 0$$

are less than one in absolute value. Then $\mathbf{X}_t$ can be expressed as an infinite autoregressive process

$$\mathbf{e}_t = \sum_{j=0}^{\infty} \mathbf{C}_j \mathbf{X}_{t-j},$$

where the $\mathbf{C}_j$ satisfy

$$\begin{aligned}
\mathbf{C}_0 &= \mathbf{I}, \\
\mathbf{C}_1 &= -\mathbf{B}_1, \\
\mathbf{C}_2 &= -\mathbf{B}_1\mathbf{C}_1 - \mathbf{B}_2, \\
&\vdots \\
\mathbf{C}_s &= -\sum_{j=1}^{q} \mathbf{B}_j \mathbf{C}_{s-j}, \qquad s = q, q+1, \ldots .
\end{aligned}$$

Proof. Reserved for the reader. ▲

It can also be demonstrated that an autoregressive moving average of the form

$$\sum_{j=0}^{p} \mathbf{A}_j \mathbf{X}_{t-j} = \sum_{j=0}^{q} \mathbf{B}_j \mathbf{e}_{t-j},$$

where the roots of

$$\left| \sum_{j=0}^{p} \mathbf{A}_j m^{p-j} \right| = 0$$

and of

$$\left| \sum_{i=0}^{q} \mathbf{B}_i m^{q-i} \right| = 0$$

are less than one in absolute value, can be given either an infinite autoregressive or an infinite moving average representation with matrix weights of the same form as the scalar weights of Theorems 2.7.1 and 2.7.2.

2.9. PREDICTION

One of the important problems in time series analysis is the following: Given n observations on a realization, predict the $(n+s)$th observation in the realization where s is a positive integer. The prediction is sometimes called the *forecast* of the $(n+s)$th observation. Because of the functional nature of a realization, prediction is also called *extrapolation*. We introduce the problem at this time because of the importance of autoregressive and moving average processes in prediction. In order to discuss alternative predictors, it is necessary to have a criterion by which the performance of a predictor is measured. The usual criterion, and the one that we adopt, is the mean square error of the predictor. For example, if $\hat{Y}_{n+s}(Y_1,\dots,Y_n)$ is the predictor of Y_{n+s} based on the n observations $Y_1, Y_2,\dots,Y_n$, then the mean square error (MSE) of the predictor is

$$\text{MSE}\{\hat{Y}_{n+s}(Y_1,\dots,Y_n)\}=E\left\{\left[Y_{n+s}-\hat{Y}_{n+s}(Y_1,\dots,Y_n)\right]^2\right\}. \qquad (2.9.1)$$

Generally, the problem of determining optimal predictors requires that the class of predictors be restricted. We shall investigate *linear* predictors.

The best linear predictor for a stationary time series with known covariance function and known mean is easily obtained.

Theorem 2.9.1. Given n observations on a realization of a stationary time series with zero mean and known covariance function, the minimum mean square error linear predictor of the $(n+s)$th, $s=1,2,\dots,$ observation is

$$\hat{Y}_{n+s}(Y_1,\dots,Y_n)=\mathbf{y}'\mathbf{b}_s, \qquad (2.9.2)$$

where

$$\mathbf{y}'=(Y_1,Y_2,\dots,Y_n),$$

$$\mathbf{b}_s=\mathbf{V}_{nn}^{+}\mathbf{V}_{ns},$$

$$\mathbf{V}_{nn}=E\{\mathbf{y}\mathbf{y}'\},$$

$$\mathbf{V}_{ns}'=\left[\gamma_Y(n+s-1),\gamma_Y(n+s-2),\dots,\gamma_Y(s)\right],$$

and

$$\mathbf{V}_{nn}^{+} \quad \text{is the Moore-Penrose generalized inverse}^4 \text{ of } \mathbf{V}_{nn}.$$

Proof. To find the minimum mean square error linear predictor, we minimize

$$E\left\{\left[Y_{n+s}-\hat{Y}_{n+s}(Y_1,\ldots,Y_n)\right]^2\right\}=E\left\{\left[Y_{n+s}-\mathbf{y}'\mathbf{b}_s\right]^2\right\}$$
$$=(1,-\mathbf{b}_s')\begin{pmatrix}\gamma_Y(0) & \mathbf{V}_{ns}' \\ \mathbf{V}_{ns} & \mathbf{V}_{nn}\end{pmatrix}\begin{pmatrix}1\\ -\mathbf{b}_s\end{pmatrix}$$

with respect to the vector of linear coefficients $\mathbf{b}_s$. Differentiating the quadratic form with respect to $\mathbf{b}_s$ and equating the derivative to zero, we find that the $\mathbf{b}_s$ that minimizes the expectation is $\mathbf{b}_s=\mathbf{V}_{nn}^{+}\mathbf{V}_{ns}$. ▲

To understand why the generalized inverse was required in Theorem 2.9.1, consider the time series

$$X_t=e_1\cos\pi t, \tag{2.9.3}$$

where e_1 is normally distributed with zero mean and variance σ^2. This time series is perfectly periodic with a period of two and covariance function $\gamma_X(h)=\sigma^2\cos\pi h$. The matrix $\mathbf{V}_{nn}$ for a sample of three observations is

$$\mathbf{V}_{33}=\begin{bmatrix}1 & -1 & 1\\ -1 & 1 & -1\\ 1 & -1 & 1\end{bmatrix}\sigma^2 \tag{2.9.4}$$

which is easily seen to be singular. Thus there are many possible linear combinations of past values that can be used to predict future values. For example, one can predict X_{t+1} by $-X_{t-1}$ or by X_{t-2} or by $-\frac{1}{2}X_{t-1}+\frac{1}{2}X_{t-2}$.

The time series of equation (2.9.3) is a member of the class of time series defined by

$$X_t=e_1\cos\lambda t+e_2\sin\lambda t, \tag{2.9.5}$$

where e_i, $i=1,2$, are independent $(0,\sigma^2)$ random variables and $\lambda\in(0,\pi]$. Given λ and two observations, it is possible to predict perfectly all future

[4] See, for example, Rao (1965, p. 25) for a discussion of the generalized inverse.

observations in a given realization. That is, for $n \geqslant 2$,

$$\begin{bmatrix} e_1 \\ e_2 \end{bmatrix} = \begin{bmatrix} \sum_{t=1}^{n} \cos^2 \lambda t & \sum_{t=1}^{n} \cos\lambda t \sin\lambda t \\ \sum_{t=1}^{n} \cos\lambda t \sin\lambda t & \sum_{t=1}^{n} \sin^2 \lambda t \end{bmatrix}^{-1} \begin{bmatrix} \sum_{t=1}^{n} X_t \cos\lambda t \\ \sum_{t=1}^{n} X_t \sin\lambda t \end{bmatrix}.$$

and all future observations can be predicted by substituting these e's into the defining equation (2.9.5).

This example illustrates that in predicting future observations of a realization from past observations of that realization, the element $\omega \in \Omega$ is fixed. In the current example, $\omega \in \Omega$ is a two-dimensional vector and can be perfectly identified for a particular realization by a few observations on that realization. In the more common situation, $\omega \in \Omega$ is infinite dimensional, and we can never determine it perfectly from a finite number of observations on the realization.

Time series for which it is possible to use the observations $Y_t, Y_{t-1}, \ldots$ of a realization to predict future observations with zero mean square error are called *deterministic*. They are also sometimes called *singular*, a term that is particularly meaningful when one looks at the covariance matrix (2.9.4). Time series that have a nonzero lower bound on the prediction error are called *regular* or *nondeterministic*.

While Theorem 2.9.1 gives a quite general solution to the problem of linear prediction, the computation of $n \times n$ generalized inverses can become inconvenient if one has a large number of predictions to make. Fortunately, the forecast equations for autoregressive and invertible moving average processes are relatively simple.

We begin with the first order autoregressive process, assuming the parameter to be known. Let the time series Y_t be defined by

$$Y_t = \rho Y_{t-1} + e_t, \qquad |\rho| < 1,$$

where the e_t are independent $(0, \sigma^2)$ random variables. We wish to construct a predictor of $Y_{n+1} = \rho Y_n + e_{n+1}$. Clearly, the "best predictor" of ρY_n is ρY_n. Therefore, we need only predict e_{n+1}. But, by assumption, e_{n+1} is independent of $Y_n, Y_{n-1}, \ldots$. Hence, the best predictor of e_{n+1} is zero, since predicting e_{n+1} is equivalent to predicting a random selection from a distribution with zero mean and finite variance. Therefore, we choose as a

predictor of Y_{n+1} at time n,

$$\hat{Y}_{n+1}(Y_1,\ldots,Y_n)=\rho Y_n. \tag{2.9.6}$$

Notice that we constructed this predictor by finding the expected value of Y_{n+1} given $Y_n, Y_{n-1},\ldots,Y_1$. Because of the nature of the time series, knowledge of $Y_{n-1}, Y_{n-2},\ldots,Y_1$ added no information beyond that contained in knowledge of Y_n. It is clear that the mean square error of this predictor is

$$E\{[Y_{n+1}-\hat{Y}_{n+1}(Y_1,\ldots,Y_n)]^2\}=E\{e_{n+1}^2\}=\sigma^2.$$

To predict more than one period into the future, we recall the representation

$$Y_{n+s}=\rho^s Y_n+\sum_{j=1}^{s}\rho^{s-j}e_{n+j}, \qquad s=1,2,\ldots. \tag{2.9.7}$$

It follows that, for independent e_t, the conditional expectation of Y_{n+s}, given $Y_1, Y_2,\ldots,Y_n$, is

$$E\{Y_{n+s}\mid Y_1,\ldots,Y_n\}=\rho^s Y_n, \tag{2.9.8}$$

and this is the best predictor for Y_{n+s}.

These ideas generalize immediately to higher order processes. Let the pth order autoregressive process be defined by

$$Y_t=\sum_{j=1}^{p}\theta_j Y_{t-j}+e_t,$$

where the roots of

$$m^p-\sum_{j=1}^{p}\theta_j m^{p-j}=0$$

are less than one in absolute value and the e_t are uncorrelated $(0,\sigma^2)$ random variables. On the basis of the arguments for the first order process we choose as the one period ahead linear predictor for the stationary pth order autoregressive process

$$\hat{Y}_{n+1}(Y_1,\ldots,Y_n)=\sum_{j=1}^{p}\theta_j Y_{n+1-j}. \tag{2.9.9}$$

It is clear that this predictor minimizes (2.9.1), since the prediction error is e_{n+1} and this is uncorrelated with Y_n and all previous Y_t's. If the e_t are independent random variables the predictor is the expected value of Y_{n+1} conditional on $Y_n, Y_{n-1}, \ldots, Y_1$. If the e_t are only uncorrelated we cannot make this statement. (See Exercise 2.16.) To obtain the best two period linear prediction, we note that

$$\begin{aligned}
Y_{n+2} &= \theta_1\left[\sum_{j=1}^{p} \theta_j Y_{n-j+1} + e_{n+1}\right] + \sum_{j=2}^{p} \theta_j Y_{n-j+2} + e_{n+2} \\
&= \theta_1\left[\hat{Y}_{n+1}(Y_1, \ldots, Y_n) + e_{n+1}\right] + \sum_{j=2}^{p} \theta_j Y_{n-j+2} + e_{n+2} \\
&= \sum_{j=1}^{p} (\theta_1\theta_j + \theta_{j+1}) Y_{n-j+1} + \theta_1 e_{n+1} + e_{n+2},
\end{aligned} \tag{2.9.10}$$

where it is understood that $\theta_{p+i}=0$ for $i=1,2,\ldots$ and, to simplify the subscripting, we have taken $p \geqslant 2$. Since e_{n+2} and e_{n+1} are uncorrelated with $(Y_1, Y_2, \ldots, Y_n)$, the two period predictor can be constructed as

$$\begin{aligned}
\hat{Y}_{n+2}(Y_1, \ldots, Y_n) &= \theta_1 \hat{Y}_{n+1}(Y_1, \ldots, Y_n) + \sum_{j=2}^{p} \theta_j Y_{n-j+2} \\
&= \sum_{j=1}^{p} (\theta_1\theta_j + \theta_{j+1}) Y_{n-j+1}.
\end{aligned}$$

In general, we can build predictors for s periods ahead by substituting the predictors for earlier periods into (2.9.9). Hence, for $p \geqslant 2$,

$$\begin{aligned}
\hat{Y}_{n+s}(Y_1, \ldots, Y_n) &= \sum_{j=1}^{s-1} \theta_j \hat{Y}_{n-j+s}(Y_1, \ldots, Y_n) + \sum_{j=s}^{p} \theta_j Y_{n-j+s}, \qquad s=2,3,\ldots,p, \\
&= \sum_{j=1}^{p} \theta_j \hat{Y}_{n-j+s}(Y_1, \ldots, Y_n), \qquad s=p+1, p+2, \ldots .
\end{aligned} \tag{2.9.11}$$

We now turn to the qth order moving average time series

$$Y_t = \sum_{j=1}^{q} \beta_j e_{t-j} + e_t, \tag{2.9.12}$$

where the roots of the characteristic equation

$$m^q + \sum_{j=1}^{q} \beta_j m^{q-j} = 0$$

are less than one in absolute value. We assume β_j, $j=1,2,\ldots,q$ are known. The pth order autoregressive process expresses Y_t as the sum of a linear combination of p previous Y's and e_t, while the qth order moving average process expresses Y_t as the sum of a linear combination of q previous e's and e_t. In both cases e_t is uncorrelated with previous Y's and uncorrelated with previous e's. Suppose we knew the e_t for $t=n,n-1,\ldots,n-q+1$ for the process (2.9.12). Then, by arguments analogous to those used for the autoregressive process, the best linear predictor for s periods ahead would be

$$\begin{aligned} \hat{Y}_{n+s}(e_1,\ldots,e_n) &= \sum_{j=s}^{q} \beta_j e_{n+s-j}, \qquad 1 \leqslant s \leqslant q, \\ &= 0, \qquad s > q. \end{aligned} \tag{2.9.13}$$

If the e_t are independent, it is clear that

$$\hat{Y}_{n+s}(e_1,\ldots,e_n) = E\{Y_{n+s} | e_1, e_2, \ldots, e_n\}.$$

Since we do not know the e_t, we must develop a predictor expressed in terms of the Y_t's. If n is small we can use the known covariance function and Theorem 2.9.1 to construct the predictor. If n is large and the roots of the characteristic equation are less than one in absolute value, we can use the autoregressive representation of Theorem 2.6.2,

$$Y_t = \sum_{j=1}^{\infty} c_j Y_{t-j} + e_t, \tag{2.9.14}$$

truncate the infinite sum at some finite $M \leqslant n$, and construct the predictions using (2.9.9) and (2.9.11). The large n predictor can also be expressed in terms of estimates of the e_t.

If e_{t-r}, $r=1,2,\ldots,q$, are known, then we can use (2.9.12) to define e_t as

$$e_t = Y_t - \sum_{j=1}^{q} \beta_j e_{t-j}.$$

Given n observations, one can approximate the e_t by $\tilde{e}_t$, where $\tilde{e}_0$, $\tilde{e}_{-1},\ldots,\tilde{e}_{-q+1}$ are taken to be zero and

$$\tilde{e}_t = Y_t - \sum_{j=1}^{q} \beta_j \tilde{e}_{t-j}, \qquad t=1,2,\ldots,n.$$

For n large relative to q the difference between $\tilde{e}_t$ and e_t will be small for $t=n,n-1,\ldots,n-q+1$. Therefore, the s period ahead predictor for large n is given by

$$\begin{aligned}\hat{Y}_{n+s}(Y_1,\ldots,Y_n) &= \sum_{j=s}^{q} \beta_j \tilde{e}_{n+s-j}, \qquad 1 \leqslant s \leqslant q,\\ &= 0, \qquad s > q.\end{aligned} \tag{2.9.15}$$

As the sample size becomes large, the variance of the one period ahead predictor (2.9.15) approaches σ^2. This is evident because

$$\begin{aligned}\lim_{m\to\infty} E\left\{\left[Y_{n+1}-\hat{Y}_{n+1}(Y_{-m},\ldots,Y_n)\right]^2\right\}\\ &= E\left\{\left[Y_{n+1}-\sum_{j=1}^{q}\beta_j e_{n+1-j}\right]^2\right\}\\ &= E\left\{e_{n+1}^2\right\}=\sigma^2.\end{aligned}$$

The procedures for autoregressive and moving average processes are easily extended to the autoregressive moving average time series defined by

$$Y_t = \sum_{j=1}^{p} \theta_j Y_{t-j} + \sum_{i=1}^{q} \beta_i e_{t-i} + e_t,$$

where the roots of

$$m^p - \sum_{j=1}^{p} \theta_j m^{p-j} = 0$$

and of

$$m^q + \sum_{i=1}^{q} \beta_i m^{q-i} = 0$$

are less than one in absolute value and the e_t are uncorrelated $(0,\sigma^2)$ random variables. For such processes the predictor for large n is given by

$$\begin{aligned}\hat{Y}_{n+s}(Y_1,\ldots,Y_n) &= \sum_{j=1}^{p} \theta_j Y_{n-j+1} + \sum_{i=1}^{q} \beta_i \tilde{e}_{n-i+1}, \qquad s=1,\\ &= \sum_{j=1}^{s-1} \theta_j \hat{Y}_{n-j+s}(Y_1,\ldots,Y_n) + \sum_{j=s}^{p} \theta_j Y_{n-j+s}\\ &\quad + \sum_{i=s}^{q} \beta_i \tilde{e}_{n-i+s}, \qquad s=2,3,\ldots,\end{aligned} \tag{2.9.16}$$

where

$$\begin{aligned}\tilde{e}_t &= 0, & t &= p, p-1, \ldots, \\ &= Y_t - \sum_{j=1}^{p} \theta_j Y_{t-j} - \sum_{i=1}^{q} \beta_i \tilde{e}_{t-i}, & t &= p+1, p+2, \ldots, n,\end{aligned}$$

and it is understood that $\theta_j = 0$ for $j > p$ and

$$\sum_{i=s}^{q} \beta_i \tilde{e}_{n-i+s} = 0, \quad s = q+1, q+2, \ldots,$$

$$\sum_{j=s}^{p} \theta_j Y_{n-j+s} = 0, \quad s = p+1, p+2, \ldots .$$

The stationary autoregressive invertible moving average, which contains the autoregressive and invertible moving average processes as special cases, can be expressed as

$$Y_t = \sum_{j=0}^{\infty} v_j e_{t-j}, \tag{2.9.17}$$

where $v_0 = 1$, $\sum_{j=0}^{\infty} |v_j| < \infty$ and the e_t are uncorrelated $(0, \sigma^2)$ random variables. The variances of the forecast errors for predictions of more than one period are readily obtained when the time series is written in this form.

Theorem 2.9.2. Given a stationary autoregressive invertible moving average with the representation (2.9.17) and the predictor,

$$\hat{Y}_{n+s}(Y_1, \ldots, Y_n) = \sum_{j=s}^{n+s-1} v_j \tilde{e}_{n-j+s},$$

where $\tilde{e}_t$, $t = 1, 2, \ldots, n$, is defined following (2.9.16), then,

$$\lim_{n \to \infty} E\left\{\left[Y_{n+s} - \hat{Y}_{n+s}(Y_1, \ldots, Y_n) - \sum_{j=0}^{s-1} v_j e_{n-j+s}\right]^2\right\} = 0.$$

Proof. The result follows immediately from Theorem 2.2.1. ▲

Using Theorem 2.9.2, we can construct the covariance matrix of the s predictors for $Y_{n+1}, Y_{n+2}, \ldots, Y_{n+s}$. Let $\mathbf{Y}'_{n,s} = [Y_{n+1}, Y_{n+2}, \ldots, Y_{n+s}]$ and let

$$\hat{\mathbf{Y}}'_{n,s} = \left[\hat{Y}_{n+1}(Y_1, \ldots, Y_n), \hat{Y}_{n+2}(Y_1, Y_2, \ldots, Y_n), \ldots, \hat{Y}_{n+s}(Y_1, \ldots, Y_n)\right].$$

Then

$$\lim_{n\to\infty} E\{(\mathbf{Y}_{n,s} - \hat{\mathbf{Y}}_{n,s})(\mathbf{Y}_{n,s} - \hat{\mathbf{Y}}_{n,s})'\}$$

$$= \begin{bmatrix} 1 & v_1 & \cdots & v_{s-1} \\ v_1 & \sum_{j=0}^{1} v_j^2 & \cdots & \sum_{j=0}^{1} v_j v_{s+j-2} \\ \vdots & \vdots & & \vdots \\ v_{s-1} & \sum_{j=0}^{1} v_j v_{s+j-2} & \cdots & \sum_{j=0}^{s-1} v_j^2 \end{bmatrix} \sigma^2.$$

Note that the predictors we have considered are unbiased; that is,

$$E\{Y_{n+s} - \hat{Y}_{n+s}(Y_1, \ldots, Y_n)\} = 0.$$

Therefore, the mean square error of the predictor is the variance. To use this information to establish confidence limits for the predictor, we require knowledge of the distribution of the e_t. If the time series is normal, confidence intervals can be constructed using normal theory. That is, a $(1-\eta)$ level confidence interval is given by

$$\hat{Y}_{n+s}(Y_1, \ldots, Y_n) \pm t_\eta \left[\operatorname{Var}\{Y_{n+s} - \hat{Y}_{n+s}(Y_1, \ldots, Y_n)\}\right]^{1/2},$$

where t_η is the value such that η of the probability of the standard normal distribution is beyond $\pm t_\eta$.

Example. In Section 6.4 we demonstrate that the U.S. Quarterly Seasonally Adjusted Unemployment Rate behaves much like the time series

$$(Y_t - 4.77) = 1.54(Y_{t-1} - 4.77) - 0.67(Y_{t-2} - 4.77) + e_t,$$

where the e_t are uncorrelated $(0, 0.115)$ random variables. Let us assume that we know that this is the proper representation. The four observations

for 1972 are 5.83, 5.77, 5.53 and 5.30. The predictions for 1973 are

$$\hat{Y}_{73-1}(Y_{72-1},\ldots,Y_{72-4}) = 4.77 + 1.54(0.53) - 0.67(0.76)$$
$$= 5.08,$$

$$\hat{Y}_{73-2}(Y_{72-1},\ldots,Y_{72-4}) = 4.77 + 1.54(0.3070) - 0.67(0.53)$$
$$= 4.89,$$

$$\hat{Y}_{73-3}(Y_{72-1},\ldots,Y_{72-4}) = 4.77 + 1.54(0.1177) - 0.67(0.3070)$$
$$= 4.75,$$

$$\hat{Y}_{73-4}(Y_{72-1},\ldots,Y_{72-4}) = 4.77 + 1.54(-0.0245) - 0.67(0.1177)$$
$$= 4.65.$$

The covariance matrix of the prediction errors can be obtained by using the representation $Y_t = \sum_{j=0}^{\infty} w_j e_{t-j}$, where the w's are defined in Theorem 2.6.1. We have $a_1 = -1.54$, $a_2 = 0.67$ and,

$$w_0 = 1,$$
$$w_1 = 1.54,$$
$$w_2 = (1.54)^2 - 0.67 = 1.702,$$
$$w_3 = 1.54(1.702) - 0.67(1.54) = 1.589.$$

Therefore, the covariance matrix of Theorem 2.9.2 is given by the product

$$0.115\mathbf{H}'\mathbf{H} = \begin{bmatrix} 0.115 & 0.177 & 0.196 & 0.183 \\ 0.177 & 0.388 & 0.478 & 0.477 \\ 0.196 & 0.478 & 0.720 & 0.789 \\ 0.183 & 0.477 & 0.789 & 1.011 \end{bmatrix},$$

where

$$\mathbf{H}' = \begin{bmatrix} 1.00 & 0 & 0 & 0 \\ 1.54 & 1.00 & 0 & 0 \\ 1.70 & 1.54 & 1.00 & 0 \\ 1.59 & 1.70 & 1.54 & 1.00 \end{bmatrix}.$$

We observe that there is a considerable increase in the variance of the predictor as s increases from 1 to 4. Also, there is a high positive correlation between the predictors.

Example. In Table 2.9.1 are seven observations from the time series

$$Y_t = 0.9Y_{t-1} + 0.7e_{t-1} + e_t,$$

where the e_t are normal independent (0, 1) random variables. The first 11 autocorrelations for this time series are, approximately, 1.00, 0.95, 0.85, 0.77, 0.69, 0.62, 0.56, 0.50, 0.45, 0.41, and 0.37. The variance of the time series is about 14.4737.

Table 2.9.1. Observations and predictions for the time series $Y_t = 0.9Y_{t-1} + 0.7e_{t-1} + e_t$

t	Observation	Prediction
1	−1.58	—
2	−2.86	—
3	−3.67	—
4	−2.33	—
5	0.42	—
6	2.87	—
7	3.43	—
8	—	3.04
9	—	2.73
10	—	2.46

To construct the linear predictor for Y_8, we obtain the seven weights by solving the system of equations

$$\begin{pmatrix} 1.00 & 0.95 & 0.85 & \cdots & 0.56 \\ 0.95 & 1.00 & 0.95 & \cdots & 0.62 \\ 0.85 & 0.95 & 1.00 & \cdots & 0.69 \\ \vdots & \vdots & \vdots & & \vdots \\ 0.56 & 0.62 & 0.69 & \cdots & 1.00 \end{pmatrix} \begin{pmatrix} b_{11} \\ b_{12} \\ b_{13} \\ \vdots \\ b_{17} \end{pmatrix} = \begin{pmatrix} 0.50 \\ 0.56 \\ 0.62 \\ \vdots \\ 0.95 \end{pmatrix}. \tag{2.9.18}$$

This system of equations gives the weights for the one period ahead forecast. The vector of weights $\mathbf{b}_1' = (0.06, -0.18, 0.32, -0.50, 0.75, -1.10, 1.59)$ is applied to $(Y_1, Y_2, Y_3, Y_4, Y_5, Y_6, Y_7)$. [In obtaining the solution more digits were used than are displayed in equation (2.9.18).] The predictor for two periods ahead is obtained by replacing the right side of (2.9.18) by $(0.45, 0.50, 0.56, 0.62, 0.69, 0.77, 0.85)'$, and the weights for the three period ahead predictor are obtained by using $(0.41, 0.45, 0.50, 0.56, 0.62, 0.69, 0.77)'$ as the right side. The predictions using these weights are displayed in Table 2.9.1.

The covariance matrix of the prediction errors is

$$E\{(\mathbf{Y}_{7,3}-\hat{\mathbf{Y}}_{7,3})(\mathbf{Y}_{7,3}-\hat{\mathbf{Y}}_{7,3})'\}=\begin{bmatrix}1.005 & 1.604 & 1.444\\ 1.604 & 3.563 & 3.907\\ 1.444 & 3.907 & 5.637\end{bmatrix},$$

where $\hat{\mathbf{Y}}_{7,3}$ and $\mathbf{Y}_{7,3}$ are defined following Theorem 2.9.2. The covariance matrix of prediction errors was obtained as $\mathbf{AVA}'$, where $\mathbf{V}$ is the 10×10 covariance matrix for 10 observations and

$$\mathbf{A}=\begin{bmatrix}0, & 0, & 1, & -1.59, & 1.10, & -0.75, & 0.50, & -0.32, & 0.18, & -0.06\\ 0, & 1, & 0, & -1.44, & 0.99, & -0.68, & 0.45, & -0.29, & 0.16, & -0.05\\ 1, & 0, & 0, & -1.29, & 0.89, & -0.61, & 0.41, & -0.26, & 0.14, & -0.05\end{bmatrix}.$$

If we had an infinite past to use in prediction the variance of the one period ahead prediction error would be one, the variance of e_t. Thus, the loss in prediction efficiency from having only seven observations is about one half of 1%. Since the time series is normal, we can construct 95% confidence intervals for the predictions. We obtain the intervals (1.08, 5.00), $(-0.97, 6.43)$, and $(-2.19, 7.11)$ for the predictions one, two, and three periods ahead, respectively.

To construct a predictor using the $\tilde{e}$-method, we set $\tilde{e}_1=0$ and calculate

$$\tilde{e}_2=Y_2-0.9Y_1-0.7\tilde{e}_1=-2.86-0.9(-1.58)=-1.438,$$

$$\tilde{e}_3=Y_3-0.9Y_2-0.7\tilde{e}_2=-3.67-0.9(-2.86)-0.7(-1.438)=-0.089,$$

$$\tilde{e}_4=-2.33-0.9(-3.67)-0.7(-0.089)=1.035,$$

$$\tilde{e}_5=0.42-0.9(-2.33)-0.7(1.035)=1.793,$$

$$\tilde{e}_6=2.87-0.9(0.42)-0.7(1.793)=1.237,$$

$$\tilde{e}_7=3.43-0.9(2.86)-0.7(1.227)=-0.003.$$

Therefore, the predictor of Y_8 is

$$\hat{Y}_8(Y_1,\ldots,Y_7)=0.9Y_7+0.7\tilde{e}_7=3.085$$

and

$$\hat{Y}_9(Y_1,\ldots,Y_7)=0.9\hat{Y}_8(Y_1,\ldots,Y_7)=2.776,$$

$$\hat{Y}_{10}(Y_1,\ldots,Y_7)=0.9\hat{Y}_9(Y_1,\ldots,Y_7)=2.498.$$

In this case the prediction using the approximate method of setting $\tilde{e}_1=0$ is quite close to the prediction obtained by the method of Theorem 2.9.1, even though the sample size is small. If the coefficient on e_{t-1} had been

closer to one, the difference would have been larger.

By Theorem 2.7.1, we may write our autoregressive moving average process as

$$Y_t = \sum_{j=0}^{\infty} v_j e_{t-j},$$

where

$$v_0 = b_0 = 1,$$

$$v_1 = b_1 - a_1 = 0.7 + 0.9 = 1.6,$$

$$v_j = -a_1 v_{j-1} = (0.9) v_{j-1}, \qquad j = 2, 3, \ldots .$$

Therefore, the large sample covariance matrix of prediction errors for predictors one, two, and three periods ahead is

$$\begin{bmatrix} 1.00 & 1.60 & 1.44 \\ 1.60 & 3.56 & 3.90 \\ 1.44 & 3.90 & 5.63 \end{bmatrix},$$

which is very close to the covariance matrix of the best predictor based on $n = 7$ observations.

REFERENCES

Sections 2.1, 2.3, 2.5–2.9. Anderson (1971), Bartlett (1966), Box and Jenkins (1970), Hannan (1970), Kendall and Stuart (1966), Pierce (1970a), Wold (1938), Yule (1927).

Section 2.2. Bartle (1964), Royden (1963), Tucker (1967).

Section 2.4. Bellman (1960), Finkbeiner (1960), Goldberg (1958), Hildebrand (1968), Kempthorne and Folks (1971), Miller (1963), (1968).

EXERCISES

1. Express the coefficients, b, of (2.4.3) in terms of the coefficients, a, of (2.4.2) for $n = 3$.
2. Given that $\{e_t\colon t \in (0, \pm 1, \pm 2, \ldots)\}$ is a sequence of independent $(0, \sigma^2)$ random variables, define

$$X_t = \sum_{i=-2}^{2} e_{t-i},$$

$$Y_t = \sum_{i=-1}^{1} X_{t-i}.$$

   (a) Derive and graph $\gamma_X(h)$ and $\gamma_Y(h)$.
   (b) Express Y_t as a moving average of e_t.
   (c) Is it possible to express e_t as a finite moving average of X_t?

3. For the second order autoregressive process, obtain an expression for $\rho(h)$ as a function of α_1 and α_2 analogous to (2.5.10).
4. Given a second order moving average time series, what are the largest and smallest possible values for $\rho(1)$? For $\rho(2)$? For $\rho(3)$?
5. Give an example of a moving average time series such that

$$\gamma(0)=2,$$
$$\gamma(1)=0,$$
$$\gamma(2)=-1,$$
$$\gamma(h)=0, \qquad |h|>2.$$

6. Find four (admittedly similar) time series with the covariance function

$$\gamma(h)=\begin{cases} 1, & h=0 \\ 0.3, & |h|=1 \\ 0, & \text{otherwise.} \end{cases}$$

7. Draw pictures illustrating the differences among the theoretical correlation functions for
   (a) A (finite) moving average process.
   (b) A (finite) autoregressive process.
   (c) A strictly periodic process of the form

$$X_t=\sum_{i=1}^{M}\{e_{1i}\cos\lambda_i t+e_{2i}\sin\lambda_i t\}, \qquad M<\infty,$$

where e_{1i} and e_{2i} are independent $(0,\sigma^2)$ random variables independent of e_{1j} and e_{2j} for all $i\neq j$, and λ_i, $i=1,2,\ldots,M$, are distinct frequencies.
8. Consider the time series $\{X_t\}$ defined by

$$X_t+\alpha_1X_{t-1}+\alpha_2X_{t-2}+\alpha_3X_{t-3}=e_t,$$

where the e_t are uncorrelated $(0,\sigma^2)$ random variables and the roots of

$$m^3+\alpha_1m^2+\alpha_2m+\alpha_3=0$$

are less than one in absolute value. Develop a system of equations defining $\gamma(0)$ as a function of the α's.
9. Let the time series Y_t be defined by

$$Y_t=\beta_0+\beta_1t+X_t, \qquad t=1,2,\ldots,$$

where

$$X_t=e_t+0.6e_{t-1},$$

$\{e_t\colon t\in(0,1,2,\ldots)\}$ is a sequence of normal independent $(0,\sigma^2)$ random variables, and β_0 and β_1 are fixed numbers. Give the mean and covariance functions for the time series.

10. The second order autoregressive time series

$$X_t + a_1 X_{t-1} + a_2 X_{t-2} = e_t,$$

where the e_t are uncorrelated $(0, \sigma^2)$ random variables and the roots of $m^2 + a_1 m + a_2 = 0$ are less than one in absolute value, can be expressed as the infinite moving average

$$X_t = \sum_{j=0}^{\infty} w_j e_{t-j}.$$

For the following three time series find and plot w_j for $j = 0, 1, \ldots, 8$ and $\rho(h)$ for $h = 0, 1, \ldots, 8$.

(a) $X_t + 1.6X_{t-1} + 0.64X_{t-2} = e_t$.
(b) $X_t - 0.4X_{t-1} - 0.45X_{t-2} = e_t$.
(c) $X_t - 1.2X_{t-1} + 0.85X_{t-2} = e_t$.

11. Given the time series

$$Y_t - 0.8Y_{t-1} = e_t + 0.7e_{t-1} + 0.6e_{t-2},$$

where the e_t are uncorrelated $(0, \sigma^2)$ random variables, find and plot $\rho(h)$ for $h = 0, 1, 2, \ldots, 8$.

12. Prove Theorem 2.1.1.

13. Find a second order autoregressive time series with complex roots whose absolute value is 0.8 that will give realizations with an apparent cycle of 20 time periods.

14. Given that X_t is a first order autoregressive process with parameter $\rho = 0.8$ use Theorem 2.9.1 to find the best one and two period ahead predictions based on three observations.

15. Let the time series X_t satisfy

$$(X_t - 4) - 0.8(X_{t-1} - 4) = e_t,$$

where the e_t are uncorrelated $(0, 7)$ random variables. Five observations from a realization of the time series are given below.

t	X_t
1	5.9
2	4.9
3	2.2
4	2.0
5	4.9

Predict X_t for $t = 6$, 7, 8, and estimate the covariance matrix for your predictions, that is, estimate the matrix $E\{(\mathbf{X}_{5,3} - \hat{\mathbf{X}}_{5,3})(\mathbf{X}_{5,3} - \hat{\mathbf{X}}_{5,3})'\}$.

16. Let Y_t be defined by

$$Y_t = \rho Y_{t-1} + e_t, \qquad |\rho| < 1,$$

where $\{e_t,\ t \in (0, \pm 1, \pm 2, \ldots)\}$ is obtained from the sequence $\{u_t\}$ of normal independent $(0, 1)$ random variables as follows:

$$e_t = \begin{cases} u_t, & t = 0, \pm 2, \pm 4, \ldots \\ 2^{-1/2}(u_{t-1}^2 - 1), & t = \pm 1, \pm 3, \ldots . \end{cases}$$

Is this a strictly stationary process? Is it a covariance stationary process? Given a sample of ten observations $(Y_1, Y_2, \ldots, Y_{10})$, what is the best linear predictor of Y_{11}? Can you construct a better predictor of Y_{11}?

17. Let Y_t be defined by

$$Y_t = e_t + \beta e_{t-1}, \qquad |\beta| < 1.$$

Show that the predictor of Y_{n+1} given in (2.9.15) is equivalent to the predictor $\hat{Y}_{n+1}(Y_1, \ldots, Y_n) = -\sum_{j=1}^{n}(-\beta)^j Y_{n-j+1}$.

18. Let $\{e_t\}$ be a sequence of independent $(0,1)$ random variables. Let the time series X_t be defined by:

(a) $X_t = e_0$.
(b) $X_t = e_t - 1.2e_{t-1} + 0.36e_{t-2}$.
(c) $X_t = e_1 \cos 0.4\pi t + e_2 \sin 0.4\pi t + 6$.
(d) $X_t = e_t + 1.2e_{t-1}$.

Assume the covariance structure is known and that we have available an infinite past for prediction, that is, $\{X_t\colon t \in (\ldots, -1, 0, 1, \ldots, n)\}$ is known. What is the variance of the prediction error for a predictor of X_{n+1} for each case above?

19. Show that if a stationary time series satisfies the difference equation

$$Y_t - Y_{t-1} = e_t,$$

then $E\{e_t^2\} = 0$.

20. Let $\{X_t\colon t \in (0, \pm 1, \pm 2, \ldots)\}$ be defined by

$$X_t + \alpha_1 X_{t-1} + \alpha_2 X_{t-2} = e_t,$$

where the e_t are uncorrelated $(0, \sigma^2)$ random variables, and the roots, r_1 and r_2, of

$$m^2 + \alpha_1 m + \alpha_2 = 0$$

are less than one in absolute value.

(a) Let r_1 and r_2 be real. Show that $X_t - r_1 X_{t-1}$ is a first order autoregressive process with parameter r_2.
(b) If the roots r_1 and r_2 form a complex conjugate pair, how would you describe the time series $Y_t = X_t - r_1 X_{t-1}$?

21. In Theorem 2.6.2 we wrote the characteristic equation as

$$m^p + b_1 m^{p-1} + b_2 m^{p-2} + \cdots + b_p = 0,$$

and required the roots to be less than one in absolute value. Show that an equivalent requirement is that the roots of

$$1 + b_1 q + b_2 q^2 + \cdots + b_p q^p = 0$$

be greater than one in absolute value.

22. Let $\Delta y_t = a_1 + a_2 \cos \omega t$, $t = 1, 2, \ldots$, where a_1, a_2 and ω are real numbers, and $y_0 = 0$. Find y_t.

23. Let $Y_t = e_t - e_{t-1}$, where the e_t are normal independent $(0, \sigma^2)$ random variables. Given n observations from a realization predict the $(n+1)$st observation in the realization. What is the

mean square error of your predictor? *Hint*: Consider the time series

$$Z_t = \sum_{j=1}^{t} Y_j$$

and predict Z_{n+1}.

24. Prove the following lemma.

Lemma. Let $\lambda > \beta \geqslant 0$ and let p be a nonnegative integer. Then there exists an M such that $(t+1)^p \beta^t < M\lambda^t$ for all $t \geqslant 0$.

25. Let

$$Y_t - 0.8\,Y_{t-1} = e_t - 0.9e_{t-4},$$

where the e_t are uncorrelated (0, 1) random variables. Using Theorem 2.7.1 obtain $E\{Y_t e_{t-4}\}$ and $E\{Y_{t-1}e_{t-4}\}$. Multiply the equation defining Y_t by Y_t and Y_{t-1} and take expectations to obtain a system of equations defining $\gamma_Y(0)$ and $\gamma_Y(1)$. Solve for the variance of Y_t.

26. Using the backward shift operator we can write the autoregressive moving average process of order (p, q) as

$$\left(\sum_{j=0}^{p} a_j \mathcal{B}^j\right) X_t = \left(\sum_{i=0}^{q} b_i \mathcal{B}^i\right) e_t.$$

Show that the coefficients of the moving average representation

$$X_t = \sum_{j=0}^{\infty} v_j e_{t-j} = \left(\sum_{j=0}^{\infty} v_j \mathcal{B}^j\right) e_t$$

given in Theorem 2.7.1 can be obtained from the quotient $(\Sigma_{j=0}^{p} a_j \mathcal{B}^j)^{-1}(\Sigma_{i=0}^{q} b_i \mathcal{B}^i)$ by long division. Obtain the autoregressive representation of Theorem 2.7.2 in the same manner.

27. Demonstrate how the nth order scalar difference equation $y_t + a_1 y_{t-1} + a_2 y_{t-2} + \cdots + a_n y_{t-n} = 0$ can be written as a first order vector difference equation of dimension n. If $\mathbf{A}$ denotes the matrix of coefficients of the resulting first order difference equation, show that the roots of $|\mathbf{A} - m\mathbf{I}| = 0$ are the same as the roots of

$$m^n + a_1 m^{n-1} + \cdots + a_n = 0.$$

28. Find $\boldsymbol{\Gamma}(h)$ for the vector autoregressive time series

$$X_{1t} = 0.9X_{1,t-1} + e_{1t}$$

$$X_{2t} = 0.8X_{1,t-1} + 0.3X_{2,t-1} + e_{2t},$$

where $\mathbf{e}_t = (e_{1t}, e_{2t})'$ is a sequence of normal independent $(\mathbf{0}, \boldsymbol{\Sigma})$ random variables and $\boldsymbol{\Sigma} = \begin{pmatrix} 2 & -1 \\ -1 & 2 \end{pmatrix}$. *Hint*: Express the covariance matrix as a linear function of the covariance functions of the two scalar autoregressive processes associated with the canonical form. Assume that we have observations $\mathbf{X}_1, \mathbf{X}_2, \ldots, \mathbf{X}_n$ with $\mathbf{X}_n' = (3, 2)$. Predict $\mathbf{X}_{n+1}$ and $\mathbf{X}_{n+2}$ and obtain the covariance matrix of the prediction errors.

29. The cobweb theory has been suggested as a model for the price behavior of some agricultural commodities. The model is composed of a demand equation and a supply

equation:

$$\text{(Demand)} \quad Q_t = \alpha_1 + \beta_1 P_t + U_{1t}$$

$$\text{(Supply)} \quad Q_t = \alpha_2 + \beta_2 P_{t-1} + U_{2t}$$

where Q_t is the quantity produced for time t, P_t is the price at time t, and $\mathbf{U}_t' = (U_{1t}, U_{2t})$ is a stationary vector time series. Assuming $\mathbf{U}_t$ is a sequence of uncorrelated $(\mathbf{0}, \mathbf{\Sigma})$ vector random variables, express the model as a vector autoregressive process. Under what conditions will the process be stationary? If $U_{1t} = e_{1t} + b_{11}e_{1,t-1}$ and $U_{2t} = e_{2t} + b_{22}e_{2,t-1}$ where $\mathbf{e}_t = (e_{1t}, e_{2t})'$ is a sequence of uncorrelated $(\mathbf{0}, \sigma^2\mathbf{I})$ vector random variables, how would you describe the vector time series $(Q_t, P_t)'$?

CHAPTER 3

Introduction to Fourier Analysis

3.1. SYSTEMS OF ORTHOGONAL FUNCTIONS—FOURIER COEFFICIENTS

In many areas of applied mathematics it is convenient to approximate a function by a linear combination of elementary functions. The reader is familiar with the Weierstrass theorem, which states that any continuous function on a compact set may be approximated by a polynomial.[1] Likewise, it may be convenient to construct a set of vectors called a *basis* such that all other vectors may be expressed as linear combinations of the elements of the basis. Often the basis vectors are constructed to be orthogonal, that is, the sum of the products of the elements of any pair is zero. The system that is of particular interest to us in the analysis of time series is the system of trigonometric polynomials.

Assume that we have a function defined on a finite number of points, N. We shall investigate the properties of the set of functions $\{\cos(2\pi mt/N), \sin(2\pi mt/N)$: $t=0,1,\ldots,N-1$ and $m=0,1,\ldots,L[N]\}$, where $L[N]$ is the largest integer less than or equal to $N/2$. The reader should observe that the points have been indexed by t running from zero to $N-1$, while m runs from zero to $L[N]$. Note that for $m=0$ the cosine is identically equal to 1. The sine function is identically zero for $m=0$ and for $m=N/2$ if N is even. Therefore, it is to be understood that these sine functions are not included in a discussion of the collection of functions. The collection of interest will always contain exactly N functions, none of which is identically zero. We shall demonstrate that these functions, defined on the integers $0,1,\ldots,N-1$, are orthogonal and we shall derive the sum of squares for each function. This constitutes a proof that the N functions defined on the integers furnish an orthogonal basis for the N-dimensional vector space.

[1]See Bartle (1964, p. 183).

Theorem 3.1.1. Given that m and r are contained in the set $\{0,1,2,\ldots,L[N]\}$, then

$$\begin{aligned}
\sum_{t=0}^{N-1} \cos\frac{2\pi m}{N}t\cos\frac{2\pi r}{N}t &= N, && m=r=0 \quad \text{or} \quad \frac{N}{2}\\
&= \frac{N}{2}, && m=r\neq 0 \quad \text{or} \quad \frac{N}{2}\\
&= 0, && m\neq r;
\end{aligned}$$

$$\sum_{t=0}^{N-1} \sin\frac{2\pi m}{N}t\cos\frac{2\pi r}{N}t=0, \qquad \forall m,r;$$

$$\begin{aligned}
\sum_{t=0}^{N-1} \sin\frac{2\pi m}{N}t\sin\frac{2\pi r}{N}t &= \frac{N}{2}, && m=r\neq 0 \quad \text{or} \quad \frac{N}{2}\\
&= 0, && m\neq r.
\end{aligned}$$

Proof. Consider first the sum of the products of two cosine functions and let

$$\begin{aligned}
S(m,r) &= \sum_{t=0}^{N-1} \cos\frac{2\pi m}{N}t\cos\frac{2\pi r}{N}t\\
&= \frac{1}{2}\sum_{t=0}^{N-1}\left[\cos\frac{2\pi t}{N}(m+r)+\cos\frac{2\pi t}{N}(m-r)\right]. \qquad (3.1.1)
\end{aligned}$$

For $m=r=0$ or $m=r=N/2$ (N even) the cosine terms on the right-hand side of (3.1.1) are always equal to one, and we have

$$S(m,r)=\frac{1}{2}\sum_{t=0}^{N-1}(1+1)=N.$$

For $m=r$, but not equal to zero and not equal to $N/2$ if N is even, the sum (3.1.1) reduces to

$$S(m,r)=\frac{1}{2}\sum_{t=0}^{N-1}\cos\frac{2\pi t}{N}2m+\frac{1}{2}N,$$

where the summation of cosines is given by

$$\sum_{t=0}^{N-1} \cos\frac{2\pi t}{N}2m = \frac{1}{2}\sum_{t=0}^{N-1}\left[e^{i(2m)2\pi t/N} + e^{-i(2m)2\pi t/N}\right].$$

The two sums are geometric series whose rates are $\exp\{i(2m)2\pi/N\}$ and $\exp\{-i(2m)2\pi/N\}$, respectively, and the first term is one in both cases. Now $i(2m)2\pi/N$ is not an integer multiple of 2π; therefore the rates are not unity. Applying the well-known formula $\Sigma_{t=0}^{N-1}\lambda^t = (1-\lambda^N)/(1-\lambda)$, the partial sum is

$$\frac{1-\left(e^{i(2m)2\pi/N}\right)^N}{1-e^{i(2m)2\pi/N}} + \frac{1-\left(e^{-i(2m)2\pi/N}\right)^N}{1-e^{-i(2m)2\pi/N}}.$$

Since $\exp\{i(2m)2\pi\} = \exp\{i2\pi\} = 1$, the numerators of the partial sums are 0, and $S(m,r)$ reduces to $N/2$. For $m \neq r$, we have

$$\begin{aligned} S(m,r) &= \frac{1}{2}\sum_{t=0}^{N-1}\left[\cos\frac{2\pi t}{N}(m+r) + \cos\frac{2\pi t}{N}(m-r)\right] \\ &= \frac{1}{4}\sum_{t=0}^{N-1}\left[e^{i(2\pi t/N)(m+r)} + e^{-i(2\pi t/N)(m+r)}\right. \\ &\qquad \left. + e^{i(2\pi t/N)(m-r)} + e^{-i(2\pi t/N)(m-r)}\right] \\ &= 0. \end{aligned}$$

The sum of products of a sine and cosine function is given by

$$\begin{aligned} &\sum_{t=0}^{N-1} \sin\frac{2\pi m}{N}t \cos\frac{2\pi r}{N}t \\ &\qquad = \frac{1}{2}\sum_{t=0}^{N-1}\left[\sin(m+r)\frac{2\pi t}{N} + \sin(m-r)\frac{2\pi t}{N}\right] = 0, \qquad \forall m,r, \end{aligned}$$

where the proof follows the same pattern as that for the product of cosines with $m = r \neq 0$. We leave the details to the reader. ▲

Having demonstrated that the N functions form an orthogonal basis, it follows that any function $f(t)$ defined on N integers can be represented by

$$f(t) = \sum_{m=0}^{L[N]} (a_m \cos\omega_m t + b_m \sin\omega_m t), \qquad t = 0, 1, \ldots, N-1, \quad (3.1.2)$$

where

$$\omega_m = \frac{2\pi m}{N}, \qquad m = 0, 1, 2, \ldots, L[N]:$$

$$a_m = \frac{2\sum_{t=0}^{N-1} f(t)\cos\omega_m t}{N}, \qquad m = 1, 2, \ldots, L[N-1],$$

$$= \frac{\sum_{t=0}^{N-1} f(t)\cos\omega_m t}{N}, \qquad m = 0, \text{ and } m = \frac{N}{2} \text{ if } N \text{ is even;}$$

$$b_m = \frac{2\sum_{t=0}^{N-1} f(t)\sin\omega_m t}{N}, \qquad m = 1, 2, \ldots, L[N-1].$$

The a_m and b_m are called *Fourier coefficients*. One way to obtain representation (3.1.2) is to find the a_m and b_m such that

$$\sum_{t=0}^{N-1} \left\{ f(t) - \sum_{m=0}^{L[N]} (a_m \cos\omega_m t + b_m \sin\omega_m t) \right\}^2 \tag{3.1.3}$$

is a minimum. Differentiating with respect to the a_j and b_j and setting the derivatives equal to zero, we obtain

$$\sum_{t=0}^{N-1} \left\{ f(t) - \sum_{m=0}^{L[N]} (a_m \cos\omega_m t + b_m \sin\,\omega_m t) \right\} \cos\omega_j t = 0,$$
$$j = 0, 1, 2, \ldots, L[N],$$

$$\sum_{t=0}^{N-1} \left\{ f(t) - \sum_{m=0}^{L[N]} (a_m \cos\omega_m t + b_m \sin\omega_m t) \right\} \sin\omega_j t = 0,$$
$$j = 1, 2, \ldots, L[N-1].$$

By the results of Theorem 3.1.1, these equations reduce to

$$\sum_{t=0}^{N-1} f(t)\cos\omega_m t = a_m \sum_{t=0}^{N-1} \cos^2\omega_m t, \qquad m = 0, 1, \ldots, L[N],$$

$$\sum_{t=0}^{N-1} f(t)\sin\omega_m t = b_m \sum_{t=0}^{N-1} \sin^2\omega_m t, \qquad m = 1, 2, \ldots, L[N-1],$$

and we have the coefficients of (3.1.2). Thus we see that the coefficients are the regression coefficients obtained by regressing the vector, $f(t)$, on the vectors $\cos\omega_m t$ and $\sin\omega_m t$. By the orthogonality of the functions, the multiple regression coefficients are the simple regression coefficients.

Because of the different sum of squares for $\cos 0t$ and $\cos\pi t$, the a's are sometimes all defined with a common divisor, $N/2$, and the first and last terms of the series modified accordingly. Often N is restricted to be odd to simplify the discussion. Specifically, for N odd, we have

$$f(t)=\frac{a_0}{2}+\sum_{m=1}^{\frac{N-1}{2}}(a_m\cos\omega_m t+b_m\sin\omega_m t),\qquad t=0,1,\dots,N-1,\quad (3.1.4)$$

where

$$a_m=\frac{2\sum_{t=0}^{N-1} f(t)\cos\omega_m t}{N},\qquad m=0,1,\dots,\frac{N-1}{2},$$

$$b_m=\frac{2\sum_{t=0}^{N-1} f(t)\sin\omega_m t}{N},\qquad m=1,2,\dots,\frac{N-1}{2}.$$

The reader will have no difficulty in identifying the definitions being used. If the leading term is given as $a_0/2$, definition (3.1.4) is being used; if not, (3.1.2) is being used.

The preceding material demonstrates that any finite vector can be represented as a linear combination of vectors defined by the sine and cosine functions. We now turn to the investigation of similar representations for functions defined on the real line.

We first consider functions defined on a finite interval of the real line. Since the interval is finite, we may code the end points in any convenient manner. When dealing with trigonometric functions, it will be most convenient to treat intervals whose length is a multiple of 2π. We shall most frequently take the inverval to be $[-\pi,\pi]$.

Definition 3.1.1. An infinite system of square integrable functions $\{\varphi_j\}_{j=0}^{\infty}$ is orthogonal on $[a,b]$ if

$$\int_a^b \varphi_j(x)\varphi_m(x)\,dx=0,\qquad j\neq m,\ j,m=0,1,\dots,$$

and

$$\int_a^b |\varphi_j(x)|^2 \, dx \neq 0, \qquad j=0,1,\ldots.$$

The following theorem states that the system of trigonometric functions is orthogonal on the interval $[-\pi,\pi]$.

Theorem 3.1.2. Given that m and j are nonnegative integers, then

$$\int_{-\pi}^{\pi} \sin mx \cos jx \, dx = 0, \qquad \forall m, j;$$

$$\begin{aligned} \int_{-\pi}^{\pi} \sin mx \sin jx \, dx &= 0, \qquad m \neq j \\ &= \pi, \qquad m=j\neq 0 \\ &= 0, \qquad m=j=0; \end{aligned}$$

$$\begin{aligned} \int_{-\pi}^{\pi} \cos mx \cos jx \, dx &= 2\pi, \qquad m=j=0 \\ &= \pi, \qquad m=j\neq 0 \\ &= 0, \qquad m\neq j. \end{aligned}$$

Proof. Reserved for the reader. ▲

The sum

$$S_n(x) = \sum_{k=0}^{n} (a_k \cos \lambda_k x + b_k \sin \lambda_k x), \tag{3.1.5}$$

$$\lambda_k = \frac{2\pi k}{T}, \qquad k=0,1,\ldots,n,$$

is called a *trigonometric polynomial* of order (or degree) n and period T. We shall take $T=2\pi$ (i.e., $\lambda_k = k$) in the sequel unless stated otherwise. If we let n increase without bound we have the infinite trigonometric series

$$S_\infty(x) = \sum_{k=0}^{\infty} (a_k \cos \lambda_k x + b_k \sin \lambda_k x). \tag{3.1.6}$$

By Theorem 3.1.2 we know that the sine and cosine functions are orthogonal on the interval $[-\pi,\pi]$, or on any interval of length 2π. We shall investigate the ability of a trigonometric polynomial to approximate a function defined on this interval.

In the finite theory of least squares we are given a vector $\mathbf{y}$ of dimension n, which we desire to approximate by a linear combination of the vectors $\mathbf{x}_k$, $k=1,2,\ldots,p$. The coefficients of the predicting equation

$$\hat{Y}_j = \sum_{k=1}^{p} b_k x_{kj}$$

are obtained by minimizing

$$\sum_{j=1}^{n}\left(Y_j - \sum_{k=1}^{p} b_k x_{kj}\right)^2$$

with respect to the b_k.

The analogous procedure for a function, $f(x)$, defined on an interval is to minimize the integral

$$\int_{-\pi}^{\pi}\left\{f(x) - \sum_{k=0}^{n}(a_k\cos\lambda_k x + b_k\sin\lambda_k x)\right\}^2 dx. \qquad (3.1.7)$$

Of course, $\{f(x)-\sum_{k=0}^{n}(a_k\cos\lambda_k x+b_k\sin\lambda_k x)\}^2$ must be integrable. In this section we shall say that a function $g(x)$ is integrable over $[a,b]$ if it is continuous, or if it has a finite number of discontinuities [at which $g(x)$ can be either bounded or unbounded], provided the improper Riemann integral exists.[2]

The reader may verify by expansion of the square and differentiation of the resultant products that the coefficients that minimize (3.1.7) are

$$a_k = \frac{1}{\pi}\int_{-\pi}^{\pi}\cos\lambda_k x\, f(x)\,dx, \qquad k=0,1,2,\ldots, \qquad (3.1.8)$$

$$b_k = \frac{1}{\pi}\int_{-\pi}^{\pi}\sin\lambda_k x\, f(x)\,dx, \qquad k=1,2,\ldots, \qquad (3.1.9)$$

[2]More general definitions could be used. Our treatment in the sequel follows closely that of Tolstov (1962), and we use his definitions.

where the first term is written as $a_0/2$. In this form the approximating sum is

$$S_n(x) = \frac{a_0}{2} + \sum_{k=1}^{n} (a_k \cos \lambda_k x + b_k \sin \lambda_k x). \qquad (3.1.10)$$

When T, the length of the interval, is not equal to 2π, λ_k is set equal to $2\pi k/T$, and we obtain the formulas

$$a_k = \frac{2}{T}\int_{-T/2}^{T/2} \cos \lambda_k x \, f(x)\, dx, \qquad k = 0, 1, 2, \ldots,$$

$$b_k = \frac{2}{T}\int_{-T/2}^{T/2} \sin \lambda_k x \, f(x)\, dx, \qquad k = 1, 2, \ldots,$$

$$S_n(x) = \frac{a_0}{2} + \sum_{k=1}^{n} (a_k \cos \lambda_k x + b_k \sin \lambda_k x). \qquad (3.1.11)$$

Observe that $\pi[(a_0^2/2) + \Sigma_{k=1}^{n}(a_k^2 + b_k^2)]$ is completely analogous to the sum of squares due to regression on n orthogonal independent variables of finite least squares theory. We repeat below the theorem analogous to the statement in finite least squares that the multiple correlation coefficient is less than or equal to one.

Theorem 3.1.3 (Bessel's inequality). Let a_k, b_k, and $S_n(x)$ be defined by (3.1.8), (3.1.9), and (3.1.10), respectively. If $f(x)$ defined on $[-\pi, \pi]$ is square integrable, then

$$\frac{1}{\pi}\int_{-\pi}^{\pi} |f(x)|^2 dx \geqslant \frac{a_0^2}{2} + \sum_{k=1}^{n} (a_k^2 + b_k^2). \qquad (3.1.12)$$

Proof. Since the square is always nonnegative,

$$0 \leqslant \int_{-\pi}^{\pi} |f(x) - S_n(x)|^2 dx = \int_{-\pi}^{\pi} \left[f^2(x) - 2f(x)S_n(x) + S_n^2(x) \right] dx$$

$$= \int_{-\pi}^{\pi} f^2(x)\, dx - 2\int_{-\pi}^{\pi} f(x) S_n(x)\, dx + \int_{-\pi}^{\pi} S_n^2(x)\, dx.$$

By the definitions of a_k and b_k and the orthogonality of the sine and

cosine functions, we have

$$\int_{-\pi}^{\pi} |f(x)-S_n(x)|^2\,dx = \int_{-\pi}^{\pi} f^2(x)\,dx - 2\left[\frac{a_0^2}{2}\pi + \pi\sum_{k=1}^{n}(a_k^2+b_k^2)\right]$$
$$+\left[\frac{a_0^2}{2}\pi + \pi\sum_{k=1}^{n}(a_k^2+b_k^2)\right]$$

and

$$\frac{1}{\pi}\int_{-\pi}^{\pi} f^2(x)\,dx \geqslant \frac{a_0^2}{2} + \sum_{k=1}^{n}(a_k^2+b_k^2).$$ ▲

Definition 3.1.2. A system of functions $\{\varphi_j(x)\}_{j=0}^{\infty}$ is *complete* if there does not exist a function, $f(x)$, such that

$$\int_a^b |f(x)|\,dx \neq 0$$

and

$$\int_a^b f(x)\varphi_j(x)\,dx = 0, \qquad j=0,1,\ldots.$$

We shall give only a few of the theorems of Fourier analysis. The following two theorems constitute a proof of completeness of the trigonometric functions. Lebesgue's proofs are given in Rogosinski (1959).

Theorem 3.1.4. If $f(x)$ is continuous on the interval $[-\pi,\pi]$, then all the Fourier coefficients are zero if and only if

$$f(x) \equiv 0.$$

Proof. Omitted. ▲

Theorem 3.1.5. If $f(x)$ defined on the interval $[-\pi,\pi]$ is integrable and if all the Fourier coefficients $\{a_k, b_k : k=0,1,2,\ldots\}$ are zero, then

$$\int_{-\pi}^{\pi} |f(x)|\,dx = 0.$$

Proof. Omitted. ▲

By Theorem 3.1.4 the Fourier coefficients of a continuous function are unique. Theorem 3.1.5 generalizes the result to absolutely integrable functions. Neither furnishes information on the nature of the convergence of the sequences $\{a_k\}$ and $\{b_k\}$.

Theorem 3.1.6 (Parseval's Theorem) answers the question of convergence for square integrable functions.

Theorem 3.1.6. If $f(x)$ defined on $[-\pi,\pi]$ is square integrable, then

$$\lim_{n\to\infty}\int_{-\pi}^{\pi}\left|f(x)-\frac{a_0}{2}-\sum_{k=1}^{n}(a_k\cos kx+b_k\sin kx)\right|^2dx=0$$

or, equivalently,

$$\frac{a_0^2}{2}+\sum_{k=1}^{\infty}\left(a_k^2+b_k^2\right)=\frac{1}{\pi}\int_{-\pi}^{\pi}|f(x)|^2dx.$$

Proof. By Bessel's inequality, the partial sum of squares of the Fourier coefficients converges for any square integrable function. (The sum of squares is monotone increasing and bounded.) Therefore, the function

$$g(x)=\frac{a_0}{2}+\sum_{k=1}^{\infty}(a_k\cos kx+b_k\sin kx),$$

where a_k and b_k are the Fourier coefficients of a square integrable function, $f(x)$, is square integrable on $[-\pi,\pi]$. Now $D(x)=f(x)-g(x)$ is square integrable and all Fourier coefficients of $D(x)$ are zero; hence, by Theorem 3.1.5,

$$\int_{-\pi}^{\pi}|f(x)-g(x)|\,dx=\int_{-\pi}^{\pi}|f(x)-g(x)|^2dx=0. \qquad \blacktriangle$$

Thus, any square integrable function defined on an interval can be approximated in mean square by a trigonometric polynomial.

As one might suspect, the convergence of the sequence of functions $S_n(x)$ at a point x_0 requires additional conditions. We shall closely follow Tolstov's (1962) text in presenting a proof of the pointwise convergence of $S_n(x_0)$ to $f(x_0)$. The following definitions are needed.

Definition 3.1.3. The *left limit* of $f(x)$ is given by

$$\lim_{\substack{x\to x_0\\ x<x_0}} f(x)=f(x_0^-)$$

provided the limit exists and is finite. The limit of $f(x)$ as $x \to x_0$, $x > x_0$, is called the *right limit* of $f(x)$ and is denoted by $f(x_0^+)$, provided it exists and is finite.

Definition 3.1.4. The *right derivative* of $f(x)$ at $x = x_0$ is defined by

$$f'(x_0^+) = \lim_{\substack{h \to 0 \\ h > 0}} \frac{f(x_0 + h) - f(x_0^+)}{h}$$

and the *left derivative* by

$$f'(x_0^-) = \lim_{\substack{h \to 0 \\ h > 0}} \frac{f(x_0 - h) - f(x_0^-)}{h}$$

provided the limits exist and are finite.

Definition 3.1.5. The function $f(x)$ has a *jump discontinuity* at the point x_0 if $f(x_0^+) \neq f(x_0^-)$.

From the definitions it is clear that the jump $f(x_0^+) - f(x_0^-)$ is finite.

Definition 3.1.6. A function $f(x)$ is *smooth* on the interval $[c, d]$ if it has a continuous derivative on the interval $[c, d]$.

Definition 3.1.7. A function $f(x)$ is *piecewise smooth* on the interval $[c, d]$ if

(i) $f(x)$ and $f'(x)$ are both continuous, or
(ii) $f(x)$ and $f'(x)$ have a finite number of jump discontinuities.

Bessel's inequality guarantees that the Fourier coefficients of any square integrable function go to zero as the frequency increases; that is,

$$\lim_{m \to \infty} a_m = \lim_{m \to \infty} b_m = 0.$$

The following lemma is a proof of the same result for a different class of functions.

Lemma 3.1.1. If $f(x)$ is a piecewise smooth function on $[c, d]$, then

$$\lim_{m \to \infty} \int_c^d f(x) \cos mx \, dx = \lim_{m \to \infty} \int_c^d f(x) \sin mx \, dx = 0.$$

Proof. Integrating by parts gives

$$\int_c^d f(x) \cos mx \, dx = \frac{1}{m} \left\{ [f(x) \sin mx]_c^d - \int_c^d f'(x) \sin mx \, dx \right\}.$$

Since both $f(x)$ and $f'(x)$ contain at most a finite number of jump discontinuities on the interval $[c,d]$, they are both bounded. Therefore, the quantity contained within the curly brackets is bounded and the result follows immediately. ▲

Using the fact that any absolutely integrable function can be approximated in the mean by a piecewise smooth function, it is possible to extend this result to any absolutely integrable function. We state this generalization without proof.

Lemma 3.1.1A. The Fourier coefficients a_m and b_m of an absolutely integrable function defined on a finite interval approach zero as $m \to \infty$.

Lemmas 3.1.2 and 3.1.3 are presented as preliminaries to the proof of Theorem 3.1.7.

Lemma 3.1.2. Define

$$A_n(u) = \frac{1}{2} + \sum_{j=1}^{n} \cos ju.$$

Then

$$\text{(i)} \qquad A_n(u) = \frac{\sin(n+1/2)u}{2\sin(u/2)} \tag{3.1.13}$$

and

$$\text{(ii)} \qquad \frac{1}{\pi}\int_0^{\pi} A_n(u)\,du = \frac{1}{\pi}\int_{-\pi}^{0} A_n(u)\,du = \frac{1}{2}. \tag{3.1.14}$$

Proof. Multiplying both sides of the definition of $A_n(u)$ by 2 $\sin(u/2)$, we have

$$\begin{aligned} 2A_n(u)\sin(u/2) &= \sin(u/2) + 2\sum_{j=1}^{n} \cos ju \sin(u/2) \\ &= \sin(u/2) + \sum_{j=1}^{n} [\sin(j+1/2)u - \sin(j-1/2)u] \\ &= \sin(n+1/2)u \end{aligned}$$

and result (i) follows. Since the integral of $\cos ju$, $j = 1, 2, \ldots$, is zero on an

interval of length 2π,

$$\int_{-\pi}^{\pi}\left[\frac{1}{2}+\sum_{j=1}^{n}\cos ju\right]du=\pi$$

and, hence,

$$\frac{1}{\pi}\int_{-\pi}^{\pi}\frac{\sin(n+1/2)u}{2\sin(u/2)}\,du=1. \tag{3.1.15}$$

Result (ii) follows since $A_n(u)$ is an even function. ▲

Lemma 3.1.3. The partial sum

$$S_n(x)=\frac{a_0}{2}+\sum_{k=1}^{n}(a_k\cos kx+b_k\sin kx)$$

may be expressed in the form

$$S_n(x)=\frac{1}{\pi}\int_{-\pi}^{\pi}f(x+u)\frac{\sin(n+1/2)u}{2\sin(u/2)}\,du, \tag{3.1.16}$$

where a_k and b_k are the Fourier coefficients of a periodic function $f(x)$ of period 2π defined in (3.1.8) and (3.1.9).

Proof. Substituting (3.1.8) and (3.1.9) into (3.1.10), we have

$$\begin{aligned}S_n(x)&=\frac{1}{2\pi}\int_{-\pi}^{\pi}f(t)\,dt+\frac{1}{\pi}\sum_{k=1}^{n}\left[\left\{\int_{-\pi}^{\pi}f(t)\cos kt\,dt\right\}\cos kx\right.\\&\qquad\left.+\left\{\int_{-\pi}^{\pi}f(t)\sin kt\,dt\right\}\sin kx\right]\\&=\frac{1}{\pi}\int_{-\pi}^{\pi}f(t)\left[\frac{1}{2}+\sum_{k=1}^{n}\cos k(t-x)\right]dt.\end{aligned}$$

Using (3.1.13) of Lemma 3.1.2, we have

$$S_n(x)=\frac{1}{\pi}\int_{-\pi}^{\pi}f(t)\frac{\sin\{(n+1/2)(t-x)\}}{2\sin\{(1/2)(t-x)\}}\,dt.$$

Setting $u=t-x$ and noting that the integral of a periodic function of period 2π over the interval $[-\pi-x,\pi-x]$ is the same as the integral over $[-\pi,\pi]$, we have

$$S_n(x)=\frac{1}{\pi}\int_{-\pi}^{\pi} f(x+u)\frac{\sin(n+1/2)u}{2\sin(u/2)}\,du. \qquad \blacktriangle$$

Formula (3.1.16) is called the *integral formula* for the partial sum of a Fourier series.

Theorem 3.1.7. Let $f(x)$ be an absolutely integrable function of period 2π. Then,

(i) At a point of continuity where $f(x)$ has a right derivative and a left derivative,

$$f(x)=\frac{a_0}{2}+\sum_{k=1}^{\infty}(a_k\cos kx+b_k\sin kx);$$

(ii) At every point of discontinuity where $f(x)$ has a right and a left derivative,

$$\frac{a_0}{2}+\sum_{k=1}^{\infty}(a_k\cos kx+b_k\sin kx)=\frac{f(x^+)+f(x^-)}{2}.$$

Proof. Consider first a point of continuity. We wish to show that the difference

$$\lim_{n\to\infty}\frac{1}{\pi}\int_{-\pi}^{\pi} f(x+u)\frac{\sin(n+1/2)u}{2\sin(u/2)}\,du-f(x) \tag{3.1.17}$$

equals zero. Multiplying (3.1.15) of Lemma 3.1.2 by $f(x)$, we have

$$f(x)=\frac{1}{\pi}\int_{-\pi}^{\pi} f(x)\frac{\sin(n+1/2)u}{2\sin(u/2)}\,du. \tag{3.1.18}$$

Therefore, the difference (3.1.17) can be written as

$$\lim_{n\to\infty}\frac{1}{\pi}\int_{-\pi}^{\pi}[f(x+u)-f(x)]\frac{\sin(n+1/2)u}{2\sin(u/2)}\,du$$

$$=\lim_{n\to\infty}\frac{1}{\pi}\int_{-\pi}^{\pi}\left(\frac{f(x+u)-f(x)}{u}\right)\left(\frac{u}{2\sin(u/2)}\right)\sin(n+1/2)u\,du.$$

Now, $u^{-1}[f(x+u)-f(x)]$ is absolutely integrable, since the existence of the left and right derivatives means that the ratio is bounded as u approaches zero. Also, $[2\sin(u/2)]^{-1}u$ is bounded. Therefore,

$$\frac{f(x+u)-f(x)}{u}\left(\frac{u}{2\sin(u/2)}\right)\overset{(\text{say})}{=}g(u)$$

is absolutely integrable. Hence, by Lemma 3.1.1A,

$$\lim_{n\to\infty}\frac{1}{\pi}\int_{-\pi}^{\pi}g(u)\sin(n+1/2)u\,du=0,$$

giving us the desired result for a point of continuity.

For a point of jump discontinuity we must prove

$$\lim_{n\to\infty}\frac{1}{\pi}\int_{-\pi}^{\pi}f(x+u)\frac{\sin(n+1/2)u}{2\sin(u/2)}\,du=\frac{f(x^+)+f(x^-)}{2}.$$

Using (3.1.14) of Lemma 3.1.2, we have

$$\frac{f(x^+)}{2}=\frac{1}{\pi}\int_{0}^{\pi}f(x^+)\frac{\sin(n+1/2)u}{2\sin(u/2)}\,du$$

and

$$\frac{f(x^-)}{2}=\frac{1}{\pi}\int_{-\pi}^{0}f(x^-)\frac{\sin(n+1/2)u}{2\sin(u/2)}\,du.$$

The same arguments on boundedness and integrability that were used for a point of continuity can be used to show that

$$\lim_{n\to\infty}\frac{1}{\pi}\int_{0}^{\pi}\frac{f(x+u)-f(x^+)}{u}\,\frac{u}{2\sin(u/2)}\sin(n+1/2)u\,du$$

$$=\lim_{n\to\infty}\frac{1}{\pi}\int_{-\pi}^{0}\frac{f(x+u)-f(x^-)}{u}\,\frac{u}{2\sin(u/2)}\sin(n+1/2)u\,du$$

$$=0.$$

▲

Theorem 3.1.8. Let $f(x)$ be a continuous periodic function of period 2π with derivative $f'(x)$ that is square integrable. Then the Fourier series of $f(x)$ converges to $f(x)$ absolutely and uniformly.

Proof. Under the assumptions we may integrate by parts to obtain

$$
\begin{aligned}
a_h &= \frac{1}{\pi}\int_{-\pi}^{\pi} f(x)\cos hx\,dx \\
&= \frac{1}{\pi h}\left[f(x)\sin hx\right]_{-\pi}^{\pi} - \frac{1}{\pi h}\int_{-\pi}^{\pi} f'(x)\sin hx\,dx \\
&= -\frac{1}{\pi h}\int_{-\pi}^{\pi} f'(x)\sin hx\,dx.
\end{aligned}
$$

Therefore, the Fourier coefficients of the function are directly related to the Fourier coefficients of the derivative by $a_h = -b_h'/h$, $b_h = a_h'/h$, where

$$
a_h' = \frac{1}{\pi}\int_{-\pi}^{\pi} f'(x)\cos hx\,dx,
$$

$$
b_h' = \frac{1}{\pi}\int_{-\pi}^{\pi} f'(x)\sin hx\,dx.
$$

The Fourier coefficients of the derivative are well defined by the assumption that $f'(x)$ is square integrable. Furthermore, by Theorem 3.1.6 (Parseval's Theorem), the series

$$
\sum_{h=1}^{\infty} \left(|a_h'|^2 + |b_h'|^2\right)
$$

converges. Since

$$
\left(|a_h'| - \frac{1}{h}\right)^2 = |a_h'|^2 - \frac{2}{h}|a_h'| + \frac{1}{h^2} \geqslant 0
$$

we have

$$
\frac{1}{h}|a_h'| + \frac{1}{h}|b_h'| \leqslant \frac{1}{2}\left(|a_h'|^2 + |b_h'|^2\right) + \frac{1}{h^2}.
$$

It follows that $\Sigma_{h=1}^{\infty}(|a_h| + |b_h|)$ converges. Now,

$$
\begin{aligned}
|a_h \cos hx + b_h \sin hx| &\leqslant |a_h \cos hx| + |b_h \sin hx| \\
&\leqslant |a_h| + |b_h|
\end{aligned}
$$

and, therefore, by the Weierstrass M-test[3] the trigonometric series

$$\frac{a_0}{2} + \sum_{h=1}^{\infty} (a_h \cos hx + b_h \sin hx)$$

converges to $f(x)$ absolutely and uniformly. ▲

As a simple consequence of the proof of Theorem 3.1.8, we have the following important result.

Corollary 3.1.8. If

$$\sum_{h=1}^{\infty} (|a_h| + |b_h|)$$

converges, then the associated trigonometric series

$$\frac{a_0}{2} + \sum_{h=1}^{\infty} (a_h \cos hx + b_h \sin hx)$$

converges absolutely and uniformly to a continuous periodic function of period 2π of which it is the Fourier series.

We are now in a position to prove a portion of Theorem 1.4.4 of Chapter 1. In the proof we require a result that will be used at later points and, therefore, we state it as a lemma.

Lemma 3.1.4. (Kronecker's lemma) If the sequence $\{a_j\}$ is such that

$$\lim_{n\to\infty} \sum_{j=0}^{n} |a_j| = A < \infty,$$

then

$$\lim_{n\to\infty} \sum_{j=0}^{n} \frac{j}{n} |a_j| = 0.$$

Proof. By assumption, given $\epsilon > 0$, there exists an N such that

$$\sum_{j=N+1}^{\infty} |a_j| < \epsilon.$$

[3]See, for example, Bartle (1964, p. 352)

Therefore, for $n>N$, we have

$$\sum_{j=0}^{n} \frac{j}{n}|a_j| < \frac{1}{n}\sum_{j=0}^{N} j|a_j| + \epsilon.$$

Clearly, for fixed N,

$$\lim_{n\to\infty} \frac{1}{n}\sum_{j=0}^{N} j|a_j| = 0,$$

and since ϵ was arbitrary, the result follows. ▲

Theorem 3.1.9. Let the correlation function $\rho(h)$ of a stationary time series be absolutely summable. Then there exists a continuous function $f(\omega)$ such that:

(i) $\rho(h)=\int_{-\pi}^{\pi} f(\omega)\cos\omega h\, d\omega$.
(ii) $f(\omega)\geqslant 0$.
(iii) $\int_{-\pi}^{\pi} f(\omega)\, d\omega = 1$.
(iv) $f(\omega)$ is an even function.

Proof. By Corollary 3.1.8

$$g(\omega)=\frac{1}{2}+\sum_{h=1}^{\infty}\rho(h)\cos h\omega$$

is a well-defined continuous function. Now, by the positive semidefinite property of the correlation function,

$$\sum_{m=1}^{n}\sum_{q=1}^{n}\rho(m-q)\cos m\omega\cos q\omega \geqslant 0$$

and

$$\sum_{m=1}^{n}\sum_{q=1}^{n}\rho(m-q)\sin m\omega\sin q\omega \geqslant 0.$$

Hence,

$$\sum_{m=1}^{n}\sum_{q=1}^{n}\rho(m-q)\left[\cos m\omega\cos q\omega+\sin m\omega\sin q\omega\right]$$
$$=\sum_{m=1}^{n}\sum_{q=1}^{n}\rho(m-q)\cos(m-q)\omega \geqslant 0.$$

Letting $m-q=h$, we have

$$\sum_{h=-(n-1)}^{n-1}\left(\frac{n-|h|}{n}\right)\rho(h)\cos h\omega \geqslant 0.$$

Now, $\rho(h)\cos h\omega$ is absolutely summable and, hence, by Lemma 3.1.4,

$$\lim_{n\to\infty}\sum_{h=-(n-1)}^{n-1}\frac{|h|}{n}\rho(h)\cos h\omega=0.$$

Therefore,

$$\lim_{n\to\infty}\sum_{h=-(n-1)}^{(n-1)}\frac{n-|h|}{n}\rho(h)\cos h\omega=\sum_{h=-\infty}^{\infty}\rho(h)\cos h\omega=2g(\omega)\geqslant 0.$$

Having shown that $g(\omega)$ satisfies conditions (i) and (ii), we need only multiply $g(\omega)$ by a constant to meet condition (iii). Since

$$\frac{1}{\pi}\int_{-\pi}^{\pi}g(\omega)\,d\omega=1,$$

the appropriate constant is π^{-1}, and we define $f(\omega)$ by

$$\begin{aligned} f(\omega)&=\frac{1}{\pi}\left[\frac{1}{2}+\sum_{h=1}^{\infty}\rho(h)\cos h\omega\right]\\ &=\frac{1}{2\pi}\sum_{h=-\infty}^{\infty}\rho(h)\cos h\omega. \end{aligned}$$

The function $f(\omega)$ is an even function, since it is the uniform limit of a sum of even functions (cosines). ▲

In Theorem 3.1.8 the square integrability of the derivative was used to demonstrate the convergence of the Fourier series. In fact, the Fourier series of a continuous function not meeting such restrictions need not converge. However, Cesàro's method may be used to recover any continuous periodic function from its Fourier series. Given the sequence $\{S_j\}_{j=1}^{\infty}$, the sequence $\{C_n\}$ defined by

$$C_n=\frac{1}{n}\sum_{j=1}^{n}S_j$$

is called the sequence of arithmetic means of $\{S_j\}$. If the sequence $\{C_n\}$ is convergent we say the sequence $\{S_j\}$ is *Cesàro summable*. If the original sequence was convergent, then $\{C_n\}$ converges.

Lemma 3.1.5. If the sequence $\{S_j\}$ converges to s, then the sequence $\{C_n\}$ converges to s.

Proof. By hypothesis, given $\epsilon > 0$, we may choose an N such that $|S_j - s| < \frac{1}{2}\epsilon$ for all $j > N$. For $n > N$, we have

$$\left|\frac{1}{n}\sum_{j=1}^{n} S_j - s\right| \leqslant \frac{1}{n}\sum_{j=1}^{N} |S_j - s| + \frac{1}{n}\sum_{j=N+1}^{n} |S_j - s|$$

$$\leqslant \frac{1}{n}\sum_{j=1}^{N} |S_j - s| + \frac{1}{2}\epsilon.$$

Since we can choose an n large enough so that the first term is less than $\frac{1}{2}\epsilon$, the result follows. ▲

Theorem 3.1.10. Let $f(\omega)$ be a continuous function of period 2π. Then the Fourier series of $f(\omega)$ is uniformly summable to $f(\omega)$ by the method of Cesàro.

Proof. The Cesàro sum is

$$C_n(\omega) = \frac{1}{n}\sum_{j=0}^{n-1}\sum_{k=0}^{j}(a_k \cos k\omega + b_k \sin k\omega)$$

$$= \frac{1}{n}\sum_{j=0}^{n-1}\frac{1}{\pi}\int_{-\pi}^{\pi} f(\omega+u)\frac{\sin(j+1/2)u}{2\sin(u/2)}\,du$$

$$= \frac{1}{n\pi}\int_{-\pi}^{\pi}\frac{f(\omega+u)}{2\sin^2(u/2)}\sum_{j=0}^{n-1}\sin(u/2)\sin(j+1/2)u\,du,$$

where we have used Lemma 3.1.3. Now,

$$\sum_{j=0}^{n-1}\sin(u/2)\sin(j+1/2)u = \frac{1}{2}\sum_{j=0}^{n-1}[\cos ju - \cos(j+1)u]$$

$$= \frac{1}{2}[1 - \cos nu]$$

$$= \sin^2(nu/2). \tag{3.1.18}$$

Therefore,

$$C_n(\omega)-f(\omega)=\frac{1}{n\pi}\int_{-\pi}^{0}[f(\omega+u)-f(\omega)]\frac{\sin^2(nu/2)}{2\sin^2(u/2)}\,du$$

$$+\frac{1}{n\pi}\int_{0}^{\pi}[f(\omega+u)-f(\omega)]\frac{\sin^2(nu/2)}{2\sin^2(u/2)}\,du, \quad (3.1.19)$$

where we have used

$$\frac{1}{\pi n}\int_{-\pi}^{\pi}\frac{\sin^2(nu/2)}{2\sin^2(u/2)}\,du=1.$$

Since $f(\omega)$ is uniformly continuous on $[-\pi,\pi]$, given any $\epsilon>0$, there is a $\delta>0$ such that

$$|f(\omega+u)-f(\omega)|<\frac{\epsilon}{2}$$

for all $|u|<\delta$ and all $\omega\in[-\pi,\pi]$. We write the second integral of (3.1.19) as

$$\frac{1}{\pi n}\int_{0}^{\delta}[f(\omega+u)-f(\omega)]\frac{\sin^2(nu/2)}{2\sin^2(u/2)}\,du$$

$$+\frac{1}{\pi n}\int_{\delta}^{\pi}[f(\omega+u)-f(\omega)]\frac{\sin^2(nu/2)}{2\sin^2(u/2)}\,du$$

$$\leqslant\frac{\epsilon}{2\pi n}\int_{0}^{\delta}\frac{\sin^2(nu/2)}{2\sin^2(u/2)}\,du$$

$$+\frac{1}{2\pi n\sin^2(\delta/2)}\int_{0}^{\pi}|f(\omega+u)-f(\omega)|\,du$$

$$\leqslant\frac{\epsilon}{4}+\frac{M}{n\sin^2(\delta/2)},$$

where M is the maximum of $|f(\omega)|$ on $[-\pi,\pi]$. A similar argument holds for the first integral of (3.1.19). Therefore, there exists an N such that (3.1.19) is less than ϵ for all $n>N$ and all $\omega\in[-\pi,\pi]$. ▲

3.2. COMPLEX REPRESENTATION OF TRIGONOMETRIC SERIES

We can represent the trigonometric series in a complex form that is somewhat more compact and that will prove useful in certain applications. Since

$$\cos\theta = \frac{\cos\theta + i\sin\theta + \cos\theta - i\sin\theta}{2} = \frac{e^{i\theta} + e^{-i\theta}}{2}$$

and

$$\sin\theta = \frac{\cos\theta + i\sin\theta - (\cos\theta - i\sin\theta)}{2i} = \frac{(-i)(e^{i\theta} - e^{-i\theta})}{2},$$

we have

$$a_k\cos kx + b_k\sin kx = a_k\left(\frac{e^{ikx} + e^{-ikx}}{2}\right) - b_k i\left(\frac{e^{ikx} - e^{-ikx}}{2}\right)$$

$$= \left(\frac{a_k - ib_k}{2}\right)e^{ikx} + \left(\frac{a_k + ib_k}{2}\right)e^{-ikx}.$$

Thus we can write the approximating sum for a function, $f(x)$, defined on $[-\pi, \pi]$, as

$$S_n(x) = \frac{a_0}{2} + \sum_{k=1}^{n}(a_k\cos kx + b_k\sin kx) = \sum_{k=-n}^{n} c_k e^{ikx},$$

where

$$c_k = \frac{a_k - ib_k}{2}, \qquad c_{-k} = c_k^* = \frac{a_k + ib_k}{2}, \qquad k = 0, 1, 2, \ldots.$$

The coefficients, c_k, are given by the integrals

$$c_k = \frac{1}{2\pi}\int_{-\pi}^{\pi} f(x)e^{-ikx}\,dx$$

$$= \frac{1}{2\pi}\int_{-\pi}^{\pi} f(x)[\cos kx - i\sin kx]\,dx$$

$$= \frac{1}{2}(a_k - ib_k). \tag{3.2.1}$$

Consider now a function X_t defined on N integers. Let us identify the integers by $\{-m, -(m-1), \ldots, -1, 0, 1, \ldots, m\}$ if N is odd and by $\{-(m-1), -(m-2), \ldots, -1, 0, 1, \ldots, m\}$ if N is even. Taking N to be even, the c_k

are given by

$$c_k = \frac{1}{N} \sum_{t=-(m-1)}^{m} X_t\, e^{-i2\pi kt/N}$$

$$= \frac{1}{N} \sum_{t=-(m-1)}^{m} X_t (\cos 2\pi kt/N - i \sin 2\pi kt/N),$$

$$k = -(m-1), \ldots, 0, \ldots, m, \tag{3.2.2}$$

and

$$X_t = \sum_{k=-(m-1)}^{m} c_k e^{i2\pi kt/N}.$$

Note that c_0 and c_m do not require separate definitions. The complex form thus makes manipulation somewhat easier, since the correct divisor need not be specified by frequency.

If N is odd and X_t is an even function (i.e., $X_t = X_{-t}$), then the coefficients, c_k, are real. Conversely, if X_t is an odd function, the coefficients c_k of equation (3.2.1) are pure imaginary.

3.3. FOURIER TRANSFORM—FUNCTIONS DEFINED ON THE REAL LINE

The results of Theorems 3.1.8 and 3.1.9 represent a special case of a more general result known as the *Fourier integral theorem*. By Theorem 3.1.8 the sequence of Fourier coefficients for a continuous periodic function with square integrable derivative can be used to construct a sequence of functions that converges to the original function. This result can be stated in a very compact form by substituting the definition of a_k and b_k into the statement of Theorem 3.1.7 to obtain

$$f(x) = \frac{a_0}{2} + \sum_{k=1}^{\infty} \left[\frac{1}{\pi} \int_{-\pi}^{\pi} f(\omega) \cos k\omega \cos kx \, d\omega \right.$$

$$\left. + \frac{1}{\pi} \int_{-\pi}^{\pi} f(\omega) \sin k\omega \sin kx \, d\omega \right]$$

$$= \frac{a_0}{2} + \sum_{k=1}^{\infty} \frac{1}{\pi} \int_{-\pi}^{\pi} f(\omega) \cos k(\omega - x) \, d\omega$$

$$= \frac{1}{2\pi} \sum_{k=-\infty}^{\infty} e^{-ikx} \int_{-\pi}^{\pi} f(\omega) e^{ik\omega} \, d\omega. \tag{3.3.1}$$

Since the reciprocal relationships are well defined, we could also write, using the notation of Theorem 3.1.9,

$$\rho(k)=\frac{1}{2\pi}\int_{-\pi}^{\pi}e^{ik\omega}\sum_{h=-\infty}^{\infty}\rho(h)e^{-ih\omega}\,d\omega.$$

We say that $\rho(k)$ and $f(x)$ form a *transform pair*. We shall associate the constant and the negative exponential with one transform and call this the *Fourier transform* or *spectral density*. The terms *spectrum* or *spectral function* are also used. Thus, in Theorem 3.1.9, the function $f(\omega)$ defined by

$$f(\omega)=\frac{1}{2\pi}\sum_{h=-\infty}^{\infty}\rho(h)e^{-i\omega h},$$

is the Fourier transform of $\rho(h)$, or the spectral density associated with $\rho(h)$.

The transform in (3.3.1) with positive exponent and no constant we call the *inverse transform* or the *characteristic function*. Thus the correlation function

$$\rho(h)=\int_{-\pi}^{\pi}f(\omega)e^{ih\omega}\,d\omega$$

is the inverse transform of $f(\omega)$ or the characteristic function of $f(\omega)$. We have mentioned several times that this is the statistical characteristic function if $f(\omega)$ is a probability density.

These definitions are merely to aid us in remembering the transform to which we have attached the constant $1/2\pi$. We trust that the reader will not be disturbed to find that our definitional placement of the constant may differ from that of authors in other fields.

Theorem 3.1.8 was for a periodic function. However, there exists a considerable body of theory applicable to integrable functions defined on the real line. For an integrable function, $f(x)$, defined on the real line we formally define the Fourier transform by

$$c(u)=\frac{1}{2\pi}\int_{-\infty}^{\infty}f(x)e^{-iux}\,dx, \tag{3.3.2}$$

where $u\in(-\infty,\infty)$. The Fourier integral theorem states that if the function, $f(x)$, meets certain regularity conditions the inverse transform of the Fourier transform is again the function. We state one version without proof. [See Tolstov, (1962, p. 188) for a proof.]

Theorem 3.3.1. Let $f(x)$ be an absolutely integrable continuous function defined on the real line with a right and a left derivative at every point. Then, for all $x \in (-\infty, \infty)$,

$$f(x) = \frac{1}{2\pi} \int_{-\infty}^{\infty} e^{iux} \int_{-\infty}^{\infty} f(t) e^{-iut} \, dt \, du. \tag{3.3.3}$$

Table 3.3.1 contains a summary of the Fourier transforms of the different types of functions we have considered. We have presented theorems for more general functions than the continuous functions described in the table. Likewise, the finite transform obviously holds for a function defined on an odd number of integers.

For the first, third, and fourth entries in the column headed "Inverse Transform" we have first listed the domain of the original function. However, the inverse transform, being a multiple of the transform of the transform, is defined for the values indicated in parentheses.

Given a continuous function defined on an interval, it may be convenient to record the function at a finite number of equally spaced points and find the Fourier transform of the N points. Time series are often created by reading "continuous" functions of time at fixed intervals. For example, river level and temperature have the appearance of continuous functions of time, but we may record for analysis the readings at only a few times during a day or season.

Let $g(x)$ be a continuous function with square integrable derivative defined on the interval $[-T, T]$. We evaluate the function at the $2m$ points

$$x_t = \left(\frac{t}{m}\right)T, \qquad t = -(m-1), -(m-2), \ldots, m-1, m.$$

The complex form of the Fourier coefficients for the vector $g(x_t)$ are given by

$$c_k^{(m)} = \frac{1}{2m} \sum_{t=-(m-1)}^{m} g(x_t) e^{-i\pi tk/m}, \qquad k = -(m-1), -(m-2), \ldots, m,$$

and $g(x_t)$ is expressible as

$$g(x_t) = \sum_{k=-(m-1)}^{m} c_k^{(m)} e^{i\pi kt/m}, \qquad t = -(m-1), -(m-2), \ldots, m. \tag{3.3.4}$$

The superscript (m) on the coefficients is to remind us that they are based on $2m$ equally spaced values of $g(x)$.

Table 3.3.1. *Summary of Fourier transforms*

Type of Function	Domain of Function	Fourier Transform	Inverse Transform
Finite sequence, X_t	$-(n-1),\ldots,0,\ldots,n$	$c_k=\frac{1}{2n}\sum_{t=-(n-1)}^{n} X_t e^{-i\pi tk/n}$ $k=-(n-1),\ldots,0,\ldots,n$	$X_t=\sum_{k=-(n-1)}^{n} c_k e^{i\pi kt/n}$ $t=-(n-1),\ldots,0,\ldots,n$ (can be extended to $t=0,\pm1,\pm2,\ldots$)
Infinite sequence, $\gamma(h)$, absolutely summable	$(0,\pm1,\pm2,\ldots)$	$f(\omega)=\frac{1}{2\pi}\sum_{h=-\infty}^{\infty}\gamma(h)e^{-i\omega h}$ $\omega\in(-\pi,\pi)$ $f(\omega)$ periodic of period 2π	$\gamma(h)=\int_{-\pi}^{\pi} f(\omega)e^{i\omega h}\,d\omega$ $h=0,\pm1,\pm2,\ldots$
Continuous piecewise smooth, $f(x)$	$[-\pi,\pi]$	$c_k=\frac{1}{2\pi}\int_{-\pi}^{\pi} f(x)e^{-ikx}\,dx$ $k=0,\pm1,\pm2,\ldots$	$f(x)=\sum_{k=-\infty}^{\infty} c_k e^{ikx}$ $x\in[-\pi,\pi]$ (can be extended as a periodic function on the real line)
Continuous periodic piecewise smooth $f(x)=f(x+T)$	$(-T/2,T/2)$ (can be extended to the real line)	$c_k=\frac{1}{T}\int_{-T/2}^{T/2} f(x)e^{-i2\pi kx/T}\,dx$ $k=0,\pm1,\pm2,\ldots$	$f(x)=\sum_{k=-\infty}^{\infty} c_k e^{i2\pi kx/T}$ $x\in(-T/2,T/2)$ (can be extended to real line)
Continuous square integrable, $f(x)$	$(-\infty,\infty)$	$c(u)=\frac{1}{2\pi}\int_{-\infty}^{\infty} f(x)e^{-iux}\,dx$ $u\in(-\infty,\infty)$	$f(x)=\int_{-\infty}^{\infty} c(u)e^{iux}\,du$ $x\in(-\infty,\infty)$

If we compute the Fourier coefficients for the function using the integral form (3.2.1), we have

$$c_k = \frac{1}{2T}\int_{-T}^{T} e^{-i(k\pi/T)x} g(x)\,dx, \qquad k=0, \pm 1, \pm 2, \ldots,$$

and the original function evaluated at the points x_t, $t=-(m-1), -(m-2), \ldots, m$, is given by

$$g(x_t) = \sum_{k=-\infty}^{\infty} c_k e^{i(k\pi/T)Tt/m}$$

$$= \sum_{k=-\infty}^{\infty} c_k e^{i\pi kt/m}.$$

By the periodicity of the complex exponential we have

$$g(x_t) = \sum_{k=-(m-1)}^{m} e^{i\pi kt/m}\big[c_k + (c_{k-2m} + c_{k+2m}) + (c_{k-4m} + c_{k+4m}) + \ldots\big]. \tag{3.3.5}$$

Equating the coefficients of $e^{i\pi kt/m}$ in (3.3.4) and (3.3.5), we have

$$c_k^{(m)} = c_k + \sum_{s=1}^{\infty} (c_{k-2ms} + c_{k+2ms}).$$

It follows that the coefficient for the kth frequency of the function defined at $2m$ points is the sum of the coefficients of the continuous function at the $k, k+2m, k-2m, k+4m, k-4m, \ldots$ frequencies. The frequencies $k \pm 2m, k \pm 4m, \ldots$ are called the *aliases* of the kth frequency.

In our representation the distance between two points is T/m. The cosine of period $2T/m$ or frequency $m/2T$ is the function of highest frequency that will be used in the Fourier transform. The frequency $m/2T$ is called the *Nyquist* frequency. The aliases of an observed frequency are frequencies obtained by adding or subtracting integer multiples of twice the Nyquist frequency.

To illustrate these ideas, let $g(x)$ defined on $[-10, 10]$ be given by

$$g(x) = 1.0\cos 2\pi\left(\frac{3}{20}\right)x + 0.5\cos \pi x + 0.4\cos 2\pi\left(\frac{13}{20}\right)x + 0.3\cos 2\pi\left(\frac{15}{20}\right)x + 0.2\cos 2\pi\left(\frac{23}{20}\right)x. \tag{3.3.6}$$

If the function is observed at the 10 points $-8, -6, \ldots, 8, 10$ and the $\{c_k^{(5)}: k=-4,-3,\ldots,4,5\}$ are computed by

$$c_k^{(5)} = \frac{1}{10} \sum_{t=-4}^{5} g\left(\frac{t}{5}10\right) e^{-i\pi t k/5},$$

we will obtain the coefficients given in Table 3.3.2.

Table 3.3.2. Computed coefficients for function (3.3.6) observed at 10 points

Period	Frequency	Cosine Coefficient $c_k^{(5)} + c_{-k}^{(5)}$	Contributing Alias Frequencies
—	0	0.5	10/20
20	1/20	0.0	—
10	2/20	0.0	—
20/3	3/20	1.6	13/20, 23/20
5	4/20	0.0	—
4	5/20	0.3	15/20

3.4. FOURIER TRANSFORM OF A CONVOLUTION

Fourier transform theory is particularly useful for certain function-of-function problems. We consider one such problem that occurs in statistics and statistical time series.

For absolutely integrable functions $f(x)$ and $g(x)$ defined on the real line, the function

$$\varphi(x) = \int_{-\infty}^{\infty} f(x-y) g(y)\, dy$$

is called the *convolution* of $f(x)$ and $g(x)$. Since $f(x)$ and $g(x)$ are absolutely integrable, $\varphi(x)$ is absolutely integrable.

We have the following theorem on the Fourier transform of a convolution.

Theorem 3.4.1. If the functions $f(x)$ and $g(x)$ are absolutely integrable, then the Fourier transform of the convolution of $f(x)$ and $g(x)$ is given by

$$\begin{aligned} c_\varphi(\omega) &= \frac{1}{2\pi} \int_{-\infty}^{\infty} \varphi(x) e^{-i\omega x}\, dx \\ &= 2\pi c_f(\omega) c_g(\omega), \end{aligned}$$

where

$$c_f(\omega)=\frac{1}{2\pi}\int_{-\infty}^{\infty} f(x)e^{-i\omega x}\,dx$$

and

$$c_g(\omega)=\frac{1}{2\pi}\int_{-\infty}^{\infty} g(x)e^{-i\omega x}\,dx$$

are the Fourier transforms of $f(x)$ and $g(x)$.

Proof. The integrals defining $c_\varphi(\omega)$, $c_f(\omega)$, and $c_g(\omega)$ are absolutely convergent. Hence,

$$\begin{aligned}
c_\varphi(\omega)&=\frac{1}{2\pi}\int_{-\infty}^{\infty} \varphi(x)e^{-i\omega x}\,dx\\
&=\frac{1}{2\pi}\int_{-\infty}^{\infty}\int_{-\infty}^{\infty} f(x-y)\,g(y)e^{-i\omega x}\,dy\,dx\\
&=\frac{1}{2\pi}\int_{-\infty}^{\infty}\int_{-\infty}^{\infty} f(x-y)e^{-i\omega(x-y)}g(y)e^{-i\omega y}\,dy\,dx\\
&=\frac{1}{2\pi}\int_{-\infty}^{\infty} f(z)e^{-i\omega z}\,dz\int_{-\infty}^{\infty} g(y)e^{-i\omega y}\,dy\\
&=2\pi c_f(\omega)c_g(\omega),
\end{aligned} \tag{3.4.1}$$

where the absolute integrability enabled us to interchange the order of integration. ▲

We discussed the convolution of two absolutely summable sequences in Section 2.2 and demonstrated that the convolution was absolutely summable.

Corollary 3.4.1.1. Given that $\{a_j\}$ and $\{b_j\}$ are absolutely summable, the Fourier transform of

$$d_m=\sum_{j=-\infty}^{\infty} a_{m-j}b_j$$

is

$$f_d(\omega)=2\pi f_a(\omega)f_b(\omega),$$

where

$$f_a(\omega) = \frac{1}{2\pi} \sum_{j=-\infty}^{\infty} a_j e^{-i\omega j}$$

and

$$f_b(\omega) = \frac{1}{2\pi} \sum_{j=-\infty}^{\infty} b_j e^{-i\omega j}.$$

Proof. The Fourier transform of $\{d_m\}$ is

$$f_d(\omega) = \frac{1}{2\pi} \sum_{m=-\infty}^{\infty} d_m e^{-i\omega m} = \frac{1}{2\pi} \sum_{m=-\infty}^{\infty} \sum_{j=-\infty}^{\infty} a_{m-j} b_j e^{-i\omega m}$$

$$= \frac{1}{2\pi} \sum_{m=-\infty}^{\infty} \sum_{j=-\infty}^{\infty} a_{m-j} b_j e^{-i\omega(m-j)} e^{-i\omega j}$$

$$= 2\pi f_a(\omega) f_b(\omega).$$ ▲

We may paraphrase these results as follows: the spectral density (Fourier transform) of a convolution is the product of the spectral densities (Fourier transforms) multiplied by 2π. That the converse is also true is clear from the proofs. We give a direct statement and proof for the product of absolutely summable sequences.

Corollary 3.4.1.2. Let $\{a_m\}$ and $\{b_m\}$ be absolutely summable and define d_m by

$$d_m = a_m b_m.$$

Then,

$$f_d(\omega) = \frac{1}{2\pi} \sum_{m=-\infty}^{\infty} d_m e^{-i\omega m} = \int_{-\pi}^{\pi} f_a(u) f_b(\omega - u)\, du$$

$$= \int_{-\pi}^{\pi} f_a(\omega - u) f_b(u)\, du,$$

where $f_a(\omega)$ and $f_b(\omega)$ are defined in Corollary 3.4.1.1.

Proof. We have

$$\begin{aligned}
f_d(\omega) &= \frac{1}{2\pi}\sum_{m=-\infty}^{\infty} d_m e^{-i\omega m} \\
&= \frac{1}{2\pi}\sum_{m=-\infty}^{\infty} a_m b_m e^{-i\omega m} \\
&= \frac{1}{2\pi}\sum_{m=-\infty}^{\infty}\left(\int_{-\pi}^{\pi} f_a(x)e^{imx}\,dx\right)b_m e^{-i\omega m} \\
&= \frac{1}{2\pi}\int_{-\pi}^{\pi} f_a(x)\sum_{m=-\infty}^{\infty} b_m e^{-im(\omega-x)}\,dx \\
&= \int_{-\pi}^{\pi} f_a(x) f_b(\omega - x)\,dx.
\end{aligned}$$

▲

REFERENCES

Jenkins and Watts (1968), Lighthill (1958), Nerlove (1964), Rogosinski (1959), Tolstov (1962), Zygmund (1959).

EXERCISES

1. Give the 12 orthogonal sine and cosine functions that furnish an orthogonal basis for the 12-dimensional vector space. Find the linear combinations of these vectors that yield the vectors $(1,0,\ldots,0),(0,1,0,\ldots,0),\ldots,(0,\ldots,0,1)$.
2. Expand the following functions defined on $[-\pi,\pi]$ in Fourier series.
   (a) $f(\omega)=\cos a\omega$, where a is not an integer.
   (b) $f(\omega)=\sin a\omega$, where a is not an integer.
   (c) $f(\omega)=e^{a|\omega|}$, where $a\neq 0$.
   (d) $f(\omega)=\begin{cases} 0, & -\pi\leqslant\omega\leqslant 0 \\ \sin\omega, & 0<\omega<\pi. \end{cases}$
3. Define $g(x)$ by

$$g(x)=\begin{cases} 0, & -\pi\leqslant x\leqslant 0 \\ 1, & 0<x<\pi. \end{cases}$$

   (a) Find the Fourier coefficients for $g(x)$.
   (b) What is the maximum value for

$$S_3(x)=\frac{1}{2}+\sum_{k=1}^{3}(a_k\cos kx+b_k\sin kx)?$$

Where does this maximum occur? What is the maximum value of $S_4(x)$? $S_5(x)$? The fact that the approximating function always overestimates the true function near the point of discontinuity is called *Gibb's Phenomenon*.

4. Prove Theorem 3.1.2.
5. Let $f(x)$ be the periodic function defined on the real line by

$$f(x)=\begin{cases} b^{-1}, & 2\pi j-b<x<2\pi j+b \\ 0, & \text{otherwise,} \end{cases}$$

where $j=0,\pm 1,\pm 2,\ldots,$ and $0<b<\pi$. Find the Fourier transform of $f(x)$.

6. Let

$$f(x)=\begin{cases} 1, & -b\leqslant x<b \\ 0, & \text{otherwise,} \end{cases}$$

where b is a positive number and $f(x)$ is defined on the real line. Find the Fourier transform of $f(x)$. Show that the limit of this transform at zero is infinity as $b\to\infty$. Show that as $b\to\infty$ the transform is bounded except at zero.

7. Let

$$\delta_n(x)=\begin{cases} 2\pi n, & -\dfrac{1}{2n}\leqslant x<\dfrac{1}{2n} \\ 0, & \text{otherwise,} \end{cases}$$

where n is a positive integer and $\delta_n(x)$ is defined on the real line. Find the Fourier transform of $\delta_n(x)$. Show that, as $n\to\infty$, the transform tends to the constant function of unit height.

8. Let the *generalized function* $\delta(x)$ represent a sequence of functions $\{\delta_n(x)\}_{n=1}^{\infty}$ such that

$$\lim_{n\to\infty}\int_{-\infty}^{\infty}\delta_n(x)f(x)\,dx=f(0),$$

where $f(x)$ is a continuous absolutely integrable function defined on the real line. Then $\delta(x)$ is called *Dirac's delta function*. Show that the sequence of functions $\{\delta_n(x)\}$ of Exercise 7 defines a generalized function.

9. Let $g_n(x)=(n/\pi)^{1/2}e^{-nx^2}$ for $x\in(-\infty,\infty)$. Show that the sequence $\{g_n(x)\}_{n=1}^{\infty}$ yields a Dirac delta function as defined in Exercise 8.

10. Let $f(x)$ defined for $x\in(-\infty,\infty)$ have the Fourier transform $c(u)$. Show that the Fourier transform of $f(ax+b)$ is $|a|^{-1}e^{ibu/a}c(u/a)$, $a\neq 0$.

11. Assume that price of a commodity is recorded on the last day of each month for a period of 144 months. The finite Fourier coefficients for the data are computed using the formulas following (3.1.2). Which coefficients will be affected if there is a weekly periodicity in prices that is perfectly represented by a sine wave of period seven days? Assume that there are 30.437 days in a month. Which coefficients will be affected if the weekly periodicity in prices can be represented by the sum of two sine waves, one of period seven days and one of period three and a half days? See Granger and Hatanaka (1964).

12. Let $f(x)$ and $g(x)$ be absolutely integrable functions, and define

$$\varphi(x)=\int_{-\infty}^{\infty}f(x-y)\,g(y)\,dy.$$

Show that $\varphi(x)$ satisfies

$$\int_{-\infty}^{\infty}|\varphi(x)|\,dx\leqslant\left(\int_{-\infty}^{\infty}|f(x)|\,dx\right)\left(\int_{-\infty}^{\infty}|g(y)|\,dy\right).$$

13. Let $f(x)$ and $g(x)$ be continuous absolutely integrable functions defined on the real line. State and prove the result analogous to Corollary 3.4.1.2 for $\psi(x)=f(x)g(x)$.

14. Give a direct proof of Corollary 3.4.1.2 for finite transforms. That is, for the two functions $\gamma(h)$ and $w(h)$ defined on the $2n-1$ integers $h=0, \pm 1, \pm 2, \ldots, \pm(n-1)$, show that

$$g(\omega_s)=\frac{1}{2n-1}\sum_{h=-(n-1)}^{n-1} w(h)\gamma(h)e^{-i\omega_s h}$$

$$=\sum_{k=-(n-1)}^{(n-1)} W(\omega_k)f(\omega_{s-k}),$$

where

$$\omega_k=\frac{2\pi k}{2n-1}, \qquad k=0, \pm 1, \pm 2, \ldots, \pm(n-1),$$

$$W(\omega_k)=\frac{1}{2n-1}\sum_{h=-(n-1)}^{n-1} w(h)e^{-i\omega_k h},$$

$$f(\omega_k)=\frac{1}{2n-1}\sum_{h=-(n-1)}^{n-1} \gamma(h)e^{-i\omega_k h}.$$

15. Let $f(\omega)$ be a nonnegative even continuous periodic function of period 2π. Show that

$$c(h)=\int_{-\pi}^{\pi} f(\omega)e^{-i\omega h}\,d\omega, \qquad h=0, \pm 1, \pm 2, \ldots,$$

is an even positive semidefinite function.

CHAPTER 4

Spectral Theory of Time Series

In Chapter 1 we discussed the correlation function as a way of characterizing a time series. In Chapter 2 we investigated representations of time series in terms of more elementary time series. These two descriptions are sometimes called descriptions in the *time domain* because of the obvious importance of the index set in the representations.

In Chapter 3 we introduced the Fourier transform of the correlation function. For certain correlation functions we demonstrated the uniqueness of the transform and showed that the correlation function is expressible as the inverse transform of the Fourier transform. The Fourier transform of the absolutely summable correlation function was called the spectral density. The spectral density furnishes another important representation of the time series. Because of the periodic nature of the trigonometric functions, the Fourier transform is often called the representation in the *frequency domain*.

4.1. THE SPECTRUM

In Chapter 1 we stated the result that the correlation function of a time series is analogous to a statistical characteristic function and may be expressed in the form

$$\rho(h)=\int_{-\pi}^{\pi} e^{i\omega h}\,dG(\omega), \tag{4.1.1}$$

where the integral is a Lebesgue-Stieltjes integral and $G(\omega)$ is a statistical distribution function. In Theorem 3.1.9 we proved this result for time series with absolutely summable covariance functions.

Since the covariance function of a stationary time series is the correlation function multiplied by the variance of the process, we have

$$\gamma(h)=\int_{-\pi}^{\pi} e^{i\omega h}\,dF(\omega), \tag{4.1.1a}$$

where

$$dF(\omega)=\gamma(0)dG(\omega).$$

Both of the functions $G(\omega)$ and $F(\omega)$ have been called the *spectral distribution function* in time series analysis. The spectral distribution function is a nondecreasing function that, for our purposes, can be assumed to be composed of the sum of two parts: an absolutely continuous portion and a step function.[1] We take (4.1.1a) as the definitional relationship between the spectral distribution function and the covariance function. The spectral distribution function is also sometimes called the *integrated spectrum*.

Let us assume $\gamma(h)$ is absolutely summable. Then, by Theorem 3.1.9, $f(\omega)$, defined by

$$\begin{aligned} f(\omega) &= \frac{1}{2\pi}\sum_{h=-\infty}^{\infty}\gamma(h)e^{-i\omega h} \\ &= \frac{1}{2\pi}\sum_{h=-\infty}^{\infty}\gamma(h)\cos\omega h, \end{aligned} \tag{4.1.2}$$

is a continuous nonnegative even function and

$$\gamma(h)=\int_{-\pi}^{\pi} f(\omega)e^{i\omega h}\,d\omega.$$

Thus for time series with absolutely summable covariance function, $dF(\omega) = f(\omega)d\omega$, where $f(\omega)$ was introduced as the spectral density function in Section 3.3.

Recall that we have taken $\{e_t\}$ to be a time series of uncorrelated $(0,\sigma^2)$ random variables. The spectral density of $\{e_t\}$ is

$$\begin{aligned} f_e(\omega) &= \frac{1}{2\pi}\sum_{h=-\infty}^{\infty}\gamma_e(h)\cos\omega h \\ &= \frac{1}{2\pi}\sigma^2, \end{aligned}$$

[1] Any statistical distribution function can be decomposed into three components: (a) a step function containing at most a countable number of finite jumps; (b) an absolutely continuous function; and (c) a "continuous singular" function. The third portion will be ignored in our treatment. See Tucker (1967, p. 15 ff.). While not formally correct, one may think of $F(\omega)$ as the sum of two parts, a step function with jumps at the points ω_j, $j=-M,-(M-1),\ldots,M-1,M$, and a function with continuous first derivative. Then the Lebesgue-Stieltjes integral, $\int g(\omega)dF(\omega)$, is the sum of two parts, $\sum_{j=-M}^{M} g(\omega_j)I(\omega_j)$ and $\int g(\omega)f(\omega)d\omega$, where $I(\omega_j)$ is the height of the jump in $F(\omega)$ at the point ω_j, $f(\omega)$ is the derivative of the continuous portion of $F(\omega)$, and $\int g(\omega)f(\omega)d\omega$ is a Riemann integral.

which is positive, continuous, and trivially periodic. Similarly,

$$\begin{aligned}\gamma_e(h) &= \int_{-\pi}^{\pi} \frac{1}{2\pi}\sigma^2 e^{i\omega h}\,d\omega \\ &= \int_{-\pi}^{\pi} \frac{1}{2\pi}\sigma^2 \cos\omega h\,d\omega \\ &= \begin{cases} \sigma^2, & h=0 \\ 0, & \text{otherwise.}\end{cases}\end{aligned}$$

As noted previously, the time series e_t is often called white noise. The reason for this description is now more apparent. The spectrum is a constant multiple of the variance and, in analogy to white light, one might say that all frequencies contribute equally to the variance.

If the spectral distribution function contains finite jumps we can visualize the spectrum containing discrete "spikes" at the jump points in the same way that we view discrete probability distributions. We now investigate a time series whose spectral distribution function is a step function.

Let a time series be composed of a finite sum of simple processes of the form (1.6.3). That is, define the time series, Y_t, by

$$Y_t = \sum_{j=0}^{M} (A_j \cos\omega_j t + B_j \sin\omega_j t), \tag{4.1.3}$$

where the A_j and B_j are random variables with zero mean and

$$\begin{aligned} E\{A_j^2\} = E\{B_j^2\} &= \sigma_j^2, & j &= 0,1,2,\ldots,M, \\ E\{B_j B_k\} = E\{A_j A_k\} &= 0, & j &\neq k, \\ E\{A_j B_k\} &= 0, & \forall j&,k, \end{aligned}$$

and ω_j, $j=0,1,\ldots,M$, are distinct frequencies contained in the interval $[-\pi,\pi]$. By (1.6.4), we have

$$\begin{aligned}\gamma_Y(h) = E\{Y_t Y_{t+h}\} &= \sum_{j=0}^{M} \sigma_j^2[\cos\omega_j t \cos\omega_j(t+h)] \\ &\quad + \sum_{j=0}^{M} \sigma_j^2[\sin\omega_j t \sin\omega_j(t+h)] \\ &= \sum_{j=0}^{M} \sigma_j^2 \cos\omega_j h. \end{aligned} \tag{4.1.4}$$

Since the function $\gamma(h)$ is composed of a finite sum of cosine functions, the graph of σ_j^2 against ω_j (or j) will give us a picture of the relative contribution of the variance associated with frequency ω_j to the variance of the time series. This is true because the variance of Y_t is given by

$$\gamma_Y(0)=\sum_{j=0}^{M}\sigma_j^2.$$

While we permitted our original frequencies in (4.1.3) to lie anywhere in the interval $[-\pi,\pi]$, it is clear that with no loss of generality we could have restricted the frequencies to $[0,\pi]$. That is, the covariance function for

$$X_t=A_j\cos(-\omega_j)t+B_j\sin(-\omega_j)t$$

is the same as that of

$$X_t=A_j\cos\omega_j t+B_j\sin\omega_j t.$$

This suggests that for a covariance function of the form $\sigma_j^2\cos\omega_j h$ we associate one half of the variance with the frequency $-\omega_j$ and one half with the frequency ω_j. To this end, we set

$$l(\omega_j)=l(-\omega_j)=\begin{cases}\frac{1}{2}\sigma_j^2, & \omega_j\neq 0\\ \sigma_j^2, & \omega_j=0.\end{cases}$$

To facilitate our representation, we assume that $\omega_0=0$ and then write the sum (4.1.4) as

$$\gamma_Y(h)=\sum_{j=-M}^{M} l(\omega_j)\cos\omega_j h. \tag{4.1.5}$$

We say that a time series with a covariance function of the form (4.1.5) has a *discrete spectrum* or a *line spectrum*. Equation (4.1.5) may be written as the Lebesgue-Stieltjes integral,

$$\gamma_Y(h)=\int_{-\pi}^{\pi}\cos\omega h\,dF(\omega)$$

$$=\int_{-\pi}^{\pi}e^{ih\omega}\,dF(\omega), \tag{4.1.5a}$$

where $F(\omega)$ is a step function with jumps of height $(1/2)\sigma_j^2$ at the points ω_j and $-\omega_j$, $\omega_j\neq 0$, and a jump of height σ_0^2 at $\omega_0=0$. By construction, the

jumps $l(\omega_j)$ are symmetric about zero, and we have expressed $\gamma_Y(h)$ in the general form (4.1.1a). We used this method of construction because the covariance function $\gamma_Y(h)=\sum_{j=0}^{M}\sigma_j^2\cos\omega_j h$ is not absolutely summable and we could not directly apply (4.1.2).

For our purposes it is sufficient for us to be able to recognize the two types of autocovariance functions and associated spectra: (1) the autocovariance function that is the sum of a finite number of cosine functions and is associated with a spectral distribution function which is a step function; and (2) the absolutely summable autocovariance function that is associated with a continuous spectral density.

Example. Let us consider an example. Assume that the covariance function is given by (4.1.4) with the variances and frequencies specified by Table 4.1.1.

Table 4.1.1. Example of variances for time series of form (4.1.3)

j	σ_j^2	ω_j
0	$\frac{1}{2}$	0
1	$\frac{1}{4}$	$-\frac{3}{4}\pi$
2	$\frac{3}{4}$	$-\frac{1}{10}\pi$
3	$\frac{5}{4}$	$\frac{3}{8}\pi$
4	$\frac{3}{4}$	$\frac{7}{8}\pi$

Defining $l(\omega_j)$ as in (4.1.5), we have

$$l(0)=\tfrac{1}{2}$$

$$l(-\tfrac{3}{4}\pi)=l(\tfrac{3}{4}\pi)=\tfrac{1}{8}$$

$$l(-\tfrac{1}{10}\pi)=l(\tfrac{1}{10}\pi)=\tfrac{3}{8}$$

$$l(-\tfrac{3}{8}\pi)=l(\tfrac{3}{8}\pi)=\tfrac{5}{8}$$

$$l(-\tfrac{7}{8}\pi)=l(\tfrac{7}{8}\pi)=\tfrac{3}{8}.$$

The original variances are plotted against frequency in Figure 4.1.1 and the line spectrum is plotted in Figure 4.1.2. The associated spectral distribution function is given in Figure 4.1.3.

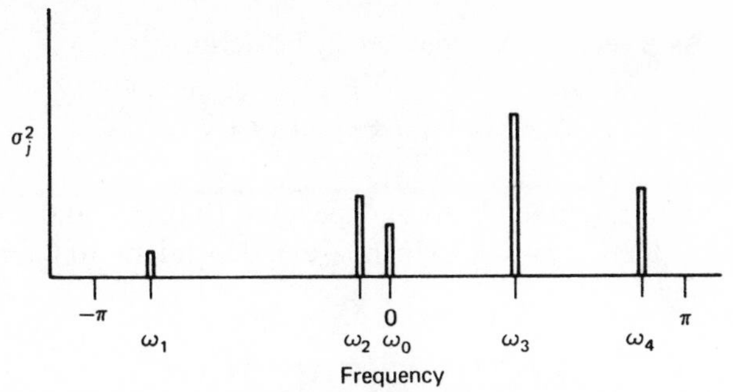

Figure 4.1.1. Graph of σ_j^2 for the time series of Table 4.1.1.

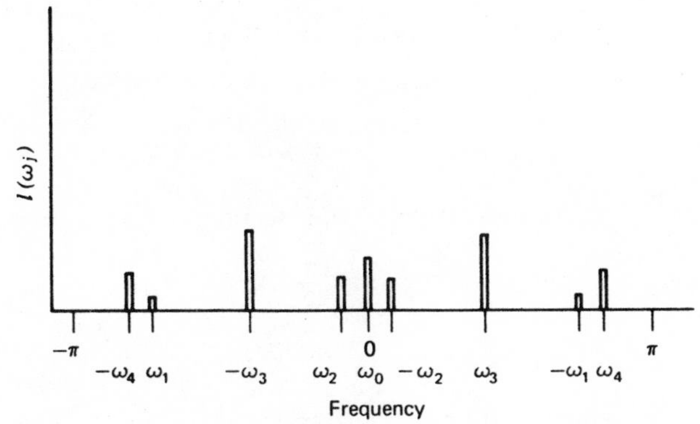

Figure 4.1.2. Graph of the line spectrum for the time series of Table 4.1.1.

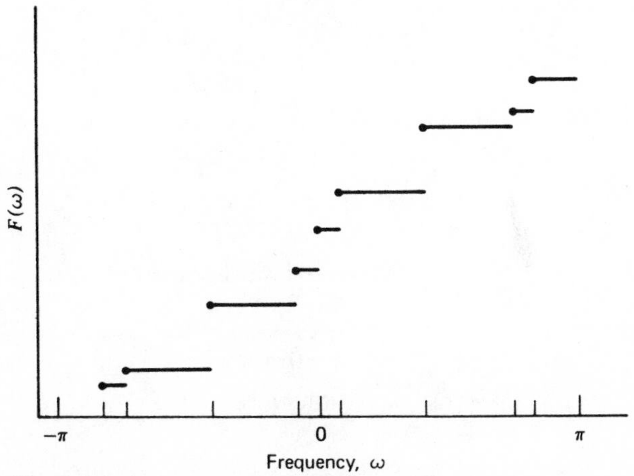

Figure 4.1.3. Spectral distribution function associated with the spectrum of Figure 4.1.2.

Example. As a second example, let X_t be defined by

$$X_t = \epsilon_1 \cos\frac{\pi}{2}t + \epsilon_2 \sin\frac{\pi}{2}t + e_t, \tag{4.1.6}$$

where the e_t, $t=0, \pm 1, \pm 2, \ldots$, are independent $(0, 0.2\pi)$ random variables independent of the ϵ_j, $j=1,2$, which are independent $(0, 0.8\pi)$ random variables. Letting

$$Y_t = \epsilon_1 \cos\frac{\pi}{2}t + \epsilon_2 \sin\frac{\pi}{2}t,$$

it follows that

$$F_X(\omega) = F_Y(\omega) + F_e(\omega),$$

$$F_e(\omega) = 0.1\pi + 0.1\omega, \qquad -\pi \leqslant \omega \leqslant \pi,$$

and

$$F_Y(\omega) = \begin{cases} 0, & -\pi \leqslant \omega < -\dfrac{\pi}{2} \\ 0.4\pi, & -\dfrac{\pi}{2} \leqslant \omega < \dfrac{\pi}{2} \\ 0.8\pi, & \dfrac{\pi}{2} \leqslant \omega \leqslant \pi. \end{cases}$$

Therefore,

$$F_X(\omega) = \begin{cases} 0.1\pi + 0.1\omega, & -\pi \leqslant \omega < -\dfrac{\pi}{2} \\ 0.5\pi + 0.1\omega, & -\dfrac{\pi}{2} \leqslant \omega < \dfrac{\pi}{2} \\ 0.9\pi + 0.1\omega, & \dfrac{\pi}{2} \leqslant \omega \leqslant \pi. \end{cases}$$

The autocovariance function of X_t is

$$\begin{aligned} \gamma_X(h) &= \int_{-\pi}^{\pi} e^{i\omega h}\, dF(\omega) \\ &= \sum_{j=-1}^{1} l(\omega_j)\cos\omega_j h + \int_{-\pi}^{\pi} 0.1\cos\omega h\, d\omega \\ &= \begin{cases} \pi, & h=0 \\ 0.8\pi\cos\dfrac{\pi}{2}h, & \text{otherwise,} \end{cases} \end{aligned}$$

where $\omega_{-1} = -\pi/2$ and $\omega_1 = \pi/2$. The reader may verify this expression by evaluating $E\{X_t X_{t+h}\}$ using the definition of X_t given in (4.1.6).

4.2 CIRCULANTS—DIAGONALIZATION OF THE COVARIANCE MATRIX OF STATIONARY PROCESSES

In this section we investigate some properties of matrices encountered in time series analysis and use these properties to obtain the matrix that will approximately diagonalize the $n \times n$ covariance matrix of n observations from a stationary time series with absolutely summable covariance function. Let the $n \times n$ covariance matrix be denoted by $\mathbf{\Gamma}$. Then,

$$\mathbf{\Gamma} = \begin{bmatrix} \gamma(0) & \gamma(1) & \gamma(2) & \cdots & \gamma(n-1) \\ \gamma(1) & \gamma(0) & \gamma(1) & \cdots & \gamma(n-2) \\ \gamma(2) & \gamma(1) & \gamma(0) & \cdots & \gamma(n-3) \\ \vdots & \vdots & \vdots & & \vdots \\ \gamma(n-1) & \gamma(n-2) & \gamma(n-3) & \cdots & \gamma(0) \end{bmatrix}.$$

It is well known[2] that for any $n \times n$ positive semidefinite covariance matrix, $\mathbf{\Gamma}$, there exists an $\mathbf{M}$ satisfying $\mathbf{M}'\mathbf{M} = \mathbf{I}$ such that

$$\mathbf{M}'\mathbf{\Gamma}\mathbf{M} = \operatorname{diag}(\lambda_1, \lambda_2, \ldots, \lambda_n),$$

where λ_i, $i = 1, 2, \ldots, n$, are the characteristic roots of $\mathbf{\Gamma}$. Our investigation will demonstrate that for large n the λ_j are approximately equal to $2\pi f(\omega_j)$, where $f(\omega)$ is the spectral density of X_t and $\omega_j = 2\pi j/n$, $j = 0, 1, 2, \ldots, n-1$. This permits us to interpret the spectral density as a multiple of the variance of the orthogonal random variables defined by the transformation $\mathbf{M}$. We initiate our study by introducing a matrix whose roots have a particularly simple representation.

A matrix of the form

$$\mathbf{\Gamma}_c = \begin{bmatrix} \gamma(0) & \gamma(1) & \cdots & \gamma(n-2) & \gamma(n-1) \\ \gamma(n-1) & \gamma(0) & \cdots & \gamma(n-3) & \gamma(n-2) \\ \gamma(n-2) & \gamma(n-1) & \cdots & \gamma(n-4) & \gamma(n-3) \\ \vdots & \vdots & & \vdots & \vdots \\ \gamma(2) & \gamma(3) & \cdots & \gamma(0) & \gamma(1) \\ \gamma(1) & \gamma(2) & \cdots & \gamma(n-1) & \gamma(0) \end{bmatrix} \tag{4.2.1}$$

is called a *circular matrix* or *circulant*. Of course, in this definition $\gamma(j)$ may be any number, but we use the covariance notation, since our

[2] See, for example, Rao (1965, p. 36).

immediate application will be to a matrix whose elements are covariances.

The characteristic roots, λ_j, and vectors, $\mathbf{x}_j$, of the matrix (4.2.1) satisfy the equation

$$\boldsymbol{\Gamma}_c \mathbf{x}_j = \lambda_j \mathbf{x}_j, \qquad j = 1, 2, \ldots, n, \tag{4.2.2}$$

and we have

$$\begin{aligned}
\gamma(0)x_{1j} + \gamma(1)x_{2j} + \cdots + \gamma(n-2)x_{n-1,j} + \gamma(n-1)x_{nj} &= \lambda_j x_{1j} \\
\gamma(n-1)x_{1j} + \gamma(0)x_{2j} + \cdots + \gamma(n-3)x_{n-1,j} + \gamma(n-2)x_{nj} &= \lambda_j x_{2j} \\
\vdots \qquad \vdots \qquad \qquad \vdots \qquad \qquad \vdots \qquad & \quad \vdots \\
\gamma(1)x_{1j} + \gamma(2)x_{2j} + \cdots + \gamma(n-1)x_{n-1,j} + \gamma(0)x_{nj} &= \lambda_j x_{nj}
\end{aligned} \tag{4.2.3}$$

where x_{kj} is the kth element of the jth characteristic vector. Let r_j be a root of the scalar equation $r^n = 1$ and set $x_{kj} = r_j^k$. The system of equations (4.2.3) becomes

$$\begin{aligned}
\gamma(0)r_j + \gamma(1)r_j^2 + \cdots + \gamma(n-2)r_j^{n-1} + \gamma(n-1)r_j^n &= \lambda_j r_j \\
\gamma(n-1)r_j + \gamma(0)r_j^2 + \cdots + \gamma(n-3)r_j^{n-1} + \gamma(n-2)r_j^n &= \lambda_j r_j^2 \\
\vdots \qquad \vdots \qquad \qquad \vdots \qquad \qquad \vdots \qquad & \quad \vdots \\
\gamma(1)r_j + \gamma(2)r_j^2 + \cdots + \gamma(n-1)r_j^{n-1} + \gamma(0)r_j^n &= \lambda_j r_j^n .
\end{aligned} \tag{4.2.4}$$

If we multiply the first equation of (4.2.4) by r_j^{n-1}, the second by r_j^{n-2}, and so forth, using $r_j^{n+k} = r_j^k$, we see that we will obtain equality for each equation if

$$\lambda_j = \sum_{h=0}^{n-1} \gamma(h) r_j^h. \tag{4.2.5}$$

The equation $r^n = 1$ has n distinct roots, $e^{i2\pi j/n}$, $j = 1, 2, \ldots, n$, which may

also be expressed as $r_j = e^{-i2\pi j/n}$, $j=0,1,\ldots,n-1$. Therefore, the n characteristic vectors associated with the n characteristic roots are given by

$$\mathbf{x}_j = \mathbf{g}_{j.}^* = n^{-1/2}\left[1, e^{-i2\pi j/n}, e^{-i2\pi 2j/n}, \ldots, e^{-i2\pi(n-1)j/n}\right]', \qquad j=0,1,\ldots,n-1, \tag{4.2.6}$$

where $\mathbf{g}_{j.}^*$ is the conjugate transpose of the row vector

$$\mathbf{g}_{j.} = n^{-1/2}\left[1, e^{i2\pi j/n}, e^{i2\pi 2j/n}, \ldots, e^{i2\pi(n-1)j/n}\right].$$

If we define the matrix $\mathbf{G}$ by

$$\mathbf{G} = \begin{bmatrix} \mathbf{g}_{0.} \\ \mathbf{g}_{1.} \\ \vdots \\ \mathbf{g}_{n-1,.} \end{bmatrix},$$

then

$$\mathbf{G}\mathbf{G}^* = \mathbf{I},$$
$$\mathbf{G}\boldsymbol{\Gamma}_c\mathbf{G}^* = \operatorname{diag}(\lambda_0, \lambda_1, \ldots, \lambda_{n-1}),$$

where $\mathbf{G}^*$ is the conjugate transpose of $\mathbf{G}$.

Setting $\gamma(1)=\gamma(n-1)$, $\gamma(2)=\gamma(n-2), \ldots, \gamma(h)=\gamma(n-h), \ldots,$ in $\boldsymbol{\Gamma}_c$, we obtain the circular symmetric matrix,

$$\boldsymbol{\Gamma}_s = \begin{bmatrix} \gamma(0) & \gamma(1) & \gamma(2) & \cdots & \gamma(2) & \gamma(1) \\ \gamma(1) & \gamma(0) & \gamma(1) & \cdots & \gamma(3) & \gamma(2) \\ \gamma(2) & \gamma(1) & \gamma(0) & \cdots & \gamma(4) & \gamma(3) \\ \vdots & \vdots & \vdots & & \vdots & \vdots \\ \gamma(1) & \gamma(2) & \gamma(3) & \cdots & \gamma(1) & \gamma(0) \end{bmatrix}. \tag{4.2.7}$$

Substituting into (4.2.5) we obtain for the roots of (4.2.7)

$$\lambda_j = \begin{cases} \displaystyle\sum_{h=-(n-1)/2}^{(n-1)/2} \gamma(h) e^{-i2\pi hj/n}, & n \text{ odd} \\ \displaystyle\sum_{h=-(n/2)+1}^{n/2} \gamma(h) e^{-i2\pi hj/n}, & n \text{ even,} \end{cases} \tag{4.2.8}$$

where we have used the periodic property, $e^{-i2\pi j(n-h)/n}=e^{i2\pi jh/n}$. Note that these roots are real, as they must be for a real symmetric matrix.

In some applications it is preferable to work with a real matrix rather than the complex matrix $\mathbf{G}$. Consider first the case of n odd. Equation (4.2.8) may also be written

$$\lambda_j=\sum_{h=-(n-1)/2}^{(n-1)/2}\gamma(h)\cos\frac{2\pi}{n}hj, \qquad j=0,1,\ldots,n-1. \tag{4.2.9}$$

Since, for $0\leqslant m\leqslant 2\pi$, $\cos m=\cos(2\pi-m)$, we see that there is a root for $j=0$ (or $j=n$) and $(n-1)/2$ roots of multiplicity two associated with $j=1,2,\ldots,(n-1)/2$. For each of these repeated roots we can find two real orthogonal vectors. These are chosen to be

$$2^{1/2}n^{-1/2}\left[1,\cos 2\pi\frac{j}{n},\cos 2\pi\frac{2j}{n},\cos 2\pi\frac{3j}{n},\ldots,\cos 2\pi\frac{(n-1)j}{n}\right]$$

and

$$2^{1/2}n^{-1/2}\left[0,\sin 2\pi\frac{j}{n},\sin 2\pi\frac{2j}{n},\sin 2\pi\frac{3j}{n},\ldots,\sin 2\pi\frac{(n-1)j}{n}\right]. \tag{4.2.10}$$

If we choose the vectors in (4.2.6) associated with the jth and $n-j$th root of one, the vectors in (4.2.10) are given by

$$2^{-1/2}(\mathbf{g}_{j.}+\mathbf{g}_{n-j,.}) \quad\text{and}\quad 2^{-1/2}i(\mathbf{g}_{j.}-\mathbf{g}_{n-j,.}).$$

Much the same pattern holds for the roots of a circular symmetric matrix of dimension $n\times n$ where n is even. There is a characteristic vector $n^{-1/2}(1,1,\ldots,1)$ associated with $j=0$ and a vector $n^{-1/2}(1,-1,1,\ldots,-1)$ associated with $j=n/2$. The remaining $(n/2)-2$ roots have multiplicity two and the roots are given by (4.2.8).

Define the orthogonal matrix $\mathbf{Q}$ by setting $n^{1/2}\,2^{-1/2}\,\mathbf{Q}'$ equal to

$$\begin{bmatrix}
2^{-1/2} & 2^{-1/2} & 2^{-1/2} & \cdots & 2^{-1/2}\\
1 & \cos 2\pi\frac{1}{n} & \cos 2\pi\frac{2}{n} & \cdots & \cos 2\pi\frac{n-1}{n}\\
0 & \sin 2\pi\frac{1}{n} & \sin 2\pi\frac{2}{n} & \cdots & \sin 2\pi\frac{n-1}{n}\\
1 & \cos 4\pi\frac{1}{n} & \cos 4\pi\frac{2}{n} & \cdots & \cos 4\pi\frac{n-1}{n}\\
\vdots & \vdots & \vdots & & \vdots\\
0 & \sin\frac{n-1}{2}2\pi\frac{1}{n} & \sin\frac{n-1}{2}2\pi\frac{2}{n} & \cdots & \sin\frac{n-1}{2}2\pi\frac{n-1}{n}
\end{bmatrix} \tag{4.2.11}$$

Note that $\mathbf{Q}$ is the matrix composed of the n characteristic vectors defined by (4.2.10) and our illustration (4.2.11) is for odd n. Let X_t be a stationary time series with covariance function $\gamma(h)$ where $\gamma(h)$ is absolutely summable. For odd n, define the $n \times n$ diagonal matrix $\mathbf{D}$ by

$$\mathbf{D} = \operatorname{diag}\{d_1, d_2, \ldots, d_n\}, \tag{4.2.12}$$

where

$$d_1 = f(0) = \frac{1}{2\pi} \sum_{h=-\infty}^{\infty} \gamma(h),$$

$$d_{2j} = d_{2j+1} = f\left(2\pi \frac{j}{n}\right) = \frac{1}{2\pi} \sum_{h=-\infty}^{\infty} \gamma(h) e^{-i2\pi hj/n}, \qquad j = 1, 2, \ldots, \frac{n-1}{2}. \tag{4.2.13}$$

For $\mathbf{\Gamma}_s$ defined by (4.2.7) the matrix $\mathbf{Q}'\mathbf{\Gamma}_s\mathbf{Q}$ is a diagonal matrix whose elements converge to $2\pi\mathbf{D}$ as n increases. This also holds for even n if the definition of $\mathbf{Q}$ is slightly modified. An additional row,

$$n^{-1/2}[1, -1, 1, \ldots, 1, -1],$$

which is the characteristic vector associated with $j = n/2$, is added to the $\mathbf{Q}'$ of (4.2.11) when n is even. The last entry in $\mathbf{D}$ for even n is

$$d_n = f(\pi) = \frac{1}{2\pi} \sum_{h=-\infty}^{\infty} \gamma(h) \cos \pi h.$$

The covariance matrix for n observations on X_t is given by

$$\mathbf{\Gamma} = \begin{bmatrix} \gamma(0) & \gamma(1) & \cdots & \gamma(n-1) \\ \gamma(1) & \gamma(0) & \cdots & \gamma(n-2) \\ \vdots & \vdots & & \vdots \\ \gamma(n-1) & \gamma(n-2) & \cdots & \gamma(0) \end{bmatrix}. \tag{4.2.14}$$

For $\gamma(h)$ absolutely summable, we now demonstrate that $\mathbf{Q}'\mathbf{\Gamma}\mathbf{Q}$ also converges to $2\pi\mathbf{D}$.

Let $\mathbf{q}_{.i} = [q_{1i}, q_{2i}, \ldots, q_{ni}]'$ be the ith column of $\mathbf{Q}$. We have, for $\mathbf{\Gamma}$ defined in (4.2.14) and $\mathbf{\Gamma}_s$ defined in (4.2.7),

$$|\mathbf{q}'_{.i}\mathbf{\Gamma}_s\mathbf{q}_{.j} - \mathbf{q}'_{.i}\mathbf{\Gamma}\mathbf{q}_{.j}| = \left| \sum_{m=1}^{M} [\gamma(m) - \gamma(n-m)] \times \sum_{k=1}^{m} [q_{ki}q_{n-m+k,j} + q_{n-m+k,i}q_{kj}] \right|, \tag{4.2.15}$$

where $M=(n-1)/2$ if n is odd and $M=(n/2)-1$ if n is even. Now (4.2.15) is less than

$$\frac{4}{n}\left\{\sum_{m=1}^{M} m|\gamma(m)|+\sum_{m=1}^{M} m|\gamma(n-m)|\right\}$$

$$\leqslant \frac{4}{n}\left\{\sum_{m=1}^{M} m|\gamma(m)|+\sum_{h=M+1}^{n} M|\gamma(h)|\right\} \tag{4.2.16}$$

since $q_{ki}q_{rj}\leqslant 2/n$ for all $k,i,r,j\in(1,2,\ldots,n)$. As n increases, the limit of the first term is zero by Lemma 3.1.4 and the limit of the second term is zero by the absolute summability of $\gamma(h)$. Therefore, the elements of $\mathbf{Q}'\mathbf{\Gamma}\mathbf{Q}$ converge to $2\pi\mathbf{D}$. We state the result as a theorem.

Theorem 4.2.1. Let $\mathbf{\Gamma}$ be the covariance matrix of n observations from a stationary time series with absolutely summable covariance function. Let $\mathbf{Q}$ be defined by (4.2.11) and take $\mathbf{D}$ to be the $n\times n$ diagonal matrix defined in (4.2.12). Then, given $\epsilon>0$, there exists an n_ϵ such that for $n>n_\epsilon$ every element of the matrix

$$\mathbf{Q}'\mathbf{\Gamma}\mathbf{Q}-2\pi\mathbf{D}$$

is less than ϵ.

Corollary 4.2.1. Let $\mathbf{\Gamma}$ be the covariance matrix of n observations from a stationary time series with covariance function that satisfies

$$\sum_{h=-\infty}^{\infty} |h|\,|\gamma(h)|=L<\infty.$$

Let $\mathbf{Q}$ and $\mathbf{D}$ be as defined in Theorem 4.2.1. Then every element of the matrix $\mathbf{Q}'\mathbf{\Gamma}\mathbf{Q}-2\pi\mathbf{D}$ is less than $4L/n$.

Proof. By (4.2.16) the difference between $\mathbf{q}_{.i}'\mathbf{\Gamma}_s\mathbf{q}_{.j}$ and $\mathbf{q}_{.i}'\mathbf{\Gamma}\mathbf{q}_{.j}$ is less than

$$\frac{4}{n}\left\{\sum_{m=1}^{M} m|\gamma(m)|+\sum_{h=M+1}^{n} M|\gamma(h)|\right\}\leqslant\frac{4}{n}\sum_{m=1}^{n} m|\gamma(m)|<\frac{2L}{n}.$$

Now $\mathbf{Q}'\mathbf{\Gamma}_s\mathbf{Q}$ is a diagonal matrix and the difference between an element of $\mathbf{Q}'\mathbf{\Gamma}_s\mathbf{Q}$ and an element of $2\pi\mathbf{D}$ is

$$\left|\sum_{h=-(n-1)/2}^{(n-1)/2}\gamma(h)e^{-i2\pi hj/n}-\sum_{h=-\infty}^{\infty}\gamma(h)e^{-i2\pi hj/n}\right|$$

$$\leqslant 2\sum_{h=(n-1)/2+1}^{\infty}|\gamma(h)|\leqslant\frac{4}{n}\sum_{h=(n+1)/2}^{\infty}h|\gamma(h)|<\frac{2L}{n}.$$

▲

We have demonstrated that, asymptotically, the **Q** defined by (4.2.11) or the **G** defined by (4.2.6) will diagonalize *all* **Γ**-matrices associated with stationary time series with absolutely summable covariance functions. That is, the transformation **Q** applied to the n observations defines n new random variables that are "nearly" uncorrelated. Each of the new random variables is a linear combination of the n original observations with weights given in (4.2.10). The variances of both the $(2j)$th and $(2j+1)$th random variables created by the transformation are approximately $2\pi f(2\pi j/n)$.

For time series with absolutely summable covariance function the variance result holds for the random variable defined for any ω, not just for integer multiples of $2\pi/n$. That is, the complex random variable

$$Y_{\omega,n} = n^{-1/2} \sum_{t=1}^{n} X_t e^{i\omega t}$$

has variance given by

$$\begin{aligned}
E\{|Y_{\omega,n}|^2\} &= \frac{1}{n} E\left\{ \sum_{t=1}^{n} X_t e^{-i\omega t} \sum_{t=1}^{n} X_t e^{i\omega t} \right\} \\
&= \frac{1}{n} E\left\{ \sum_{t=1}^{n} \sum_{j=1}^{n} X_t X_j e^{-i\omega(t-j)} \right\} \\
&= \frac{1}{n} \sum_{t=1}^{n} \sum_{j=1}^{n} \gamma(t-j) e^{-i\omega(t-j)} \\
&= \sum_{h=-(n-1)}^{(n-1)} \frac{n-|h|}{n} \gamma(h) e^{-i\omega h}.
\end{aligned}$$

By Lemma 3.1.4,

$$\lim_{n\to\infty} E\{|Y_{\omega,n}|^2\} = 2\pi f_X(\omega).$$

We shall return to this point in Chapter 7 and investigate functions of the random variables created by applying the transformation (4.2.10) or (4.2.6) to a finite number of observations from a realization.

4.3. THE SPECTRAL DENSITY OF MOVING AVERAGE AND AUTOREGRESSIVE TIME SERIES

By the Fourier integral theorem we know that the Fourier transform of an absolutely summable sequence is unique and that the sequence is given once again by the inverse transform of the transform. Therefore, we expect

the spectral density of a particular kind of time series, say a finite moving average, to have a distinctive form. We shall see that this is the case; the spectral densities of autoregressive and moving average processes are easily recognizable.

Stationary moving average, stationary autoregressive, and stationary autoregressive moving average time series are contained within the representation

$$X_t = \sum_{j=-\infty}^{\infty} b_j e_{t-j}, \tag{4.3.0}$$

where

$$\sum_{j=-\infty}^{\infty} |b_j| < \infty$$

and the e_t are uncorrelated $(0, \sigma^2)$ random variables. Therefore, if we obtain the spectral density of a time series of the form (4.3.0), we shall have the spectral density for such types of time series. In fact, it is convenient to treat the case of an infinite weighted sum of a more general time series.

Theorem 4.3.1. Let X_t be a stationary time series with an absolutely summable covariance function and let $\{a_j\}_{j=-\infty}^{\infty}$ be absolutely summable. Then the spectral density of

$$Y_t = \sum_{j=-\infty}^{\infty} a_j X_{t-j}$$

is

$$f_Y(\omega) = (2\pi)^2 f_X(\omega) f_a(\omega) f_a^*(\omega), \tag{4.3.1}$$

where $f_X(\omega)$ is the spectral density of X_t,

$$f_a(\omega) = \frac{1}{2\pi} \sum_{j=-\infty}^{\infty} a_j e^{-\iota\omega j}$$

is the Fourier transform of a_j, and

$$f_a^*(\omega) = \frac{1}{2\pi} \sum_{j=-\infty}^{\infty} a_j e^{\iota\omega j}$$

is the complex conjugate of the Fourier transform of a_j.

Proof. Given the absolute summability, we may interchange integration and summation and calculate the expectation term by term. Therefore, letting $E\{X_t\}=0$, we have

$$E\{Y_tY_{t+h}\}=E\left\{\left(\sum_{j=-\infty}^{\infty} a_jX_{t-j}\right)\left(\sum_{k=-\infty}^{\infty} a_kX_{t-k+h}\right)\right\}$$

$$=E\left\{\sum_{j=-\infty}^{\infty} a_j \sum_{k=-\infty}^{\infty} a_kX_{t-j}X_{t-k+h}\right\}$$

$$=\sum_{j=-\infty}^{\infty} a_j \sum_{k=-\infty}^{\infty} a_k\gamma_X(h-k+j).$$

If we set $p=h-k+j$, then

$$\gamma_Y(h)=\sum_{j=-\infty}^{\infty} a_j \sum_{p=-\infty}^{\infty} a_{j+h-p}\gamma_X(p). \tag{4.3.2}$$

Since $\gamma_Y(h)$ is absolutely summable, we may write

$$f_Y(\omega)=\frac{1}{2\pi}\sum_{h=-\infty}^{\infty} e^{-\iota\omega h}\gamma_Y(h)$$

$$=\frac{1}{2\pi}\sum_{h=-\infty}^{\infty} e^{-\iota\omega h}\sum_{j=-\infty}^{\infty} a_j \sum_{p=-\infty}^{\infty} a_{j+h-p}\gamma_X(p)$$

$$=\frac{1}{2\pi}\sum_{h=-\infty}^{\infty}\sum_{j=-\infty}^{\infty}\sum_{p=-\infty}^{\infty} a_ja_{j+h-p}\gamma_X(p)e^{\iota\omega j}e^{-\iota\omega p}e^{-\iota\omega(j+h-p)}$$

$$=\frac{1}{2\pi}\sum_{j=-\infty}^{\infty} a_je^{\iota\omega j}\sum_{p=-\infty}^{\infty}\gamma_X(p)e^{-\iota\omega p}\sum_{s=-\infty}^{\infty} a_se^{-\iota\omega s},$$

where we have made the transformation $s=j+h-p$. ▲

The theorem is particularly useful, and we shall make repeated application of the result. If the X_t of the theorem are assumed to be uncorrelated random variables we obtain the spectral density of a moving average time series.

Corollary 4.3.1.1. The spectral density of the moving average process

$$X_t=\sum_{j=-\infty}^{\infty} a_je_{t-j},$$

where $\{e_t\}$ is a sequence of uncorrelated $(0,\sigma^2)$ random variables and the sequence $\{a_j\}$ is absolutely summable, is

$$f_X(\omega)=\frac{\sigma^2}{2\pi}\left(\sum_{j=-\infty}^{\infty} a_j e^{-ij\omega}\right)\left(\sum_{j=-\infty}^{\infty} a_j e^{ij\omega}\right). \tag{4.3.3}$$

Proof. The result follows immediately from Theorem 4.3.1. ▲

Let us use (4.3.3) to calculate the spectral density of the finite moving average

$$X_t=\sum_{j=0}^{p} a_j e_{t-j}. \tag{4.3.4}$$

We have

$$f_X(\omega)=\frac{\sigma^2}{2\pi}\left(\sum_{j=0}^{p} a_j e^{-ij\omega}\right)\left(\sum_{j=0}^{p} a_j e^{ij\omega}\right)$$

$$=\frac{\sigma^2}{2\pi}\left(a_0+a_1e^{-i\omega}+a_2e^{-i2\omega}+\cdots+a_pe^{-ip\omega}\right)$$

$$\times\left(a_0+a_1e^{i\omega}+a_2e^{i2\omega}+\cdots+a_pe^{ip\omega}\right)$$

$$=\frac{\sigma^2}{2\pi}\Big[\left(a_0^2+a_1^2+a_2^2+\cdots+a_p^2\right)$$

$$+(a_0a_1+a_1a_2+a_2a_3+\cdots+a_{p-1}a_p)e^{-i\omega}$$

$$+(a_0a_1+a_1a_2+a_2a_3+\cdots+a_{p-1}a_p)e^{i\omega}$$

$$+(a_0a_2+a_1a_3+\cdots+a_{p-2}a_p)e^{-i2\omega}$$

$$+\cdots+a_0a_pe^{-ip\omega}+a_0a_pe^{ip\omega}\Big]$$

$$=\frac{1}{2\pi}\Big[\gamma_X(0)+\gamma_X(1)e^{-i\omega}+\gamma_X(1)e^{i\omega}+\gamma_X(2)e^{-i2\omega}$$

$$+\cdots+\gamma_X(p)e^{-ip\omega}+\gamma_X(p)e^{ip\omega}\Big]$$

$$=\frac{1}{2\pi}\left[\sum_{h=-p}^{p}\gamma_X(h)e^{-i\omega h}\right]. \tag{4.3.5}$$

Equation (4.3.5) is a proof of the corollary for finite p and serves to reinforce our confidence in the general result. The spectral density of a finite moving average may, alternatively, be expressed in terms of the roots of the auxiliary equation. The quantity

$$\sum_{j=0}^{p} a_j e^{-i\omega j}$$

is seen to be a polynomial in $e^{-i\omega}$. Therefore, it may be written as

$$\sum_{j=0}^{p} a_j e^{-i\omega j} = e^{-i\omega p} \sum_{j=0}^{p} a_j e^{i\omega(p-j)} = e^{-i\omega p} \prod_{j=1}^{p} \left(e^{i\omega} - m_j\right),$$

where $a_0 = 1$ and the m_j are the roots of

$$m^p + a_1 m^{p-1} + a_2 m^{p-2} + \cdots + a_p = 0.$$

It follows that the spectral density of the $\{X_t\}$ defined by (4.3.4) is

$$f_X(\omega) = \frac{\sigma^2}{2\pi}\left[\prod_{j=1}^{p}\left(e^{-i\omega} - m_j\right)\right]\left[\prod_{j=1}^{p}\left(e^{i\omega} - m_j\right)\right]. \tag{4.3.6}$$

We can also use Corollary 4.3.1.1 to compute the spectral density of the first order autoregressive process

$$X_t = \sum_{j=0}^{\infty} \rho^j e_{t-j}, \qquad |\rho| < 1.$$

The transform of the weights $a_j = \rho^j$ is given by

$$g_a(\omega) = \frac{1}{2\pi} \sum_{j=0}^{\infty} \rho^j e^{-ij\omega}$$

$$= \frac{1}{2\pi} \sum_{j=0}^{\infty} \left(\rho e^{-i\omega}\right)^j = \frac{1}{2\pi} \frac{1}{1 - \rho e^{-i\omega}}$$

and the complex conjugate by

$$g_a^*(\omega) = \frac{1}{2\pi} \sum_{j=0}^{\infty} \rho^j e^{ij\omega} = \frac{1}{2\pi} \frac{1}{1 - \rho e^{i\omega}}.$$

Therefore,

$$\begin{aligned} f_X(\omega) &= (2\pi)^2 f_e(\omega)\, g_a(\omega)\, g_a^*(\omega) \\ &= \frac{\sigma^2}{2\pi}\left(\frac{1}{1-\rho e^{-\iota\omega}}\right)\left(\frac{1}{1-\rho e^{\iota\omega}}\right) \\ &= \frac{\sigma^2}{2\pi}\,\frac{1}{1-2\rho\cos\omega+\rho^2}, \end{aligned} \tag{4.3.7}$$

since $f_e(\omega)$, the spectral density of the uncorrelated sequence $\{e_t\}$, is $\sigma^2/2\pi$ for all ω.

We note that we could also find the spectral density of X_t by setting $Y_t = e_t$ in (4.3.1) and using the known spectral density of e_t. Applying this approach to the pth order autoregressive time series, we have the following corollary.

Corollary 4.3.1.2. The spectral density of the pth order autoregressive time series $\{X_t\}$ defined by

$$\sum_{j=0}^{p} \alpha_j X_{t-j} = e_t,$$

where $\alpha_0 = 1$, $\alpha_p \neq 0$, the e_t are uncorrelated $(0,\sigma^2)$ random variables, and the roots $\{m_j\colon j=1,2,\ldots,p\}$ of

$$R(m) = \sum_{j=0}^{p} \alpha_j m^{p-j} = 0$$

are less than one in absolute value, is given by

$$\begin{aligned} f_X(\omega) &= \frac{\sigma^2}{2\pi}\left[\left(\sum_{j=0}^{p} \alpha_j e^{-\iota\omega j}\right)\left(\sum_{j=0}^{p} \alpha_j e^{\iota\omega j}\right)\right]^{-1} \\ &= \frac{\sigma^2}{2\pi}\prod_{j=1}^{p} \frac{1}{\left(1-2m_j\cos\omega+m_j^2\right)}. \end{aligned} \tag{4.3.8}$$

Proof. Apply Theorem 4.3.1 to the finite moving average time series

$$e_t = \sum_{j=0}^{p} \alpha_j X_{t-j}$$

to obtain

$$(2\pi)^{-1}\sigma^2 = f_e(\omega) = T(\omega) f_X(\omega),$$

where

$$\begin{aligned} T(\omega) &= \left[\sum_{j=0}^{p} \alpha_j e^{-i\omega j}\right]\left[\sum_{j=0}^{p} \alpha_j e^{i\omega j}\right] \\ &= \left[e^{i\omega p} \sum_{j=0}^{p} \alpha_j e^{-i\omega j}\right]\left[e^{-i\omega p} \sum_{j=0}^{p} \alpha_j e^{i\omega j}\right] \\ &= R(e^{i\omega}) R(e^{-i\omega}). \end{aligned}$$

Neither $R(e^{i\omega})$ nor $R(e^{-i\omega})$ can be zero, since the roots of $R(m)$ cannot have an absolute value of one. Therefore, $T(\omega)$ is not equal to zero for any ω, and we are able to write

$$f_X(\omega) = f_e(\omega) T^{-1}(\omega).$$

This is the first representation given in (4.3.8). The alternative form for $f_X(\omega)$ follows by writing $T(\omega)$ in the factored form

$$\begin{aligned} T(\omega) &= \left|\prod_{j=1}^{p} \left(1 - e^{-i\omega} m_j\right)\right|^2 \\ &= \prod_{j=1}^{p} \left(1 - 2m_j \cos\omega + m_j^2\right). \end{aligned}$$

▲

We can also write (4.3.8) in the factored form,

$$f_X(\omega) = \frac{\sigma^2}{2\pi}\left[\prod_{j=1}^{p}\left(e^{-i\omega} - m_j\right)\prod_{j=1}^{p}\left(e^{i\omega} - m_j\right)\right]^{-1}. \tag{4.3.9}$$

The spectral density of the autoregressive moving average process is sometimes called a rational spectral density because it is the ratio of polynomials in $e^{i\omega}$.

Corollary 4.3.1.3. Let an autoregressive moving average process of

order (p,q) be defined by

$$\sum_{j=0}^{p} \alpha_j X_{t-j} = \sum_{k=0}^{q} \beta_k e_{t-k},$$

where $\alpha_0 = \beta_0 = 1$, $\alpha_p \neq 0$, $\beta_q \neq 0$, the e_t are uncorrelated $(0,\sigma^2)$ random variables, and the roots $\{m_j: j=1,2,\ldots,p\}$ of

$$R(m) = \sum_{j=0}^{p} \alpha_j m^{p-j} = 0$$

are less than one in absolute value. Then the spectral density of $\{X_t\}$ is given by

$$f_X(\omega) = \frac{\sigma^2}{2\pi} \frac{\left(\sum_{k=0}^{q} \beta_k e^{i\omega k}\right)\left(\sum_{k=0}^{q} \beta_k e^{-i\omega k}\right)}{\left(\sum_{j=0}^{p} \alpha_j e^{i\omega j}\right)\left(\sum_{j=0}^{p} \alpha_j e^{-i\omega j}\right)}$$

$$= \frac{\sigma^2}{2\pi} \frac{\prod_{k=1}^{q} \left(e^{i\omega} - r_k\right) \prod_{k=1}^{q} \left(e^{-i\omega} - r_k\right)}{\prod_{j=1}^{p} \left(e^{i\omega} - m_j\right) \prod_{j=1}^{p} \left(e^{-i\omega} - m_j\right)}, \tag{4.3.10}$$

where $\{r_k: k=1,2,\ldots,q\}$ are the roots of

$$\sum_{k=0}^{q} \beta_k r^{q-k} = 0.$$

Proof. Reserved for the reader. ▲

In Theorem 2.6.4 we demonstrated that a finite moving average process with no roots of unit absolute value always has a representation with characteristic equation whose roots are less than one in absolute value. In Theorem 2.6.3 we obtained a moving average representation for a time series with $|\rho(1)| < 0.5$ and $\rho(h) = 0$, $h = \pm 2, \pm 3, \ldots$. We can now extend Theorem 2.6.3 to time series with a finite number of nonzero covariances.

Theorem 4.3.2. Let the stationary time series X_t have zero mean and spectral density

$$f(\omega) = \frac{1}{2\pi} \sum_{h=-q}^{q} \gamma(h) e^{-i\omega h},$$

where $f(\omega)$ is strictly positive, $\gamma(h)$ is the covariance function of X_t, and $\gamma(q)\neq 0$. Then X_t has the representation

$$X_t=\sum_{j=1}^{q}\beta_j e_{t-j}+e_t=\sum_{j=0}^{q}\beta_j e_{t-j},$$

where the e_t are uncorrelated $(0,\sigma^2)$ random variables with $\sigma^2=\gamma(0)\times[\Sigma_{j=0}^{q}\beta_j^2]^{-1}$ and the roots of

$$m^q+\sum_{j=1}^{q}\beta_j m^{q-j}=0$$

are less than one in absolute value.

Proof. The function

$$A(z)=z^q[\gamma(q)]^{-1}\sum_{h=-q}^{q}\gamma(h)z^h=[\gamma(q)]^{-1}\sum_{h=0}^{2q}\gamma(h-q)z^h$$

is a polynomial of degree $2q$ in z with unit coefficient for z^{2q}. Therefore, it can be written in the factored form

$$A(z)=\prod_{r=1}^{2q}(z-m_r),$$

where the m_r are the roots of $A(z)=0$. If $A(m_r)=0$, then

$$\sum_{h=-q}^{q}\gamma(h)m_r^h=0,$$

and, by symmetry,

$$\sum_{h=-q}^{q}\gamma(h)m_r^{-h}=0.$$

Since $f(\omega)$ is strictly positive none of the roots are of unit absolute value. Also, the coefficients are real, so that any complex roots occur as conjugate pairs. Therefore, we can arrange the roots in q pairs, $(m_r,m_{-r})=(m_r,m_r^{-1})$, where multiple roots are repeated the proper number of times. We let m_r, $r=1,2,\ldots,q$, denote the roots less than one in absolute value.

Using these roots we define β_j, $j=0,1,2,\ldots,q$, by

$$\prod_{r=1}^{q}(z-m_r)=\sum_{j=0}^{q}\beta_j z^{q-j}.$$

Therefore,

$$\begin{aligned}
z^{-q}\gamma(q)A(z)&=z^{-q}\gamma(q)\prod_{r=1}^{q}(z-m_r)\prod_{r=1}^{q}(z-m_r^{-1})\\
&=\gamma(q)\prod_{r=1}^{q}(z-m_r)\prod_{r=1}^{q}(1-z^{-1}m_r^{-1})\\
&=\gamma(q)\left(\sum_{j=0}^{q}\beta_j z^{j}\right)\left(\sum_{j=0}^{q}\beta_j z^{-j}\right)\prod_{r=1}^{q}(-m_r)^{-1}.
\end{aligned}$$

It follows that the expression in $e^{-i\omega}$ defining $f(\omega)$ can be written as

$$f(\omega)=2\pi\sigma^2|f_\beta(\omega)|^2,$$

where $f_\beta(\omega)$ is the Fourier transform of $\{\beta_j\}$ and $\sigma^2=\gamma(0)(\Sigma_{j=0}^{q}\beta_j^2)^{-1}$. Now define the sequence of random variables $\{e_t\}$ by $e_t=\Sigma_{j=0}^{\infty}c_jX_{t-j}$, where the sequence $\{c_j\}$ is defined in Theorem 2.6.2. These c_j are such that $X_t=\Sigma_{j=0}^{q}\beta_j e_{t-j}$, which means that $|f_c(\omega)f_\beta(\omega)|^2=(2\pi)^{-4}$. Therefore, the spectral density of e_t is

$$\begin{aligned}
f_e(\omega)&=(2\pi)^2|f_c(\omega)|^2 f(\omega)\\
&=(2\pi)^2|f_c(\omega)|^2\left[\frac{\sigma^2|2\pi f_\beta(\omega)|^2}{2\pi}\right]\\
&=\frac{\sigma^2}{2\pi},
\end{aligned}$$

and the e_t are uncorrelated with variance σ^2. ▲

By our earlier results in Fourier series we know that a continuous periodic function can be approximated by a trigonometric polynomial. This means that a time series with a continuous spectral density is "very nearly" a moving average process and also "very nearly" an autoregressive process.

Theorem 4.3.3. Let $g(\omega)$ be a nonnegative even continuous periodic

function of period 2π. Then, given an $\epsilon > 0$, there is a time series with the representation

$$X_t = \sum_{j=0}^{q} \beta_j e_{t-j}$$

such that $|f_X(\omega) - g(\omega)| < \epsilon$ for all $\omega \in [-\pi, \pi]$ where $\beta_0 = 1$, q is a finite integer, the e_t are uncorrelated $(0, \sigma^2)$ random variables, and

$$\sigma^2 = \left(\sum_{j=0}^{q} \beta_j^2 \right)^{-1} \int_{-\pi}^{\pi} g(\omega)\, d\omega.$$

Proof. The result is trivial if $g(\omega) \equiv 0$; hence, we assume $g(\omega) > 0$ for some ω. Given $\epsilon > 0$, let $\delta = 2\epsilon G (4\pi M + 3G)^{-1}$ and set

$$\begin{aligned} C(\omega) &= g(\omega), && \text{if } g(\omega) > \delta \\ &= \delta, && \text{otherwise,} \end{aligned}$$

where

$$M = \max_{\omega} g(\omega)$$

$$G = \int_{-\pi}^{\pi} g(\omega)\, d\omega.$$

Then by Theorem 3.1.10, there is a q such that $|f_Y(\omega) - C(\omega)| < \frac{1}{2}\delta$ for all $\omega \in [-\pi, \pi]$, where

$$f_Y(\omega) = \frac{1}{2\pi} \sum_{h=-q}^{q} \gamma(h) e^{-i\omega h},$$

$$\gamma(h) = q^{-1}(q - |h|) a_h,$$

and

$$a_h = \int_{-\pi}^{\pi} e^{i\omega h} C(\omega)\, d\omega.$$

By Exercise 4.14 $\gamma(h)$ is positive semidefinite, and $f_Y(\omega)$ is of the same form as the spectral density in Theorem 4.3.2. Therefore, we may write

$$f_Y(\omega)=(2\pi)^{-1}\tau^2\left|\sum_{j=0}^{q}\beta_j e^{-i\omega j}\right|^2,$$

where $\beta_0=1$, the roots of $\sum_{j=0}^{q}\beta_j m^{q-j}$ are less than one in absolute value, and $\tau^2=[\sum_{j=0}^{q}\beta_j^2]^{-1}\int_{-\pi}^{\pi}C(\omega)\,d\omega$. Setting

$$f_X(\omega)=(2\pi)^{-1}\sigma^2\left|\sum_{j=0}^{q}\beta_j e^{-i\omega j}\right|^2,$$

where

$$\sigma^2=\tau^2\left[\int_{-\pi}^{\pi}f_Y(\omega)\,d\omega\right]^{-1}\int_{-\pi}^{\pi}g(\omega)\,d\omega,$$

we obtain the desired result. ▲

Theorem 4.3.4. Let $g(\omega)$ be a nonnegative even continuous periodic function of period 2π. Then, given an $\epsilon>0$, there is a time series with the representation

$$\sum_{j=0}^{p}\alpha_j Y_{t-j}=e_t$$

such that $|f_Y(\omega)-g(\omega)|<\epsilon$ for all $\omega\in[-\pi,\pi]$, where $\alpha_0=1$, p is a finite integer, the roots of

$$\sum_{j=0}^{p}\alpha_j m^{p-j}=0$$

are less than one in absolute value, and the e_t are uncorrelated $(0,\sigma^2)$ random variables.

Proof. The result is trivial if $g(\omega)\equiv 0$. Assuming $g(\omega)>0$ for some ω define

$$\begin{aligned} d(\omega)&=g^{-1}(\omega), && \text{if } g(\omega)>\tfrac{1}{2}\epsilon \\ &=2\epsilon^{-1}, && \text{otherwise.} \end{aligned}$$

Also define $G = \max g(\omega)$ and let $0 < \delta < \epsilon[G\{2G+\epsilon\}]^{-1}$. Then, by Theorem 3.1.10, there is a finite p such that

$$\left| \sum_{h=-p}^{p} C_h e^{-i\omega h} - d(\omega) \right| < \delta$$

for all $\omega \in [-\pi, \pi]$. By Theorem 4.3.2,

$$\sum_{h=-p}^{p} C_h e^{-i\omega h} = M \sum_{j=0}^{p} \alpha_j e^{-i\omega j} \sum_{j=0}^{p} \alpha_j e^{i\omega j},$$

where $\alpha_0 = 1$, M is a constant, and the roots of $\Sigma_{j=0}^{p} \alpha_j m^{p-j} = 0$ are less than one in absolute value. Hence, by defining

$$f_Y(\omega) = M^{-1}\left[\left(\sum_{j=0}^{p} \alpha_j e^{-i\omega j}\right)\left(\sum_{j=0}^{p} \alpha_j e^{i\omega j}\right)\right]^{-1}$$

and $\sigma^2 = 2\pi M^{-1}$, we have the conclusion. ▲

Operations on a time series are sometimes called *filtering*. In some areas, particularly in electronics, physical devices are constructed to filter time series. Examples include the modification of radio waves and radar signals. The creation of a linearly weighted sum is the application of a *linear filter*, the weights being identified as the filter. If, as in (4.3.0), the weights are constant over time, the filter is called *time invariant*.[3] Obviously such operations change the behavior of the time series and, in practice, this is often the objective. By investigating the properties of the filter, one is able to make statements about the change in behavior.

It is clear that moving average and autoregressive processes are obtained from a white noise process by the application of a linear time invariant filter. Thus, using this terminology, we have been studying the effects of a linear time invariant filter on the spectral density of a time series.

To further illustrate the effects of the application of a linear filter, consider the simple time series

$$X_t = 2\alpha\cos(\omega t + \beta)$$

$$= \alpha\left[e^{i(\omega t+\beta)} + e^{-i(\omega t+\beta)}\right]$$

[3] We may define the time invariant property as follows. If the output of a filter g applied to X_t is Y_t, then the filter is time invariant if $g(X_{t+h}) = Y_{t+h}$ for all integer h.

and apply the absolutely summable linear filter a_j to obtain

$$Y_t = \alpha \sum_{j=-\infty}^{\infty} a_j \left[e^{i\{\omega(t-j)+\beta\}} + e^{-i\{\omega(t-j)+\beta\}} \right]$$

$$= \alpha e^{i(\omega t+\beta)} \left\{ \sum_{j=-\infty}^{\infty} a_j e^{-i\omega j} \right\} + \alpha e^{-i(\omega t+\beta)} \left\{ \sum_{j=-\infty}^{\infty} a_j e^{i\omega j} \right\}.$$

We may express the Fourier transform of a_j in a general complex form

$$2\pi f_a(\omega) = \sum_{j=-\infty}^{\infty} a_j e^{-i\omega j} \overset{\text{(say)}}{=} u(\omega) + iv(\omega)$$

$$= \psi(\omega)\left[\cos\varphi(\omega) + i\sin\varphi(\omega)\right] = \psi(\omega)e^{i\varphi(\omega)}, \qquad (4.3.11)$$

where

$$\psi(\omega) = \left\{ [u(\omega)]^2 + [v(\omega)]^2 \right\}^{1/2}$$

$$\varphi(\omega) = \tan^{-1}\left[\frac{v(\omega)}{u(\omega)} \right].$$

The reader should note the convention used in defining the sign of $v(\omega)/u(\omega)$; $v(\omega)$ is the coefficient of i (not $-i$) in the original transform.

In this notation, the filtered cosine wave becomes

$$Y_t = \alpha\psi(\omega)\left[e^{i\{\omega t+\beta+\varphi(\omega)\}} + e^{-i\{\omega t+\beta+\varphi(\omega)\}} \right]$$

$$= 2\alpha\psi(\omega)\cos\{\omega t + \beta + \varphi(\omega)\}.$$

Thus, the process Y_t is also a cosine function with the original period, but with amplitude $2\alpha\psi(\omega)$ and phase angle $\beta + \varphi(\omega)$. The function $\psi(\omega)$ is called the *gain* of the filter and the function $\varphi(\omega)$ is the *phase angle* (or simply *phase*) of the filter. The rationale for these terms is clear when one observes the effect of the filter on the cosine time series. The transform $2\pi f_a(\omega)$ is called the *frequency response function* or the *transfer function* of the filter. From (4.3.1) we know that the spectrum of a time series, Y_t, created by linearly filtering a time series, X_t, is the spectrum of the original series, $f_X(\omega)$, multiplied by the squared gain of the filter. The squared gain is sometimes called the *power transfer function*.

A particularly simple filter is the perfect *delay*. Let $\{X_t\colon t \in (0, \pm 1, \pm 2, \ldots)\}$ be a stationary time series. The time series delayed or lagged by τ

periods is defined by

$$Y_t = X_{t-\tau},$$

where the filter is defined by $a_\tau = 1$ and $a_j = 0$, $j \neq \tau$. Hence the frequency response function of the filter is $e^{-i\omega\tau} = \cos\omega\tau - i\sin\omega\tau$, the gain is $(\cos^2\omega\tau + \sin^2\omega\tau)^{1/2} = 1$, and the phase angle is $\tan^{-1}(-\sin\omega\tau/\cos\omega\tau) = -\omega\tau$. A cosine wave of frequency ω and corresponding period $2\pi\omega^{-1}$ completes $(2\pi)^{-1}\omega\tau$ cycles during a "time period" of length τ. Thus the phase of such a cosine wave lagged τ periods is shifted by the quantity $(2\pi)^{-1}\omega\tau$.

4.4. VECTOR PROCESSES

The spectral representation of vector time series follows in a straightforward manner from that of scalar time series. We denote the cross covariance of two zero mean stationary time series X_{jt} and X_{mt} by $\gamma_{jm}(h) = \{\Gamma(h)\}_{jm} = E\{X_{jt}X_{m,t+h}\}$ and assume that $\{\gamma_{jm}(h)\}$ is absolutely summable. Then,

$$\frac{1}{2\pi}\sum_{h=-\infty}^{\infty} \gamma_{jm}(h)e^{-i\omega h} = f_{jm}(\omega) \tag{4.4.1}$$

is a continuous periodic function of ω, which we call the *cross spectral function* of X_{jt} and X_{mt}. Since $\gamma_{jm}(h)$ may not be symmetric about 0, $f_{jm}(\omega)$ is, in general, a complex valued function. As such, it can be written as

$$f_{jm}(\omega) = c_{jm}(\omega) - iq_{jm}(\omega),$$

where $c_{jm}(\omega)$ and $q_{jm}(\omega)$ are real valued functions of ω. The function $c_{jm}(\omega)$ is called the *coincident spectral density* or simply the *cospectrum*. The function $q_{jm}(\omega)$ is called the *quadrature spectral density*. The function $c_{jm}(\omega)$ is the cosine portion of the transform and is an even function of ω, and $q_{jm}(\omega)$ is the sine portion and is an odd function of ω. Thus we may define these quantities as the transforms

$$c_{jm}(\omega) = \frac{1}{2\pi}\sum_{h=-\infty}^{\infty} \tfrac{1}{2}\left[\gamma_{jm}(h) + \gamma_{jm}(-h)\right]e^{-i\omega h},$$

$$-iq_{jm}(\omega) = \frac{1}{2\pi}\sum_{h=-\infty}^{\infty} \tfrac{1}{2}\left[\gamma_{jm}(h) - \gamma_{jm}(-h)\right]e^{-i\omega h},$$

or, in real terms,

$$c_{jm}(\omega)=\frac{1}{2\pi}\sum_{h=-\infty}^{\infty}\tfrac{1}{2}\left[\gamma_{jm}(h)+\gamma_{jm}(-h)\right]\cos\omega h,$$

$$q_{jm}(\omega)=\frac{1}{2\pi}\sum_{h=-\infty}^{\infty}\tfrac{1}{2}\left[\gamma_{jm}(h)-\gamma_{jm}(-h)\right]\sin\omega h. \tag{4.4.2}$$

By the Fourier integral theorem,

$$\gamma_{jm}(h)=\int_{-\pi}^{\pi}e^{\iota\omega h}f_{jm}(\omega)\,d\omega.$$

If we let $\boldsymbol{f}(\omega)$ denote the matrix with typical element $f_{jm}(\omega)$, we have the matrix representations

$$\boldsymbol{\Gamma}(h)=\int_{-\pi}^{\pi}e^{\iota\omega h}\boldsymbol{f}(\omega)\,d\omega \tag{4.4.3}$$

$$\boldsymbol{f}(\omega)=\frac{1}{2\pi}\sum_{h=-\infty}^{\infty}e^{-\iota\omega h}\boldsymbol{\Gamma}(h). \tag{4.4.4}$$

For a general stationary time series we can write

$$\boldsymbol{\Gamma}(h)=\int_{-\pi}^{\pi}e^{\iota\omega h}\,d\boldsymbol{F}(\omega) \tag{4.4.5}$$

in complete analogy to (4.1.1a).

Let us investigate some of the properties of the matrix $\boldsymbol{f}(\omega)$.

Definition 4.4.1 A square complex valued matrix $\mathbf{B}$ is called a *Hermitian matrix* if it is equal to its conjugate transpose, that is,

$$\mathbf{B}=\mathbf{B}^*,$$

where the jmth element of $\mathbf{B}^*$ is the complex conjugate of the mjth element of $\mathbf{B}$.

Definition 4.4.2 A Hermitian matrix, $\mathbf{B}$, is *positive definite* if for any complex vector $\mathbf{w}$, such that $\mathbf{w}^*\mathbf{w}>0$,

$$\mathbf{w}^*\mathbf{B}\mathbf{w}>0,$$

and it is *positive semidefinite* if

$$\mathbf{w}^*\mathbf{B}\mathbf{w}\geqslant 0.$$

Lemma 4.4.1. For stationary vector time series of dimension k satisfying

$$\sum_{h=-\infty}^{\infty} |\gamma_{jm}(h)| < \infty$$

for $j,m=1,2,\ldots,k$, the matrix $\boldsymbol{f}(\omega)$ is a positive semidefinite Hermitian matrix for all ω in $[-\pi,\pi]$.

Proof. The matrix $\boldsymbol{f}(\omega)$ is Hermitian since, for all ω in $[-\pi,\pi]$ and for $j,m=1,2,\ldots,k$,

$$\begin{aligned} f^*_{jm}(\omega) &= \frac{1}{2\pi}\sum_{h=-\infty}^{\infty} \gamma^*_{jm}(h)e^{i\omega h} \\ &= \frac{1}{2\pi}\sum_{h=-\infty}^{\infty} \gamma_{mj}(-h)e^{i\omega h} \\ &= \frac{1}{2\pi}\sum_{r=-\infty}^{\infty} \gamma_{mj}(r)e^{-i\omega r} \\ &= f_{mj}(\omega). \end{aligned}$$

Consider the complex valued time series

$$Z_t = \boldsymbol{\alpha}'\mathbf{X}_t = \sum_{j=1}^{k} \alpha_j X_{jt},$$

where $\boldsymbol{\alpha}'=(\alpha_1,\alpha_2,\ldots,\alpha_k)$ is a vector of arbitrary complex numbers. The autocovariance function of Z_t is given by

$$\gamma_Z(h) = \sum_{j=1}^{k}\sum_{r=1}^{k} \alpha_j\alpha_r^*\gamma_{jr}(h).$$

Now $\gamma_Z(h)$ is positive semidefinite and, therefore,

$$\sum_{m=1}^{n}\sum_{q=1}^{n} \gamma_Z(m-q)e^{-im\omega}e^{iq\omega} \geqslant 0$$

and

$$\sum_{h=-(n-1)}^{(n-1)} \frac{n-|h|}{n}\left[\sum_{j=1}^{k}\sum_{r=1}^{k} \alpha_j\alpha_r^*\gamma_{jr}(h)\right]e^{-ih\omega} \geqslant 0.$$

Taking the limit as $n\to\infty$, we have

$$\sum_{j=1}^{k}\sum_{r=1}^{k}\alpha_j\alpha_r^* f_{jr}(\omega)\geqslant 0,$$

which establishes that $\boldsymbol{f}(\omega)$ is positive semidefinite. ▲

It follows immediately from Lemma 4.4.1 that the determinant of any two by two matrix of the form

$$\begin{pmatrix} f_{jj}(\omega) & f_{jm}(\omega)\\ f_{mj}(\omega) & f_{mm}(\omega)\end{pmatrix}$$

is nonnegative. Hence,

$$\begin{aligned}|f_{jm}(\omega)|^2=f_{jm}(\omega)f_{jm}^*(\omega)&=f_{jm}(\omega)f_{mj}(\omega)\\ &\leqslant f_{jj}(\omega)f_{mm}(\omega).\end{aligned}\tag{4.4.6}$$

The quantity

$$\mathcal{K}_{jm}^2(\omega)=\frac{|f_{jm}(\omega)|^2}{f_{jj}(\omega)f_{mm}(\omega)}=\frac{c_{jm}^2(\omega)+q_{jm}^2(\omega)}{f_{jj}(\omega)f_{mm}(\omega)}\tag{4.4.7}$$

is called the *squared coherency function*. The spectral density may be zero at certain frequencies, in which case $\mathcal{K}^2(\omega)$ is of the form 0/0. We adopt the convention of assigning zero to the coherency in such situations. The inequality (4.4.6), sometimes written as

$$\mathcal{K}_{jm}^2(\omega)\leqslant 1,\tag{4.4.8}$$

is called the *coherency inequality*.

To further appreciate the properties of $\boldsymbol{f}(\omega)$ consider the time series

$$\begin{aligned}X_{1t}&=A_1\cos rt+B_1\sin rt\\ X_{2t}&=A_2\cos rt+B_2\sin rt,\end{aligned}\tag{4.4.9}$$

where $r\in(0,\pi)$ and $(A_1,B_1,A_2,B_2)'$ is distributed as a multivariate normal with zero mean and covariance matrix

$$\begin{bmatrix}\sigma_{11} & 0 & \sigma_{13} & \sigma_{14}\\ 0 & \sigma_{11} & -\sigma_{14} & \sigma_{13}\\ \sigma_{13} & -\sigma_{14} & \sigma_{33} & 0\\ \sigma_{14} & \sigma_{13} & 0 & \sigma_{33}\end{bmatrix}.$$

Then,

$$\gamma_{11}(h) = \sigma_{11}\cos rh,$$

$$\gamma_{22}(h) = \sigma_{33}\cos rh,$$

$$\begin{aligned}\gamma_{12}(h) &= E\{A_1A_2[\cos rt]\cos r(t+h) + A_1B_2[\cos rt]\sin r(t+h) \\ &\quad + B_1A_2[\sin rt]\cos r(t+h) + B_1B_2[\sin rt]\sin r(t+h)\} \\ &= \sigma_{13}\cos rh + \sigma_{14}\sin rh.\end{aligned}$$

The matrix analog of the spectral distribution function introduced in Section 4.1 is

$$\boldsymbol{F}(\omega) = \begin{pmatrix} F_{11}(\omega) & F_{12}(\omega) \\ F_{21}(\omega) & F_{22}(\omega) \end{pmatrix},$$

where $F_{12}(\omega)$ is the cross spectral distribution function. For the example (4.4.9), $F_{11}(\omega)$ is a step function with jumps of height $(1/2)\sigma_{11}$ at $\pm r$, $F_{22}(\omega)$ is a step function with jumps of height $(1/2)\sigma_{33}$ at $\pm r$, $F_{12}(\omega)$ is a complex valued function where $\mathrm{Re}\{F_{12}(\omega)\}$ is a step function with a jump of height $(1/2)\sigma_{13}$ at $\pm r$, and $\mathrm{Im}\{F_{12}(\omega)\}$ is a step function with a jump of $(1/2)\sigma_{14}$ at $-r$ and a jump of $-(1/2)\sigma_{14}$ at r. Since the elements of $\boldsymbol{F}(\omega)$ are pure jump functions, we have a pure line spectrum. The real portion of the cross line spectrum is one half of the covariance between the coefficients of the cosine portions of the two time series, which is also one half of the covariance between the coefficients of the sine portions. The absolute value of the imaginary portion of the cross line spectrum is one half of the covariance between the coefficient of the cosine portion of the first time series and the coefficient of the sine portion of the second time series. This is one half of the negative of the covariance between the coefficient of the sine of the first time series and the coefficient of the cosine of the second. To consider the point further, we write the time series X_{2t} as a sine wave,

$$X_{2t} = \psi\sin(rt + \varphi),$$

where

$$\psi = (A_2^2 + B_2^2)^{1/2}$$

$$\varphi = \tan^{-1}\frac{A_2}{B_2}.$$

Let X_{3t} be the cosine with the same amplitude and phase,

$$\begin{aligned} X_{3t} &= \psi \cos(rt + \varphi) \\ &= B_2 \cos rt - A_2 \sin rt. \end{aligned}$$

It follows that X_{2t} is uncorrelated with X_{3t}. The covariance between X_{1t} and X_{2t} is

$$\begin{aligned} E\{X_{1t}X_{2t}\} &= E\{(A_1 \cos rt + B_1 \sin rt)(A_2 \cos rt + B_2 \sin rt)\} \\ &= \sigma_{13}(\cos^2 rt + \sin^2 rt) = \sigma_{13} \end{aligned}$$

and the covariance between X_{1t} and X_{3t} is

$$\begin{aligned} E\{X_{1t}X_{3t}\} &= E\{(A_1 \cos rt + B_1 \sin rt)(B_2 \cos rt - A_2 \sin rt)\} \\ &= \sigma_{14}. \end{aligned}$$

The covariance between X_{1t} and X_{2t} is proportional to the real part of the cross spectrum at r. The covariance between X_{1t} and X_{3t} is proportional to the imaginary portion of the cross spectrum at r. The fact that X_{3t} is X_{2t} "shifted" by a phase angle of $\pi/2$ explains why the real portion of the cross spectrum is called the cospectrum and the imaginary part is called the quadrature spectrum. The squared coherency is the multiple correlation between X_{1t} and the pair X_{2t}, X_{3t}, that is,

$$\mathcal{K}_{12}^2(r) = \frac{\sigma_{13}^2 + \sigma_{14}^2}{\sigma_{33}\sigma_{11}}.$$

We now introduce some cross spectral quantities useful in the analysis of input-output systems. Let the bivariate time series $(X_t, Y_t)'$ have absolutely summable covariance function. We may then write the cross spectral function as

$$f_{XY}(\omega) = A_{XY}(\omega) e^{i\varphi_{XY}(\omega)}, \tag{4.4.10}$$

where

$$\varphi_{XY}(\omega) = \tan^{-1}\left[\frac{-q_{XY}(\omega)}{c_{XY}(\omega)}\right]$$

$$A_{XY}(\omega) = \left[c_{XY}^2(\omega) + q_{XY}^2(\omega)\right]^{1/2}.$$

We use the convention of setting $\varphi_{XY}(\omega)=0$ when both $c_{XY}(\omega)$ and $q_{XY}(\omega)$ are zero. The quantity $\varphi_{XY}(\omega)$ is called the *phase spectrum* and $A_{XY}(\omega)$ is called the *cross amplitude spectrum*. The *gain* of Y_t over X_t is defined by

$$\psi_{XY}(\omega)=\frac{A_{XY}(\omega)}{f_{XX}(\omega)} \tag{4.4.11}$$

for those ω where $f_{XX}(\omega)>0$.

Let us assume that an absolutely summable linear filter is applied to an input time series, X_t, with zero mean, absolutely summable covariance function and everywhere positive spectral density to yield an output time series, Y_t, where

$$Y_t=\sum_{j=-\infty}^{\infty} a_j X_{t-j}. \tag{4.4.12}$$

The cross covariance function is

$$\gamma_{XY}(h)=E\{X_t Y_{t+h}\}=\sum_{j=-\infty}^{\infty} a_j \gamma_{XX}(h-j), \qquad h=0,\pm 1,\pm 2,\ldots,$$

and, by Corollary 3.4.1.1, the cross spectral function is

$$f_{XY}(\omega)=2\pi f_a(\omega) f_{XX}(\omega), \tag{4.4.13}$$

where

$$f_a(\omega)=\frac{1}{2\pi}\sum_{j=-\infty}^{\infty} a_j e^{-\iota\omega j},$$

$$f_{XX}(\omega)=\frac{1}{2\pi}\sum_{h=-\infty}^{\infty} \gamma_{XX}(h) e^{-\iota\omega h}.$$

It follows that $f_{XY}(\omega)/f_{XX}(\omega)$ is the transfer function of the filter.

Recall that the phase $\varphi(\omega)$ and gain $\psi(\omega)$ of the filter $\{a_j\}$ were defined by

$$2\pi f_a(\omega)=\psi(\omega)e^{\iota\varphi(\omega)}.$$

Since $f_{XY}(\omega)$ is the product of $2\pi f_a(\omega)$ and $f_{XX}(\omega)$, where $f_{XX}(\omega)$ is a real function of ω, it follows that the phase spectrum is the same as the phase of

the filter. The cross amplitude spectrum is the product of the spectrum of the input time series and the gain of the filter. That is,

$$A_{XY}(\omega)=|f_{XY}(\omega)|=f_{XX}(\omega)|2\pi f_a(\omega)| \tag{4.4.14}$$

and the gain of Y_t over X_t is simply the gain of the filter.

By Theorem 4.3.1, the spectral density of Y_t is

$$f_{YY}(\omega)=(2\pi)^2 f_a(\omega)f_a^*(\omega)f_{XX}(\omega)$$

and it follows that the squared coherency, for $|f_a(\omega)|>0$, is

$$\mathcal{K}_{XY}^2(\omega)=\frac{|f_{XY}(\omega)|^2}{f_{YY}(\omega)f_{XX}(\omega)}=\frac{|2\pi f_a(\omega)f_{XX}(\omega)|^2}{(2\pi)^2|f_a(\omega)|^2|f_{XX}(\omega)|^2}=1.$$

This is an interesting result, because it shows that the squared coherency between an output time series created by the application of a linear filter and the original time series is one at all frequencies. The addition of an error (noise) time series to the output will produce a coherency less than one in a linear system. For example, consider the bivariate time series

$$\begin{aligned}X_{1t}&=\beta X_{1,t-1}+e_{1t}\\X_{2t}&=\alpha_1X_{1t}+\alpha_2X_{1,t-1}+e_{2t},\end{aligned} \tag{4.4.16}$$

where $|\beta|<1$ and $\{(e_{1t},e_{2t})'\}$ is a sequence of uncorrelated vector random variables with $E\{e_{1t}^2\}=\sigma_{11}$, $E\{e_{2t}^2\}=\sigma_{22}$, and $E\{e_{1t}e_{2,t+h}\}=0$ for all t and h.

The input time series X_{1t} is a first order autoregressive time series and the output X_{2t} is a linear function of X_{1t}, $X_{1,t-1}$ and e_{2t}. The autocovariance and cross covariance functions are, therefore, easily computed:

$$\gamma_{11}(h)=E\{X_{1t}X_{1,t+h}\}=\frac{\beta^{|h|}}{1-\beta^2}\sigma_{11},$$

$$\gamma_{22}(h)=\begin{cases}(\alpha_1^2+\alpha_2^2)\gamma_{11}(0)+2\alpha_1\alpha_2\gamma_{11}(1)+\sigma_{22}, & h=0\\(\alpha_1^2+\alpha_2^2)\gamma_{11}(h)+\alpha_1\alpha_2[\gamma_{11}(h+1)+\gamma_{11}(h-1)], & h\neq0,\end{cases}$$

$$\gamma_{12}(h)=E\{X_{1t}X_{2,t+h}\}=\alpha_1\gamma_{11}(h)+\alpha_2\gamma_{11}(h-1).$$

The cross covariance and cross correlation functions for parameters $\alpha_1 = 0.5$, $\alpha_2 = 1.0$, $\gamma_{11}(0) = 1.0$, $\sigma_{22} = 0.5$, and $\beta = 0.8$ are displayed in Table 4.4.1.

Table 4.4.1. Autocovariance and cross covariance functions of the time series defined in (4.4.16)

h	$\gamma_{11}(h)$	$\gamma_{22}(h)$	$\gamma_{12}(h)$	$\rho_{12}(h)$
−6	0.262	0.596	0.341	0.213
−5	0.328	0.745	0.426	0.267
−4	0.410	0.932	0.532	0.333
−3	0.512	1.165	0.666	0.417
−2	0.640	1.456	0.832	0.521
−1	0.800	1.820	1.040	0.651
0	1.000	2.550	1.300	0.814
1	0.800	1.820	1.400	0.877
2	0.640	1.456	1.120	0.701
3	0.512	1.165	0.896	0.561
4	0.410	0.932	0.717	0.449
5	0.328	0.745	0.573	0.359
6	0.262	0.596	0.459	0.287

The spectral matrix has elements

$$f_{11}(\omega) = g_{11}(\omega)\left(1 - \beta e^{-i\omega}\right)^{-1}\left(1 - \beta e^{i\omega}\right)^{-1},$$

$$f_{22}(\omega) = \left(\alpha_1 + \alpha_2 e^{-i\omega}\right)\left(\alpha_1 + \alpha_2 e^{i\omega}\right) f_{11}(\omega) + g_{22}(\omega),$$

$$f_{12}(\omega) = \left(\alpha_1 + \alpha_2 e^{-i\omega}\right) f_{11}(\omega),$$

$$f_{21}(\omega) = f_{12}^*(\omega) = \left(\alpha_1 + \alpha_2 e^{i\omega}\right) f_{11}(\omega),$$

where $g_{11}(\omega)$ is the spectral density of e_{1t} and $g_{22}(\omega)$ is the spectral density of e_{2t}; that is,

$$g_{11}(\omega) = (2\pi)^{-1}\sigma_{11},$$

$$g_{22}(\omega) = (2\pi)^{-1}\sigma_{22}.$$

The cospectrum and quadrature spectral density are

$$c_{12}(\omega) = (\alpha_1 + \alpha_2 \cos\omega) f_{11}(\omega),$$

$$q_{12}(\omega) = \alpha_2\{\sin\omega\} f_{11}(\omega),$$

the phase spectrum and cross amplitude spectrum are

$$\varphi_{12}(\omega) = \tan^{-1}\left[\frac{-\alpha_2 \sin\omega}{(\alpha_1 + \alpha_2 \cos\omega)}\right],$$

$$A_{12}(\omega) = f_{11}(\omega)\left[\alpha_1^2 + 2\alpha_1\alpha_2\cos\omega + \alpha_2^2\right]^{1/2},$$

and the squared coherency is

$$\mathcal{K}_{12}^2(\omega) = \frac{|2\pi f_\alpha(\omega)|^2 |f_{11}(\omega)|^2}{f_{11}(\omega)\left[|2\pi f_\alpha(\omega)|^2 f_{11}(\omega) + g_{22}(\omega)\right]}$$

$$= \frac{1}{1+\eta(\omega)},$$

where

$$\eta(\omega) = \frac{g_{22}(\omega)}{|2\pi f_\alpha(\omega)|^2 f_{11}(\omega)}.$$

Since $g_{22}(\omega)$ is positive at all frequencies, the squared coherency is strictly less than one at all frequencies. The quantity $\eta(\omega)$ is sometimes called the *noise to signal ratio* in physical applications. Although the presence of noise reduces the squared coherency, the ratio of the cross spectrum to the spectrum of the input series still gives the transfer function of the filter. This is because the time series e_{2t} is uncorrelated with the input X_{1t}. The quantity

$$f_{22}(\omega)\left[1 - \mathcal{K}_{12}^2(\omega)\right] = f_{22}(\omega) - f_{21}(\omega) f_{11}^{-1}(\omega) f_{12}(\omega)$$

is sometimes called the *error spectral density* or *error spectrum*. We see that for a model such as (4.4.16),

$$g_{22}(\omega) = f_{22}(\omega) - f_{21}(\omega) f_{11}^{-1}(\omega) f_{12}(\omega).$$

For the example of Table 4.4.1 the elements of the spectral matrix are

$$f_{11}(\omega) = \frac{0.36}{2\pi(1-0.8e^{-i\omega})(1-0.8e^{i\omega})}$$

$$f_{22}(\omega) = \frac{(0.5+1.0e^{-i\omega})(0.5+1.0e^{i\omega})(0.36)}{(2\pi)(1-0.8e^{-i\omega})(1-0.8e^{i\omega})} + \frac{0.5}{2\pi}$$

$$= \frac{1.23(1-0.17e^{-i\omega})(1-0.17e^{i\omega})}{2\pi(1-0.8e^{-i\omega})(1-0.8e^{i\omega})}$$

$$f_{12}(\omega) = \frac{(0.5+1.0e^{-i\omega})(0.36)}{2\pi(1-0.8e^{-i\omega})(1-0.8e^{i\omega})}.$$

It is interesting that the spectral density of X_{2t} is that of an autoregressive moving average (1, 1) process. Note that if X_{2t} had been defined as the simple sum of X_{1t} and e_{2t}, the spectral density would also have been that of an autoregressive moving average (1, 1) process but with different parameters.

The noise to signal ratio is

$$\eta(\omega) = \frac{g_{22}(\omega)}{|2\pi f_\alpha(\omega)|^2 f_{11}(\omega)} = \frac{(1-0.8e^{-i\omega})(1-0.8e^{i\omega})}{(0.72)(0.5+e^{-i\omega})(0.5+e^{i\omega})}$$

and the squared coherency is

$$\mathcal{K}_{12}^2(\omega) = \frac{0.45+0.36\cos\omega}{1.27-0.44\cos\omega}.$$

Let us consider the example a bit further. The input time series X_{1t} is autocorrelated. Let us filter both the input and output with the same filter, choosing the filter so that the input time series becomes a sequence of uncorrelated random variables. Thus we define

$$X_{3t} = X_{1t} - \beta X_{1,t-1} = e_{1t}$$
$$X_{4t} = X_{2t} - \beta X_{2,t-1}$$

and it follows that

$$X_{4t} = \alpha_1(X_{1t} - \beta X_{1,t-1}) + \alpha_2(X_{1,t-1} - \beta X_{1,t-2}) + e_{2t} - \beta e_{2,t-1}$$
$$= \alpha_1 e_{1t} + \alpha_2 e_{1,t-1} + e_{2t} - \beta e_{2,t-1}.$$

The cross covariance function of X_{3t} and X_{4t} then has a particularly simple form:

$$\gamma_{34}(h)=\begin{cases}\alpha_1\sigma_{11}, & h=0\\ \alpha_2\sigma_{11}, & h=1\\ 0, & \text{otherwise.}\end{cases}$$

By transforming the input series to white noise, the cross covariance function is transformed into the coefficients of the function (or linear filter) that defines X_{4t} as a function of X_{3t}. The spectral matrix of X_{3t} and X_{4t} has elements

$$f_{33}(\omega)=\frac{\sigma_{11}}{2\pi},$$

$$f_{44}(\omega)=\left(\alpha_1+\alpha_2 e^{-i\omega}\right)\left(\alpha_1+\alpha_2 e^{i\omega}\right)\frac{\sigma_{11}}{2\pi}$$

$$+\left(1-\beta e^{-i\omega}\right)\left(1-\beta e^{i\omega}\right)\frac{\sigma_{22}}{2\pi},$$

$$f_{34}(\omega)=\left(\alpha_1+\alpha_2 e^{-i\omega}\right)\frac{\sigma_{11}}{2\pi}.$$

The reader may verify that

$$\varphi_{34}(\omega)=\tan^{-1}\left[\frac{-\alpha_2\sin\omega}{(\alpha_1+\alpha_2\cos\omega)}\right]$$

and

$$A_{34}(\omega)=\frac{\sigma_{11}}{2\pi}\left[\alpha_1^2+2\alpha_1\alpha_2\cos\omega+\alpha_2^2\right]^{1/2}.$$

As we expected from (4.4.11) and (4.4.13), the phase spectrum is unchanged by the transformation, since we transformed both input and output with the same filter. The filter changed the input spectrum and, as a result, that portion of the cross amplitude spectrum associated with the input was altered. Also the error spectral density

$$f_{44}(\omega)-f_{43}(\omega)f_{33}^{-1}(\omega)f_{34}(\omega)=\left(1-\beta e^{-i\omega}\right)\left(1-\beta e^{i\omega}\right)\frac{\sigma_{22}}{2\pi}$$

is that of a moving average with parameter $-\beta$. The reader may verify that $\mathcal{K}_{34}^2(\omega)$ is the same as $\mathcal{K}_{12}^2(\omega)$.

We have introduced and illustrated the ideas of phase spectrum and amplitude spectrum using the input-output model. Naturally these quantities can be computed, in much the same manner as we compute the correlation for a bivariate normal distribution, without recourse to this model. The same is true of squared coherency and error spectrum, which have immediate generalizations to higher dimensions.

The effect of the application of a matrix filter to a vector time series is summarized in Theorem 4.4.1.

Theorem 4.4.1. Let $\mathbf{X}_t$ be a real k-dimensional stationary time series with absolutely summable covariance matrix and let $\{\mathbf{A}_j\}_{j=-\infty}^{\infty}$ be a sequence of real $k \times k$ matrices that is absolutely summable. Then the spectral density of

$$\mathbf{Y}_t = \sum_{j=-\infty}^{\infty} \mathbf{A}_j' \mathbf{X}_{t-j}$$

is

$$\boldsymbol{f}_{\mathbf{Y}}(\omega) = (2\pi)^2 \boldsymbol{f}_{\mathbf{A}}^*(\omega) \boldsymbol{f}_{\mathbf{X}}(\omega) \boldsymbol{f}_{\mathbf{A}}(\omega),$$

where

$$\boldsymbol{f}_{\mathbf{X}}(\omega) = \frac{1}{2\pi} \sum_{h=-\infty}^{\infty} \boldsymbol{\Gamma}_{\mathbf{X}}(h) e^{-\iota\omega h},$$

$$\boldsymbol{f}_{\mathbf{A}}(\omega) = \frac{1}{2\pi} \sum_{j=-\infty}^{\infty} \mathbf{A}_j e^{-\iota\omega j},$$

and $\boldsymbol{f}_{\mathbf{A}}^*(\omega)$ is the conjugate transpose of $\boldsymbol{f}_{\mathbf{A}}(\omega)$,

$$\boldsymbol{f}_{\mathbf{A}}^*(\omega) = \frac{1}{2\pi} \sum_{j=-\infty}^{\infty} \mathbf{A}_j' e^{\iota\omega j}.$$

Proof. The proof parallels that of Theorem 4.3.1. We have

$$\begin{aligned}\boldsymbol{f}_{\mathbf{Y}}(\omega) &= \frac{1}{2\pi} \sum_{h=-\infty}^{\infty} \sum_{j=-\infty}^{\infty} \sum_{s=-\infty}^{\infty} \mathbf{A}_j' \boldsymbol{\Gamma}_{\mathbf{X}}(h-s+j) \mathbf{A}_s e^{-\iota\omega h} \\ &= \frac{1}{2\pi} \sum_{h=-\infty}^{\infty} \sum_{j=-\infty}^{\infty} \sum_{s=-\infty}^{\infty} \mathbf{A}_j' e^{\iota\omega j} \boldsymbol{\Gamma}_{\mathbf{X}}(h-s+j) e^{-\iota\omega(h-s+j)} \mathbf{A}_s e^{-\iota\omega s}\end{aligned}$$

and the result follows. ▲

As in the scalar case, the spectral matrix of autoregressive and moving average time series follows immediately from Theorem 4.4.1.

Corollary 4.4.1.1. The spectral density of the moving average process

$$\mathbf{X}_t = \sum_{j=-\infty}^{\infty} \mathbf{B}_j \mathbf{e}_{t-j},$$

where $\{\mathbf{e}_t\}$ is a sequence of uncorrelated $(\mathbf{0}, \boldsymbol{\Sigma})$ vector random variables and the sequence $\{\mathbf{B}_j\}$ is absolutely summable, is

$$f_{\mathbf{X}}(\omega) = \frac{1}{2\pi}\left(\sum_{j=-\infty}^{\infty} \mathbf{B}_j e^{ij\omega}\right)\boldsymbol{\Sigma}\left(\sum_{j=-\infty}^{\infty} \mathbf{B}_j' e^{-ij\omega}\right)$$

$$= (2\pi)^2 f_{\mathbf{B}'}^{*}(\omega)\boldsymbol{\Sigma} f_{\mathbf{B}'}(\omega).$$

Corollary 4.4.1.2. Define the vector autoregressive process $\mathbf{X}_t$ by

$$\sum_{j=0}^{p} \mathbf{A}_j \mathbf{X}_{t-j} = \mathbf{e}_t,$$

where $\mathbf{A}_0 = \mathbf{I}$, $\mathbf{A}_p \neq \mathbf{0}$, the $\mathbf{e}_t$ are uncorrelated $(\mathbf{0}, \boldsymbol{\Sigma})$ vector random variables and the roots of

$$\left|\sum_{j=0}^{p} \mathbf{A}_j m^{p-j}\right| = 0$$

are less than one in absolute value. Then the spectral density of $\mathbf{X}_t$ is

$$f_{\mathbf{X}}(\omega) = 2\pi\left[f_{\mathbf{A}'}^{*}(\omega)\right]^{-1}\boldsymbol{\Sigma}\left[f_{\mathbf{A}'}(\omega)\right]^{-1},$$

where

$$f_{\mathbf{A}'}(\omega) = \frac{1}{2\pi}\sum_{j=0}^{p} \mathbf{A}_j' e^{-i\omega j}.$$

4.5. MEASUREMENT ERROR—SIGNAL DETECTION

In any statistical model the manner in which the "errors" enter the model is very important. In models of the "input-output" or "independent variable-dependent variable" form, measurement error in the output or dependent variable is relatively easy to handle. On the other hand,

measurement error in the input or independent variable typically introduces additional complexity into the analysis. In the simple regression model with normal errors the presence of normal measurement error in the independent variable requires additional information, such as knowledge of the variance of the measurement error, before one can estimate the slope of the regression line.

In time series analysis the distinction between independent variable and dependent variable may be blurred. For example, in predicting a future observation in the realization, the past observations play the role of independent variables while the future observation plays the role of (unknown) dependent variable. As a result, considerable care is required in specifying and treating measurement errors in such analyses. One important problem where errors of observation play a central role is the estimation of the values of an underlying time series that is observed with measurement error.

To introduce the problem, let $\{X_t: t \in (0, \pm 1, \pm 2, \ldots)\}$ be a stationary time series with zero mean. Because of measurement error we do not observe X_t directly. Instead, we observe

$$Y_t = X_t + u_t, \tag{4.5.1}$$

where $\{u_t: t \in (0, \pm 1, \pm 2, \ldots)\}$ is a time series with u_t independent of X_j for all t, j. The u_t are the measurement errors or the noise in the system. A problem of *signal measurement* or signal detection is the construction of an estimator of X_t given a realization on Y_t (or a portion of the realization). We assume the covariance functions of X_t and u_t are known.

We first consider the problem in the time domain and restrict ourselves to finding a linear filter that minimizes the mean square error of our estimator of X_t. Thus we desire weights $\{a_j: j = -L, -(L-1), \ldots, M-1, M\}$, for $L \geqslant 0$ and $M \geqslant 0$ fixed, such that

$$E\left\{\left(X_t - \sum_{j=-L}^{M} a_j Y_{t-j}\right)^2\right\} = E\left\{\left[X_t - \sum_{j=-L}^{M} a_j (X_{t-j} + u_{t-j})\right]^2\right\} \tag{4.5.2}$$

is a minimum. We set the derivatives of (4.5.2) with respect to the a_j equal to zero and obtain the system of equations

$$\sum_{j=-L}^{M} a_j[\gamma_{XX}(j-r) + \gamma_{uu}(j-r)] = \gamma_{XX}(r),$$

$$r = -L, -(L-1), \ldots, M-1, M. \tag{4.5.3}$$

For modest choices of L and M this system of linear equations is easily solved for the a_j. The mean square error of the estimator can then be obtained from (4.5.2).

To obtain the set of weights one would use if one had available the entire realization, we investigate the problem in the frequency domain. This permits us to establish the bound on the mean square error and to use this bound to evaluate the performance of a filter with a finite number of weights. We assume that u_t and X_t have continuous spectral densities with bounded derivatives. The spectral density of Y_t and the crosss spectral density follow immediately from the independence:

$$f_{YY}(\omega) = f_{XX}(\omega) + f_{uu}(\omega),$$

$$f_{XY}(\omega) = f_{XX}(\omega).$$

We assume that $f_{YY}(\omega)$ is strictly positive and that $f_{YY}^{-1}(\omega)$ has bounded first derivative.

We shall search in the class of absolutely summable filters $\{a_j\}$ for that filter such that our estimator of X_t,

$$\hat{X}_t = \sum_{j=-\infty}^{\infty} a_j Y_{t-j},$$

will have minimum mean square error. We write

$$f_{\hat{X}\hat{X}}(\omega) = (2\pi)^2 f_a(\omega) f_a^*(\omega) f_{YY}(\omega)$$

$$f_{X\hat{X}}(\omega) = \frac{1}{2\pi} \sum_{h=-\infty}^{\infty} \gamma_{X\hat{X}}(h) e^{-ih\omega}$$

$$= 2\pi f_a(\omega) f_{XY}(\omega).$$

Now the mean square error is

$$E\{(X_t - \hat{X}_t)^2\} = E\left\{\left(X_t - \sum_{j=-\infty}^{\infty} a_j Y_{t-j}\right)^2\right\}, \tag{4.5.4}$$

and $W_t = X_t - \sum_{j=-\infty}^{\infty} a_j Y_{t-j}$ is a stationary time series with a spectral density, say $f_{WW}(\omega)$. Hence, the variance of W_t is

$$\gamma_{WW}(0) = E\{W_t^2\} = \int_{-\pi}^{\pi} f_{WW}(\omega)\, d\omega$$

$$= \int_{-\pi}^{\pi} \left[f_{XX}(\omega) - 2\pi\{f_a(\omega) f_{XY}(\omega) + f_a^*(\omega) f_{YX}(\omega)\} \right.$$

$$\left. + (2\pi)^2 f_a(\omega) f_a^*(\omega) f_{YY}(\omega) \right] d\omega. \tag{4.5.5}$$

We have converted the problem of finding the a_j that minimize (4.5.4) to the problem of finding the $f_a(\omega)$ that minimizes (4.5.5). This problem retains the same form and appearance as the classical regression problem with $f_a(\omega)$ playing the role of the vector of coefficients. Therefore, we take as our candidate solution

$$f_a(\omega) = \frac{1}{2\pi} \frac{f_{YX}(\omega)}{f_{YY}(\omega)}, \tag{4.5.6}$$

which gives

$$\begin{aligned}\gamma_{WW}(0) &= \int_{-\pi}^{\pi} \left[f_{XX}(\omega) - \frac{f_{XY}(\omega) f_{XY}^*(\omega)}{f_{YY}(\omega)} \right] d\omega \\ &= \int_{-\pi}^{\pi} f_{XX}(\omega) \left[1 - \mathcal{K}_{XY}^2(\omega) \right] d\omega. \end{aligned} \tag{4.5.7}$$

The weights a_j are given by the inverse transform of $f_a(\omega)$,

$$a_j = \int_{-\pi}^{\pi} f_a(\omega) e^{i\omega j}\, d\omega, \qquad j = 0, \pm 1, \pm 2, \ldots .$$

To demonstrate that these weights yield the minimum value for the mean square error, we consider an alternative to (4.5.6):

$$f_b(\omega) = \frac{f_{YX}(\omega)}{2\pi f_{YY}(\omega)} + \frac{f_d(\omega)}{2\pi f_{YY}(\omega)},$$

where $f_b(\omega)$ must be in the class of functions such that the integral defining the mean square error exists. The mean square error is then

$$\begin{aligned}&\int_{-\pi}^{\pi} \left[f_{XX}(\omega) - 2\pi \left\{ f_b(\omega) f_{YX}^*(\omega) + f_b^*(\omega) f_{YX}(\omega) \right\} \right. \\ &\qquad \left. + (2\pi)^2 f_b(\omega) f_b^*(\omega) f_{YY}(\omega) \right] d\omega \\ &= \int_{-\pi}^{\pi} \left[f_{XX}(\omega) - \frac{|f_{XY}(\omega)|^2}{f_{YY}(\omega)} + \frac{|f_d(\omega)|^2}{f_{YY}(\omega)} \right] d\omega. \end{aligned} \tag{4.5.8}$$

Since $|f_d(\omega)|^2 / f_{YY}(\omega)$ is nonnegative, we conclude that the $f_a(\omega)$ of (4.5.6) yields the minimum value for (4.5.5).

Note that $2\pi f_a(\omega) = f_{YY}^{-1}(\omega) f_{YX}(\omega)$ is a real valued symmetric function of

ω because $f_{YX}(\omega)=f_{XX}(\omega)$ is a real valued symmetric function of ω. Therefore, the weights a_j are also symmetric about zero.

We summarize in Theorem 4.5.1.

Theorem 4.5.1. Let $\{X_t: t\in(0,\pm 1,\pm 2,\ldots)\}$ and $\{u_t: t\in(0,\pm 1,\pm 2,\ldots)\}$ be independent zero mean stationary time series and define $Y_t=X_t+u_t$. Let $f_{YY}(\omega)$, $f_{YY}^{-1}(\omega)$, and $f_{XX}(\omega)$ be continuous with bounded first derivatives. Then the best linear filter for extracting X_t from a realization of Y_t is given by

$$a_j=\int_{-\pi}^{\pi} f_a(\omega)e^{\iota\omega j}\,d\omega,$$

where

$$f_a(\omega)=\frac{f_{YX}(\omega)}{2\pi f_{YY}(\omega)}.$$

Furthermore, the mean square error of $\sum_{j=-\infty}^{\infty} a_j Y_{t-j}$ as an estimator for X_t is

$$\int_{-\pi}^{\pi}\left\{f_{XX}(\omega)-\left[f_{YY}(\omega)\right]^{-1}|f_{XY}(\omega)|^2\right\}d\omega$$

$$=\int_{-\pi}^{\pi} f_{XX}(\omega)\left[1-\mathcal{K}_{XY}^2(\omega)\right]d\omega.$$

Proof. Since the derivatives of $f_{YY}^{-1}(\omega)$ and $f_{YX}(\omega)$ are bounded, the derivative of $f_{YY}^{-1}(\omega)f_{YX}(\omega)$ is square integrable. Therefore, by Theorem 3.1.8, the Fourier coefficients of $f_{YY}^{-1}(\omega)f_{YX}(\omega)$, the a_j, are absolutely summable and, by Theorem 2.2.1, $\sum_{j=-\infty}^{\infty} a_j Y_{t-j}$ is a well-defined random variable. That $\{a_j\}$ is the best linear filter follows from (4.5.8), and the mean square error of the filter follows from (4.5.7). ▲

Example. To illustrate the ideas of this section, we use some data on the sediment suspended in the water of the Des Moines River at Boone, Iowa. A portion of the data obtained by daily sampling of the water during 1973 are displayed in Table 4.5.1. The data are the logarithm of the parts per million of suspended sediment. Since the laboratory determinations are made on a small sample of water collected from the river, the readings can be represented as

$$Y_t=X_t+u_t$$

Table 4.5.1. Logarithm of sediment suspended in Des Moines River, Boone, Iowa, 1973.

Daily Observations	Smoothed Observations Two-sided Filter	Smoothed Observations One-sided Filter
5.44	—	—
5.38	—	—
5.43	5.40	—
5.22	5.26	5.26
5.28	5.27	5.28
5.21	5.22	5.22
5.23	5.25	5.23
5.33	5.37	5.31
5.58	5.63	5.53
6.18	6.08	6.04
6.16	6.14	6.10
6.07	6.13	6.04
6.56	6.38	6.42
5.93	5.96	5.99
5.70	5.69	5.74
5.36	5.39	5.42
5.17	5.24	5.22
5.35	5.36	5.33
5.51	5.51	5.47
5.80	5.67	5.72
5.29	5.36	5.37
5.28	5.29	5.30
5.27	—	5.27
5.17	—	5.19

SOURCE. U. S. Department of Interior Geological Survey—Water Resources Division, Sediment Concentration Notes, Des Moines River, Boone, Iowa.

where Y_t is the recorded value, X_t is the "true" average sediment in the river water, and u_t is the measurement error introduced by sampling and laboratory determination. Assume that X_t can be represented as a first order autoregressive process

$$X_t - 5.28 = 0.81(X_{t-1} - 5.28) + e_t$$

where the e_t are independent $(0, 0.172)$ random variables. Assume further that u_t is a sequence of independent $(0, 0.053)$ random variables independent of X_j for all t, j.

To construct a filter $\{a_{-2}, a_{-1}, a_0, a_1, a_2\}$ that will best estimate X_t using

the observations $\{Y_{t-2}, Y_{t-1}, Y_t, Y_{t+1}, Y_{t+2}\}$, we solve the system of equations

$$\begin{bmatrix} 0.553 & 0.405 & 0.328 & 0.266 & 0.215 \\ 0.405 & 0.553 & 0.405 & 0.328 & 0.266 \\ 0.328 & 0.405 & 0.553 & 0.405 & 0.328 \\ 0.266 & 0.328 & 0.405 & 0.553 & 0.405 \\ 0.215 & 0.266 & 0.328 & 0.405 & 0.553 \end{bmatrix} \begin{bmatrix} a_{-2} \\ a_{-1} \\ a_0 \\ a_1 \\ a_2 \end{bmatrix} = \begin{bmatrix} 0.328 \\ 0.405 \\ 0.500 \\ 0.405 \\ 0.328 \end{bmatrix}$$

to obtain

$$\mathbf{a} = \{a_{-2}, a_{-1}, a_0, a_1, a_2\} = \{0.023, 0.120, 0.702, 0.120, 0.023\}.$$

The mean square error of the filtered time series $5.28 + \Sigma_{j=-2}^{2} a_j(Y_{t-j} - 5.28)$ as an estimator of X_t is $0.500 - \mathbf{a}[0.328, 0.405, 0.500, 0.405, 0.328]' = 0.0372$. The data transformed by this filter are displayed in the second column of Table 4.5.1. Note that the filtered data are "smoother" in that the variance of changes from one period to the next is smaller for the filtered data than for the original data.

We now obtain a one-sided filter that can be used to estimate the most recent value of X_t using only the most recent and the four preceding values of Y_t. The estimator of X_t is given by

$$5.28 + b_0(Y_t - 5.28) + b_1(Y_{t-1} - 5.28) + b_2(Y_{t-2} - 5.28) + b_3(Y_{t-3} - 5.28) + b_4(Y_{t-4} - 5.28),$$

where

$$\begin{bmatrix} b_0 \\ b_1 \\ b_2 \\ b_3 \\ b_4 \end{bmatrix} = \begin{bmatrix} 0.553 & 0.405 & 0.328 & 0.266 & 0.215 \\ 0.405 & 0.553 & 0.405 & 0.328 & 0.266 \\ 0.328 & 0.405 & 0.553 & 0.405 & 0.328 \\ 0.266 & 0.328 & 0.405 & 0.553 & 0.405 \\ 0.215 & 0.266 & 0.328 & 0.405 & 0.553 \end{bmatrix}^{-1} \begin{bmatrix} 0.500 \\ 0.405 \\ 0.328 \\ 0.266 \\ 0.215 \end{bmatrix} = \begin{bmatrix} 0.790 \\ 0.134 \\ 0.023 \\ 0.005 \\ 0.000 \end{bmatrix}.$$

This one-sided estimator of X_t has a mean square error of 0.0419. The filtered data using this filter are given in the last column of Table 4.5.1.

To obtain the minimum value for the mean square error of the two-sided

filter, we evaluated (4.5.7), where

$$f_{XX}(\omega)=\frac{0.172}{2\pi(1-0.81e^{-i\omega})(1-0.81e^{i\omega})},$$

$$f_{YY}(\omega)=\frac{0.172}{2\pi(1-0.81e^{-i\omega})(1-0.81e^{i\omega})}+\frac{0.053}{2\pi}$$

$$=\frac{0.2524(1-0.170e^{i\omega})(1-0.170e^{-i\omega})}{2\pi(1-0.81e^{-i\omega})(1-0.81e^{i\omega})},$$

obtaining

$$\gamma_{WW}(0)=\int_{-\pi}^{\pi}f_{XX}(\omega)\,d\omega-\int_{-\pi}^{\pi}\frac{(0.172)^2\,d\omega}{2\pi(0.2524)|1-0.170e^{-i\omega}|^2|1-0.81e^{-i\omega}|^2}$$
$$=0.500-0.463=0.037.$$

The infinite sequence of weights is given by the inverse transform of

$$\frac{0.172}{2\pi(0.2524)(1-0.17e^{-i\omega})(1-0.17e^{i\omega})},$$

which yields

$$(\ldots,0.1193,0.7017,0.1193,0.0203,0.0034,\ldots).$$

While it is possible to use spectral methods to evaluate the minimum mean square error of a one-sided filter [see, for example, Yaglom (1962, p. 97)], we examine the problem in a slightly different manner. Since Y_t is an autoregressive moving average (1, 1) process, the methods of Section 2.9 can be used to obtain a one period ahead predictor $\hat{Y}_t(Y_{t-1}, Y_{t-2}, \ldots)$ based on an infinite past. As $Y_t = X_t + u_t$, where u_t is a sequence of independent random variables, the best predictor of X_t based on Y_{t-1}, $Y_{t-2}, \ldots$ must be the same as the best predictor of Y_t based on Y_{t-1}, $Y_{t-2}, \ldots$. Furthermore, given the predictor, the partial correlation between any Y_{t-j}, $j>0$, and X_t is zero. Therefore, to obtain the best filter for X_t using Y_t, $Y_{t-1}, \ldots$, we find the optimal linear combination of

$\hat{Y}_t(Y_{t-1}, Y_{t-2}, \ldots)$ and Y_t. Denote the linear combination by $c_0 Y_t + c_1 \hat{Y}_t(Y_{t-1}, Y_{t-2}, \ldots)$, where

$$\begin{pmatrix} c_0 \\ c_1 \end{pmatrix} = \begin{pmatrix} 0.5530 & 0.3006 \\ 0.3006 & 0.3006 \end{pmatrix}^{-1} \begin{pmatrix} 0.5000 \\ 0.3006 \end{pmatrix} = \begin{pmatrix} 0.7900 \\ 0.2100 \end{pmatrix},$$

$$\begin{pmatrix} 0.5530 & 0.3006 \\ 0.3006 & 0.3006 \end{pmatrix}$$

is the covariance matrix of $[Y_t, \hat{Y}_t(Y_{t-1}, Y_{t-2}, \ldots)]$, and $(0.5000, 0.3006)$ is the vector of covariances between $[Y_t, \hat{Y}_t(Y_{t-1}, Y_{t-2}, \ldots)]$ and X_t. It follows that the minimum mean square error for a one-sided predictor is 0.0419.

REFERENCES

Sections 4.1, 4.3–4.5. Anderson (1971), Blackman and Tukey (1959), Brillinger (1975), Granger and Hatanaka (1964), Hannan (1960), (1970), Harris (1967), Jenkins and Watts (1968), Kendall and Stuart (1966), Koopmans (1974), Whittle (1963), Yaglom (1962).

Section 4.2. Amemiya and Fuller (1967), Bellman (1960), Wahba (1968).

EXERCISES

1. Which of the following functions is the spectral density of a stationary time series? Explain why or why not.
   (a) $f(\omega) = 1 - (1/2)\omega^2, \quad -\pi \leqslant \omega \leqslant \pi.$
   (b) $f(\omega) = 1 + (1/2)\omega, \quad -\pi \leqslant \omega \leqslant \pi.$
   (c) $f(\omega) = 476 + \cos 13\omega, \quad -\pi \leqslant \omega \leqslant \pi.$

2. Let $\{X_t: t \in (0, \pm 1, \pm 2, \ldots)\}$ be defined by $X_t = e_t + 0.4 e_{t-1}$. Compute the autocovariance function $\gamma(h)$ and the spectral density $f(\omega)$, given that the e_t are independently and identically distributed $(0, \sigma^2)$ random variables.

3. Give the spectral density for the time series defined by

$$X_t - \beta X_{t-1} = e_t + \alpha_1 e_{t-1} + \alpha_2 e_{t-2}, \qquad t = 0, \pm 1, \pm 2, \ldots,$$

where $|\beta| < 1$, the roots of $m^2 + \alpha_1 m + \alpha_2 = 0$ are less than one in absolute value, and the e_t are uncorrelated $(0, \sigma^2)$ random variables.

4. Let

$$X_t + \alpha_1 X_{t-1} + \alpha_2 X_{t-2} = Z_t$$

and

$$Z_t + \beta_1 Z_{t-1} + \beta_2 Z_{t-2} = e_t,$$

where the e_t are uncorrelated $(0,\sigma^2)$ random variables, and the roots of both $m^2+\alpha_1 m+\alpha_2=0$ and $r^2+\beta_1 r+\beta_2=0$ are less than one in absolute value. Give an expression for the spectral density of X_t. How do you describe the time series X_t?

5. Find the covariance function and spectral distribution function for the time series

$$X_t=u_1\cos t+u_2\sin t+Y_t,$$

where $(u_1,u_2)'$ is distributed as a bivariate normal random variable with zero mean and diagonal covariance matrix diag$\{2,2\}$, $Y_t=e_t-e_{t-1}$, and the e_t are independent (0,3) random variables, independent of $(u_1,u_2)'$.

6. Given the following spectral distribution function:

$$\begin{aligned} F_X(\omega)&=\pi+\omega, && -\pi\leqslant\omega<\frac{-\pi}{2}\\ &=5\pi+\omega, && \frac{-\pi}{2}\leqslant\omega<\frac{\pi}{2}\\ &=9\pi+\omega, && \frac{\pi}{2}\leqslant\omega\leqslant\pi. \end{aligned}$$

What is the variance of X_t? What is the spectral distribution function of X_t-X_{t-1}? Is there a k such that X_t-X_{t-k} will have a continuous spectral distribution function?

7. Prove Corollary 4.3.1.3.

8. Let $\{e_t\colon t\in(0,\pm1,\pm2,\ldots)\}$ be a time series of uncorrelated $(0,\sigma^2)$ random variables. Let $X_t=e_{t-2}+0.5e_{t-3}$. Give the covariance matrix, $\boldsymbol{\Gamma}(h)$, and the spectral matrix $\boldsymbol{f}(\omega)$ for $(X_t,e_t)'$.

9. Let $\{e_{1t}\}$ and $\{e_{2t}\}$ be two independent sequences of uncorrelated random variables with variances 0.34 and 0.50 respectively. Let

$$X_{1t}=0.8X_{1t-1}+e_{1t}$$
$$X_{2t}=X_{1t}+e_{2t}.$$

Find $\gamma_{22}(h)$, $\gamma_{12}(h)$, $f_{12}(\omega)$, and $f_{22}(\omega)$.

10. The complex vector random variable $\mathbf{X}$ of dimension q is distributed as a complex multivariate normal if the real vector $[(\mathrm{Re}\,\mathbf{X})', (\mathrm{Im}\,\mathbf{X})']'$ is distributed as a multivariate normal with mean $[(\mathrm{Re}\,\boldsymbol{\mu})', (\mathrm{Im}\,\boldsymbol{\mu})']'$ and covariance matrix

$$\frac{1}{2}\begin{pmatrix}\mathrm{Re}\,\boldsymbol{\Sigma} & -\mathrm{Im}\,\boldsymbol{\Sigma}\\ \mathrm{Im}\,\boldsymbol{\Sigma} & \mathrm{Re}\,\boldsymbol{\Sigma}\end{pmatrix},$$

where $\boldsymbol{\mu}=E\{\mathbf{X}\}$ and $\boldsymbol{\Sigma}$ is a positive semidefinite Hermitian matrix. Let $(A_1,B_1,A_2,B_2)'$ be the multivariate normal random variable defined following (4.4.9). Show that $\mathbf{X}=(A_1+\iota B_1, A_2+\iota B_2)'$ is distributed as a complex bivariate normal random variable. Give the Hermitian matrix $\boldsymbol{\Sigma}$.

11. Let $\{a_j\}_{j=-\infty}^{\infty}$ and $\{b_s\}_{s=-\infty}^{\infty}$ be absolutely summable, and let $\{X_t\colon t\in(0,\pm1,\pm2,\ldots)\}$ be a stationary time series with absolutely summable covariance function. Consider the following filtering operation: (a) apply the filter $\{a_j\}$ to the time series X_t to obtain a time series Z_t and then (b) apply the filter $\{b_j\}$ to Z_t to obtain a time series Y_t. What is the transfer function of this filtering operation? Express the spectral density of Y_t as a function of the spectral density of X_t and of the transfer function.

12. Let $\{a_j\}$, $\{b_j\}$, and $\{X_t\}$ be as defined in Exercise 11, and define

$$Y_t = \sum_{j=-\infty}^{\infty} (a_j + b_j) X_{t-j}.$$

Express the spectral density of Y_t as a function of the spectral density of X_t and the transfer functions of $\{a_j\}$ and $\{b_j\}$.

13. Let X_t and Y_t be defined by

$$X_t = 0.9X_{t-1} + e_t$$

$$Y_t = X_t + u_t,$$

where $\{e_t\}$ is a sequence of normal independent (0, 1) random variables independent of the sequence $\{u_t\}$. Let u_t satisfy the difference equation

$$u_t + 0.5u_{t-1} = v_t,$$

where $\{v_t\}$ is a sequence of normal independent (0, 0.3) random variables. Assuming that only Y_t is observed, construct the filter $\{a_j: j = -2, -1, 0, 1, 2\}$ so that

$$\sum_{j=-2}^{2} a_j Y_{t-j}$$

is the minimum mean square error estimator of X_t. Construct the one-sided filter $\{b_j: j = 0, 1, \ldots, 5\}$ to estimate X_t. How does the mean square error of these filters compare with the lower bound for linear filters?

14. Let $f(\omega)$ be an even nonnegative continuous periodic function of period 2π. Let

$$a(h) = \int_{-\pi}^{\pi} f(\omega) e^{-i\omega h} \, d\omega.$$

Show that, for q a positive integer,

$$\gamma(h) = \begin{cases} \dfrac{q - |h|}{q} a(h), & h = 0, \pm 1, \pm 2, \ldots, \pm q \\ 0, & \text{otherwise} \end{cases}$$

is the covariance function of a stationary time series. (*Hint*: See Exercise 3.15 and Theorem 3.1.10.)

15. Let X_t be a time series with continuous spectral density $f_X(\omega)$. Let Y_t be a time series satisfying

$$\sum_{j=0}^{p} \alpha_j Y_{t-j} = e_t,$$

$$|f_Y(\omega) - f_X(\omega)| < \epsilon,$$

where the e_t are uncorrelated $(0, \sigma^2)$ random variables, $\alpha_0 = 1$, and Y_t is defined by Theorem 4.3.4.

(a) Show that $|\gamma_X(h) - \gamma_Y(h)| < 2\pi\epsilon$ for all h.

(b) Let $f_X(\omega)$ be strictly positive. Prove that given $\epsilon > 0$ there is a p and a set $\{\alpha_j$:

$j = 0, 1, \ldots, p\}$ with $\alpha_0 = 1$ such that the time series Z_t defined by

$$Z_t = \sum_{j=0}^{p} \alpha_j X_{t-j}$$

satisfies

$$\left| f_Z(\omega) - \frac{\sigma^2}{2\pi} \right| < \epsilon$$

for all ω, where $f_Z(\omega)$ is the spectral density of Z_t and $\gamma_Z(0) = \sigma^2$. Show that

$$\sum_{h=1}^{\infty} \gamma_Z^2(h) \leqslant 2\pi\epsilon^2.$$

(c) Let $f_X(\omega)$ be strictly positive. Show that, given $\epsilon > 0$, one may define two autoregressive time series Y_{1t} and Y_{2t} with spectral densities

$$f_{Y_1}(\omega) = \frac{\sigma_1^2}{2\pi \left| \sum_{j=0}^{p_1} \alpha_{1j} e^{-i\omega j} \right|^2}$$

$$f_{Y_2}(\omega) = \frac{\sigma_2^2}{2\pi \left| \sum_{j=0}^{p_2} \alpha_{2j} e^{-i\omega j} \right|^2}$$

such that

$$f_X(\omega) - \epsilon \leqslant f_{Y_1}(\omega) \leqslant f_X(\omega) \leqslant f_{Y_2}(\omega) \leqslant f_X(\omega) + \epsilon.$$

(d) For the three time series defined in part (c) prove that

$$\operatorname{Var}\left\{ \sum_{t=1}^{n} a_t Y_{1t} \right\} \leqslant \operatorname{Var}\left\{ \sum_{t=1}^{n} a_t X_t \right\} \leqslant \operatorname{Var}\left\{ \sum_{t=1}^{n} a_t Y_{2t} \right\}$$

for any fixed real numbers $\{a_t \colon t = 1, 2, \ldots, n\}$.

16. Let

$$X_{1t} = e_t - 0.8 e_{t-1}$$

$$X_{2t} = u_t - 0.9 u_{t-4}$$

where $\{e_t\}$ is a sequence of independent $(0,1)$ random variables independent of $\{u_t\}$ a sequence of independent $(0,6)$ random variables. Express

$$Y_t = X_{1t} + X_{2t}$$

as a moving average process.

17. Let Y_t be defined by

$$Y_t = S_t + Z_t,$$

$$S_t = 0.9S_{t-4} + u_t,$$

$$Z_t = 0.8Z_{t-1} + e_t,$$

where the sequence $\{(e_t, u_t)'\}$ is a sequence of normal independent $(\mathbf{0}, \mathbf{\Sigma})$ random variables with $\mathbf{\Sigma} = \text{diag}(0.1, 0.6)$. Construct the optimum filter $\{a_j: j = -9, -8, \ldots, 8, 9\}$ to estimate S_t where the estimator is defined by $\Sigma_{j=-9}^{9} a_j Y_{t-j}$. Construct the best one sided filter $\{b_j: j = 0, 1, \ldots, 8, 9\}$ to estimate S_t. Compare the mean square error of these filters with the lower bound for linear filters.

CHAPTER 5

Some Large Sample Theory

So far, we have been interested in ways of representing time series and describing their properties. In most practical situations we have a portion of a realization, or of several realizations, and we wish a description (an estimate of the parameters) of the time series.

Most of the presently available results on the estimation of the covariance function, the parameters of autoregressive and moving average processes, and the spectral density rest on large sample theory. Therefore, we shall present some results in large sample statistics.

5.1. ORDER IN PROBABILITY

Concepts of relative magnitude or *order of magnitude* are useful in investigating limiting behavior of random variables. We first define the concepts of order as used in real analysis. Let $\{a_n\}_{n=1}^{\infty}$ be a sequence of real numbers and $\{g_n\}_{n=1}^{\infty}$ be a sequence of positive real numbers.

Definition 5.1.1. We say a_n is of smaller order than g_n and write

$$a_n = o(g_n)$$

if

$$\lim_{n\to\infty} \frac{a_n}{g_n} = 0.$$

Definition 5.1.2. We say a_n is at most of order g_n and write

$$a_n = O(g_n)$$

if there exists a real number M such that $g_n^{-1}|a_n| \leqslant M$ for all n.

The properties of Lemma 5.1.1 are easily established using the definitions of order and the properties of limits.

Lemma 5.1.1. Let $\{a_n\}$ and $\{b_n\}$ be sequences of real numbers. Let $\{f_n\}$ and $\{g_n\}$ be sequences of positive real numbers.

(i) If $a_n = o(f_n)$ and $b_n = o(g_n)$, then

$$a_n b_n = o(f_n g_n),$$

$$|a_n|^s = o(f_n^s) \quad \text{for} \quad s > 0,$$

$$a_n + b_n = o(\max\{f_n, g_n\}).$$

(ii) If $a_n = O(f_n)$ and $b_n = O(g_n)$, then

$$a_n b_n = O(f_n g_n),$$

$$|a_n|^s = O(f_n^s) \quad \text{for} \quad s \geqslant 0,$$

$$a_n + b_n = O(\max\{f_n, g_n\}).$$

(iii) If $a_n = o(f_n)$ and $b_n = O(g_n)$, then

$$a_n b_n = o(f_n g_n).$$

Proof. Reserved for the reader. ▲

The concepts of order when applied to random variables are closely related to *convergence in probability.*

Definition 5.1.3. The sequence of random variables $\{X_n\}$ converges in probability to the random variable X, and we write

$$p\lim X_n = X,$$

(the *probability limit* of X_n is X) if, for every $\epsilon > 0$,

$$\lim_{n\to\infty} P\{|X_n - X| > \epsilon\} = 0.$$

An equivalent definition is that for every $\epsilon > 0$ and $\delta > 0$ there exists an N such that for all $n > N$,

$$P\{|X_n - X| > \epsilon\} < \delta.$$

The notation

$$X_n \xrightarrow{P} X$$

is also frequently used to indicate that X_n converges in probability to X.

For sequences of random variables, definitions of *order in probability* were introduced by Mann and Wald (1943b). Let $\{X_n\}$ be a sequence of random variables and $\{g_n\}$ a sequence of positive real numbers.

Definition 5.1.4. We say X_n is of smaller order in probability than g_n and write

$$X_n = o_p(g_n)$$

if

$$p\lim \frac{X_n}{g_n} = 0.$$

Definition 5.1.5. We say X_n is at most of order in probability g_n and write

$$X_n = O_p(g_n)$$

if, for every $\epsilon > 0$, there exists a positive real number M_ϵ such that

$$P\{|X_n| \geqslant M_\epsilon g_n\} \leqslant \epsilon$$

for all n.

If $X_n = O_p(g_n)$, we sometimes say that X_n is *bounded in probability* by g_n. We define a vector random variable to be $O_p(g_n)$ if every element of the vector is $O_p(g_n)$ as follows.

Definition 5.1.6. If $\mathbf{X}_n$ is a k-dimensional random variable, $\mathbf{X}_n$ is at most of order in probability g_n, and we write

$$\mathbf{X}_n = O_p(g_n)$$

if, for every $\epsilon > 0$, there exists a positive real number M_ϵ such that

$$P\{|X_{jn}| \geqslant M_\epsilon g_n\} \leqslant \epsilon, \qquad j = 1, 2, \ldots, k,$$

for all n.

We say $\mathbf{X}_n$ is of smaller order in probability than g_n and write

$$\mathbf{X}_n = o_p(g_n)$$

if, for every $\epsilon>0$ and $\delta>0$ there exists an N such that for all $n>N$,

$$P\{|X_{jn}|>\epsilon g_n\}<\delta, \qquad j=1,2,\dots,k.$$

Note that k might be a function of n and $\mathbf{X}_n$ could still satisfy the definition. However, it is clear that the M_ϵ of the definition is a function of ϵ only (and not of n).

A matrix random variable may be viewed as a vector random variable with the elements displayed in a particular manner, or as a collection of vector random variables. Therefore, we shall define the order of matrix random variables in an analogous manner.

Definition 5.1.7. A $k\times r$ matrix, $\mathbf{B}_n$, of random variables is at most of order in probability g_n, and we write

$$\mathbf{B}_n=O_p(g_n)$$

if, for every $\epsilon>0$, there exists a positive real number M_ϵ such that

$$P\{|b_{ijn}|\geqslant M_\epsilon g_n\}\leqslant\epsilon, \qquad i=1,2,\dots,k, \qquad j=1,2,\dots,r,$$

for all n, where the b_{ijn} are the elements of $\mathbf{B}_n$. We say that $\mathbf{B}_n$ is of smaller order in probability than g_n and write

$$\mathbf{B}_n=o_p(g_n)$$

if, for every $\epsilon>0$ and $\delta>0$ there exists an N such that for all $n>N$,

$$P\{|b_{ijn}|>\epsilon g_n\}<\delta, \qquad i=1,2,\dots,k, \qquad j=1,2,\dots,r.$$

For real numbers $a_i, i=1,2,\dots,n$, we know that $|\Sigma_{i=1}^n a_i|\leqslant\Sigma_{i=1}^n|a_i|$. The following lemma, resting on this property, furnishes a useful bound on the probability that the absolute value of a sum of random variables exceeds a given number.

Lemma 5.1.2. Let $\mathbf{X}_i$, $i=1,2,\dots,n$, be k-dimensional random variables. Then, for every $\epsilon>0$,

$$P\left\{\left|\sum_{i=1}^n \mathbf{X}_i\right|\geqslant\epsilon\right\}\leqslant\sum_{i=1}^n P\left\{|\mathbf{X}_i|>\frac{\epsilon}{n}\right\}.$$

Proof. Let $\epsilon>0$ be arbitrary. We see that if $\Sigma_{i=1}^n|\mathbf{X}_i|\geqslant\epsilon$, then $|\mathbf{X}_i|\geqslant\epsilon/n$

for at least one $i \in \{1,2,\ldots,n\}$. Therefore,

$$P\left\{\left|\sum_{i=1}^{n} \mathbf{X}_i\right| \geqslant \epsilon\right\} \leqslant P\left\{\sum_{i=1}^{n} |\mathbf{X}_i| \geqslant \epsilon\right\}$$

$$\leqslant \sum_{i=1}^{n} P\left\{|\mathbf{X}_i| \geqslant \frac{\epsilon}{n}\right\}. \qquad \blacktriangle$$

Definition 5.1.3 applies for vector random variables of fixed dimension as well as for scalar random variables, if it is understood that $|\mathbf{X}_n - \mathbf{X}|$ is the common Euclidean distance.

Lemma 5.1.3. Let $\mathbf{X}_n$ be a k-dimensional random variable such that

$$p\lim X_{jn} = X_j, \qquad j = 1,2,\ldots,k,$$

where X_{jn} is the jth element of $\mathbf{X}_n$. Then, for k fixed,

$$p\lim \mathbf{X}_n = \mathbf{X}.$$

Proof. By hypothesis, for each j and for every $\epsilon > 0$ and $\delta > 0$, there exists an integer N_j such that for all $n > N_j$

$$P\left\{|X_{jn} - X_j| > k^{-1/2}\epsilon\right\} \leqslant k^{-1}\delta.$$

Let N be the maximum of $\{N_1, N_2, \ldots, N_k\}$. Using Lemma 5.1.2, we have

$$P\{|\mathbf{X}_n - \mathbf{X}| > \epsilon\} \leqslant \sum_{j=1}^{k} P\left\{|X_{jn} - X_j| > k^{-1/2}\epsilon\right\} \leqslant \delta$$

for $n > N$. ▲

The proof of Lemma 5.1.3 should also help to make it clear that if k is not fixed, then the fact that $\mathbf{X}_n = o_p(1)$ does not necessarily imply that $p\lim|\mathbf{X}_n - \mathbf{X}| = 0$. The vector random variable composed of n entries all equal to $n^{-1/2}$ furnishes a counterexample for $k = n$.

We shall demonstrate later (Theorems 5.1.5 and 5.1.6) that operations valid for order are also valid for order in probability. Since it is relatively easy to establish the properties analogous to those of Lemma 5.1.1, we do so at this time.

Lemma 5.1.4. Let $\{f_n\}$ and $\{g_n\}$ be sequences of positive real numbers, and let $\{X_n\}$ and $\{Y_n\}$ be sequences of random variables.

(i) If $X_n = o_p(f_n)$ and $Y_n = o_p(g_n)$, then

$$X_n Y_n = o_p(f_n g_n),$$

$$|X_n|^s = o_p(f_n^s) \quad \text{for} \quad s > 0,$$

$$X_n + Y_n = o_p(\max\{f_n, g_n\}).$$

(ii) If $X_n = O_p(f_n)$ and $Y_n = O_p(g_n)$, then

$$X_n Y_n = O_p(f_n g_n),$$

$$|X_n|^s = O_p(f_n^s) \quad \text{for} \quad s \geqslant 0,$$

$$X_n + Y_n = O_p(\max\{f_n, g_n\}).$$

(iii) If $X_n = o_p(f_n)$ and $Y_n = O_p(g_n)$, then

$$X_n Y_n = o_p(f_n g_n).$$

Proof. We investigate only part i, leaving parts ii and iii as an exercise. By arguments similar to those of Lemma 5.1.3, $|X_n Y_n| > f_n g_n$ implies that $|X_n / f_n| > 1$ or (and) $|Y_n / g_n| > 1$. By hypothesis, given $\epsilon > 0$ and $\delta > 0$, there is an N such that

$$P\{|X_n| > \epsilon f_n\} < \frac{1}{2}\delta,$$

$$P\{|Y_n| > \epsilon g_n\} < \frac{1}{2}\delta$$

for $n > N$. Therefore,

$$\begin{aligned} P\{|X_n Y_n| > \epsilon^2 f_n g_n\} &\leqslant P\left\{\left|\frac{X_n}{f_n}\right| > \epsilon \quad \text{or} \quad \left|\frac{Y_n}{g_n}\right| > \epsilon\right\} \\ &\leqslant P\left\{\left|\frac{X_n}{f_n}\right| > \epsilon\right\} + P\left\{\left|\frac{Y_n}{g_n}\right| > \epsilon\right\} \\ &< \delta \end{aligned}$$

for $n > N$.

The second equality in part i follows from

$$P\{|X_n| > \epsilon f_n\} = P\{|X_n|^s > \epsilon^s f_n^s\},$$

which holds for all $\epsilon > 0$.

Let $q_n = \max\{f_n, g_n\}$. Given $\epsilon > 0$ and $\delta > 0$, there exists an n such that

$$P\left\{|X_n| > \frac{1}{2}\epsilon q_n\right\} < \frac{1}{2}\delta,$$

$$P\left\{|Y_n| > \frac{1}{2}\epsilon q_n\right\} < \frac{1}{2}\delta$$

for $n > N$. Hence, the third result of part i follows by Lemma 5.1.2. ▲

One of the most useful tools for establishing the order in probability of random variables is Chebyshev's inequality.

Theorem 5.1.1 (Chebyshev's inequality). Let $r > 0$, let X be a random variable such that $E\{|X|^r\} < \infty$, and let $F(x)$ be the distribution function of X. Then, for every $\epsilon > 0$ and finite A,

$$P\{|X - A| \geqslant \epsilon\} \leqslant \frac{E\{|X - A|^r\}}{\epsilon^r}.$$

Proof. Let us denote by S the set of x for which $|x - A| \geqslant \epsilon$ and by $\tilde{S}$ the set of x for which $|x - A| < \epsilon$. Then,

$$\int |x - A|^r \, dF(x) = \int_S |x - A|^r \, dF(x) + \int_{\tilde{S}} |x - A|^r \, dF(x)$$

$$\geqslant \epsilon^r \int_S dF(x) = \epsilon^r P\{|X - A| \geqslant \epsilon\}. \qquad ▲$$

It follows from Chebyshev's inequality that any random variable with finite variance is bounded in probability by the square root of its second moment about the origin.

Corollary 5.1.1.1. Let $\{X_n\}$ be a sequence of random variables and $\{a_n\}$ a sequence of positive real numbers such that

$$E\{X_n^2\} = O(a_n^2).$$

Then

$$X_n = O_p(a_n).$$

Proof. By assumption there exists an M_1 such that

$$E\{X_n^2\} < M_1^2 a_n^2$$

for all n. By Chebyshev's inequality, for any $M_2 > 0$,

$$P\{|X_n| \geqslant M_2 a_n\} \leqslant \frac{E\{X_n^2\}}{M_2^2 a_n^2}.$$

Hence, given $\epsilon > 0$, we choose $M_2 \geqslant M_1 \epsilon^{-1/2}$, and the result follows. ▲

If the sequence $\{X_n\}$ has zero mean or a mean whose order is less than or equal to the order of the standard error, then the order in probability of the sequence is the order of the standard error.

Corollary 5.1.1.2. Let the sequence of random variables $\{X_n\}$ satisfy

$$E\{(X_n - E\{X_n\})^2\} = O(a_n^2)$$

and

$$E\{X_n\} = O(a_n),$$

where $\{a_n\}$ is a sequence of positive real numbers. Then,

$$X_n = O_p(a_n).$$

Proof. By the assumptions and by property ii of Lemma 5.1.1,

$$E\{X_n^2\} = E\{(X_n - E\{X_n\})^2\} + (E\{X_n\})^2 = O(a_n^2),$$

and the result follows by Corollary 5.1.1.1. ▲

Let the probability limits of two sequences of random variables be defined. We now demonstrate that the sequences have a common probability limit if the probability limit of the sequence of differences is zero.

Theorem 5.1.2. Let $\{X_n\}$ and $\{Y_n\}$ be sequences of random variables such that

$$p\lim|X_n - Y_n| = 0.$$

If there exists a random variable X such that $p\lim X_n = X$, then

$$p\lim Y_n = X.$$

Proof. Given $\epsilon > 0$ and $\delta > 0$, there exists, by hypothesis, an N such that for $n > N$,

$$P\left\{|Y_n - X_n| \geqslant \frac{\epsilon}{2}\right\} \leqslant \frac{\delta}{2}$$

and

$$P\left\{|X_n - X| \geqslant \frac{\epsilon}{2}\right\} \leqslant \frac{\delta}{2}.$$

But, by Lemma 5.1.2 for $n > N$,

$$\begin{aligned} P\{|Y_n - X| \geqslant \epsilon\} &\leqslant P\left\{|Y_n - X_n| \geqslant \frac{\epsilon}{2}\right\} + P\left\{|X_n - X| \geqslant \frac{\epsilon}{2}\right\} \\ &\leqslant \delta. \end{aligned}$$

▲

Definition 5.1.8. For $r \geqslant 1$, the sequence of random variables $\{X_n\}$ is said to *converge in rth mean* if $E\{|X_n|^r\} < \infty$ for all n and

$$E\{|X_n - X_m|^r\} \to 0$$

as $n \to \infty$ and $m \to \infty$. If $\{X_n\}$ converges to X in rth mean, we denote this by writing

$$X_n \xrightarrow{r} X.$$

Using Chebyshev's inequality it is easy to demonstrate that convergence in rth mean implies convergence in probability.

Theorem 5.1.3. Let $\{X_n\}$ be a sequence of random variables with finite rth moments. If there exists a random variable X such that $X_n \xrightarrow{r} X$, then $X_n \xrightarrow{P} X$.

Proof. Given $\epsilon > 0$,

$$P\{|X_n - X| > \epsilon\} \leqslant \frac{E\{|X_n - X|^r\}}{\epsilon^r}$$

by Chebyshev's inequality. For $\delta > 0$ there is, by hypothesis, an integer $N = N(\epsilon, \delta)$ such that for all $n > N$,

$$E\{|X_n - X|^r\} < \delta\epsilon^r$$

and, therefore, for $n > N$,

$$P\{|X_n - X| > \epsilon\} < \delta. \qquad \blacktriangle$$

One useful consequence is Corollary 5.1.3.1, which can be paraphrased as follows. If the sequence of differences of two sequences of random variables converges in squared mean to zero, then the two sequences of random variables have a common probability limit if the limit exists.

Corollary 5.1.3.1. Let $\{X_n\}$ and $\{Y_n\}$ be sequences of random variables such that

$$\lim_{n\to\infty} E\{(X_n - Y_n)^2\} = 0.$$

If there exists a random variable Y such that $p\lim Y_n = Y$, then $p\lim X_n = Y$.

Proof. By Theorem 5.1.3, we have that $p\lim(X_n - Y_n) = 0$. The conclusion follows by Theorem 5.1.2. ▲

Corollary 5.1.3.2. If the sequence of random variables $\{X_n\}$ is such that

$$\lim_{n\to\infty} E\{X_n\} = \mu$$

and

$$\lim_{n\to\infty} E\{[X_n - E\{X_n\}]^2\} = 0,$$

then $p\lim X_n = \mu$.

Proof. The proof follows directly by letting the constants μ and $\{E\{X_n\}\}$ be, respectively, the Y and $\{Y_n\}$ of Corollary 5.1.3.1. ▲

Since we often work with functions of sequences of random variables, the following theorem is very important. The theorem states that if the function $\mathbf{g}(\mathbf{x})$ is continuous, then "the probability limit of the function is the function of the probability limit."

Theorem 5.1.4. Let $\{\mathbf{X}_n\}$ be a sequence of real valued k-dimensional random variables such that $p\lim \mathbf{X}_n = \mathbf{X}$. Let $\mathbf{g}(\mathbf{x})$ be a function mapping the real k-dimensional vector $\mathbf{x}$ into a real p-dimensional space. Let $\mathbf{g}(\mathbf{x})$ be continuous. Then $p\lim \mathbf{g}(\mathbf{X}_n) = \mathbf{g}(\mathbf{X})$.

Proof. Given $\epsilon > 0$ and $\delta > 0$, let A be a closed and bounded k-dimensional set such that

$$P\{\mathbf{X} \in A\} \geqslant 1 - \frac{1}{2}\delta.$$

Since $\mathbf{g}(\mathbf{x})$ is continuous, it is uniformly continuous on A, and there exists a δ_ϵ such that

$$|\mathbf{g}(\mathbf{x}_1)-\mathbf{g}(\mathbf{x}_2)|<\epsilon$$

if $|\mathbf{x}_1-\mathbf{x}_2|<\delta_\epsilon$ and $\mathbf{x}_1$ is in A. Since $p\lim \mathbf{X}_n=\mathbf{X}$, there exists an N such that for $n>N$,

$$P\{|\mathbf{X}_n-\mathbf{X}|>\delta_\epsilon\}<\frac{1}{2}\delta.$$

Therefore, for $n>N$,

$$\begin{aligned}
P\{|\mathbf{g}(\mathbf{X}_n)-\mathbf{g}(\mathbf{X})|>\epsilon\} &= P\{|\mathbf{g}(\mathbf{X}_n)-\mathbf{g}(\mathbf{X})|>\epsilon \mid \mathbf{X}\not\in A\}P\{\mathbf{X}\not\in A\} \\
&\quad + P\{|\mathbf{g}(\mathbf{X}_n)-\mathbf{g}(\mathbf{X})|>\epsilon \mid \mathbf{X}\in A\}P\{\mathbf{X}\in A\} \\
&\leqslant P\{\mathbf{X}\not\in A\}+P\{|\mathbf{X}_n-\mathbf{X}|>\delta_\epsilon \mid \mathbf{X}\in A\}P\{\mathbf{X}\in A\} \\
&\leqslant P\{\mathbf{X}\not\in A\}+P\{|\mathbf{X}_n-\mathbf{X}|>\delta_\epsilon\} \\
&<\delta. \qquad \blacktriangle
\end{aligned}$$

Theorem 5.1.4 can be extended to functions that are continuous except on a set D, where $P\{\mathbf{X}\in D\}=0$. See, for example, Tucker (1967, p. 104).

Mann and Wald (1943b) demonstrated that the algebra of the common order relationships holds for order in probability. The following two theorems are similar to a paraphrase of Mann and Wald's result given by Pratt (1959). The proof follows more closely that of Mann and Wald, however.

Theorem 5.1.5. Let $\{\mathbf{X}_n\}$ be a sequence of k-dimensional random variables with elements $\{X_{jn}: j=1,2,\ldots,k\}$, and let $\{\mathbf{r}_n\}$ be a sequence of k-dimensional vectors with positive real elements $\{r_{jn}: j=1,2,\ldots,k\}$ such that

$$X_{jn}=O_p(r_{jn}), \qquad j=1,2,\ldots,t,$$
$$X_{jn}=o_p(r_{jn}), \qquad j=t+1,t+2,\ldots,k.$$

Let $g_n(\mathbf{x})$ be a sequence of real valued (Borel measurable) functions defined on k-dimensional Euclidian space and let $\{s_n\}$ be a sequence of positive real numbers. Let $\{\mathbf{a}_n\}$ be a nonrandom sequence of k-dimensional vectors. If

$$g_n(\mathbf{a}_n)=O(s_n)$$

for all sequences $\{\mathbf{a}_n\}$ such that

$$a_{jn} = O(r_{jn}), \qquad j=1,2,\ldots,t,$$

$$a_{jn} = o(r_{jn}), \qquad j=t+1,t+2,\ldots,k,$$

then

$$g_n(\mathbf{X}_n) = O_p(s_n).$$

Proof. Set $\epsilon > 0$. By assumption there exist real numbers $M_1, M_2, \ldots, M_t$ and sequences $\{M_{jn}\}$, $j=t+1,t+2,\ldots,k$, such that $\lim_{n\to\infty} M_{jn} = 0$ and

$$P\{|X_{jn}| \geqslant M_j r_{jn}\} < \frac{\epsilon}{k}, \qquad j=1,2,\ldots,t,$$

$$P\{|X_{jn}| \geqslant M_{jn} r_{jn}\} < \frac{\epsilon}{k}, \qquad j=t+1,t+2,\ldots,k$$

for all n. Let $\{A_n\}$ be a sequence of k-dimensional sets defined by

$$A_n = \{(y_1, y_2, \ldots, y_k) : -M_j r_{jn} \leqslant y_j \leqslant M_j r_{jn} \quad \text{for} \quad 1 \leqslant j \leqslant t$$
$$\text{and } -M_{jn} r_{jn} \leqslant y_j \leqslant M_{jn} r_{jn} \quad \text{for} \quad t+1 \leqslant j \leqslant k\}.$$

Then, for $\mathbf{a} \in A_n$, there exists an M such that $|g_n(\mathbf{a})| < M s_n$. Hence,

$$\frac{|g_n(\mathbf{X}_n)|}{s_n} < M$$

for all $\mathbf{X}_n$ contained in A_n, and the result follows, since the A_n were constructed so that $P\{\mathbf{X}_n \in A_n\} > 1 - \epsilon$ for all n. ▲

Theorem 5.1.6. If we replace

$$g_n(\mathbf{a}_n) = O(s_n)$$

by

$$g_n(\mathbf{a}_n) = o(s_n)$$

in the hypothesis of Theorem 5.1.5, we may replace

$$g_n(\mathbf{X}_n) = O_p(s_n)$$

by

$$g_n(\mathbf{X}_n) = o_p(s_n)$$

in the conclusion.

Proof. The set A_n is constructed exactly as in the proof of Theorem 5.1.5. There then exists a sequence $\{b_n\}$ such that $\lim_{n\to\infty} b_n = 0$, and

$$\frac{|g_n(\mathbf{a})|}{s_n} < b_n$$

for $\mathbf{a}$ contained in A_n. Therefore,

$$\frac{g_n(\mathbf{X}_n)}{s_n} < b_n$$

for all $\mathbf{X}_n$ contained in A_n and the result follows from the construction of the A_n. ▲

Corollary 5.1.5. Let $\{X_n\}$ be a sequence of scalar random variables such that

$$X_n = a + O_p(r_n),$$

where $r_n \to 0$ as $n \to \infty$. If $g(x)$ is a function with s continuous derivatives at $x = a$, then

$$g(X_n) = g(a) + g^{(1)}(a)(X_n - a)$$

$$+ \cdots + \frac{1}{(s-1)!} g^{(s-1)}(a)(X_n - a)^{s-1} + O_p(r_n^s),$$

where $g^{(j)}(a)$ is the jth derivative of $g(x)$ evaluated at $x = a$.

Proof. Since the statement holds for a sequence of real numbers, the result follows from Theorem 5.1.5. A direct proof can be obtained by expanding $g(x)$ in a Taylor series with remainder

$$\frac{g^{(s)}(b)(X_n - a)^s}{s!},$$

where b is between X_n and a. Since $g^{(s)}(x)$ is continuous at a, for n

sufficiently large, $g^{(s)}(b)$ is bounded in probability [i.e., $g^{(s)}(b) = O_p(1)$]. Therefore,

$$\frac{g^{(s)}(b)(X_n - a)^s}{s!} = O_p(r_n^s). \qquad \blacktriangle$$

Corollary 5.1.6. If

$$X_n = a + O_p(r_n)$$

in the hypothesis of Corollary 5.1.5 is replaced by

$$X_n = a + o_p(r_n),$$

then the remainder $O_p(r_n^s)$ is replaced by $o_p(r_n^s)$.

Proof. The proof is nearly identical to that of Corollary 5.1.5. ▲

Note that the condition on the derivative defining the remainder can be weakened. As we saw in the proof of Corollary 5.1.5, we need only that $g^{(s)}(b)$ is bounded in probability.

The corollaries generalize immediately to vector random variables. For example, let

$$\mathbf{X}_n = \mathbf{a} + o_p(r_n),$$

where $\mathbf{X}_n = (X_{1n}, X_{2n}, \ldots, X_{kn})'$, $\mathbf{a} = (a_1, a_2, \ldots, a_k)'$, and $r_n \to 0$ as $n \to \infty$. Let $g(\mathbf{x})$ be a real valued function defined on k-dimensional Euclidian space with continuous partial derivatives of order three at $\mathbf{a}$; then, for example,

$$g(\mathbf{X}_n) = g(\mathbf{a}) + \sum_{j=1}^{k} \frac{\partial g(\mathbf{a})}{\partial x_j}(X_{jn} - a_j)$$

$$+ \sum_{j=1}^{k} \sum_{i=1}^{k} \frac{1}{2!} \frac{\partial^2 g(\mathbf{a})}{\partial x_j \partial x_i}(X_{jn} - a_j)(X_{in} - a_i) + o_p(r_n^3),$$

where

$\dfrac{\partial g(\mathbf{a})}{\partial x_j}$ is the partial derivative of $g(\mathbf{x})$ with respect to x_j evaluated at $\mathbf{x} = \mathbf{a}$

and

$\dfrac{\partial^2 g(\mathbf{a})}{\partial x_j \partial x_i}$ is the second partial derivative of $g(\mathbf{x})$ with respect to x_j and x_i evaluated at $\mathbf{x} = \mathbf{a}$.

5.2. CONVERGENCE IN DISTRIBUTION

In the preceding section we discussed conditions under which a sequence of random variables converges in probability to a limit random variable. A second type of convergence important in statistics is the convergence of a sequence of distribution functions to a limit function. The classical example of such convergence is given by the central limit theorem wherein the sequence of distribution functions converges pointwise to the normal distribution function.

Definition 5.2.1. If $\{X_n\}$ is a sequence of random variables with distribution functions $\{F_{X_n}(x)\}$, then $\{X_n\}$ is said to converge in distribution (or in law) to the random variable X with distribution function $F_X(x)$, and we write $X_n \xrightarrow{\mathcal{L}} X$, if

$$\lim_{n\to\infty} F_{X_n}(x) = F_X(x)$$

at all x for which $F_X(x)$ is continuous.

Note that the sequence of distribution functions is converging to a function that is itself a distribution function. Some authors define this type of convergence by saying the sequence $\{F_{X_n}(x)\}$ converges *completely* to $F_X(x)$. Thus our symbolism

$$X_n \xrightarrow{\mathcal{L}} X$$

is understood to mean that $F_{X_n}(x)$ converges to the distribution function of the random variable X. The notation

$$F_{X_n}(x) \xrightarrow{C} F_X(x)$$

is also used.

Theorem 5.2.1. Let $\{X_n\}$ and $\{Y_n\}$ be sequences of random variables such that

$$p\lim(X_n - Y_n) = 0.$$

If there exists a random variable X such that

$$X_n \xrightarrow{\mathcal{L}} X,$$

then

$$Y_n \xrightarrow{\mathcal{L}} X.$$

Proof. Let W and Z be random variables with distribution functions $F_W(w)$ and $F_Z(z)$, respectively, and fix $\epsilon>0$ and $\delta>0$. We first show that

$$P\{|Z-W|>\epsilon\}\leqslant\delta$$

implies that

$$F_Z(z-\epsilon)-\delta\leqslant F_W(z)\leqslant F_Z(z+\epsilon)+\delta,$$

for all z. This result holds, since

$$\begin{aligned}F_Z(z-\epsilon)-F_W(z)&=P\{Z\leqslant z-\epsilon\}-P\{W\leqslant z\}\\&\leqslant P\{Z\leqslant z-\epsilon\}-P\{Z\leqslant z-\epsilon \text{ and } W\leqslant z\}\\&=P\{Z\leqslant z-\epsilon \text{ and } W>z\}\\&\leqslant P\{(W-Z)>\epsilon\}\\&\leqslant P\{|W-Z|>\epsilon\}\leqslant\delta,\end{aligned}$$

and, in a similar manner,

$$\begin{aligned}F_W(z)-F_Z(z+\epsilon)&=P\{W\leqslant z\}-P\{Z\leqslant z+\epsilon\}\\&\leqslant P\{W\leqslant z\}-P\{Z\leqslant z+\epsilon \text{ and } W\leqslant z\}\\&=P\{W\leqslant z \text{ and } Z>z+\epsilon\}\\&\leqslant P\{|Z-W|>\epsilon\}\leqslant\delta.\end{aligned}$$

Let x_0 be a continuity point for $F_X(x)$. Then, given $\delta>0$, there is an $\eta>0$ such that $|F_X(x)-F_X(x_0)|<\frac{1}{4}\delta$ for $|x-x_0|\leqslant\eta$ and $F_X(x)$ is continuous at $x_0-\eta$ and at $x_0+\eta$. Furthermore, for this δ and η, there is an N_1 such that, for $n>N_1$,

$$P\{|X_n-Y_n|>\eta\}<\frac{1}{2}\delta$$

and, therefore,

$$F_{X_n}(x-\eta)-\frac{1}{2}\delta\leqslant F_{Y_n}(x)\leqslant F_{X_n}(x+\eta)+\frac{1}{2}\delta$$

for all x. Also, there is an N_2 such that, for $n>N_2$,

$$|F_{X_n}(x_0-\eta)-F_X(x_0-\eta)|<\frac{1}{4}\delta$$

and

$$|F_{X_n}(x_0+\eta)-F_X(x_0+\eta)|<\frac{1}{4}\delta.$$

Therefore, given the continuity point x_0 and $\delta>0$, there is an $\eta>0$ and an $N=\max(N_1,N_2)$ such that for $n>N$,

$$\begin{aligned}F_X(x_0)-\delta &< F_X(x_0-\eta)-\frac{3}{4}\delta < F_{X_n}(x_0-\eta)-\frac{1}{2}\delta\\ &\leqslant F_{Y_n}(x_0)\leqslant F_{X_n}(x_0+\eta)+\frac{1}{2}\delta < F_X(x_0+\eta)+\frac{3}{4}\delta\\ &< F_X(x_0)+\delta.\end{aligned}$$ ▲

As a corollary we have the result that convergence in probability implies convergence in law.

Corollary 5.2.1.1. Let $\{X_n\}$ be a sequence of random variables. If there exists a random variable X such that $p\lim X_n=X$, then $X_n \xrightarrow{\mathfrak{L}} X$.

Corollary 5.2.1.2. Let $\{X_n\}$ and X be random variables such that $p\lim X_n=X$. If $g(x)$ is a continuous function, then the distribution of $g(X_n)$ converges to the distribution of $g(X)$.

Proof. This follows immediately since, by Theorem 5.1.4,

$$p\lim g(X_n)=g(X).$$ ▲

We state the following two important theorems without proof.

Theorem 5.2.2 (Helly-Bray). If $\{F_n(\mathbf{x})\}$ is a sequence of distribution functions over k-dimensional Euclidean space $\mathfrak{R}^{(k)}$ such that $F_n(\mathbf{x}) \xrightarrow{C} F(\mathbf{x})$, then

$$\int g(\mathbf{x})\,dF_n(\mathbf{x})\rightarrow\int g(\mathbf{x})\,dF(\mathbf{x}) \quad \text{as} \quad n\rightarrow\infty$$

for every bounded continuous function $g(\mathbf{x})$.

Theorem 5.2.3. Let $\{F_n(\mathbf{x})\}$ be a sequence of distribution functions over $\mathfrak{R}^{(k)}$ with corresponding characteristic functions $\{\varphi_n(\mathbf{u})\}$.

(i) If $F_n(\mathbf{x}) \xrightarrow{C} F(\mathbf{x})$, then $\varphi_n(\mathbf{u})\rightarrow\varphi(\mathbf{u})$ at all $\mathbf{u}\in\mathfrak{R}^{(k)}$, where $\varphi(\mathbf{u})$ is the characteristic function associated with $F(\mathbf{x})$.

(ii) Continuity Theorem. If $\varphi_n(\mathbf{u})$ converges pointwise to a function

$\varphi(\mathbf{u})$ that is continuous at $(0,0,\ldots,0)\in\mathcal{R}^{(k)}$, then $\varphi(\mathbf{u})$ is the characteristic function of a distribution function $F(\mathbf{x})$ and $F_n(\mathbf{x})\xrightarrow{C}F(\mathbf{x})$.

Theorem 5.2.4. Let $\{\mathbf{X}_n\}$ be a sequence of k-dimensional random variables with distribution functions $\{F_{\mathbf{X}_n}(\mathbf{x})\}$ such that $F_{\mathbf{X}_n}(\mathbf{x})\xrightarrow{C}F_{\mathbf{X}}(\mathbf{x})$ and let T be a continuous mapping from $\mathcal{R}^{(k)}$ to $\mathcal{R}^{(p)}$; then

$$F_{T(\mathbf{X}_n)}(\mathbf{y})\xrightarrow{C}F_{T(\mathbf{X})}(\mathbf{y}).$$

Proof. By the Helly-Bray theorem, the characteristic function of $T(\mathbf{X}_n)$ converges to the characteristic function of $T(\mathbf{X})$, and the result follows. ▲

Theorem 5.2.5. Let $\{\mathbf{X}_n\}$ and $\{\mathbf{Y}_n\}$ be two sequences of k-dimensional random variables such that $\mathbf{X}_n$ is independent of $\mathbf{Y}_n$ for all n. If there exist random variables $\mathbf{X}$ and $\mathbf{Y}$ such that $F_{\mathbf{X}_n}(\mathbf{x})\xrightarrow{C}F_{\mathbf{X}}(\mathbf{x})$ and $F_{\mathbf{Y}_n}(\mathbf{y})\xrightarrow{C}F_{\mathbf{Y}}(\mathbf{y})$, then

$$F_{\mathbf{X}_n\mathbf{Y}_n}(\mathbf{x},\mathbf{y})\xrightarrow{C}F_{\mathbf{X}}(\mathbf{x})F_{\mathbf{Y}}(\mathbf{y}).$$

Proof. The characteristic function of $(\mathbf{X}_n',\mathbf{Y}_n')'$ is given by

$$\int\int e^{i\mathbf{u}'\mathbf{x}}e^{i\mathbf{v}'\mathbf{y}}\,dF_{\mathbf{X}_n\mathbf{Y}_n}(\mathbf{x},\mathbf{y})=\int e^{i\mathbf{u}'\mathbf{x}}\,dF_{\mathbf{X}_n}(\mathbf{x})\int e^{i\mathbf{v}'\mathbf{y}}\,dF_{\mathbf{Y}_n}(\mathbf{y})$$
$$=\varphi_{\mathbf{X}_n}(\mathbf{u})\varphi_{\mathbf{Y}_n}(\mathbf{v}).$$

Now, by the Helly-Bray theorem, $\varphi_{\mathbf{X}_n}(\mathbf{u})\to\varphi_{\mathbf{X}}(\mathbf{u})$ and $\varphi_{\mathbf{Y}_n}(\mathbf{v})\to\varphi_{\mathbf{Y}}(\mathbf{v})$. Therefore,

$$\varphi_{\mathbf{XY}}(\mathbf{u},\mathbf{v})=\lim_{n\to\infty}\varphi_{\mathbf{X}_n}(\mathbf{u})\varphi_{\mathbf{Y}_n}(\mathbf{v})=\varphi_{\mathbf{X}}(\mathbf{u})\varphi_{\mathbf{y}}(\mathbf{v}).$$

By the continuity theorem, this implies that $F_{\mathbf{X}_n\mathbf{Y}_n}(\mathbf{x},\mathbf{y})\xrightarrow{C}F_{\mathbf{XY}}(\mathbf{x},\mathbf{y})$, where $F_{\mathbf{XY}}(\mathbf{x},\mathbf{y})$ is the distribution function of independent random variables associated with the characteristic function $\varphi_{\mathbf{XY}}(\mathbf{u},\mathbf{v})$. ▲

From Corollary 5.2.1.1, we know that convergence in probability implies convergence in law. For the special case wherein a sequence of random variables converges in law to a constant random variable, the converse is also true.

Lemma 5.2.1. Let $\{\mathbf{Y}_n\}$ be a sequence of p-dimensional random variables with corresponding distribution functions $\{F_{\mathbf{Y}_n}(\mathbf{y})\}$. Let $\mathbf{Y}$ be a p-dimensional random variable with distribution function $F_{\mathbf{Y}}(\mathbf{y})$ such that $P\{\mathbf{Y}=\mathbf{b}\}=1$, $\mathbf{b}$ is a constant vector, and

$$F_{\mathbf{Y}_n}(\mathbf{y}) \xrightarrow{C} F_{\mathbf{Y}}(\mathbf{y}).$$

Then, given $\epsilon>0$, there exists an N such that, for $n>N$,

$$P\{|\mathbf{Y}_n-\mathbf{b}|\geqslant\epsilon\}<\epsilon.$$

Proof. Let $B=\{\mathbf{y}: y_1>b_1-\epsilon/p, y_2>b_2-\epsilon/p,\ldots,y_p>b_p-\epsilon/p\}$. Then $F_{\mathbf{Y}}(\mathbf{y})=0$ on the complement of B. Fix $\epsilon>0$. As $F_{\mathbf{Y}_n}(\mathbf{y}) \xrightarrow{C} F_{\mathbf{Y}}(\mathbf{y})$, there exists an N_0 such that, for $n>N_0$,

$$F_{\mathbf{Y}_n}(\mathbf{g}_1)=F_{\mathbf{Y}_n}\left(b_1-\frac{\epsilon}{p},b_2+\frac{\epsilon}{p},\ldots,b_p+\frac{\epsilon}{p}\right)<\frac{\epsilon}{2p},$$

$$F_{\mathbf{Y}_n}(\mathbf{g}_2)=F_{\mathbf{Y}_n}\left(b_1+\frac{\epsilon}{p},b_2-\frac{\epsilon}{p},\ldots,b_p+\frac{\epsilon}{p}\right)<\frac{\epsilon}{2p},$$

$$\vdots$$

$$F_{\mathbf{Y}_n}(\mathbf{g}_p)=F_{\mathbf{Y}_n}\left(b_1+\frac{\epsilon}{p},b_2+\frac{\epsilon}{p},\ldots,b_p-\frac{\epsilon}{p}\right)<\frac{\epsilon}{2p}.$$

There also exists an N_1 such that, for $n>N_1$, $1-F_{\mathbf{Y}_n}(b_1+\epsilon/p, b_2+\epsilon/p,\ldots,b_p+\epsilon/p)<\epsilon/2$. Therefore, for $n>\max(N_0,N_1)$,

$$P\left\{b_1-\frac{\epsilon}{p}\leqslant y_1\leqslant b_1+\frac{\epsilon}{p}, b_2-\frac{\epsilon}{p}\leqslant y_2\leqslant b_2+\frac{\epsilon}{p},\ldots,\right.$$

$$\left. b_p-\frac{\epsilon}{p}\leqslant y_p\leqslant b_p+\frac{\epsilon}{p}\right\}$$

$$\geqslant F_{\mathbf{Y}_n}\left(b_1+\frac{\epsilon}{p},b_2+\frac{\epsilon}{p},\ldots,b_p+\frac{\epsilon}{p}\right)-\sum_{i=1}^{p}F_{\mathbf{Y}_n}(\mathbf{g}_i)\geqslant 1-\epsilon,$$

and it follows that

$$P\{|\mathbf{Y}_n-\mathbf{b}|\geqslant\epsilon\}<\epsilon. \qquad \blacktriangle$$

Theorem 5.2.6. Let $\{(\mathbf{X}_n',\mathbf{Y}_n')'\}$ be a sequence of $(k+p)$-dimensional

random variables where $\mathbf{X}_n$ is k-dimensional. Let the sequence of joint distribution functions be denoted by $\{F_{\mathbf{X}_n\mathbf{Y}_n}(\mathbf{x},\mathbf{y})\}$ and the sequences of marginal distribution functions by $\{F_{\mathbf{X}_n}(\mathbf{x})\}$ and $\{F_{\mathbf{Y}_n}(\mathbf{y})\}$. If there exists a k-dimensional random variable $\mathbf{X}$ and a p-dimensional random variable $\mathbf{Y}$ such that $F_{\mathbf{X}_n}(\mathbf{x}) \xrightarrow{C} F_{\mathbf{X}}(\mathbf{x})$ and $F_{\mathbf{Y}_n}(\mathbf{y}) \xrightarrow{C} F_{\mathbf{Y}}(\mathbf{y})$, where $P\{\mathbf{Y}=\mathbf{b}\}=1$ and $\mathbf{b}=(b_1,b_2,\ldots,b_p)'$ is a constant vector, then

$$F_{\mathbf{X}_n\mathbf{Y}_n}(\mathbf{x},\mathbf{y}) \xrightarrow{C} F_{\mathbf{XY}}(\mathbf{x},\mathbf{y}).$$

Proof. Now $P\{\mathbf{Y}=\mathbf{b}\}=1$ implies that

$$F_{\mathbf{X}}(\mathbf{x}) = F_{\mathbf{XY}}(\mathbf{x},\mathbf{b}),$$

that $F_{\mathbf{XY}}(\mathbf{x},\mathbf{y})=0$ if any element of $\mathbf{y}$ is less than the corresponding element of $\mathbf{b}$, and that $F_{\mathbf{XY}}(\mathbf{x},\mathbf{y})=F_{\mathbf{X}}(\mathbf{x})$ if every element of $\mathbf{y}$ is greater than or equal to the corresponding element of $\mathbf{b}$. Fix $\epsilon>0$ and consider a point $(\mathbf{x}_0,\mathbf{y}_0)$ where at least one element of $\mathbf{y}_0$ is less than the corresponding element of $\mathbf{b}$ by an amount ϵ. Then $F_{\mathbf{XY}}(\mathbf{x}_0,\mathbf{y}_0)=0$. However, there is an N_0 such that, for $n>N_0$,

$$F_{\mathbf{X}_n\mathbf{Y}_n}(\mathbf{x}_0,\mathbf{y}_0)<\frac{\epsilon}{2}$$

by Lemma 5.2.1. Let $(\mathbf{x}_0,\mathbf{y}_1)$ be a continuity point of $F_{\mathbf{XY}}(\mathbf{x},\mathbf{y})$ where every element of $\mathbf{y}_1$ exceeds the corresponding element of $\mathbf{b}$ by $\epsilon/p^{1/2}>0$. Because $F_{\mathbf{X}_n}(\mathbf{x}) \xrightarrow{C} F_{\mathbf{X}}(\mathbf{x})$ and $F_{\mathbf{Y}_n}(\mathbf{y}) \xrightarrow{C} F_{\mathbf{Y}}(\mathbf{y})$, we can choose N_1 such that, for $n \geqslant N_1$, $|F_{\mathbf{X}_n}(\mathbf{x}_0)-F_{\mathbf{X}}(\mathbf{x}_0)|<\epsilon/2$ and $|F_{\mathbf{Y}_n}(\mathbf{y}_1)-F_{\mathbf{Y}}(\mathbf{y}_1)|<\epsilon/2$. Hence,

$$\begin{aligned}|F_{\mathbf{X}_n\mathbf{Y}_n}(\mathbf{x}_0,\mathbf{y}_1)-F_{\mathbf{XY}}(\mathbf{x}_0,\mathbf{y}_1)| &\leqslant |F_{\mathbf{X}_n\mathbf{Y}_n}(\mathbf{x}_0,\mathbf{y}_1)-F_{\mathbf{X}_n}(\mathbf{x}_0)| \\ &\quad + |F_{\mathbf{X}_n}(\mathbf{x}_0)-F_{\mathbf{X}}(\mathbf{x}_0)| \\ &< 1-F_{\mathbf{Y}_n}(\mathbf{y}_1)+\frac{\epsilon}{2}<\epsilon.\end{aligned}$$

▲

Utilizing Theorems 5.2.4 and 5.2.6, we obtain the following corollary.

Corollary 5.2.6.1. Let $\{\mathbf{X}_n\}$ and $\{\mathbf{Y}_n\}$ be two sequences of k-dimensional random variables. If there exists a k-dimensional random variable $\mathbf{Y}$

and a fixed vector $\mathbf{b}$ such that $\mathbf{Y}_n \xrightarrow{\mathcal{L}} \mathbf{Y}$ and $\mathbf{X}_n \xrightarrow{\mathcal{L}} \mathbf{b}$; then,

(i) $\mathbf{X}_n + \mathbf{Y}_n \xrightarrow{\mathcal{L}} \mathbf{b} + \mathbf{Y}$

(ii) $\mathbf{X}_n'\mathbf{Y}_n \xrightarrow{\mathcal{L}} \mathbf{b}'\mathbf{Y}$.

Corollary 5.2.6.2. Let $\{\mathbf{Y}_n\}$ be a sequence of k-dimensional random variables and let $\{\mathbf{A}_n\}$ be a sequence of $k \times k$ random matrices. If there exists a random vector $\mathbf{Y}$ and a fixed nonsingular matrix $\mathbf{A}$ such that $\mathbf{Y}_n \xrightarrow{\mathcal{L}} \mathbf{Y}$, $\mathbf{A}_n \xrightarrow{\mathcal{L}} \mathbf{A}$, then

$$\mathbf{A}_n^{-1}\mathbf{Y}_n \xrightarrow{\mathcal{L}} \mathbf{A}^{-1}\mathbf{Y}.$$

5.3 CENTRAL LIMIT THEOREMS

The exact distributions of many statistics encountered in practice have not been obtained. Fortunately, many statistics in the class of continuous functions of means or of sample moments converge in distribution to normal random variables. We give without proof the following central limit theorem.

Theorem 5.3.1 (Lindeberg central limit theorem). Let $\{Z_t\colon t = 1, 2, \ldots\}$ be a sequence of independent random variables with distribution functions $\{F_t(z)\}$. Let $E\{Z_t\} = \mu_t$, $E\{(Z_t - \mu_t)^2\} = \sigma_t^2$, and assume

$$\lim_{n\to\infty} V_n^{-1} \sum_{t=1}^{n} \int_{|z-\mu_t| > \epsilon V_n^{1/2}} (z - \mu_t)^2 \, dF_t(z) = 0 \tag{5.3.1}$$

for all $\epsilon > 0$, where $V_n = \Sigma_{t=1}^n \sigma_t^2$. Then,

$$V_n^{-1/2} \sum_{t=1}^{n} (Z_t - \mu_t) \xrightarrow{\mathcal{L}} N(0, 1),$$

where $N(0, 1)$ denotes the normal distribution with mean zero and variance one.

A form of the central limit theorem whose conditions are often more easily verified is the following theorem.

Theorem 5.3.2 (Liapounov central limit theorem). Let $\{Z_t\colon t = 1, 2, \ldots\}$ be a sequence of independent random variables with distribution functions

$\{F_t(z)\}$. Let $E\{Z_t\}=\mu_t$, $E\{(Z_t-\mu_t)^2\}=\sigma_t^2$, and $V_n=\sum_{t=1}^n\sigma_t^2$. If

$$\lim_{n\to\infty}\frac{\sum_{t=1}^{n}\int|z-\mu_t|^{2+\delta}\,dF_t(z)}{V_n^{1+\delta/2}}=0$$

for some $\delta>0$, then

$$V_n^{-1/2}\sum_{t=1}^{n}(Z_t-\mu_t)\xrightarrow{\mathcal{L}}N(0,1).$$

Proof. Let $\delta>0$ and $\epsilon>0$, and define the set A_ϵ by

$$A_\epsilon=\left\{z\colon |z-\mu_t|>\epsilon V_n^{1/2}\right\}.$$

Then

$$\frac{1}{V_n}\sum_{t=1}^{n}\int_{A_\epsilon}(z-\mu_t)^2\,dF_t(z)\leqslant\frac{1}{V_n\left(\epsilon V_n^{1/2}\right)^{\delta}}\sum_{t=1}^{n}\int_{A_\epsilon}|z-\mu_t|^{2+\delta}\,dF_t(z)$$

$$\leqslant\frac{1}{V_n^{1+\delta/2}\epsilon^{\delta}}\sum_{t=1}^{n}\int|z-\mu_t|^{2+\delta}\,dF_t(z)$$

which, by hypothesis, goes to zero as $n\to\infty$. Therefore, the condition on the $2+\delta$ moment implies the condition of the Lindeberg theorem. ▲

The reader is referred to the texts of Tucker (1967), Gnedenko (1967), and Loève (1963) for discussions of these theorems.

For a proof of the following extension of the central limit theorems to the multivariate case, see Varadarajan (1958).

Theorem 5.3.3. Let $\{\mathbf{Z}_n\colon n=1,2,\dots\}$ be a sequence of k-dimensional random variables with distribution functions $\{F_{\mathbf{Z}_n}(\mathbf{z})\}$. Let $F_{X_n}(x)$ be the distribution function of $X_n=\boldsymbol{\lambda}'\mathbf{Z}_n$, where $\boldsymbol{\lambda}$ is a fixed vector. A necessary and sufficient condition for $F_{\mathbf{Z}_n}(\mathbf{z})$ to converge to the k-variate distribution function $F_{\mathbf{Z}}(\mathbf{z})$ is that $F_{X_n}(x)$ converge to a limit for each $\boldsymbol{\lambda}$.

In most of our applications of Theorem 5.3.3, each $F_{X_n}(x)$ will be converging to a normal distribution function and, hence, the vector random variable $\mathbf{Z}_n$ will converge in distribution to a multivariate normal.

5.4. APPROXIMATING A SEQUENCE OF EXPECTATIONS

Taylor expansions and the order in probability concepts introduced in Section 5.1 are very important in investigating the limiting behavior of sample statistics. Care must be taken, however, in understanding the meaning of statements about such behavior. To illustrate, consider the sequence $\{\bar{x}_n^{-1}\}$, where $\bar{x}_n$ is the mean of n normal independent $(\mu, 1)$ random variables, $\mu \neq 0$. By Corollary 5.1.5, we may write

$$\bar{x}_n^{-1} = \mu^{-1} - \mu^{-2}(\bar{x}_n - \mu) + \mu^{-3}(\bar{x}_n - \mu)^2 - \mu^{-4}(\bar{x}_n - \mu)^3 + O_p(n^{-2}) \quad (5.4.1)$$

and

$$n^{1/2}(\bar{x}_n^{-1} - \mu^{-1}) = -n^{1/2}\mu^{-2}(\bar{x}_n - \mu) + O_p(n^{-1/2}).$$

It follows that

$$p\lim n^{1/2}\left[(\bar{x}_n^{-1} - \mu^{-1}) + \mu^{-2}(\bar{x}_n - \mu)\right] = 0.$$

Therefore, by Theorem 5.2.1, the limiting distribution of $n^{1/2}(\bar{x}_n^{-1} - \mu^{-1})$ is the same as the limiting distribution of $-n^{1/2}\mu^{-2}(\bar{x}_n - \mu)$. The distribution of $-n^{1/2}\mu^{-2}(\bar{x}_n - \mu)$ is $N(0, \mu^{-4})$ for all n and it follows that the limiting distribution of $n^{1/2}(\bar{x}_n^{-1} - \mu^{-1})$ is $N(0, \mu^{-4})$.

On the other hand, it can be demonstrated that $E\{\bar{x}_n^{-1}\}$ exists for *no* finite n. Since the expectation of $(\bar{x}_n^{-1} - \mu^{-1})$ is not defined, it is clear that one cannot speak of the sequence of expectations $\{E\{\bar{x}_n^{-1}\}\}$, and it is incorrect to say that $E\{\bar{x}_n^{-1}\}$ converges to μ^{-1}.

The example illustrates that a random variable, Y_n, may converge in probability and hence in distribution to a random variable, Y, where Y possesses finite moments even though $E\{Y_n\}$ is not defined. If we know that Y_n has finite moments of order $r > 1$, we may be able to determine that the sequence of expectations differs from a given sequence by an amount of specified order. The conditions required to permit such statements are typically more stringent than those required to obtain convergence in distribution. In this section we investigate these conditions and develop approximations to the expectation of functions of mean or "meanlike" statistics. In preparation for that investigation we consider the expectations of integer powers of sample means of random variables with zero population means. Let $(\bar{x}_n, \bar{y}_n, \bar{z}_n, \bar{w}_n)'$ be a vector of sample means computed from a random sample selected from a distribution function

with zero mean vector and finite fourth moments. Then,

$$E\{\bar{x}_n\bar{y}_n\bar{z}_n\} = \frac{1}{n^3}E\left\{\sum_i\sum_j\sum_k X_iY_jZ_k\right\}$$

$$= \frac{1}{n^3}E\left\{\sum_i X_iY_iZ_i + \sum_{i\neq j} X_iY_jZ_j + \sum_{i\neq j} X_jY_iZ_j + \sum_{i\neq j} X_jY_jZ_i + \sum_{i\neq j\neq k} X_iY_jZ_k\right\}$$

$$= \frac{1}{n^2}E\{XYZ\},$$

where $E\{XYZ\}$ is the expectation of the product of the original random variables. Similarly,

$$E\{\bar{x}_n\bar{y}_n\bar{z}_n\bar{w}_n\} = \frac{1}{n^4}E\left\{\sum_i\sum_j\sum_k\sum_m X_iY_jZ_kW_m\right\}$$

$$= \frac{1}{n^4}E\left\{\sum_i X_iY_iZ_iW_i + \sum_{i\neq j}(X_iY_jZ_jW_j + X_jY_iZ_jW_j + X_jY_jZ_iW_j + X_jY_jZ_jW_i) + \sum_{i\neq j}(X_iY_iZ_jW_j + X_iY_jZ_iW_j + X_iY_jZ_jW_i) + \sum_{i\neq j\neq k}(X_iY_iZ_jW_k + X_iY_jZ_iW_k + X_iY_jZ_kW_i + X_jY_iZ_iW_k + X_jY_iZ_kW_i + X_jY_kZ_iW_i) + \sum_{i\neq j\neq k\neq m} X_iY_jZ_kW_m\right\}$$

$$= \frac{1}{n^3}E\{XYZW\} + \frac{n-1}{n^3}(\sigma_{xy}\sigma_{zw} + \sigma_{xz}\sigma_{yw} + \sigma_{xw}\sigma_{yz}),$$

where σ_{xy} is the covariance between X and Y, σ_{zw} is the covariance between Z and W, and so forth.

We note that the expectation of a product of either three or four means is $O(n^{-2})$. This is an example of the general result that we state as a

theorem. Our proof follows closely that of Hansen, Hurwitz, and Madow (1953).

Theorem 5.4.1. Let $\bar{\mathbf{x}}_n = (\bar{x}_{1n}, \bar{x}_{2n}, \ldots, \bar{x}_{mn})'$ be the mean of a random sample of n vector random variables selected from a distribution function with mean vector zero and finite Bth moment. Consider the sequence $\{\bar{\mathbf{x}}_n\}_{n=1}^{\infty}$ and let $b_1, b_2, \ldots, b_m$ be nonnegative integers such that $B = \sum_{i=1}^{m} b_i$. Then,

$$E\{\bar{x}_{1n}^{b_1}\bar{x}_{2n}^{b_2}\ldots\bar{x}_{mn}^{b_m}\} = O(n^{-B/2}), \quad \text{if } B \text{ is even}$$

$$= O(n^{-(B+1)/2}), \qquad \text{if } B \text{ is odd.}$$

Proof. Now $E\{\bar{x}_{1n}^{b_1}\bar{x}_{2n}^{b_2}\ldots\bar{x}_{mn}^{b_m}\}$ can be expanded into a sum of terms such as

$$n^{-B}X_{1i_1}X_{1i_2}\ldots X_{1i_{b_1}}X_{2i_{b_1+1}}X_{2i_{b_1+2}}\ldots X_{2i_{b_1+b_2}}\ldots X_{mi_{B-b_m+1}}X_{mi_{B-b_m+2}}\ldots X_{mi_B}.$$

If there is a subscript matched by no other subscript the expected value of the product is zero. (Recall that in the four variable case this included terms of the form $X_iY_jZ_jW_j$, $X_iY_jZ_kZ_k$ and $X_iY_jZ_kZ_r$.) If every subscript agrees with at least one other subscript, we group the terms with common subscripts to obtain, say, H groups. The expected value is then the product of the H expected values. The sum contains $n(n-1)\ldots(n-H+1)$ terms for a particular configuration of H different subscripts. The order of $n^{-B}[n(n-1)\ldots(n-H+1)]$ is n^{-B+H} and will be maximized if we choose H as large as possible. If B is even, the largest H that gives a nonzero expectation is $B/2$, in which case we have $B/2$ groups, each containing two indexes. If B is odd the largest H that gives a nonzero expectation is $(B-1)/2$, in which case we have $(B-1)/2-1$ groups of two and one group of three. ▲

The following lemma is used in the proof of the principal results of this section.

Lemma 5.4.1. Let $\{\mathbf{X}_n\}$ be a sequence of k-dimensional random variables with corresponding distribution functions $\{F_n(\mathbf{x})\}$ such that

$$\int |x_i - \mu_i|^s \, dF_n(\mathbf{x}) = O(a_n^s), \qquad i = 1, 2, \ldots, k,$$

where the integral is over $\mathcal{R}^{(k)}$, $a_n > 0$, s is a positive integer, $E\{\mathbf{X}_n\} = \boldsymbol{\mu} = (\mu_1, \mu_2, \ldots, \mu_k)'$, $|\mathbf{x}| = [\sum_{i=1}^{k} x_i^2]^{1/2}$ is the Euclidean norm, and

$$\lim_{n\to\infty} a_n = 0.$$

Then,

$$\int |x_1-\mu_1|^{p_1}|x_2-\mu_2|^{p_2}\dots|x_k-\mu_k|^{p_k}dF_n(\mathbf{x})=O(1),$$

where the p_i, $i=1,2,\dots,k$, are nonnegative real numbers satisfying

$$\sum_{i=1}^{k} p_i \leqslant s.$$

Proof. Without loss of generality, set all $\mu_i=0$. Define $A=[-1,1]$, and let $I_A(x)$ be the indicator function with value one for $x\in A$ and zero for $x\notin A$. Then, for $0\leqslant q\leqslant s$,

$$|x_i|^q\leqslant I_A(x_i)+|x_i|^s,$$

so that

$$\int |x_i|^q dF_n(\mathbf{x})\leqslant\int[I_A(x_i)+|x_i|^s]dF_n(\mathbf{x})\leqslant 1+O(a_n^s)=O(1),$$

where the integrals are over $\mathfrak{R}^{(k)}$. By the Hölder inequality,

$$\int |x_1|^{p_1}|x_2|^{p_2}\dots|x_k|^{p_k}dF_n(\mathbf{x})$$
$$\leqslant\left[\int|x_1|^r dF_n(\mathbf{x})\right]^{p_1/r}\left[\int|x_2|^r dF_n(\mathbf{x})\right]^{p_2/r}\dots\left[\int|x_k|^r dF_n(\mathbf{x})\right]^{p_k/r}$$
$$=O(1),$$

where

$$r=\sum_{i=1}^{k} p_i. \qquad \blacktriangle$$

Theorem 5.4.2. Let $\{X_n\}$ be a sequence of real valued random variables with corresponding distribution functions $\{F_n(x)\}$, and let $\{f_n(x)\}$ be a sequence of real valued functions. Assume that for some positive integers s and N_0:

(i) $\int|x-\mu|^{2s}dF_n(x)=a_n^{2s}$, where $a_n\to 0$ as $n\to\infty$.
(ii) $\int|f_n(x)|^2 dF_n(x)=O(1)$.
(iii) $f_n^{(s)}(x)$ is continuous in x over a closed and bounded interval S

for n greater than N_0, where $f_n^{(j)}(x)$ denotes the jth derivative of $f_n(x)$ evaluated at x and $f_n^{(0)}(x) \equiv f_n(x)$.

(iv) μ is an interior point of S.

(v) There is a K such that, for $n > N_0$,

$$|f_n^{(s)}(x)| \leq K, \qquad \text{for all } x \in S,$$

and

$$|f_n^{(r)}(\mu)| \leq K, \qquad \text{for } r = 0, 1, \ldots, s-1.$$

Then,

$$\int f_n(x)\, dF_n(x) = f_n(\mu) + Q(n,s) + O(a_n^s),$$

where

$$\begin{aligned} Q(n,s) &= 0, & s &= 1 \\ &= \sum_{j=1}^{s-1} \frac{1}{j!} f_n^{(j)}(\mu) \int (x-\mu)^j\, dF_n(x), & s &> 1. \end{aligned}$$

Proof. See the proof of Theorem 5.4.3. ▲

Theorem 5.4.3. Let $\{\mathbf{X}_n\}$ be a sequence of k-dimensional random variables with corresponding distribution functions $\{F_n(\mathbf{x})\}$ and let $\{f_n(\mathbf{x})\}$ be a sequence of functions mapping $\mathcal{R}^{(k)}$ into $\mathcal{R}$. Let $\delta \in (0, \infty)$ and define $\alpha = \delta^{-1}(1+\delta)$. Assume that for some positive integers s and N_0:

(i) $\int |\mathbf{x} - \boldsymbol{\mu}|^{\alpha s}\, dF_n(\mathbf{x}) = a_n^{\alpha s}$, where $a_n \to 0$ as $n \to \infty$.

(ii) $\int |f_n(\mathbf{x})|^{1+\delta}\, dF_n(\mathbf{x}) = O(1)$.

(iii) $f_n^{(i_1, \ldots, i_s)}(\mathbf{x})$ is continuous in $\mathbf{x}$ over a closed and bounded sphere S for all n greater than N_0, where

$$f_n^{(i_1, \ldots, i_r)}(\mathbf{x}_0) = \frac{\partial^r}{\partial x_{i_1} \cdots \partial x_{i_r}} f_n(\mathbf{x}) \bigg|_{\mathbf{x} = \mathbf{x}_0}.$$

(iv) $\boldsymbol{\mu}$ is an interior point of S.

(v) There is a finite number K such that, for $n > N_0$,

$$\left| f_n^{(i_1, \ldots, i_s)}(\mathbf{x}) \right| \leq K, \qquad \text{for all } \mathbf{x} \in S,$$

$$\left| f_n^{(i_1, \ldots, i_r)}(\boldsymbol{\mu}) \right| \leq K, \qquad \text{for } r = 1, 2, \ldots, s-1,$$

and

$$|f_n(\mu)| \leq K.$$

Then,

$$\int f_n(\mathbf{x})\,dF_n(\mathbf{x}) = f_n(\mu) + \sum_{j=1}^{s-1} \frac{1}{j!} \int D^j f_n(\mu)(\mathbf{x}-\mu)^j\,dF_n(\mathbf{x}) + O(a_n^s),$$

where

$$D^r f_n(\mu)(\mathbf{x}-\mu)^r = \sum_{i_1=1}^{k} \sum_{i_2=1}^{k} \cdots \sum_{i_r=1}^{k} f_n^{(i_1,\ldots,i_r)}(\mu) \prod_{j=1}^{r} (x_{i_j} - \mu_{i_j})$$

and, for $s=1$, it is understood that

$$\int f_n(\mathbf{x})\,dF_n(\mathbf{x}) = f_n(\mu) + O(a_n).$$

The result also holds if we replace (ii) with the condition that the $f_n(\mathbf{x})$ are uniformly bounded for n sufficiently large and assume that (i), (iii), (iv), and (v) hold for $\alpha = 1$.

Proof. We consider only those n greater than N_0. By Taylor's theorem there is a sequence of functions $\{\mathbf{Y}_n\}$ mapping S into S such that

$$f_n(\mathbf{x}) = f_n(\mu) + I_S(\mathbf{x}) \sum_{j=1}^{s-1} \frac{1}{j!} D^j f_n(\mu)(\mathbf{x}-\mu)^j + R_n(\mathbf{x}),$$

where

$$I_S(\mathbf{x}) = \begin{cases} 1, & \text{if } \mathbf{x} \in S \\ 0, & \text{otherwise} \end{cases}$$

and

$$R_n(\mathbf{x}) = \begin{cases} (s!)^{-1} D^s f_n(\mathbf{Y}_n(\mathbf{x}))(\mathbf{x}-\mu)^s, & \mathbf{x} \in S \\ f_n(\mathbf{x}) - f_n(\mu), & \text{otherwise.} \end{cases}$$

For $\mathbf{x} \in S$ we have $\mathbf{Y}_n(\mathbf{x}) \in S$ so that

$$|R_n(\mathbf{x})| \leqslant (s!)^{-1} \sum_{i_1=1}^{k} \cdots \sum_{i_s=1}^{k} \left\{ \left| f_n^{(i_1,\ldots,i_s)}(\mathbf{Y}_n(\mathbf{x})) \right| |x_{i_1} - \mu_{i_1}| \cdots |x_{i_s} - \mu_{i_s}| \right\}$$

$$\leqslant (s!)^{-1} \sum_{i_1=1}^{k} \cdots \sum_{i_s=1}^{k} K |\mathbf{x} - \boldsymbol{\mu}|^s$$

$$= (s!)^{-1} K k^s |\mathbf{x} - \boldsymbol{\mu}|^s.$$

Thus,

$$\int_S |R_n(\mathbf{x})| \, dF_n(\mathbf{x}) \leqslant (s!)^{-1} K k^s \int_S |\mathbf{x} - \boldsymbol{\mu}|^s \, dF_n(\mathbf{x})$$
$$= O(a_n^s).$$

Now, for finite δ and letting $\tilde{S}$ denote the complement of S,

$$\int_{\tilde{S}} |R_n(\mathbf{x})| \, dF_n(\mathbf{x})$$

$$= \int I_{\tilde{S}}(\mathbf{x}) |f_n(\mathbf{x}) - f_n(\boldsymbol{\mu})| \, dF_n(\mathbf{x})$$

$$\leqslant \left[\int |f_n(\mathbf{x}) - f_n(\boldsymbol{\mu})|^{1+\delta} \, dF_n(\mathbf{x}) \right]^{1/(1+\delta)} \left[\int I_{\tilde{S}}(\mathbf{x}) \, dF_n(\mathbf{x}) \right]^{\delta/(1+\delta)}$$

by Hölder's inequality. By Theorem 5.1.1 (Chebyshev's inequality),

$$\int I_{\tilde{S}}(\mathbf{x}) \, dF_n(\mathbf{x}) \leqslant M \int |\mathbf{x} - \boldsymbol{\mu}|^{\alpha s} \, dF_n(\mathbf{x}) = O(a_n^{\alpha s}), \quad \text{for some } M > 0.$$

Therefore, for $\delta \in (0, \infty)$,

$$\int |R_n(\mathbf{x})| \, dF_n(\mathbf{x}) = O(a_n^s).$$

This result also holds for $\alpha = 1$ and $f_n(\mathbf{x})$ uniformly bounded (by $K^\dagger$, say), because then,

$$\int I_{\tilde{S}}(\mathbf{x}) |f_n(\mathbf{x}) - f_n(\boldsymbol{\mu})| \, dF_n(\mathbf{x}) \leqslant (2K^\dagger) \int I_{\tilde{S}}(\mathbf{x}) \, dF_n(\mathbf{x}) = O(a_n^s).$$

We now have that, for $s>1$,

$$\int f_n(\mathbf{x})\,dF_n(\mathbf{x}) = f_n(\boldsymbol{\mu}) + \int \sum_{j=1}^{s-1} (j!)^{-1} D^j f_n(\boldsymbol{\mu})(\mathbf{x}-\boldsymbol{\mu})^j\,dF_n(\mathbf{x})$$

$$-\int_{\tilde{S}} \sum_{j=1}^{s-1} (j!)^{-1} D^j f_n(\boldsymbol{\mu})(\mathbf{x}-\boldsymbol{\mu})^j\,dF_n(\mathbf{x}) + O(a_n^s).$$

However,

$$\int_{\tilde{S}} \sum_{j=1}^{s-1} (j!)^{-1} D^j f_n(\boldsymbol{\mu})(\mathbf{x}-\boldsymbol{\mu})^j\,dF_n(\mathbf{x})$$

$$= \int_{\tilde{S}} \sum_{j=1}^{s-1} (j!)^{-1} \sum_{i_1=1}^{k} \cdots \sum_{i_j=1}^{k} f_n^{(i_1,\ldots,i_j)}(\boldsymbol{\mu}) \prod_{r=1}^{j} (x_{i_r}-\mu_{i_r})\,dF_n(\mathbf{x})$$

$$\leqslant \int_{\tilde{S}} Kk^s\left[\,|\mathbf{x}-\boldsymbol{\mu}|^s+1\,\right] dF_n(\mathbf{x})$$

$$= O(a_n^s). \qquad \blacktriangle$$

Theorems 5.4.2 and 5.4.3 require that $\int |f_n(\mathbf{x})|^{1+\delta}\,dF_n(\mathbf{x})$ be bounded for all n sufficiently large. The following theorem gives sufficient conditions for a sequence of integrals $\int |f_n(\mathbf{x})|\,dF_n(\mathbf{x})$ to be $O(1)$.

Theorem 5.4.4. Let $\{f_n(\mathbf{x})\}$ be a sequence of real valued (measurable) functions and let $\{\mathbf{X}_n\}$ be a sequence of k-dimensional random variables with corresponding distribution functions $\{F_n(\mathbf{x})\}$. Assume that:

(i) $|f_n(\mathbf{x})| \leqslant K_1$ for $\mathbf{x}\in\bar{S}$, where S is a bounded open set containing μ, $\bar{S}$ is the closure of S, and K_1 is a finite constant.

(ii) $|f_n(\mathbf{x})| \leqslant Y(\mathbf{x})n^p$ for some $p>0$ and for a function $Y(\cdot)$ such that $\int |Y(\mathbf{x})|^\gamma\,dF_n(\mathbf{x}) = O(1)$ for some γ, $1<\gamma<\infty$.

(iii) $\int |\mathbf{x}-\boldsymbol{\mu}|^r\,dF_n(\mathbf{x}) = O(n^{-\eta p})$ for a positive integer r and an η such that $1/\eta + 1/\gamma = 1$.

Then

$$\int |f_n(\mathbf{x})|\,dF_n(\mathbf{x}) = O(1).$$

The result also holds for $\eta=1$ given that (i), (iii), and $|f_n(\mathbf{x})| \leqslant K_2 n^p$ hold for all $\mathbf{x}$, some $p>0$, and K_2 a finite constant.

Proof. Let $\delta > 0$ be such that

$$A = \{\mathbf{x}: |\mathbf{x} - \boldsymbol{\mu}| \leqslant \delta\} \subset S.$$

By Chebyshev's inequality,

$$\begin{aligned} P\{\mathbf{X}_n \in \tilde{A}\} &= P\{|\mathbf{X}_n - \boldsymbol{\mu}| > \delta\} \\ &\leqslant \delta^{-r} \int |\mathbf{x} - \boldsymbol{\mu}|^r \, dF_n(\mathbf{x}) \\ &= O(n^{-\eta p}). \end{aligned}$$

For $1 < \gamma < \infty$,

$$\begin{aligned} \int |f_n(\mathbf{x})| \, dF_n(\mathbf{x}) &= \int_A |f_n(\mathbf{x})| \, dF_n(\mathbf{x}) + \int_{\tilde{A}} |f_n(\mathbf{x})| \, dF_n(\mathbf{x}) \\ &\leqslant \int_A K_1 \, dF_n(\mathbf{x}) + n^p \int_{\tilde{A}} |Y(\mathbf{x})| \, dF_n(\mathbf{x}) \\ &\leqslant K_1 + n^p \int I_{\tilde{A}}(\mathbf{x}) |Y(\mathbf{x})| \, dF_n(\mathbf{x}) \\ &\leqslant K_1 + n^p \left[\int I_{\tilde{A}}(\mathbf{x}) \, dF_n(\mathbf{x}) \right]^{1/\eta} \left[\int |Y(\mathbf{x})|^\gamma \, dF_n(\mathbf{x}) \right]^{1/\gamma} \\ &= K_1 + n^p [P\{\mathbf{X}_n \in \tilde{A}\}]^{1/\eta} [O(1)] \\ &= O(1). \end{aligned}$$

For $|f_n(\mathbf{x})| \leqslant K_2 n^p$ and $\eta = 1$, we have

$$\begin{aligned} \int |f_n(\mathbf{x})| \, dF_n(\mathbf{x}) &\leqslant K_1 + K_2 n^p \int I_{\tilde{A}}(\mathbf{x}) \, dF_n(\mathbf{x}) \\ &= O(1). \end{aligned}$$

▲

Example. To illustrate some of the ideas of this section, we consider the problem of estimating $\log \mu$, where $\mu > 0$ is the mean of a random variable with finite sixth moment. Let $\bar{x}_n$ be the mean of a simple random sample of n such observations. Since $\log x$ is not defined for $x \leqslant 0$, we consider the estimator

$$\begin{aligned} f_n(\bar{x}_n) &= \log \bar{x}_n, & \text{if } \bar{x}_n > \frac{1}{n} \\ &= -\log n, & \text{if } \bar{x}_n \leqslant \frac{1}{n}. \end{aligned}$$

Suppose that we desire an approximation to the sequence of expectations of $f_n(\bar{x}_n)$ to order n^{-1}. We first apply Theorem 5.4.2 with $a_n = O(n^{-1/2})$ and $s=3$. By Theorem 5.4.1, we have that

$$E\left\{(\bar{x}_n-\mu)^{2s}\right\}=O(n^{-s}), \qquad s=1,2,3, \tag{5.4.2}$$

so that condition i of Theorem 5.4.2 is met. To establish condition ii, we demonstrate that $|f_n(x)|^2$ satisfies the conditions of Theorem 5.4.4. Since $f_n(x)$ is continuous for $x>0$, condition i of Theorem 5.4.4 is satisfied. For $p=1$, $\gamma=\eta=2$, and

$$\begin{aligned} Y(x) &= 1, \qquad x<1 \\ &= x, \qquad \text{otherwise,} \end{aligned}$$

we have that $|f_n(x)|^2 < nY(x)$ and $\int |Y(x)|^2 \, dF_n(x) = O(1)$. This is condition ii of Theorem 5.4.4; from equation (5.4.2) we see that condition iii also holds for $r=4$, 5, or 6. Therefore, the conditions of Theorem 5.4.4 are satisfied and the integral of $|f_n(x)|^2$ is order one.

Since the third derivative of log x is continuous for positive x, we may choose an $N_0 > \mu^{-1}$ and an interval containing μ such that $f_n^{(3)}(x)$ is continuous on that interval for all $n > N_0$. To illustrate, let $\mu = 0.1$ and take the interval S to be $[0.05, 0.15]$. Then, for $n > 20$, $f_n^{(3)}(x)$ is continuous on S. Therefore, conditions iii, iv, and v of Theorem 5.4.2 are also met, and we may write

$$\begin{aligned} E\{f_n(\bar{x}_n)\} &= \log\mu + \frac{1}{\mu}E\{\bar{x}_n-\mu\} \\ &\quad - \frac{1}{2\mu^2}E\left\{(\bar{x}_n-\mu)^2\right\} + O(n^{-3/2}) \\ &= \log\mu - \frac{\sigma^2}{2n\mu^2} + O(n^{-3/2}), \end{aligned}$$

where σ^2 is the variance of the observations. To order n^{-1}, the bias in $f_n(\bar{x}_n)$ as an estimator of log μ is $-(2n\mu^2)^{-1}\sigma^2$.

Since $\bar{x}_n$ possesses finite fifth moments, we can decrease the above remainder term to $O(n^{-2})$ by carrying the expansion to one more term and using the more general form of Theorem 5.4.3. By equation (5.4.2), condition i of Theorem 5.4.3 holds for $s=4$, $\alpha=3/2$, and $a_n = O(n^{-1/2})$, so that $\delta=2$ by definition. Condition ii is established by demonstrating that

$|f_n(x)|^3$ satisfies the conditions of Theorem 5.4.4. Defining

$$Y(x)=\begin{cases}1, & x<1\\ x^{3/2}, & x\geqslant 1\end{cases}$$

and letting $p=3/2$, we see that $|f_n(x)|^3\leqslant Y(x)n^p$ and

$$\int |Y(x)|^2\,dF_n(x)\leqslant \int |1+x^{3/2}|^2\,dF_n(x)=O(1),$$

as $\bar{x}_n$ has finite third moments. Therefore, we are using $\gamma=\eta=2$ and $\eta p=3$. Since $\int |x-\mu|^5\,dF_n(x)=O(n^{-3})$ and $|f_n(x)|^3$ is continuous, the conditions of Theorem 5.4.4 are met. The sixth derivative of log x is continuous for positive x, and we can find an N_0 and a set S such that conditions iii, iv, and v of Theorem 5.4.3 are also met. Because $f_n^{(3)}(\mu)=O(1)$ and $E\{(\bar{x}_n-\mu)^3\}=O(n^{-2})$, we have

$$\begin{aligned}E\{f_n(\bar{x}_n)\}&=\log\mu-(2n\mu^2)^{-1}\sigma^2\\&\quad+\frac{1}{3!}f_n^{(3)}(\mu)E\{(\bar{x}_n-\mu)^3\}+O(n^{-2})\\&=\log\mu-(2n\mu^2)^{-1}\sigma^2+O(n^{-2}).\end{aligned}$$

5.5. GAUSS-NEWTON ESTIMATION OF NONLINEAR PARAMETERS

The expected value of a time series is sometimes expressible as a nonlinear function of unknown parameters and observable functions of time. For example, we might have for $\{Y_t\colon t\in(0,1,2,\ldots)\}$,

$$E\{Y_t\}=\sum_{j=0}^{t}\lambda^j x_{t-j},$$

where $\{x_t\}$ is a sequence of constants and λ is unknown. The estimation of a parameter such as λ is considerably more complicated than the estimation of a parameter that enters the expected value function in a linear manner. A number of methods of treating the nonlinear estimation problem are available in the statistical literature. We shall consider one, called the Gauss-Newton procedure.

We assume the model

$$Y_t=f(\mathbf{x}_t;\boldsymbol{\theta}^0)+e_t,\qquad t=1,2,\ldots, \tag{5.5.1}$$

where the e_t are independent $(0,\sigma^2)$ random variables, $\boldsymbol{\theta}^0$ is a k-dimensional parameter contained in the parameter space Θ, $\mathbf{x}_t$ is an m-dimensional vector contained in the space χ, and $f(\mathbf{x}_t;\boldsymbol{\theta})$ has continuous third derivatives with respect to $\boldsymbol{\theta}$ for all $\boldsymbol{\theta}\in\Theta$ and $\mathbf{x}_t\in\chi$.

The derivatives of the function $f(\mathbf{x}_t;\boldsymbol{\theta})$ are very important in the treatment of this problem, and we introduce a shorthand notation for them. Let $f^{(j)}(\mathbf{x}_t;\hat{\boldsymbol{\theta}})$ denote the first partial derivative of $f(\mathbf{x};\boldsymbol{\theta})$ with respect to the jth element of $\boldsymbol{\theta}$ evaluated at the point $\boldsymbol{\theta}=\hat{\boldsymbol{\theta}}$, $\mathbf{x}=\mathbf{x}_t$. Likewise, let $f^{(js)}(\mathbf{x}_t;\hat{\boldsymbol{\theta}})$ denote the second partial derivative of $f(\mathbf{x};\boldsymbol{\theta})$ with respect to the jth and sth elements of $\boldsymbol{\theta}$ evaluated at the point $\boldsymbol{\theta}=\hat{\boldsymbol{\theta}}$, $\mathbf{x}=\mathbf{x}_t$. The third partial derivative, $f^{(jrs)}(\mathbf{x}_t;\hat{\boldsymbol{\theta}})$, is similarly defined.

An important special case of estimation for this problem occurs when one is able to obtain an initial consistent estimator of the parameter vector $\boldsymbol{\theta}^0$. Most of our applications fall into this category, and we consider this situation in some detail.

We assume:

1. The vectors, $\{\mathbf{x}_t\}$, form a fixed sequence.
2. There is an open set S such that $S\subset\Theta$, $\boldsymbol{\theta}^0\in S$, and

$$\lim_{n\to\infty} n^{-1}\mathbf{F}'(\boldsymbol{\theta})\mathbf{F}(\boldsymbol{\theta})=\mathbf{B}(\boldsymbol{\theta})$$

is nonsingular for all $\theta\in S$, where $\mathbf{F}(\boldsymbol{\theta})$ is the $n\times k$ matrix with tjth element given by $f^{(j)}(\mathbf{x}_t;\boldsymbol{\theta})$.
3. $\lim_{n\to\infty} n^{-1}\mathbf{G}'(\boldsymbol{\theta})\mathbf{G}(\boldsymbol{\theta})=\mathbf{L}(\boldsymbol{\theta})$ uniformly in $\boldsymbol{\theta}$ on the closure of S, $\bar{S}$, where the elements of $\mathbf{L}(\boldsymbol{\theta})$ are continuous functions of $\boldsymbol{\theta}$ on $\bar{S}$ and $\mathbf{G}(\boldsymbol{\theta})$ is an $n\times(1+k+k^2+k^3)$ matrix with tth row given by

$$\big[f(\mathbf{x}_t;\boldsymbol{\theta}),f^{(1)}(\mathbf{x}_t;\boldsymbol{\theta}),f^{(2)}(\mathbf{x}_t;\boldsymbol{\theta}),\ldots,f^{(k)}(\mathbf{x}_t;\boldsymbol{\theta}),f^{(11)}(\mathbf{x}_t;\boldsymbol{\theta}),$$
$$f^{(12)}(\mathbf{x}_t;\boldsymbol{\theta}),\ldots,f^{(1k)}(\mathbf{x}_t;\boldsymbol{\theta}),f^{(21)}(\mathbf{x}_t;\boldsymbol{\theta}),\ldots,f^{(kk)}(\mathbf{x}_t;\boldsymbol{\theta}),$$
$$f^{(111)}(\mathbf{x}_t;\boldsymbol{\theta}),\ldots,f^{(kkk)}(\mathbf{x}_t;\boldsymbol{\theta})\big].$$

4. There is available an initial estimator of $\boldsymbol{\theta}^0$, say $\hat{\boldsymbol{\theta}}$, that satisfies $(\hat{\boldsymbol{\theta}}-\boldsymbol{\theta}^0)=O_p(a_n)$, where $\lim_{n\to\infty}a_n=0$.

The principle of least squares would lead us to choose as our estimator of $\boldsymbol{\theta}^0$ that $\boldsymbol{\theta}$ in Θ that minimizes

$$Q(\boldsymbol{\theta})=\frac{1}{n}\sum_{t=1}^{n}\left[Y_t-f(\mathbf{x}_t;\boldsymbol{\theta})\right]^2.$$

A Taylor's series expansion of $f(\mathbf{x}_t; \boldsymbol{\theta}^0)$ about $\hat{\boldsymbol{\theta}}$ gives

$$f(\mathbf{x}_t; \boldsymbol{\theta}^0) = f(\mathbf{x}_t; \hat{\boldsymbol{\theta}}) + \sum_{j=1}^{k} f^{(j)}(\mathbf{x}_t; \hat{\boldsymbol{\theta}})(\theta_j^0 - \hat{\theta}_j) + d(\mathbf{x}_t; \hat{\boldsymbol{\theta}}), \tag{5.5.2}$$

where θ_j^0 and $\hat{\theta}_j$ are the jth elements of $\boldsymbol{\theta}^0$ and $\hat{\boldsymbol{\theta}}$, respectively,

$$d(\mathbf{x}_t; \hat{\boldsymbol{\theta}}) = \frac{1}{2} \sum_{r=1}^{k} \sum_{j=1}^{k} f^{(jr)}(\mathbf{x}_t; \ddot{\boldsymbol{\theta}})(\theta_j^0 - \hat{\theta}_j)(\theta_r^0 - \hat{\theta}_r),$$

and $\ddot{\boldsymbol{\theta}}$ is on the line segment joining $\hat{\boldsymbol{\theta}}$ and $\boldsymbol{\theta}^0$. On the basis of equation (5.5.2), we consider the modified sum of squares

$$\begin{aligned}\hat{Q}(\boldsymbol{\theta}) &= \frac{1}{n} \sum_{t=1}^{n} \left\{ Y_t - f(\mathbf{x}_t; \hat{\boldsymbol{\theta}}) - \sum_{j=1}^{k} f^{(j)}(\mathbf{x}_t; \hat{\boldsymbol{\theta}})[\theta_j - \hat{\theta}_j] \right\}^2 \\ &= n^{-1}\{\mathbf{w} - \mathbf{F}(\hat{\boldsymbol{\theta}})[\boldsymbol{\theta} - \hat{\boldsymbol{\theta}}]\}'\{\mathbf{w} - \mathbf{F}(\hat{\boldsymbol{\theta}})[\boldsymbol{\theta} - \hat{\boldsymbol{\theta}}]\},\end{aligned}$$

where $\mathbf{w}$ is the $n \times 1$ vector with tth element given by

$$w_t = Y_t - f(\mathbf{x}_t; \hat{\boldsymbol{\theta}}).$$

Minimizing $\hat{Q}(\boldsymbol{\theta})$ with respect to $\boldsymbol{\delta} = (\boldsymbol{\theta} - \hat{\boldsymbol{\theta}})$ leads to

$$\tilde{\boldsymbol{\theta}} = \hat{\boldsymbol{\theta}} + \tilde{\boldsymbol{\delta}} \tag{5.5.3}$$

as an estimator of $\boldsymbol{\theta}^0$, where

$$\tilde{\boldsymbol{\delta}} = [\mathbf{F}'(\hat{\boldsymbol{\theta}})\mathbf{F}(\hat{\boldsymbol{\theta}})]^{-1}\mathbf{F}'(\hat{\boldsymbol{\theta}})\mathbf{w}.$$

We call $\tilde{\boldsymbol{\theta}}$ the *one step Gauss-Newton estimator.*

We can write,

$$\begin{aligned}\tilde{\boldsymbol{\delta}} &= [\mathbf{F}'(\hat{\boldsymbol{\theta}})\mathbf{F}(\hat{\boldsymbol{\theta}})]^{-1}\mathbf{F}'(\hat{\boldsymbol{\theta}})[\mathbf{f}(\boldsymbol{\theta}^0) - \mathbf{f}(\hat{\boldsymbol{\theta}}) + \mathbf{e}] \\ &= [\mathbf{F}'(\hat{\boldsymbol{\theta}})\mathbf{F}(\hat{\boldsymbol{\theta}})]^{-1}\mathbf{F}'(\hat{\boldsymbol{\theta}})[\mathbf{F}(\hat{\boldsymbol{\theta}})\boldsymbol{\delta}^0 + \mathbf{e}] \\ &\quad + [\mathbf{F}'(\hat{\boldsymbol{\theta}})\mathbf{F}(\hat{\boldsymbol{\theta}})]^{-1}\mathbf{R}(\hat{\boldsymbol{\theta}}),\end{aligned}$$

where $\mathbf{f}(\boldsymbol{\theta})$ is the $n \times 1$ vector with tth element $f(\mathbf{x}_t; \boldsymbol{\theta})$, $\mathbf{e}$ is the $n \times 1$ vector

with tth element e_t, $\boldsymbol{\delta}^0 = \boldsymbol{\theta}^0 - \hat{\boldsymbol{\theta}}$, and the jth element of $n^{-1}\mathbf{R}(\hat{\boldsymbol{\theta}})$ is given by

$$\frac{1}{2n}\sum_{r=1}^{k}\sum_{s=1}^{k}\sum_{t=1}^{n} f^{(j)}(\mathbf{x}_t;\hat{\boldsymbol{\theta}})f^{(rs)}(\mathbf{x}_t;\ddot{\boldsymbol{\theta}})\left[\theta_r^0 - \hat{\theta}_r\right]\left[\theta_s^0 - \hat{\theta}_s\right],$$

$\ddot{\boldsymbol{\theta}}$ between[1] $\hat{\boldsymbol{\theta}}$ and $\boldsymbol{\theta}^0$.

The elements of $\mathbf{L}(\boldsymbol{\theta})$ are bounded on $\bar{S}$ and, given $\epsilon_1 > 0$, there exists an N such that

$$\frac{1}{n}\sum_{t=1}^{n}\left[f^{(j)}(\mathbf{x}_t;\boldsymbol{\theta})\right]^2$$

and

$$\frac{1}{n}\sum_{t=1}^{n}\left[f^{(rs)}(\mathbf{x}_t;\boldsymbol{\theta})\right]^2$$

differ from the respective elements of $\mathbf{L}(\boldsymbol{\theta})$ by less than ϵ_1 for all $\boldsymbol{\theta} \in \bar{S}$ and all $n > N$. As $\hat{\boldsymbol{\theta}} - \boldsymbol{\theta}^0 = O_p(a_n)$, given $\epsilon_2 > 0$, there is an N_1 such that, for $n > N_1$, $P\{\hat{\boldsymbol{\theta}} \in \bar{S}\} > 1 - \epsilon_2$. Therefore,

$$\frac{1}{2n}\sum_{t=1}^{n} f^{(j)}(\mathbf{x}_t;\hat{\boldsymbol{\theta}})f^{(rs)}(\mathbf{x}_t;\ddot{\boldsymbol{\theta}}) = O_p(1), \qquad j,r,s = 1,2,\ldots,k,$$

and

$$n^{-1}\mathbf{R}(\hat{\boldsymbol{\theta}}) = O_p(a_n^2).$$

For sufficiently large n, $\mathbf{F}'(\boldsymbol{\theta})\mathbf{F}(\boldsymbol{\theta})$ is nonsingular for all $\boldsymbol{\theta}$ in S. It follows that

$$\frac{1}{n}\mathbf{F}'(\hat{\boldsymbol{\theta}})\mathbf{F}(\hat{\boldsymbol{\theta}}) = \frac{1}{n}\mathbf{F}'(\boldsymbol{\theta}^0)\mathbf{F}(\boldsymbol{\theta}^0) + O_p(a_n) \tag{5.5.4}$$

and

$$p\lim\left[\frac{1}{n}\mathbf{F}'(\hat{\boldsymbol{\theta}})\mathbf{F}(\hat{\boldsymbol{\theta}})\right]^{-1} = \mathbf{B}^{-1}(\boldsymbol{\theta}^0), \tag{5.5.5}$$

[1]When we use the word "between" in this context we mean that $\ddot{\boldsymbol{\theta}}$ lies on the line segment joining $\hat{\boldsymbol{\theta}}$ and $\boldsymbol{\theta}^0$.

where the jsth element of $\mathbf{B}(\boldsymbol{\theta}^0)$ is

$$\{\mathbf{B}(\boldsymbol{\theta}^0)\}_{js} = \lim_{n\to\infty}\left[n^{-1}\sum_{t=1}^{n} f^{(j)}(\mathbf{x}_t;\boldsymbol{\theta}^0)f^{(s)}(\mathbf{x}_t;\boldsymbol{\theta}^0)\right].$$

Therefore,

$$[\mathbf{F}'(\hat{\boldsymbol{\theta}})\mathbf{F}(\hat{\boldsymbol{\theta}})]^{-1}\mathbf{R}(\hat{\boldsymbol{\theta}}) = O_p(a_n^2). \tag{5.5.6}$$

The jth element of $n^{-1}\mathbf{F}'(\hat{\boldsymbol{\theta}})\mathbf{e}$ is

$$\begin{aligned} n^{-1}\sum_{t=1}^{n} f^{(j)}(\mathbf{x}_t;\hat{\boldsymbol{\theta}})e_t &= n^{-1}\sum_{t=1}^{n} f^{(j)}(\mathbf{x}_t;\boldsymbol{\theta}^0)e_t \\ &\quad + n^{-1}\sum_{t=1}^{n}\sum_{s=1}^{k} f^{(js)}(\mathbf{x}_t;\boldsymbol{\theta}^0)[\hat{\theta}_s-\theta_s^0]e_t \\ &\quad + (2n)^{-1}\sum_{t=1}^{n}\sum_{s=1}^{k}\sum_{r=1}^{k} f^{(jsr)}(\mathbf{x}_t;\boldsymbol{\theta}^\dagger)[\hat{\theta}_s-\theta_s^0][\hat{\theta}_r-\theta_r^0]e_t \\ &= n^{-1}\sum_{t=1}^{n} f^{(j)}(\mathbf{x}_t;\boldsymbol{\theta}^0)e_t + O_p(\max\{a_n^2, a_n n^{-1/2}\}), \end{aligned} \tag{5.5.7}$$

where $\boldsymbol{\theta}^\dagger$ is between $\hat{\boldsymbol{\theta}}$ and $\boldsymbol{\theta}^0$.

We summarize our discussion in a theorem.

Theorem 5.5.1. Assume that the nonlinear model (5.5.1) and assumptions 1 to 4 hold. If $\tilde{\boldsymbol{\theta}}$ is the one step Gauss-Newton estimator of (5.5.3), then

$$(\tilde{\boldsymbol{\theta}}-\boldsymbol{\theta}^0) = [\mathbf{F}'(\boldsymbol{\theta}^0)\mathbf{F}(\boldsymbol{\theta}^0)]^{-1}\mathbf{F}'(\boldsymbol{\theta}^0)\mathbf{e} + O_p(\max\{a_n^2, a_n n^{-1/2}\}).$$

Proof. Write

$$\tilde{\boldsymbol{\theta}}-\boldsymbol{\theta}^0 = [\mathbf{F}'(\hat{\boldsymbol{\theta}})\mathbf{F}(\hat{\boldsymbol{\theta}})]^{-1}\mathbf{F}'(\hat{\boldsymbol{\theta}})\mathbf{e} + [\mathbf{F}'(\hat{\boldsymbol{\theta}})\mathbf{F}(\hat{\boldsymbol{\theta}})]^{-1}\mathbf{R}(\hat{\boldsymbol{\theta}})$$

and apply (5.5.6) and (5.5.7) to obtain

$$\tilde{\boldsymbol{\theta}}-\boldsymbol{\theta}^0 = [\mathbf{F}'(\hat{\boldsymbol{\theta}})\mathbf{F}(\hat{\boldsymbol{\theta}})]^{-1}\mathbf{F}'(\boldsymbol{\theta}^0)\mathbf{e} + O_p(\max\{a_n^2, a_n n^{-1/2}\}).$$

The result follows from (5.5.4) and (5.5.5). ▲

If the order in probability of the error in the initial estimator is no smaller than $n^{-1/2}$, and we carry out a step of the Gauss-Newton procedure, the order of the error in the new estimator is no larger than that in the original estimator. It will be an unusual situation in which the Gauss-Newton procedure does not lead to an estimator whose asymptotic properties are at least as good as those of the initial estimator.

Corollary 5.5.1. Assume that the model of Theorem 5.5.1 holds and that $a_n^2 = o(n^{-1/2})$. If, for some $\delta > 0$, $E\{|e_t|^{2+\delta}\}$ and $f^{(j)}(\mathbf{x}_t; \boldsymbol{\theta}^0)$, $j = 1, 2, \ldots, k$, are uniformly bounded, then

$$n^{1/2}(\tilde{\boldsymbol{\theta}} - \boldsymbol{\theta}^0) \xrightarrow{\mathcal{L}} N\left(\mathbf{0}, \sigma^2 \mathbf{B}^{-1}(\boldsymbol{\theta}^0)\right)$$

as $n \to \infty$, where $\mathbf{B}(\boldsymbol{\theta})$ is defined in assumption 2.

Proof. Consider the linear combination

$$n^{1/2}\boldsymbol{\lambda}' n^{-1}\left[\mathbf{F}'(\boldsymbol{\theta}^0)\mathbf{F}(\boldsymbol{\theta}^0)\right](\tilde{\boldsymbol{\theta}} - \boldsymbol{\theta}^0) = n^{-1/2} \sum_{t=1}^{n} \sum_{j=1}^{k} \lambda_j f^{(j)}(\mathbf{x}_t; \boldsymbol{\theta}^0) e_t + o_p(1),$$

where $\boldsymbol{\lambda}$ is a fixed k-dimensional vector. Since $f^{(j)}(\mathbf{x}_t; \boldsymbol{\theta}^0)$ and $E\{|e_t|^{2+\delta}\}$ are bounded,

$$\lim_{n \to \infty} \frac{\displaystyle\sum_{t=1}^{n} E\left\{\left|\sum_{j=1}^{k} \lambda_j f^{(j)}(\mathbf{x}_t; \boldsymbol{\theta}^0) e_t\right|^{2+\delta}\right\}}{\left\{\displaystyle\sum_{t=1}^{n} \left|\sum_{j=1}^{k} \lambda_j f^{(j)}(\mathbf{x}_t; \boldsymbol{\theta}^0)\right|^2\right\}^{1+\delta/2} \sigma^{2+\delta}} = 0$$

and the conditions of the Liapounov central limit theorem are satisfied. As $\boldsymbol{\lambda}$ was arbitrary, the multivariate extension follows by Theorem 5.3.3. ▲

To estimate the variance of the limiting distribution of $n^{1/2}(\tilde{\boldsymbol{\theta}} - \boldsymbol{\theta}^0)$ we must estimate $\mathbf{B}^{-1}(\boldsymbol{\theta}^0)$ and σ^2. Clearly, $[n^{-1}\mathbf{F}'(\hat{\boldsymbol{\theta}})\mathbf{F}(\hat{\boldsymbol{\theta}})]^{-1}$ or $[n^{-1}\mathbf{F}'(\tilde{\boldsymbol{\theta}})\mathbf{F}(\tilde{\boldsymbol{\theta}})]^{-1}$ furnishes a consistent estimator for $\mathbf{B}^{-1}(\boldsymbol{\theta}^0)$.

We now demonstrate that

$$s^2 = \frac{1}{n-k} \sum_{t=1}^{n} \left[Y_t - f(\mathbf{x}_t; \tilde{\boldsymbol{\theta}})\right]^2 \tag{5.5.8}$$

is a consistent estimator for σ^2. We have

$$s^2 = \frac{1}{n-k}\sum_{t=1}^{n}\left[e_t - \{f(\mathbf{x}_t;\tilde{\boldsymbol{\theta}}) - f(\mathbf{x}_t;\boldsymbol{\theta}^0)\}\right]^2$$

$$= \frac{1}{n-k}\sum_{t=1}^{n}\left[e_t - \sum_{j=1}^{k} f^{(j)}(\mathbf{x}_t;\boldsymbol{\theta}^0)(\tilde{\theta}_j - \theta_j^0)\right]^2$$

$$+ R_2(\mathbf{x}_t;\tilde{\boldsymbol{\theta}}),$$

where

$$R_2(\mathbf{x}_t;\tilde{\boldsymbol{\theta}}) = \frac{1}{n-k}\sum_{t=1}^{n}\sum_{j=1}^{k}\sum_{s=1}^{k}\left[f^{(j)}(\mathbf{x}_t;\ddot{\boldsymbol{\theta}})f^{(s)}(\mathbf{x}_t;\ddot{\boldsymbol{\theta}})\right.$$

$$\left. - f^{(j)}(\mathbf{x}_t;\boldsymbol{\theta}^0)f^{(s)}(\mathbf{x}_t;\boldsymbol{\theta}^0)\right](\tilde{\theta}_j - \theta_j^0)(\tilde{\theta}_s - \theta_s^0)$$

$$- \frac{1}{n-k}\sum_{t=1}^{n}\left[e_t - \{f(\mathbf{x}_t;\ddot{\boldsymbol{\theta}}) - f(\mathbf{x}_t;\boldsymbol{\theta}^0)\}\right]$$

$$\times \sum_{j=1}^{k}\sum_{s=1}^{k} f^{(js)}(\mathbf{x}_t;\ddot{\boldsymbol{\theta}})(\tilde{\theta}_j - \theta_j^0)(\tilde{\theta}_s - \theta_s^0)$$

$$= O_p\left(\max\{a_n^3, n^{-1/2}a_n^2\}\right),$$

and $\ddot{\boldsymbol{\theta}}$ is on the line segment joining $\tilde{\boldsymbol{\theta}}$ and $\boldsymbol{\theta}^0$. By Theorem 5.5.1,

$$\tilde{\boldsymbol{\theta}} - \boldsymbol{\theta}^0 = \left[\mathbf{F}'(\boldsymbol{\theta}^0)\mathbf{F}(\boldsymbol{\theta}^0)\right]^{-1}\mathbf{F}'(\boldsymbol{\theta}^0)\mathbf{e} + O_p\left(\max\{a_n^2, a_n n^{-1/2}\}\right)$$

and, therefore,

$$s^2 = \frac{1}{n-k}\mathbf{e}'\left(\mathbf{I} - \mathbf{F}(\boldsymbol{\theta}^0)\left[\mathbf{F}'(\boldsymbol{\theta}^0)\mathbf{F}(\boldsymbol{\theta}^0)\right]^{-1}\mathbf{F}'(\boldsymbol{\theta}^0)\right)\mathbf{e} + O_p\left(\max\{a_n^3, n^{-1/2}a_n^2\}\right). \tag{5.5.9}$$

The leading term in (5.5.9) is the familiar residual sum of squares of linear regression. Hence, using the matrix $[\mathbf{F}'(\hat{\boldsymbol{\theta}})\mathbf{F}(\hat{\boldsymbol{\theta}})]^{-1}$ and the mean square s^2, all of the standard linear regression theory holds approximately for $\tilde{\boldsymbol{\theta}}$. The conclusion of Theorem 5.5.1 can be obtained under considerably relaxed assumptions. In later chapters we shall obtain the conclusions of Theorem 5.5.1 and Corollary 5.5.1 for random $\mathbf{x}_t$.

While the theorem demonstrates that the one step estimator is asymptotically unchanged by additional iteration if $\hat{\boldsymbol{\theta}}-\boldsymbol{\theta}^0=o_p(n^{-1/4})$, it is often advisable to iterate the procedure. For a particular sample we are not guaranteed that iteration of (5.5.3) and (5.5.4), using $\tilde{\boldsymbol{\theta}}$ of the previous step as the initial estimator, will converge. Therefore, if one iterates, the estimator $\tilde{\boldsymbol{\theta}}$ should be replaced at each step by

$$\tilde{\boldsymbol{\theta}}_\nu=\hat{\boldsymbol{\theta}}+\nu\tilde{\boldsymbol{\delta}},$$

where $\hat{\boldsymbol{\theta}}$ is the estimator of the previous step and $\nu\in(0,1]$ is chosen so that $n^{-1}\Sigma_{t=1}^n[Y_t-f(\mathbf{x}_t;\tilde{\boldsymbol{\theta}}_\nu)]^2$ is less than $n^{-1}\Sigma_{t=1}^n[Y_t-f(\mathbf{x}_t;\hat{\boldsymbol{\theta}})]^2$, and so that $\tilde{\boldsymbol{\theta}}_\nu\in\Theta$.

If the e_t are normal independent $(0,\sigma^2)$ random variables, then, under the assumptions of Theorem 5.5.1, it is possible to demonstrate that the maximum likelihood estimator and the Gauss-Newton estimator have the same limiting distribution.

Example. To illustrate the Gauss-Newton procedure, consider the model

$$Y_t=\theta_0+\theta_1x_{t1}+\theta_1^2x_{t2}+e_t, \tag{5.5.10}$$

where the e_t are normal independent $(0,\sigma^2)$ random variables. While the superscript 0 was used on $\boldsymbol{\theta}$ in our derivation to identify the true parameter value, it is not a common practice to use that notation when discussing a particular application of the procedure. Observations generated by this model are given in Table 5.5.1. To obtain initial estimators of θ_0 and θ_1, we ignore the nonlinear restriction and regress Y_t on $x_{t0}\equiv 1$, x_{t1}, and x_{t2}. This gives the regression equation

$$\hat{Y}_t=0.877+1.262x_{t1}+1.150x_{t2}.$$

Table 5.5.1. Data and regression variables used in the estimation of the parameters of model (5.5.10)

t	Y_t	x_{t0}	x_{t1}	x_{t2}	w_t	$x_{t1}+2\hat{\theta}_1x_{t2}$
1	9	1	2	4	−0.777	12.100
2	19	1	7	8	−3.464	27.199
3	11	1	1	9	−5.484	23.724
4	14	1	3	7	−1.821	20.674
5	9	1	7	0	−0.714	7.000
6	3	1	0	2	−1.065	5.050

Assuming that $n^{-1}\Sigma x_{t1}^2$, $n^{-1}\Sigma x_{t2}^2$, and $n^{-1}\Sigma x_{t1}x_{t2}$ converge to form a positive definite matrix, the coefficients for x_{t0} and x_{t1} are estimators for θ_0 and θ_1, respectively, with errors $O_p(n^{-1/2})$. We note that $(1.150)^{1/2}$ is also a consistent estimate of θ_1. Using 0.877 and 1.262 as initial estimators, we compute

$$w_t = Y_t - 0.877 - 1.262x_{t1} - 1.593x_{t2}.$$

The rows of $\mathbf{F}(\hat{\boldsymbol{\theta}})$ are given by

$$\mathbf{F}_{t.}(\hat{\boldsymbol{\theta}}) = (1, x_{t1} + 2\hat{\theta}_1 x_{t2}), \qquad t = 1, 2, \ldots, 6.$$

Regressing w_t on x_{t0} and $x_{t1} + 2\hat{\theta}_1 x_{t2}$, we obtain

$$\tilde{\boldsymbol{\delta}} = \begin{pmatrix} 0.386 \\ -0.163 \end{pmatrix}$$

$$\tilde{\boldsymbol{\theta}} = \begin{pmatrix} 0.877 \\ 1.262 \end{pmatrix} + \begin{pmatrix} 0.386 \\ -0.163 \end{pmatrix} = \begin{pmatrix} 1.263 \\ 1.099 \end{pmatrix}.$$

The changes in the initial estimates are moderately large and, therefore, we decide to carry out a second iteration. Using $\tilde{\boldsymbol{\theta}}$ as an initial estimate, we construct the regression variables of Table 5.5.2. Carrying out the regression, we obtain

$$\tilde{\boldsymbol{\theta}}_{(2)} = \begin{pmatrix} 1.263 \\ 1.099 \end{pmatrix} + \begin{pmatrix} -0.019 \\ -0.008 \end{pmatrix} = \begin{pmatrix} 1.244 \\ 1.091 \end{pmatrix},$$

$$[\mathbf{F}'(\tilde{\boldsymbol{\theta}})\mathbf{F}(\tilde{\boldsymbol{\theta}})]^{-1} = \begin{pmatrix} 0.7918 & -0.0436 \\ -0.0436 & 0.0030 \end{pmatrix},$$

where the columns of $\mathbf{F}(\tilde{\boldsymbol{\theta}})$ are the last two columns of Table 5.5.2, and the estimated variance is $s^2 = 1.73$. Since the estimated changes are small, we

Table 5.5.2. Variables used in second iteration of estimation for model (5.5.10)

t	w_t	x_{t0}	$x_{t1} + 2.198x_{t2}$
1	0.707	1	10.793
2	0.379	1	24.585
3	−2.234	1	20.784
4	0.984	1	18.387
5	0.043	1	7.000
6	−0.679	1	4.396

accept $\tilde{\boldsymbol{\theta}}_{(2)}$ as our final estimate. The estimated standard errors for our estimators of θ_0 and θ_1 are 1.17 and 0.073, respectively. The standard errors are obtained as part of the output of most regression programs when w_t is regressed on x_{t0} and $x_{t1}+2.198x_{t2}$.

5.6. INSTRUMENTAL VARIABLES

In many applications, estimators of the parameters of an equation of the regression type are desired, but the classical assumption that the matrix of explanatory variables is fixed is violated. Some of the columns of the matrix of explanatory variables may be measured with error and (or) may be generated by a stochastic mechanism such that the assumption that the error in the equation is independent of the explanatory variables becomes suspect.

Let us assume that we have the model

$$\mathbf{y}=\boldsymbol{\Phi}\boldsymbol{\beta}+\mathbf{X}\boldsymbol{\lambda}+\mathbf{z}, \tag{5.6.1}$$

where $\boldsymbol{\beta}$ is a $k_1\times 1$ vector of unknown parameters, $\boldsymbol{\lambda}$ is a $k_2\times 1$ vector of unknown parameters, $\mathbf{y}$ is an $n\times 1$ vector, $\boldsymbol{\Phi}$ is an $n\times k_1$ matrix, $\mathbf{X}$ is an $n\times k_2$ matrix, and $\mathbf{z}$ is an $n\times 1$ vector of unknown random variables with zero mean. The matrix $\boldsymbol{\Phi}$ is fixed, but the elements of $\mathbf{X}$ may contain a random component that is correlated with $\mathbf{z}$.

Estimators obtained by ordinary least squares may be seriously biased because of the correlation between $\mathbf{z}$ and $\mathbf{X}$. If information is available on variables that do not enter the equation, it may be possible to use these variables to obtain consistent estimators of the parameters. Such variables are called *instrumental variables*. The instrumental variables must be correlated with the variables entering the matrix $\mathbf{X}$ but not with the random components of the model. Assume that we have observations on k_3 instrumental variables, $k_3\geqslant k_2$. We denote the $n\times k_3$ matrix of observations on the instrumental variables by $\boldsymbol{\psi}$ and assume the elements of $\boldsymbol{\psi}$ are fixed. We express $\mathbf{X}$ as a linear combination of $\boldsymbol{\Phi}$ and $\boldsymbol{\psi}$:

$$\begin{aligned}\mathbf{X}&=\boldsymbol{\Phi}\boldsymbol{\delta}_1+\boldsymbol{\psi}\boldsymbol{\delta}_2+\mathbf{w}\\&=(\boldsymbol{\Phi}:\boldsymbol{\psi})\boldsymbol{\delta}+\mathbf{w},\end{aligned} \tag{5.6.2}$$

where

$$\boldsymbol{\delta}=\begin{pmatrix}\boldsymbol{\delta}_1\\ \boldsymbol{\delta}_2\end{pmatrix}=\begin{pmatrix}\boldsymbol{\Phi}'\boldsymbol{\Phi} & \boldsymbol{\Phi}'\boldsymbol{\psi}\\ \boldsymbol{\psi}'\boldsymbol{\Phi} & \boldsymbol{\psi}'\boldsymbol{\psi}\end{pmatrix}^{-1}\begin{pmatrix}E\{\boldsymbol{\Phi}'\mathbf{X}\}\\ E\{\boldsymbol{\psi}'\mathbf{X}\}\end{pmatrix}.$$

Note that the residuals $\mathbf{w}$ follow from the definition of $\boldsymbol{\delta}$. Therefore, $\mathbf{w}$ may be a sum of random and fixed components, but the fixed component is orthogonal to $\boldsymbol{\Phi}$ and $\boldsymbol{\psi}$ by construction.

The instrumental variable estimators we consider are obtained by regressing $\mathbf{X}$ on $\boldsymbol{\Phi}$ and $\boldsymbol{\psi}$, computing the estimated values $\hat{\mathbf{X}}$ from this regression, and then replacing $\mathbf{X}$ by $\hat{\mathbf{X}}$ in the regression equation

$$\mathbf{y}=\boldsymbol{\Phi}\boldsymbol{\beta}+\mathbf{X}\boldsymbol{\lambda}+\mathbf{z}.$$

The instrumental variable estimators of $\boldsymbol{\beta}$ and $\boldsymbol{\lambda}$ are given by the regression of $\mathbf{y}$ on $\boldsymbol{\Phi}$ and $\hat{\mathbf{X}}$:

$$\hat{\boldsymbol{\theta}}=\begin{pmatrix}\hat{\boldsymbol{\beta}}\\ \hat{\boldsymbol{\lambda}}\end{pmatrix}=\begin{pmatrix}\boldsymbol{\Phi}'\boldsymbol{\Phi} & \boldsymbol{\Phi}'\hat{\mathbf{X}}\\ \hat{\mathbf{X}}'\boldsymbol{\Phi} & \hat{\mathbf{X}}'\hat{\mathbf{X}}\end{pmatrix}^{-1}\begin{pmatrix}\boldsymbol{\Phi}'\mathbf{y}\\ \hat{\mathbf{X}}'\mathbf{y}\end{pmatrix}, \tag{5.6.3}$$

where

$$\hat{\mathbf{X}}=(\boldsymbol{\Phi}:\boldsymbol{\psi})\left[(\boldsymbol{\Phi}:\boldsymbol{\psi})'(\boldsymbol{\Phi}:\boldsymbol{\psi})\right]^{-1}(\boldsymbol{\Phi}:\boldsymbol{\psi})'\mathbf{X}.$$

To investigate the properties of this estimator we make the assumptions:

1. $\mathbf{Q}_n=\mathbf{D}_{1n}^{-1}(\boldsymbol{\Phi}:\boldsymbol{\psi})'(\boldsymbol{\Phi}:\boldsymbol{\psi})\,\mathbf{D}_{1n}^{-1}$ is a nonsingular matrix for all $n>k_1+k_3$, and $\lim_{n\to\infty}\mathbf{Q}_n=\mathbf{Q}$, where $\mathbf{Q}$ is nonsingular and $\mathbf{D}_{1n}$ is a diagonal matrix whose elements are the square roots of the diagonal elements of $(\boldsymbol{\Phi}:\boldsymbol{\psi})'(\boldsymbol{\Phi}:\boldsymbol{\psi})$.
2. $\mathbf{M}_n$ is nonsingular for all $n>k_1+k_3$, and

$$\lim_{n\to\infty}\mathbf{M}_n=\mathbf{M},$$

where $\mathbf{M}$ is a positive definite matrix,

$$\mathbf{M}_n=\mathbf{D}_{2n}^{-1}(\boldsymbol{\Phi}:\bar{\mathbf{X}})'(\boldsymbol{\Phi}:\bar{\mathbf{X}})\mathbf{D}_{2n}^{-1},$$

$$\bar{\mathbf{X}}=(\boldsymbol{\Phi}:\boldsymbol{\psi})\boldsymbol{\delta},$$

and $\mathbf{D}_{2n}$ is a diagonal matrix whose elements are the square roots of the diagonal elements of $(\boldsymbol{\Phi}:\bar{\mathbf{X}})'(\boldsymbol{\Phi}:\bar{\mathbf{X}})$.
3. $\lim_{n\to\infty}\mathbf{R}_n=\mathbf{R}$, where $\mathbf{R}$ is finite and

$$\mathbf{R}_n=E\left\{\mathbf{D}_{2n}^{-1}\left[\boldsymbol{\Phi}:\bar{\mathbf{X}}\right]\mathbf{z}\mathbf{z}'\left[\boldsymbol{\Phi}:\bar{\mathbf{X}}\right]\mathbf{D}_{2n}^{-1}\right\}.$$

4. (a) $\lim_{n\to\infty}\mathbf{B}_n=\mathbf{B}$, where $\mathbf{B}$ is finite and

$$\mathbf{B}_n=E\left\{\mathbf{D}_{1n}^{-1}(\boldsymbol{\Phi}:\boldsymbol{\psi})'\mathbf{z}\mathbf{z}'(\boldsymbol{\Phi}:\boldsymbol{\psi})\mathbf{D}_{1n}^{-1}\right\}.$$

(b) $\lim_{n\to\infty}\mathbf{G}_{nij}=\mathbf{G}_{ij}$, $i,j=1,2,\ldots,k_1+k_2$, where $\mathbf{G}_{ij}$ is finite,

$$\mathbf{G}_{nij}=E\left\{\mathbf{D}_{1n}^{-1}[\mathbf{\Phi}:\boldsymbol{\psi}]'\mathbf{w}_{.i}\mathbf{w}'_{.j}[\mathbf{\Phi}:\boldsymbol{\psi}]\mathbf{D}_{1n}^{-1}\right\},$$

and $\mathbf{w}_{.i}$ is the ith column of the matrix $\mathbf{w}$.

5. (a) $\lim_{n\to\infty}d_{jnii}=\infty, j=1,2,\ i=1,2,\ldots,k_1+k_{4-j}$, where d_{jnii} is the ith diagonal element of $\mathbf{D}_{jn}$.

(b)
$$\lim_{n\to\infty}\frac{\varphi_{ni}^2}{\sum_{t=1}^{n}\varphi_{ti}^2}=0, \qquad i=1,2,\ldots,k_1,$$

$$\lim_{n\to\infty}\frac{\psi_{nj}^2}{\sum_{t=1}^{n}\psi_{tj}^2}=0, \qquad j=1,2,\ldots,k_3,$$

where φ_{ti} is the tith element of $\mathbf{\Phi}$ and ψ_{tj} is the tjth element of $\boldsymbol{\psi}$.

Theorem 5.6.1. Let the model of (5.6.1) to (5.6.2) and assumptions 1 through 4 and 5(a) hold. Then,

$$\mathbf{D}_{2n}(\hat{\boldsymbol{\theta}}-\boldsymbol{\theta})=\mathbf{M}_n^{-1}\mathbf{D}_{2n}^{-1}(\mathbf{\Phi}:\overline{\mathbf{X}})'\mathbf{z}+o_p(1),$$

where

$$\hat{\boldsymbol{\theta}}=\left[(\mathbf{\Phi}:\hat{\mathbf{X}})'(\mathbf{\Phi}:\hat{\mathbf{X}})\right]^{-1}(\mathbf{\Phi}:\hat{\mathbf{X}})'\mathbf{y}$$

$$\hat{\mathbf{X}}=(\mathbf{\Phi}:\boldsymbol{\psi})\hat{\boldsymbol{\delta}},$$

and

$$\hat{\boldsymbol{\delta}}=\left[(\mathbf{\Phi}:\boldsymbol{\psi})'(\mathbf{\Phi}:\boldsymbol{\psi})\right]^{-1}(\mathbf{\Phi}:\boldsymbol{\psi})'\mathbf{X}.$$

Proof. Define $\hat{\mathbf{M}}_n$ to be the matrix $\mathbf{M}_n$ with $\overline{\mathbf{X}}$ replaced by $\hat{\mathbf{X}}$. By assumption 4, the variance of $\mathbf{D}_{1n}(\hat{\boldsymbol{\delta}}-\boldsymbol{\delta})$ is of order one, and by assumption 3, the variance of $\mathbf{D}_{2n}^{-1}(\mathbf{\Phi}:\overline{\mathbf{X}})'\,\mathbf{z}$ is of order one. Thus,

$$\begin{aligned}\hat{\mathbf{M}}_n-\mathbf{M}_n&=\mathbf{D}_{2n}^{-1}(\mathbf{\Phi}:\hat{\mathbf{X}})'(\mathbf{\Phi}:\hat{\mathbf{X}})\mathbf{D}_{2n}^{-1}-\mathbf{D}_{2n}^{-1}(\mathbf{\Phi}:\overline{\mathbf{X}})'(\mathbf{\Phi}:\overline{\mathbf{X}})\mathbf{D}_{2n}^{-1}\\
&=\mathbf{D}_{2n}^{-1}(\mathbf{\Phi}:\overline{\mathbf{X}})'\left[\mathbf{0}:\ (\mathbf{\Phi}:\boldsymbol{\psi})\mathbf{D}_{1n}^{-1}\mathbf{D}_{1n}(\hat{\boldsymbol{\delta}}-\boldsymbol{\delta})\right]\mathbf{D}_{2n}^{-1}\\
&\quad+\mathbf{D}_{2n}^{-1}\left[\mathbf{0}:\ (\mathbf{\Phi}:\boldsymbol{\psi})\mathbf{D}_{1n}^{-1}\mathbf{D}_{1n}(\hat{\boldsymbol{\delta}}-\boldsymbol{\delta})\right]'(\mathbf{\Phi}:\overline{\mathbf{X}})\mathbf{D}_{2n}^{-1}\\
&\quad+\mathbf{D}_{2n}^{-1}\left[\mathbf{0}:\ (\mathbf{\Phi}:\boldsymbol{\psi})\mathbf{D}_{1n}^{-1}\mathbf{D}_{1n}(\hat{\boldsymbol{\delta}}-\boldsymbol{\delta})\right]'\left[\mathbf{0}:\ (\mathbf{\Phi}:\boldsymbol{\psi})\mathbf{D}_{1n}^{-1}\mathbf{D}_{1n}(\hat{\boldsymbol{\delta}}-\boldsymbol{\delta})\right]\mathbf{D}_{2n}^{-1}\\
&=o_p(1).\end{aligned}$$

Similarly,

$$\mathbf{D}_{2n}^{-1}(\mathbf{\Phi}:\hat{\mathbf{X}})'\mathbf{z}-\mathbf{D}_{2n}^{-1}(\mathbf{\Phi}:\overline{\mathbf{X}})'\mathbf{z}=o_p(1).$$

Using

$$\begin{aligned}\mathbf{y}&=(\mathbf{\Phi}:\mathbf{X})\boldsymbol{\theta}+\mathbf{z}\\&=\mathbf{\Phi}\boldsymbol{\beta}+(\hat{\mathbf{X}}+\hat{\mathbf{w}})\boldsymbol{\lambda}+\mathbf{z}\end{aligned}$$

and

$$(\mathbf{\Phi}:\hat{\mathbf{X}})'\hat{\mathbf{w}}=\mathbf{0},$$

we have

$$\begin{aligned}\mathbf{D}_{2n}(\hat{\boldsymbol{\theta}}-\boldsymbol{\theta})&=\hat{\mathbf{M}}_n^{-1}\mathbf{D}_{2n}^{-1}(\mathbf{\Phi}:\hat{\mathbf{X}})'\mathbf{z}=\left[\mathbf{M}_n^{-1}+o_p(1)\right]\mathbf{D}_{2n}^{-1}(\mathbf{\Phi}:\overline{\mathbf{X}})'\mathbf{z}+o_p(1)\\&=\mathbf{M}_n^{-1}\mathbf{D}_{2n}^{-1}(\mathbf{\Phi}:\overline{\mathbf{X}})'\mathbf{z}+o_p(1).\end{aligned}$$

▲

Since $\lim_{n\to\infty}\mathbf{M}_n=\mathbf{M}$, we can also write

$$\mathbf{D}_{2n}(\hat{\boldsymbol{\theta}}-\boldsymbol{\theta})=\mathbf{M}^{-1}\mathbf{D}_{2n}^{-1}(\mathbf{\Phi}:\overline{\mathbf{X}})'\mathbf{z}+o_p(1). \tag{5.6.3}$$

In many applications $n^{-1/2}\ \mathbf{D}_{1n}$ and $n^{-1/2}\mathbf{D}_{2n}$ have finite nonsingular limits. In these cases $n^{1/2}$ can be used as the normalizing factor and the remainder in (5.6.3) is $O_p(n^{-1/2})$.

It follows from Theorem 5.6.1 that the sampling behavior of $\mathbf{D}_{2n}(\hat{\boldsymbol{\theta}}-\boldsymbol{\theta})$ is approximately that of $\mathbf{M}_n^{-1}\mathbf{D}_{2n}^{-1}(\mathbf{\Phi}:\overline{\mathbf{X}})'\mathbf{z}$, which has variance $\mathbf{M}_n^{-1}\mathbf{R}_n\mathbf{M}_n^{-1}$, where $\mathbf{R}_n$ was defined in assumption 3. In some situations it is reasonable to assume the elements of $\mathbf{z}$ are independent $(0,\sigma_z^2)$ random variables.

Corollary 5.6.1. Let model (5.6.1) to (5.6.2) and assumptions 1 to 5 hold with the elements of $\mathbf{z}$ independently distributed $(0,\sigma_z^2)$ random variables with $E\{Z_t^4\}=\eta\sigma_z^4$. Then,

$$\mathbf{D}_{2n}(\hat{\boldsymbol{\theta}}-\boldsymbol{\theta})\xrightarrow{\mathcal{L}}N\left(\mathbf{0},\mathbf{M}\sigma_z^2\right).$$

Furthermore, a consistent estimator for σ_z^2 is

$$s_z^2=\frac{1}{n-k_1-k_2}\hat{\mathbf{z}}'\hat{\mathbf{z}}, \tag{5.6.4}$$

where $\hat{\mathbf{z}}=\mathbf{y}-(\mathbf{\Phi}:\mathbf{X})\hat{\boldsymbol{\theta}}$.

Proof. A proof is not presented at this time. The normality result is a special case of Theorem 6.3.4 of Chapter 6. That s_z^2 is a consistent estimator of σ_z^2 can be demonstrated by the arguments of Theorem 9.8.3. ▲

Our analysis treated $\mathbf{\Phi}$ and $\boldsymbol{\psi}$ as fixed. Theorem 5.6.1 will hold for $\mathbf{\Phi}$ and (or) $\boldsymbol{\psi}$ random, provided the second moments exist, the probability limits analogous to the limits of assumptions 1, 2, and 5 exist, and the error terms in $\mathbf{y}$ and $\mathbf{X}$ are independent of $\mathbf{\Phi}$ and $\boldsymbol{\psi}$.

A discussion of instrumental variables particularly applicable when k_3 is larger than k_2 is given in Sargan (1958). A model where the method of instrumental variables is appropriate will be discussed in Chapter 9.

Example. To illustrate the method of instrumental variables, we use some data collected in an animal feeding experiment. Twenty-four lots of pigs were fed six different rations characterized by the percent of protein in the ration. The remainder of the ration was primarily carbohydrate from corn. We simplify by calling this remainder corn. Twelve of the lots were weighed after two weeks and 12 after four weeks. The logarithms of the gain and of the feed consumed are given in Table 5.6.1. We consider the model

$$G_i = \beta_0 + \beta_1 P_i + \beta_2 C_i + Z_i, \qquad i = 1, 2, \ldots, 24,$$

where G_i is the logarithm of gain, C_i is the logarithm of corn consumed, P_i is the logarithm of protein consumed, and Z_i is the random error for the ith lot.

It is clear that neither corn nor protein but, instead, their ratio, is fixed by the experimental design. The observations on corn and protein for a particular ration are constrained to lie on a ray through the origin with slope corresponding to the ratio of the percentages of the two items in the ration. The logarithms of these observations will lie on parallel lines. Since $C_i - P_i$ is fixed, we rewrite the model as

$$G_i = \beta_0 + (\beta_1 + \beta_2) P_i + \beta_2 (C_i - P_i) + Z_i.$$

In terms of the notation of (5.6.1), $G_i = Y_i$ and $P_i = X_i$. Candidates for ψ_i are functions of $(C_i - P_i)$ and of time on feed. As one simple model for the protein consumption we suggest

$$P_i = \delta_0 + \delta_1 (C_i - P_i) + \delta_2 t_i + \delta_3 (t_i - 3)(C_i - P_i) + W_i,$$

where t_i is the time on feed of the ith lot. The ordinary least squares estimate of the equation is

$$\hat{P}_i = \underset{(0.48)}{4.45} - \underset{(0.09)}{0.95} (C_i - P_i) + \underset{(0.15)}{0.46}\, t_i - \underset{(0.09)}{0.02} (t_i - 3)(C_i - P_i),$$

Table 5.6.1. Gain and feed consumed by 24 lots of pigs

Lot i	Time on Feed Weeks	Log Gain G	Log Corn C	Log Protein P	$\hat{P}$	$\hat{Z}$
1	2	4.477	5.366	3.465	3.601	−0.008
2	2	4.564	5.488	3.587	3.601	−0.042
3	2	4.673	5.462	3.647	3.682	0.035
4	2	4.736	5.598	3.783	3.682	−0.036
5	2	4.718	5.521	3.787	3.757	−0.033
6	2	4.868	5.580	3.846	3.757	0.059
7	2	4.754	5.516	3.858	3.828	−0.043
8	2	4.844	5.556	3.898	3.828	0.007
9	2	4.836	5.470	3.884	3.895	0.035
10	2	4.828	5.463	3.877	3.895	0.034
11	2	4.745	5.392	3.876	3.961	−0.026
12	2	4.852	5.457	3.941	3.961	0.017
13	4	5.384	6.300	4.399	4.453	−0.021
14	4	5.493	6.386	4.485	4.453	0.003
15	4	5.513	6.350	4.535	4.537	0.001
16	4	5.583	6.380	4.565	4.537	0.041
17	4	5.545	6.314	4.580	4.616	0.013
18	4	5.613	6.368	4.634	4.616	0.028
19	4	5.687	6.391	4.733	4.690	0.028
20	4	5.591	6.356	4.698	4.690	−0.033
21	4	5.591	6.288	4.702	4.760	−0.015
22	4	5.700	6.368	4.782	4.760	0.015
23	4	5.700	6.355	4.839	4.828	−0.019
24	4	5.656	6.332	4.816	4.828	−0.041

SOURCE. Data courtesy of Research Department, Moorman Manufacturing Company. The data are a portion of a larger experiment conducted by the Moorman Manufacturing Company in 1974.

where the numbers in parentheses are the estimated standard errors of the regression coefficients. If the W_i are independent $(0, \sigma^2)$ random variables, the usual regression assumptions are satisfied. The interaction term contributes very little to the regression, but the time coefficient is highly significant. This supports assumption 2 because it suggests that the partial correlation between P_i and t_i after adjusting for $C_i - P_i$ is not zero. This, in turn, implies that the matrix M_n is nonsingular. The $\hat{P}$-values for this regression are given in Table 5.6.1. Regressing G_i on $\hat{P}_i$ and $(C_i - P_i)$, we obtain

$$\hat{G}_i = 0.49 + 0.98 P_i + 0.31(C_i - P_i).$$

In this problem it is reasonable to treat the Z's as independent random variables. We also assume that they have common variance. The estimated

residuals are shown in the last column of Table 5.6.1. These must be computed directly as

$$G_i - 0.49 - 0.98P_i - 0.31(C_i - P_i).$$

The residuals obtained in the second round regression computations are $G_i - 0.49 - 0.98\hat{P}_i - 0.31(C_i - P_i)$ and are inappropriate for the construction of variance estimates. From the $\hat{Z}$'s we obtain

$$s_z^2 = (21)^{-1} \sum_{i=1}^{24} \hat{Z}_i^2 = 0.0010.$$

The inverse of the matrix used in computing the estimates is

$$\begin{bmatrix} 14.73 & -1.32 & -5.37 \\ -1.32 & 0.23 & 0.21 \\ -5.37 & 0.21 & 2.62 \end{bmatrix},$$

and it follows that the estimated standard errors of the estimates are (0.121), (0.015), and (0.051), respectively.

REFERENCES

Section 5.1. Chernoff (1956), Chung (1968), Mann and Wald (1943b), Pratt (1959), Tucker (1967).

Sections 5.2, 5.3. Chung (1968), Cramér (1946), Gnedenko (1967), Loéve (1963), Rao (1965a), Tucker (1967), Varadarajan (1958), Wilks (1962).

Section 5.4. Cramér (1946), DeGracie and Fuller (1972), Hansen, Hurwitz, and Madow (1953), Hatanaka (1973).

Section 5.5. Gallant (1971), Hartley (1961), Hartley and Booker (1965), Jennrich (1969), Malinvaud (1970b).

Section 5.6. Basmann (1957), Sargan (1958), (1959), Theil (1961).

EXERCISES

1. Let $\{a_n\}$ and $\{b_n\}$ be sequences of real numbers and let $\{f_n\}$ and $\{g_n\}$ be sequences of positive real numbers such that $a_n = O(f_n)$ and $b_n = O(g_n)$. Show that:
   (a) $|a_n|^s = O(f_n^s)$, $s > 0$.
   (b) $a_n b_n = O(f_n g_n)$.

2. Let $\{f_n\}$, $\{g_n\}$, and $\{r_n\}$ be sequences of positive real numbers, and let $\{X_n\}$, $\{Y_n\}$, and $\{Z_n\}$ be sequences of random variables such that $X_n = O_p(f_n)$, $Y_n = O_p(g_n)$, and $Z_n = o_p(r_n)$. Without recourse to Theorems 5.1.5 and 5.1.6, show that:
   (a) $|X_n|^s = O_p(f_n^s)$, $s > 0$.

(b) $X_nY_n = O_p(f_ng_n)$.
(c) $X_n + Y_n = O_p(\max\{f_n, g_n\})$.
(d) $X_nZ_n = o_p(f_nr_n)$.
(e) If $g_n/r_n = o(1)$, then $Y_n + Z_n = o_p(r_n)$.

3. Let $X_1, \ldots, X_n$ and $Y_1, \ldots, Y_n$ be two independent random samples, each of size n. Let the X_i be $N(0,4)$ and the Y_i be $N(2,9)$. Find the order in probability of the following statistics.
(a) $\bar{X}_n$. (d) $\bar{Y}_n^2$.
(b) $\bar{Y}_n$. (e) $\bar{X}_n\bar{Y}_n$.
(c) $\bar{X}_n^3$. (f) $(\bar{X}_n + 1)$.

Find the most meaningful expression (i.e., the smallest quantity for which the statement is true).

4. Prove that $\text{plim}_{n\to\infty}\hat{\theta} = \theta^0$ implies that there exists a sequence of positive real numbers $\{a_n\}$ such that $\lim_{n\to\infty} a_n = 0$ and $\hat{\theta} - \theta^0 = O_p(a_n)$.

5. Let $\{a_t\}$ be a sequence of constants satisfying $\sum_{t=1}^{\infty}|a_t| < \infty$; also let $\{X_{tn}: t = 1,2,\ldots,n;\ n = 1,2,\ldots\}$ be a triangular array of random variables such that $E\{X_{tn}^2\} = O(b_n^2)$, $t = 1,2,\ldots,n$, where $\lim_{n\to\infty} b_n = 0$. Prove

$$\sum_{t=1}^{n} a_t X_{tn} = O_p(b_n).$$

6. Let $\bar{x}_n$ be distributed as a normal $(\mu, \sigma^2/n)$ random variable with $\mu \neq 0$, and define $Y_n = \bar{x}_n^3 - \bar{x}_n^{-2}$. Expand Y_n in a Taylor's series through terms of $O_p(n^{-1})$. Find the expectation of these terms.

7. Let $F_1, F_2, \ldots, F_M$ be a finite collection of distribution functions with finite variances and common mean μ. A sequence of random variables $\{X_t: t \in (1,2,\ldots)\}$ is created by randomly choosing, for each t, one of the distribution functions and then making a random selection from the chosen distribution. Show that, as $n\to\infty$,

$$\frac{\sum_{t=1}^{n} X_t - n\mu}{\left(\sum_{t=1}^{n}\sigma_t^2\right)^{1/2}} \xrightarrow{\mathcal{L}} N(0,1),$$

where σ_t^2 is the variance of the distribution chosen for index t. Does

$$\frac{(1/n)\sum_{t=1}^{n} X_t - \mu}{\left((1/M)\sum_{j=1}^{M}\sigma_j^2\right)^{1/2}} \xrightarrow{\mathcal{L}} N(0,1),$$

where σ_j^2 is the variance of the jth distribution, $j = 1,2,\ldots,M$?

8. Let X_i be normal independent (μ, σ^2) random variables and let $\bar{x}_n = n^{-1}\sum_{i=1}^{n} X_i$.
(a) It is known that $\mu \neq 0$. How would you approximate the distribution of $\bar{x}_n^2$ in large samples?
(b) It is known that $\mu = 0$. How would you approximate the distribution of $\bar{x}_n^2$ in large samples?

Explain and justify your answers giving the normalizing constants necessary to produce nondegenerate limiting distributions.

9. Let $\bar{x}_n$ be distributed as a normal $(\mu, \sigma^2/n)$ random variable, $\mu > 0$.
   (a) Find $E\{Y_n\}$ to order $(1/n)$, where

$$Y_n = \operatorname{sgn} \bar{x}_n |\bar{x}_n|^{1/2}$$

   and $\operatorname{sgn} \bar{x}_n$ denotes the sign of $\bar{x}_n$.
   (b) Find $E\{(Y_n - \mu^{1/2})^2\}$ to order $(1/n)$.
   (c) What is the limiting distribution of $n^{1/2}(Y_n - \mu^{1/2})$?

10. Let $\bar{x}_n$ be distributed as a normal $(\mu, \sigma^2/n)$ random variable, $\mu \neq 0$, and define

$$Z_n = \frac{n\bar{x}_n}{n\bar{x}_n^2 + \sigma^2}.$$

Show that $E\{\bar{x}_n^{-1}\}$ is not defined but that $E\{Z_n\}$ is defined. Show further that Z_n satisfies the conditions of Theorem 5.4.3 and find the expectation of Z_n through terms of $O(1/n)$.

11. Let $(X, Y)'$ be distributed as a bivariate normal random variable with mean $(\mu_x, \mu_y)'$ and covariance matrix

$$\begin{pmatrix} \sigma_x^2 & \sigma_{xy} \\ \sigma_{xy} & \sigma_y^2 \end{pmatrix}.$$

Given a sample of size n from this bivariate population, derive an estimator for the product, $\mu_x \mu_y$, that is unbiased to $O(1/n)$.

12. Assume the model

$$Y_t = \alpha + \lambda e^{-\beta x_t} + u_t,$$

where the u_t are distributed as normal independent $(0, \sigma^2)$ random variables. We have available the following data:

t	Y_t	x_t
1	47.3	0.0
2	87.0	0.4
3	120.1	0.8
4	130.4	1.6
5	58.8	0.0
6	111.9	1.0
7	136.5	2.0
8	132.0	4.0
9	68.8	0.0
10	138.1	1.5
11	145.7	3.0
12	143.0	5.9

where Y is yield of corn and x is applied nitrogen. Given the initial values $\hat{\alpha} = 143$, $\hat{\lambda} = -85$, $\hat{\beta} = 1.20$, carry out two iterations of the Gauss-Newton procedure. Using the estimates obtained at the second iteration, estimate the covariance matrix of the estimator.

13. In the illustration of Section 5.5 only the coefficient of x_{t1} was used in constructing an initial estimate of θ_1. Identifying the original equation as

$$Y_t = \theta_0 + \theta_1 x_{t1} + \alpha^2 x_{t2} + e_t,$$

construct an estimator for the covariance matrix of $(\hat{\theta}_0, \hat{\theta}_1, \hat{\alpha})$, where the coefficients $\hat{\theta}_0, \hat{\theta}_1$, and $\hat{\alpha}^2$ are the ordinary least squares estimates. Using this covariance matrix, find the λ that minimizes the estimated variance of

$$\lambda \hat{\theta}_1 + (1-\lambda)\hat{\alpha}$$

as an estimator of θ_1. Use the estimated linear combination as an initial estimate in the Gauss-Newton procedure.

14. Assuming that the e_t are normal independent $(0, \sigma^2)$ random variables, obtain the likelihood function associated with model (5.5.1). By evaluating the expectations of the second partial derivatives of the likelihood function with respect to the parameters, demonstrate that the asymptotic covariance matrix of the maximum likelihood estimator is the same as that given in Theorem 5.5.1.

15. Let Y_t satisfy the model

$$Y_t = \theta_0 + \theta_1 e^{-\theta_2 t} + e_t, \qquad t = 1, 2, \ldots,$$

where $\theta_2 > 0$ and e_t is a sequence of normal independent $(0, \sigma^2)$ random variables. Does this model satisfy assumptions 1, 2 and 3 of Section 5.5? Would the model with the t in the exponent replaced by x_t where

$$\{x_t\} = \{1, 2, 3, 4, 1, 2, 3, 4, \ldots\}$$

satisfy the three assumptions?

16. An experiment is conducted to study the relationship between the phosphate content of the leaf and the yield of grain for the corn plant. In the experiment different levels of phosphate fertilizer were applied to the soil of 20 experimental plots. The leaf phosphate and the grain yield of the corn were recorded for each plot. Denoting yield by Y, applied phosphate by A, and leaf phosphate by P, the sums of squares and cross products matrix for A, P, and Y were computed as

$$\begin{pmatrix} \Sigma A^2 & \Sigma AP & \Sigma AY \\ \Sigma PA & \Sigma P^2 & \Sigma PY \\ \Sigma YA & \Sigma YP & \Sigma Y^2 \end{pmatrix} = \begin{pmatrix} 69{,}600 & 16{,}120 & 3{,}948 \\ 16{,}120 & 8{,}519 & 1{,}491 \\ 3{,}948 & 1{,}491 & 739 \end{pmatrix}.$$

The model

$$Y_t = \beta P_t + v_t, \qquad t = 1, 2, \ldots, 20,$$

where the v_t are normal independent $(0, \sigma^2)$ random variables, is postulated. Estimate β by the method of instrumental variables using A_t as the instrumental variable. Estimate the standard error of your coefficients. Compare your estimate to the ordinary least squares estimate.

CHAPTER 6

Estimation of the Mean and Autocorrelations

In this chapter we shall derive some large sample results for the sampling behavior of the estimated mean, covariances, and autocorrelations.

6.1. ESTIMATION OF THE MEAN

Consider a stationary time series, X_t, with mean μ, which we desire to estimate. If it were possible to obtain a number of independent realizations, then the average of the realization averages would converge in mean square to μ as the number of realizations increased. That is, given m samples of n observations each,

$$\operatorname{Var}\left\{\frac{1}{m}\sum_{j=1}^{m}\bar{x}_{(j)}\right\} \leqslant \frac{1}{m}\gamma_X(0);$$

where $\bar{x}_{(j)} = n^{-1}\Sigma_{t=1}^{n} X_{(j)t}$ is the mean of the n observations from the jth realization.

However, in many areas of application, it is difficult or impossible to obtain multiple realizations. For example, most economic time series constitute a single realization. Therefore, the question becomes whether or not we can use the average of a single realization to estimate the mean. Clearly, the sample mean, $\bar{x}_n$, is an unbiased estimator for the mean of a covariance stationary time series. If the mean square error of the sample mean as an estimator of the population mean approaches zero as the number of observations included in the mean increases, we say that the time series is *ergodic* for the mean. We now investigate conditions under which the time series is ergodic for the mean. Theorem 6.1.1 demonstrates that the sample mean may be a consistent estimator for nonstationary time

series if the nonstationarity is of a transient nature. The theorem follows Parzen (1962).

Theorem 6.1.1. Let $\{X_t: t \in (1,2,\ldots)\}$ be a time series satisfying

$$\lim_{t\to\infty} E\{X_t\} = \mu,$$

$$\lim_{n\to\infty} \operatorname{Cov}\{\bar{x}_n, X_n\} = 0,$$

where

$$\bar{x}_n = \frac{1}{n}\sum_{t=1}^{n} X_t.$$

Then,

$$\lim_{n\to\infty} E\{(\bar{x}_n - \mu)^2\} = 0.$$

Proof. Now

$$E\{(\bar{x}_n - \mu)^2\} = \operatorname{Var}\{\bar{x}_n\} + \left(\frac{1}{n}\sum_{t=1}^{n} E\{X_t\} - \mu\right)^2$$

where the second term on the right converges to zero by Lemma 3.1.5. Furthermore,

$$\begin{aligned}
\operatorname{Var}\{\bar{x}_n\} &= \frac{1}{n^2}\sum_{t=1}^{n}\sum_{j=1}^{n} \operatorname{Cov}\{X_t, X_j\} \\
&= \frac{2}{n^2}\sum_{t=1}^{n}\sum_{j=1}^{t} \operatorname{Cov}\{X_t, X_j\} - \frac{1}{n^2}\sum_{t=1}^{n} \operatorname{Var}\{X_t\} \\
&= \frac{2}{n^2}\sum_{t=1}^{n} t \operatorname{Cov}\{\bar{x}_t, X_t\} - \frac{1}{n^2}\sum_{t=1}^{n} \operatorname{Var}\{X_t\} \\
&\leqslant \frac{2}{n}\sum_{t=1}^{n} |\operatorname{Cov}\{\bar{x}_t, X_t\}|,
\end{aligned}$$

which also converges to zero by Lemma 3.1.5. ▲

Corollary 6.1.1.1. Let $\{X_t\}$ be a stationary time series whose covariance function $\gamma(h)$ converges to zero as h gets large. Then,

$$\lim_{n\to\infty} \operatorname{Var}\{\bar{x}_n\} = 0.$$

Proof. By Lemma 3.1.5, the convergence of $\gamma(h)$ to zero implies that $\operatorname{Cov}\{\bar{x}_n, X_n\} = (1/n)\sum_{h=0}^{n-1}\gamma(h)$ converges to zero and the result follows by Theorem 6.1.1. ▲

Corollary 6.1.1.2. A stationary time series with absolutely summable covariance function is ergodic for the mean. Furthermore,

$$\lim_{n\to\infty} n\operatorname{Var}\{\bar{x}_n\} = \sum_{h=-\infty}^{\infty} \gamma(h).$$

Proof. The assumption of absolute summability implies

$$\lim_{h\to\infty} \gamma(h) = 0$$

and ergodicity follows from Corollary 6.1.1.1. We have

$$n\operatorname{Var}\{\bar{x}_n\} = \frac{1}{n}\sum_{j=1}^{n}\sum_{t=1}^{n}\gamma(t-j) = \frac{1}{n}\sum_{h=-(n-1)}^{n-1}(n-|h|)\gamma(h)$$

and $n\operatorname{Var}\{\bar{x}_n\}$ converges to the stated result by Lemma 3.1.4. ▲

Theorem 6.1.2. If the spectral density of a stationary time series X_t is continuous, then

$$\lim_{n\to\infty} n\operatorname{Var}\{\bar{x}_n\} = 2\pi f(0), \tag{6.1.1}$$

where $f(0)$ is the spectral density of X_t evaluated at zero.

Proof. By Theorem 3.1.10 the Fourier series of a continuous periodic function is uniformly summable by the method of Cesàro. The autocovariances $\gamma(k)$ are equal to π times the a_k of that theorem. Therefore,

$$\begin{aligned} 2\pi f(0) &= \lim_{n\to\infty} 2\pi\left[\frac{a_0}{2} + \frac{1}{n}\sum_{r=2}^{n}\sum_{k=1}^{r-1} a_k\right] \\ &= \lim_{n\to\infty}\left[\gamma(0) + \frac{2}{n}\sum_{r=2}^{n}\sum_{k=1}^{r-1}\gamma(k)\right] \\ &= \lim_{n\to\infty}\left[\gamma(0) + \frac{2}{n}\sum_{r=1}^{n}(n-r)\gamma(r)\right] \\ &= \lim_{n\to\infty} n\operatorname{Var}\{\bar{x}_n\}. \end{aligned}$$

▲

Since the absolute summability of the covariance function implies that $f(\omega)$ is continuous, it follows that (6.1.1) holds for a time series with absolutely summable covariance function. Thus, for a wide class of time series, the sample mean has a variance that is declining at the rate n^{-1}. In large samples the variance is approximately the spectral density evaluated at zero multiplied by $2\pi n^{-1}$.

To investigate the efficiency of the sample mean as an estimator of μ, we write

$$Y_t = \mu + Z_t$$

where Z_t is a time series with zero mean, and we define $\mathbf{V}$ to be the covariance matrix for n observations on Y_t. Thus,

$$\mathbf{V} = E\{\mathbf{z}\mathbf{z}'\},$$

where $\mathbf{z}$ is the column vector of n observations on Z_t. If the covariance matrix is known and nonsingular the best linear unbiased estimator of the mean is given by

$$\hat{\mu} = (\mathbf{1}'\mathbf{V}^{-1}\mathbf{1})^{-1}\mathbf{1}'\mathbf{V}^{-1}\mathbf{y}, \tag{6.1.2}$$

where $\mathbf{1}$ is a column vector composed of n ones and $\mathbf{y}$ is the vector of n observations on Y_t. The variance of $\hat{\mu}$ is

$$\operatorname{Var}\{\hat{\mu}\} = (\mathbf{1}'\mathbf{V}^{-1}\mathbf{1})^{-1}, \tag{6.1.3}$$

whereas the variance of $\bar{y}_n = n^{-1}\Sigma_{t=1}^{n} Y_t$ is

$$\operatorname{Var}\{\bar{y}_n\} = n^{-2}\mathbf{1}'\mathbf{V}\mathbf{1}. \tag{6.1.4}$$

Let Y_t be a pth order autoregressive process defined by

$$(Y_t - \mu) + \sum_{j=1}^{p} \alpha_j (Y_{t-j} - \mu) = e_t, \tag{6.1.5}$$

where the e_t are uncorrelated $(0, \sigma^2)$ random variables, and the roots of the characteristic equation are less than one in absolute value. For known α_j, the estimator (6.1.2) can be constructed by transforming the observations into a sequence of uncorrelated constant variance observations. Using the

Gram-Schmidt orthogonalization procedure we obtain

$$
\begin{aligned}
W_1 &= \delta_{11} Y_1 \\
W_2 &= \delta_{22} Y_2 - \delta_{21} Y_1 \\
&\vdots \\
W_p &= \delta_{pp} Y_p - \sum_{j=1}^{p-1} \delta_{p,p-j} Y_{p-j} \\
W_t &= Y_t + \sum_{j=1}^{p} \alpha_j Y_{t-j}, \qquad t = p+1, p+2, \ldots, n,
\end{aligned}
\tag{6.1.6}
$$

where $\delta_{11} = \gamma^{-1/2}(0)\sigma$, $\delta_{22} = [\{1-\rho^2(1)\}\gamma(0)]^{-1/2}\sigma$, $\delta_{21} = \rho(1)[\{1-\rho^2(1)\}\gamma(0)]^{-1/2}\sigma$, and so forth. The expected values are

$$
\begin{aligned}
E\{W_1\} &= \delta_{11}\mu \\
E\{W_2\} &= (\delta_{22} - \delta_{21})\mu \\
&\vdots \\
E\{W_p\} &= \left(\delta_{pp} - \sum_{j=1}^{p-1} \delta_{p,p-j}\right)\mu \\
E\{W_t\} &= \left(1 + \sum_{j=1}^{p} \alpha_j\right)\mu, \qquad t = p+1, p+2, \ldots, n.
\end{aligned}
$$

In matrix notation we let $\mathbf{T}$ denote the transformation defined in (6.1.6). Then $\mathbf{T}'\mathbf{T}\sigma^{-2} = \mathbf{V}^{-1}$, $E\{\mathbf{T}\mathbf{y}\} = \mathbf{T}\mathbf{1}\mu$, and

$$
\hat{\mu} = (\mathbf{1}'\mathbf{T}'\mathbf{T}\mathbf{1})^{-1}\mathbf{1}'\mathbf{T}'\mathbf{T}\mathbf{y}. \tag{6.1.7}
$$

With the aid of this transformation, we can demonstrate that the sample mean has the same asymptotic efficiency as the best linear unbiased estimator.

Theorem 6.1.3. Let Y_t be the stationary pth order autoregressive process defined in (6.1.5). Then,

$$
\lim_{n\to\infty} n \operatorname{Var}\{\bar{y}_n\} = \lim_{n\to\infty} n \operatorname{Var}\{\hat{\mu}\},
$$

where $\hat{\mu}$ is defined in (6.1.2).

Proof. Without loss of generality, we let $\sigma^2 = 1$. The spectral density of Y_t is then

$$f_Y(\omega) = \frac{1}{2\pi}\left[\sum_{j=0}^{p} \alpha_j e^{-i\omega j} \sum_{j=0}^{p} \alpha_j e^{i\omega j}\right]^{-1},$$

where $\alpha_0 = 1$. With the exception of $2p^2$ terms in the upper left and lower right corners of $\mathbf{T}'\mathbf{T}$, the elements of $\mathbf{V}^{-1}$ are given by

$$v^{ir} = \begin{cases} \sum_{r=0}^{p-h} \alpha_r \alpha_{r+h}, & |i-r| = h \leq p \\ 0, & \text{otherwise.} \end{cases} \tag{6.1.8}$$

The values of the elements v^{ir} in (6.1.8) depend only on $|i-r|$ and we recognize $\sum_{r=0}^{p-|h|} \alpha_r \alpha_{r+|h|} = \gamma_m(h)$, say, as the covariance function of a pth order moving average. Therefore, for $n > 2p$,

$$\begin{aligned}
\frac{1}{n}\mathbf{1}'\mathbf{V}^{-1}\mathbf{1} &= \frac{1}{n}\sum_{i=1}^{n}\sum_{r=1}^{n} v^{ir} \\
&= \frac{1}{n}\sum_{i=1}^{n}\sum_{r=1}^{n} \gamma_m(i-r) + \frac{1}{n}\sum_{i=1}^{p}\sum_{r=1}^{p}\left[v^{ir} - \gamma_m(i-r)\right] \\
&\quad + \frac{1}{n}\sum_{i=n-p+1}^{n}\sum_{r=n-p+1}^{n}\left[v^{ir} - \gamma_m(i-r)\right] \\
&= \frac{1}{n}\sum_{i=1}^{n}\sum_{r=1}^{n} \gamma_m(i-r) + O(n^{-1}).
\end{aligned}$$

It follows that

$$\begin{aligned}
\lim_{n\to\infty} n \operatorname{Var}\{\hat{\mu}\} &= \lim_{n\to\infty}\left[\frac{1}{n}\sum_{i=1}^{n}\sum_{r=1}^{n}\gamma_m(i-r)\right]^{-1} = \left|\sum_{j=0}^{p}\alpha_j\right|^{-2} \\
&= 2\pi f_Y(0) = \lim_{n\to\infty} n \operatorname{Var}\{\bar{y}_n\}. \qquad \blacktriangle
\end{aligned}$$

Using the fact that a general class of spectral densities can be approximated by the spectral density of an autoregressive process (see Theorem 4.3.4), Grenander and Rosenblatt (1957) have shown that the mean, and

certain other linear estimators, have the same asymptotic efficiency as the generalized least squares estimator for time series with spectral densities in that class.

6.2. ESTIMATORS OF THE AUTOCOVARIANCE AND AUTOCORRELATION FUNCTIONS

While the sample mean is a natural estimator to consider for the mean of a stationary time series, a number of estimators have been proposed for the covariance function. If the mean is known and, without loss of generality, taken to be zero, the estimator

$$\tilde{\gamma}(h)=\frac{1}{n-h}\sum_{t=1}^{n-h} X_t X_{t+h} \tag{6.2.1}$$

is seen to be the mean of $n-h$ observations from the time series, say,

$$Z_{th}=X_t X_{t+h}.$$

For stationary time series, $E\{Z_{th}\}=\gamma_X(h)$ for all t and it follows that $\tilde{\gamma}(h)$ is an unbiased estimator of $\gamma(h)$. In most practical situations the mean is unknown and must be estimated. We list below two possible estimators of the covariance function when the mean is estimated. In both expressions h is taken to be greater than or equal to zero.

$$\gamma^{\dagger}(h)=\frac{1}{n-h}\sum_{t=1}^{n-h}(X_t-\bar{x}_n)(X_{t+h}-\bar{x}_n) \tag{6.2.2}$$

$$\hat{\gamma}(h)=\frac{1}{n}\sum_{t=1}^{n-h}(X_t-\bar{x}_n)(X_{t+h}-\bar{x}_n) \tag{6.2.3}$$

It is clear that these estimators differ by factors that become small at the rate n^{-1}. Unlike $\tilde{\gamma}(h)$, neither of the estimators is unbiased. The bias is given in Theorem 6.2.2 of this section.

The estimator, $\hat{\gamma}(h)$, can be shown to have smaller mean square error than $\gamma^{\dagger}(h)$ for certain types of time series. This estimator also has the advantage of guaranteeing positive definiteness of the estimated covariance function. In most of our applications we shall use the estimator $\hat{\gamma}(h)$.

As one might expect, the variances of the estimated autocovariances are much more complicated than that of the mean. The theorem we present is

due to Bartlett (1946). A result needed in the proof will be useful in later sections, and we state it as a lemma.

Lemma 6.2.1. Let $\{\delta_j\}_{j=-\infty}^{\infty}$ and $\{c_j\}_{j=-\infty}^{\infty}$ be two absolutely summable sequences. Then, for fixed integers r, h and d, $d \geqslant 0$,

$$\lim_{n\to\infty} \frac{1}{n} \sum_{s=1}^{n+d} \sum_{t=1}^{n} \delta_{s-t+r} c_{s-t+h} = \sum_{p=-\infty}^{\infty} \delta_{p+r} c_{p+h}$$

$$= \sum_{p=-\infty}^{\infty} \delta_p c_{p+h-r}.$$

Proof. Let $p = s - t$. Then,

$$\begin{aligned}
\frac{1}{n} \sum_{s=1}^{n+d} \sum_{t=1}^{n} \delta_{s-t+r} c_{s-t+h} &= \frac{1}{n} \sum_{p=0}^{n-1} \sum_{s=p+1}^{n} \delta_{p+r} c_{p+h} \\
&\quad + \frac{1}{n} \sum_{p=-(n-1)}^{-1} \sum_{s=1}^{n+p} \delta_{p+r} c_{p+h} \\
&\quad + \frac{1}{n} \sum_{s=n+1}^{n+d} \sum_{t=1}^{n} \delta_{s-t+r} c_{s-t+h} \\
&= \sum_{p=-(n-1)}^{n-1} \frac{(n-|p|)}{n} \delta_{p+r} c_{p+h} \\
&\quad + \frac{1}{n} \sum_{s=n+1}^{n+d} \sum_{t=1}^{n} \delta_{s-t+r} c_{s-t+h}.
\end{aligned}$$

Now,

$$\lim_{n\to\infty} \frac{1}{n} \sum_{s=n+1}^{n+d} \sum_{t=1}^{n} \delta_{s-t+r} c_{s-t+h} \leqslant \lim_{n\to\infty} \frac{1}{n} \sum_{p=1}^{n+d} d|\delta_{p+r}||c_{p+h}| = 0$$

with the inequality resulting from the inclusion of additional terms and the introduction of absolute values. ▲

Theorem 6.2.1. Let the time series $\{X_t\}$ be defined by

$$X_t = \sum_{j=-\infty}^{\infty} \alpha_j e_{t-j},$$

where the sequence $\{\alpha_j\}$ is absolutely summable, and the e_t are independent $(0,\sigma^2)$ random variables with $E\{e_t^4\}=\eta\sigma^4$. Then, for fixed h and q, $(h \geqslant q \geqslant 0)$,

$$\lim_{n\to\infty} (n-q)\operatorname{Cov}\{\tilde{\gamma}(h),\tilde{\gamma}(q)\}$$
$$=(\eta-3)\gamma(h)\gamma(q) + \sum_{p=-\infty}^{\infty} [\gamma(p)\gamma(p-h+q)+\gamma(p+q)\gamma(p-h)], \tag{6.2.4}$$

where $\tilde{\gamma}(h)$ is defined in (6.2.1).

Proof. Using

$$E\{e_t e_u e_v e_w\}=\begin{cases}\eta\sigma^4, & t=u=v=w\\ \sigma^4, & \text{if subscripts are equal in pairs but not all equal}\\ 0, & \text{otherwise,}\end{cases}$$

it follows that

$$E\{X_t X_{t+h} X_{t+h+p} X_{t+h+p+q}\}=(\eta-3)\sigma^4 \sum_{j=-\infty}^{\infty} \alpha_j\alpha_{j+h}\alpha_{j+h+p}\alpha_{j+h+p+q}$$
$$+\gamma(h)\gamma(q)+\gamma(h+p)\gamma(p+q)$$
$$+\gamma(h+p+q)\gamma(p). \tag{6.2.5}$$

Thus,

$$E\{\tilde{\gamma}(h)\tilde{\gamma}(q)\}-\gamma(h)\gamma(q)$$
$$=\frac{1}{(n-h)(n-q)}E\left\{\sum_{s=1}^{n-q}\sum_{t=1}^{n-h} X_t X_{t+h} X_s X_{s+q}\right\}-\gamma(h)\gamma(q)$$
$$=\frac{(\eta-3)\sigma^4}{(n-h)(n-q)}\sum_{s=1}^{n-q}\sum_{t=1}^{n-h}\sum_{j=-\infty}^{\infty}\alpha_j\alpha_{j+h}\alpha_{j+s-t}\alpha_{j+s-t+q}$$
$$+\frac{1}{(n-h)(n-q)}\sum_{s=1}^{n-q}\sum_{t=1}^{n-h}[\gamma(s-t)\gamma(s-t-h+q)$$
$$+\gamma(s-t+q)\gamma(s-t-h)]. \tag{6.2.6}$$

Applying Lemma 6.2.1 to equation (6.2.6), we have

$$\lim_{n\to\infty}(n-q)\operatorname{Cov}\{\tilde{\gamma}(h),\tilde{\gamma}(q)\}=(\eta-3)\sigma^4\sum_{j=-\infty}^{\infty}\sum_{p=-\infty}^{\infty}\alpha_j\alpha_{j+h}\alpha_{j+p}\alpha_{j+p+q}$$

$$+\sum_{p=-\infty}^{\infty}[\gamma(p)\gamma(p-h+q)$$

$$+\gamma(p+q)\gamma(p-h)]. \qquad \blacktriangle$$

For normal time series $\eta=3$, and we have

$$\operatorname{Cov}\{\tilde{\gamma}(h),\tilde{\gamma}(q)\}\doteq\frac{1}{n-q}\sum_{p=-\infty}^{\infty}[\gamma(p)\gamma(p-h+q)+\gamma(p+q)\gamma(p-h)].$$

Corollary 6.2.1.1. Given a time series, $\{e_t\colon t\in(0,\pm1,\pm2,\ldots)\}$, where the e_t are normal independent $(0,\sigma^2)$ random variables, then, for $h,q\geqslant 0$,

$$\operatorname{Cov}\{\tilde{\gamma}_e(h),\tilde{\gamma}_e(q)\}=\begin{cases}\dfrac{2\sigma^4}{n}, & h=q=0\\[2ex] \dfrac{\sigma^4}{n-h}, & h=q\neq 0\\[2ex] 0, & \text{otherwise.}\end{cases}$$

Proof. Reserved for the reader. ▲

In Theorem 6.2.1 the estimator was constructed assuming the mean to be known. As we have mentioned, the estimation of the unknown mean introduces a bias into the estimated covariance. However, the variance formulas presented in Theorem 6.2.1 remain valid approximations for the estimator defined in equation (6.2.2).

Theorem 6.2.2. Given fixed $h\geqslant q\geqslant 0$ and a time series X_t satisfying the assumptions of Theorem 6.2.1,

$$E\{\hat{\gamma}(h)-\gamma(h)\}=-\frac{|h|}{n}\gamma(h)-\frac{n-|h|}{n}\operatorname{Var}\{\bar{x}_n\}+O(n^{-2})$$

and

$$\lim_{n\to\infty}\frac{n^2}{(n-h)}\operatorname{Cov}\{\hat{\gamma}(h),\hat{\gamma}(q)\}$$

$$=(\eta-3)\gamma(h)\gamma(q)+\sum_{p=-\infty}^{\infty}[\gamma(p)\gamma(p-h+q)+\gamma(p+q)\gamma(p-h)].$$

Proof. From the definition of $\hat{\gamma}(h)$, we obtain

$$\hat{\gamma}(h)=\frac{1}{n}\left[\sum_{t=1}^{n-h} X_t X_{t+h}-\bar{x}_n \sum_{t=1}^{n-h}(X_t+X_{t+h})+(n-h)\bar{x}_n^2\right]$$

$$=\frac{n-h}{n}\tilde{\gamma}(h)+\frac{\bar{x}_n}{n}\left(\sum_{t=1}^{h} X_t+\sum_{t=n-h+1}^{n} X_t-2h\bar{x}_n\right)-\frac{(n-h)}{n}\bar{x}_n^2.$$

Now,

$$\left|E\left\{\bar{x}_n\left(\sum_{t=1}^{h} X_t+\sum_{t=n-h+1}^{n} X_t-2h\bar{x}_n\right)\right\}\right| \leqslant \frac{2h}{n}\sum_{j=-\infty}^{\infty}|\gamma(j)|$$

and we have the first result. Since, by an application of (6.2.5) $\operatorname{Var}\{\bar{x}_n^2\} = O(n^{-2})$, the second conclusion also follows. ▲

Using Theorems 6.2.1 and 6.2.2 and the results of Chapter 5, we can approximate the mean and variance of the estimated correlation function. If the mean is known, we consider the estimated autocorrelation

$$\tilde{r}(h)=\frac{\tilde{\gamma}(h)}{\tilde{\gamma}(0)}$$

and, if the mean is unknown, the estimator

$$\hat{r}(h)=\frac{\hat{\gamma}(h)}{\hat{\gamma}(0)}. \tag{6.2.7}$$

If the denominator of the estimator is zero, we define the estimator to be zero.

Theorem 6.2.3. Let the time series $\{X_t\}$ be defined by

$$X_t=\sum_{j=-\infty}^{\infty}\alpha_j e_{t-j},$$

where the sequence $\{\alpha_j\}$ is absolutely summable, and the e_t are indepen-

dent $(0,\sigma^2)$ random variables with $E\{e_t^6\}=\eta\sigma^6$. Then, for fixed h and q,

$$\begin{aligned}\operatorname{Cov}\{\tilde{r}(h),\tilde{r}(q)\} &= [\gamma(0)]^{-2}[\operatorname{Cov}\{\tilde{\gamma}(h),\tilde{\gamma}(q)\}-\rho(h)\operatorname{Cov}\{\tilde{\gamma}(0),\tilde{\gamma}(q)\} \\ &\quad -\rho(q)\operatorname{Cov}\{\tilde{\gamma}(0),\tilde{\gamma}(h)\}+\rho(h)\rho(q)\operatorname{Var}\{\tilde{\gamma}(0)\}]+O(n^{-2}) \\ &= \frac{1}{n}\sum_{p=-\infty}^{\infty}[\rho(p)\rho(p-h+q)+\rho(p+q)\rho(p-h) \\ &\quad -2\rho(q)\rho(p)\rho(p-h)-2\rho(h)\rho(p)\rho(p-q) \\ &\quad +2\rho(h)\rho(q)\rho^2(p)]+O(n^{-2}), \end{aligned} \tag{6.2.8}$$

$$\operatorname{Cov}\{\hat{r}(h),\hat{r}(q)\}=\operatorname{Cov}\{\tilde{r}(h),\tilde{r}(q)\}+O(n^{-2}),$$

$$\begin{aligned}E\{\tilde{r}(h)-\rho(h)\} &= [\gamma(0)]^{-2}[\rho(h)\operatorname{Var}\{\tilde{\gamma}(0)\}-\operatorname{Cov}\{\tilde{\gamma}(0),\tilde{\gamma}(h)\}] \\ &\quad +O(n^{-2}),\end{aligned}$$

$$\begin{aligned}E\{\hat{r}(h)\} &= \frac{n-h}{n}\rho(h)-[\gamma(0)]^{-1}[1-\rho(h)]\operatorname{Var}\{\bar{x}_n\} \\ &\quad +[\gamma(0)]^{-2}[\rho(h)\operatorname{Var}\{\tilde{\gamma}(0)\}-\operatorname{Cov}\{\hat{\gamma}(h),\hat{\gamma}(0)\}]+O(n^{-2}).\end{aligned}$$

Proof. The estimated autocorrelations are bounded and are differentiable functions of the estimated covariances on a closed set containing the true parameter vector as an interior point. Furthermore, the derivatives are bounded on that set. Hence, the conditions of Theorem 5.4.3 are met with $\{\tilde{\gamma}(h),\tilde{\gamma}(q),\tilde{\gamma}(0)\}$ playing the role of $\{\mathbf{X}_n\}$ of that theorem. Since the function $\tilde{r}(h)$ is bounded, we take $\alpha=1$.

Expanding $[\tilde{r}(h)-\rho(h)][\tilde{r}(q)-\rho(q)]$ through third order terms and using Theorem 5.4.1 to establish that the expected value of the third order moments is $O(n^{-2})$, we have result (6.2.8). The remaining results are established in a similar manner. ▲

For the first order autoregressive time series $X_t=\rho X_{t-1}+e_t$ it is relatively easy to evaluate the variances of the estimated autocorrelations. We have,

for $h>0$,

$$\operatorname{Var}\{\hat{r}(h)\} \doteq \frac{n-h}{n^2} \sum_{p=-\infty}^{\infty} \{\rho^{|2p|}+\rho^{|p+h|+|p-h|}-4\rho^{|h|}\rho^{|p|+|p-h|}+2\rho^{|2h|}\rho^{|2p|}\}$$

$$= \frac{n-h}{n^2}\left\{\frac{(1+\rho^2)(1-\rho^{2h})}{1-\rho^2}-2h\rho^{2h}\right\}. \tag{6.2.9}$$

We note that for large h,

$$\operatorname{Var}\{\hat{r}(h)\} \doteq \frac{n-h}{n^2}\frac{1+\rho^2}{1-\rho^2}. \tag{6.2.10}$$

For a time series where the correlations approach zero rapidly, the variance of $\hat{r}(h)$ for large h can be approximated by the first term of (6.2.8). That is, for such a time series and for h such that $\rho(h)\doteq 0$, we have

$$\operatorname{Var}\{\hat{r}(h)\} \doteq \frac{1}{n}\sum_{p=-\infty}^{\infty}\rho^2(p). \tag{6.2.11}$$

We are particularly interested in the behavior of the estimated autocorrelations for a time series of independent random variables, since this is often a working hypothesis in time series analysis.

Corollary 6.2.3. Let $\{e_t\}$ be a sequence of independent $(0,\sigma^2)$ random variables with sixth moment $\eta\sigma^6$. Then, for $h \geqslant q > 0$,

$$E\{\hat{r}(h)\} = -\frac{n-h}{n(n-1)} + O(n^{-2}),$$

$$\operatorname{Cov}\{\hat{r}(h),\hat{r}(q)\} = \frac{n-h}{n^2} + O(n^{-2}), \qquad h=q\neq 0,$$

$$= O(n^{-2}), \qquad \text{otherwise.}$$

Proof. Omitted. ▲

For large n the bias in $\hat{r}(h)$ is negligible. However, it is easy to reduce the bias in the null case of independent random variables. It is suggested that the estimator

$$\hat{\rho}(h) = \hat{r}(h) + \frac{n-h}{(n-1)^2}\left(1-[\hat{r}(h)]^2\right) \tag{6.2.12}$$

be used for hypothesis testing when only a small number of observations is available. For time series of the type specified in Corollary 6.2.3 and $h,q>0$,

$$E\{\hat{\rho}(h)\}=O(n^{-2}),$$

$$\begin{aligned}\operatorname{Cov}\{\hat{\rho}(h),\hat{\rho}(q)\} &= \frac{n-h}{n^2}+O(n^{-2}), && h=q>0,\\ &= O(n^{-2}), && \text{otherwise.}\end{aligned}$$

In the next section we prove that the $\hat{r}(h)$ and $\hat{\rho}(h)$ are approximately normally distributed. The approximate distribution of the autocorrelations will be adequate for most purposes, but we mention one exact distributional result. A statistic closely related to the first order autocorrelation is the von Neumann ratio:[1]

$$d_v=\frac{\sum_{t=2}^{n}(X_t-X_{t-1})^2}{\sum_{t=1}^{n}(X_t-\bar{x}_n)^2}. \tag{6.2.13}$$

We see that

$$\begin{aligned}d_v &= \frac{\sum_{t=2}^{n}(X_t-\bar{x}_n)^2-2\sum_{t=2}^{n}(X_t-\bar{x}_n)(X_{t-1}-\bar{x}_n)+\sum_{t=1}^{n-1}(X_t-\bar{x}_n)^2}{\sum_{t=1}^{n}(X_t-\bar{x}_n)^2}\\ &= 2\left[1-\hat{r}(1)-\frac{(X_1-\bar{x}_n)^2+(X_n-\bar{x}_n)^2}{2\sum_{t=1}^{n}(X_t-\bar{x}_n)^2}\right].\end{aligned}$$

If the X_t are normal independent (μ,σ^2) random variables, it is possible to show that $E\{d_v\}=2$. Therefore, $r_v=\frac{1}{2}(d_v-2)$ is an unbiased estimator of zero in that case. Von Neumann (1941) and Hart (1942) have given the exact distribution of r_v under the assumption that X_t is a sequence of normal independent (μ,σ^2) random variables. Tables of percentage points are given in Hart (1942) and in Anderson (1971, p. 345). Inspection of

[1] The ratio is sometimes defined with the multiplier $n/(n-1)$.

these tables demonstrates that the percentage points of $t_v = r_v(n+1)^{1/2}(1-r_v^2)^{-1/2}$ are approximately those of Student's t with $n+3$ degrees of freedom for n greater than 10.

Clearly the distribution of $\hat{\rho}(1)[1-\hat{\rho}^2(1)]^{-1/2}(n+1)^{1/2}$ where $\hat{\rho}(1)$ is defined in (6.2.12) is close to the distribution of t_v and, therefore, may also be approximated by Student's t with $n+3$ degrees of freedom when the observations are independent normal random variables.

Kendall and Stuart (1966) and Anderson (1971) present discussions of the distributional theory of statistics such as d_v.

6.3. SOME CENTRAL LIMIT THEOREMS FOR STATIONARY TIME SERIES

The results of this chapter have, so far, been concerned with the mean and variance of certain sample statistics computed from a single realization. In order to perform tests of hypotheses or set confidence limits for the underlying parameters, some distribution theory is required.

That the mean of a time series composed of a finite moving average of independent random variables is, in the limit, normally distributed is a simple extension of the central limit theorems of Section 5.3.

Theorem 6.3.1. Let $\{X_t: t\in(0,\pm1,\pm2,\ldots)\}$ be defined by

$$X_t = \mu + \sum_{j=0}^{m} b_j e_{t-j},$$

where $b_0 = 1$, $\sum_{j=0}^{m} b_j \neq 0$, and the e_t are independent $(0,\sigma^2)$ random variables with distribution functions $F_t(e)$ such that, for $\epsilon > 0$,

$$\lim_{n\to\infty} n^{-1}\sigma^{-2}\sum_{t=1}^{n}\int_{|e|>\epsilon\sigma n^{1/2}} e^2\,dF_t(e) = 0.$$

Then,

$$n^{1/2}(\bar{x}_n-\mu) \xrightarrow{\mathcal{L}} N\left[0,\sigma^2\left[\sum_{j=0}^{m} b_j\right]^2\right].$$

Proof. We have

$$n^{1/2}(\bar{x}_n-\mu)=n^{-1/2}\sum_{t=1}^{n}(X_t-\mu)$$

$$=n^{-1/2}\sum_{t=1}^{n}(e_t+b_1e_{t-1}+b_2e_{t-2}+\cdots+b_me_{t-m})$$

$$=n^{-1/2}\sum_{j=0}^{m}b_j\sum_{t=1}^{n}e_t+n^{-1/2}\sum_{s=1}^{m}\sum_{j=s}^{m}b_je_{1-s}$$

$$-n^{-1/2}\sum_{s=0}^{m-1}\sum_{j=s+1}^{m}b_je_{n-s}.$$

It is clear that both $n^{-1/2}\sum_{s=0}^{m-1}\sum_{j=s+1}^{m}b_je_{n-s}$ and $n^{-1/2}\sum_{s=1}^{m}\sum_{j=s}^{m}b_je_{1-s}$ converge in probability to zero as n increases. As the e_t satisfy the assumptions of the Lindeberg central limit theorem, $n^{-1/2}\sum_{t=1}^{n}e_t$ converges in distribution to a normal random variable with mean zero and variance σ^2. ▲

If $\sum_{j=0}^{m}b_j=0$, then $\bar{x}_n-\mu$ has a variance that approaches zero at a rate faster than n^{-1}. The reader can verify this by considering, for example, the time series $X_t=\mu+e_t-e_{t-1}$. In such a case the theorem holds in the sense that $n^{1/2}(\bar{x}_n-\mu)$ is converging to the singular (zero variance) normal distribution.

Moving average time series of the type investigated in Theorem 6.3.1 are special cases of a more general class of time series called *m-dependent*.

Definition 6.3.1. The sequence of random variables $\{Z_t\colon t\in(0,\pm1,\pm2,\ldots)\}$ is said to be m-dependent if $s-r>m$, where m is a positive integer, implies that the two sets

$$(\ldots,Z_{r-2},Z_{r-1},Z_r),\qquad(Z_s,Z_{s+1},Z_{s+2},\ldots)$$

are independent.

We give a theorem for such time series due to Hoeffding and Robbins (1948). We assume bounded third moments in the next two theorems to simplify the presentation, but the reader may replace this assumption by a weaker condition.

Theorem 6.3.2. Let $\{Z_t\colon t\in(1,2,\ldots)\}$ be a sequence of m-dependent random variables with $E\{Z_t\}=0$, $E\{Z_t^2\}=\sigma_t^2<\infty$, $E\{|Z_t|^3\}\leqslant\beta^3$. Let the

limit

$$\lim_{p\to\infty} p^{-1}\sum_{j=1}^{p} A_{t+j}=A,$$

$A\neq 0$, be uniform for $t=1,2,\ldots,$ where

$$A_t=E\{Z_{t+m}^2\}+2\sum_{j=1}^{m} E\{Z_{t+m-j}Z_{t+m}\}.$$

Then,

$$n^{-1/2}\sum_{t=1}^{n} Z_t \xrightarrow{\mathcal{L}} N(0,A).$$

Proof. Fix an α, $0<\alpha<1/4$, and let k be the largest integer less than n^{α} and p be the largest integer less than n/k. Define

$$Y_i=\sum_{j=1}^{k-m} Z_{(i-1)k+j}, \qquad i=1,2,\ldots,p,$$

$$S_p=n^{-1/2}\sum_{i=1}^{p} Y_i.$$

Then the difference

$$D_n=n^{-1/2}\sum_{t=1}^{n} Z_t-S_p=n^{-1/2}\left\{\sum_{i=1}^{p-1}\left(\sum_{j=1}^{m} Z_{ki-m+j}\right)+\sum_{t=pk-m+1}^{n} Z_t\right\}.$$

For n large enough so that $k>2m$, the sums $\sum_{j=1}^{m} Z_{ki-m+j}$, $i=1,2,\ldots,p$, are independent. By the assumption on the third moments, we have

$$\operatorname{Var}\left\{\sum_{j=1}^{m} Z_{ki-m+j}\right\}\leqslant m^2\beta^2,$$

$$\operatorname{Var}\left\{\sum_{t=pk-m+1}^{n} Z_t\right\}\leqslant (k+m)^2\beta^2.$$

It follows that

$$\operatorname{Var}\{D_n\} \leqslant n^{-1}\beta^2\{m^2(p-1)+(k+m)^2\} = o(1),$$

and $n^{-1/2}\sum_{t=1}^{n} Z_t$ converges in mean square to S_p. Since Z_t is correlated only with $Z_{t-m}, Z_{t-m+1}, \ldots, Z_{t+m-1}, Z_{t+m}$, the addition of a Z_t to a sum containing the m preceding terms, $Z_{t-1}, Z_{t-2}, \ldots, Z_{t-m}$, increases the variance by the amount A_{t-m}. Therefore,

$$\operatorname{Var}\{Y_i\} = \operatorname{Var}\left\{\sum_{j=1}^{k-m} Z_{(i-1)k+j}\right\} = \operatorname{Var}\left\{\sum_{j=1}^{m} Z_{(i-1)k+j}\right\} + \sum_{j=1}^{k-2m} A_{(i-1)k+j}$$

and

$$\operatorname{Var}\{S_p\} = \frac{1}{n}\sum_{i=1}^{p}\operatorname{Var}\{Y_i\}$$

$$= \frac{1}{n}\sum_{i=1}^{p}\operatorname{Var}\left\{\sum_{j=1}^{m} Z_{(i-1)k+j}\right\} + \frac{1}{n}\sum_{i=1}^{p}\sum_{j=1}^{k-2m} A_{(i-1)k+j}.$$

Since $\operatorname{Var}\{\sum_{j=1}^{m} Z_{(i-1)k+j}\} \leqslant m^2\beta^2$, and $p^{-1}\sum_{j=1}^{p} A_{t+j}$ converges uniformly, we have

$$\lim_{p\to\infty}\frac{1}{n}\sum_{i=1}^{p}\operatorname{Var}\{Y_i\} = \lim_{p\to\infty} p^{-1}\sum_{i=1}^{p}\operatorname{Var}(k^{-1/2}Y_i) = A.$$

The third moment of $k^{-1/2}Y_i$ is

$$E\{|k^{-1/2}Y_i|^3\} = k^{-3/2}\,E\left\{\left|\sum_{j=1}^{k-m} Z_{(i-1)k+j}\right|^3\right\} \leqslant k^{3/2}\beta^3.$$

Since

$$\lim_{p\to\infty}\frac{\sum_{i=1}^{p} E\{|k^{-1/2}Y_i|^3\}}{\left(\sum_{i=1}^{p} E\{|k^{-1/2}Y_i|^2\}\right)^{3/2}} \leqslant \lim_{p\to\infty}\frac{p(k^{3/2}\beta^3)}{(pA)^{3/2}} = 0,$$

$p^{-1/2}\sum_{i=1}^{p} k^{-1/2}Y_i$ satisfies the conditions of Liapounov's central limit theorem, and the result follows. ▲

For the mth order moving average,

$$A \equiv A_t = \gamma(0) + 2\sum_{j=1}^{m} \gamma(j)$$

$$= \sum_{j=0}^{m} b_j^2\sigma^2 + 2\sum_{j=1}^{m}\sum_{s=0}^{m-j} b_s b_{s+j}\sigma^2$$

$$= \left(\sum_{j=0}^{m} b_j\right)^2 \sigma^2$$

which agrees with Theorem 6.3.1.

The results of Theorems 6.3.1 and 6.3.2 may be generalized to infinite moving averages of independent random variables. Our proofs follow those of Diananda (1953) and Anderson (1959, 1971). We first state a lemma required in the proofs of the primary theorems.

Lemma 6.3.1. Let the random variables ξ_n with distribution functions $F_{\xi_n}(z)$ be defined by

$$\xi_n = S_{kn} + D_{kn}$$

for $k = 1, 2, \ldots$ and $n = 1, 2, \ldots$. Let

$$\operatorname*{plim}_{k\to\infty} D_{kn} = 0$$

uniformly in n,

$$F_{S_{kn}}(z) \xrightarrow{C} F_{\psi_k}(z) \quad \text{as} \quad n\to\infty,$$

and

$$F_{\psi_k}(z) \xrightarrow{C} F_{\xi}(z) \quad \text{as} \quad k\to\infty.$$

Then,

$$F_{\xi_n}(z) \xrightarrow{C} F_{\xi}(z) \quad \text{as} \quad n\to\infty.$$

Proof. Let z_0 be a point of continuity of $F_\xi(z)$ and fix $\delta > 0$. There is an $\epsilon > 0$ such that

$$|F_\xi(z) - F_\xi(z_0)| < \frac{\delta}{4}$$

for z in $[z_0-\epsilon, z_0+\epsilon]$ and $F_\xi(z)$ is continuous at $z_0-\epsilon$ and at $z_0+\epsilon$. By hypothesis, there exists a K_0 such that

$$|F_{\psi_k}(z_0-\epsilon)-F_\xi(z_0-\epsilon)|<\frac{\delta}{4},$$

$$|F_{\psi_k}(z_0+\epsilon)-F_\xi(z_0+\epsilon)|<\frac{\delta}{4}$$

for $k>K_0$. This means that

$$|F_{\psi_k}(z)-F_\xi(z_0)|<\frac{2\delta}{4}$$

for $k>K_0$ and $z_0-\epsilon\leqslant z\leqslant z_0+\epsilon$.

Now there is a K_1 such that for $k\geqslant K_1$,

$$P\{|\xi_n-S_{kn}|>\epsilon\}<\frac{\delta}{4}$$

for all n. By the arguments used in the proof of Theorem 5.2.1, this implies that

$$F_{S_{kn}}(z-\epsilon)-\frac{\delta}{4}\leqslant F_{\xi_n}(z)\leqslant F_{S_{kn}}(z+\epsilon)+\frac{\delta}{4}$$

for all z.

Fix k at the maximum of K_0 and K_1. Let z_1 and z_2 be continuity points of $F_{\psi_k}(z)$ such that $z_0-\epsilon\leqslant z_1<z_0$ and $z_0<z_2\leqslant z_0+\epsilon$. Then there exists an N such that for $n>N$,

$$|F_{S_{kn}}(z_1)-F_{\psi_k}(z_1)|<\frac{\delta}{4}$$

$$|F_{S_{kn}}(z_2)-F_{\psi_k}(z_2)|<\frac{\delta}{4}$$

Therefore, for $n>N$,

$$\begin{aligned}F_\xi(z_0)-\delta &\leqslant F_{\psi_k}(z_1)-\frac{2\delta}{4}\leqslant F_{S_{kn}}(z_1)-\frac{\delta}{4}\\ &\leqslant F_{\xi_n}(z_0)\leqslant F_{S_{kn}}(z_2)+\frac{\delta}{4}\leqslant F_{\psi_k}(z_2)+\frac{2\delta}{4}\\ &\leqslant F_\xi(z_0)+\delta.\end{aligned}$$

▲

Theorem 6.3.3. Let X_t be a time series defined by

$$X_t = \sum_{j=0}^{\infty} \alpha_j e_{t-j},$$

where $\sum_{j=0}^{\infty}|\alpha_j| < \infty$, $\sum_{j=0}^{\infty}\alpha_j \neq 0$, and the e_t are independent $(0,\sigma^2)$ random variables with finite third moment, β. Then,

$$n^{-1/2}\sum_{t=1}^{n} X_t \xrightarrow{\mathcal{L}} N\left(0, \sum_{h=-\infty}^{\infty} \gamma_X(h)\right),$$

where

$$\sum_{h=-\infty}^{\infty} \gamma_X(h) = \left(\sum_{j=0}^{\infty} \alpha_j\right)^2 \sigma^2.$$

Proof. Let

$$Y_{tk} = \sum_{j=0}^{k} \alpha_j e_{t-j},$$

$$W_{tk} = \sum_{j=k+1}^{\infty} \alpha_j e_{t-j},$$

and define the normalized sums

$$S_{kn} = n^{-1/2} \sum_{t=1}^{n} Y_{tk}$$

$$D_{kn} = n^{-1/2} \sum_{t=1}^{n} W_{tk}.$$

For a fixed k, W_{tk} is a stationary time series such that

$$\gamma_W(h) = \sum_{j=k+1}^{\infty} \alpha_j \alpha_{j+|h|}\sigma^2, \qquad h = 0, \pm 1, \pm 2\ldots .$$

It follows that

$$\begin{aligned}
\operatorname{Var}\{D_{kn}\} &= \frac{1}{n} \sum_{h=-(n-1)}^{n-1} (n-|h|)\gamma_W(h) \\
&\leqslant \sigma^2 \sum_{j=k+1}^{\infty} \alpha_j^2 + 2\sigma^2 \sum_{h=1}^{n-1} \sum_{j=k+1}^{\infty} |\alpha_j \alpha_{j+h}| \\
&\leqslant \sigma^2 \left(\sum_{j=k+1}^{\infty} |\alpha_j| \right)^2.
\end{aligned}$$

Therefore, by Chebyshev's inequality,

$$\operatorname*{plim}_{k\to\infty} D_{kn}=0$$

uniformly in n. For fixed k, Y_{tk} is a finite moving average with finite third moment. Hence, S_{kn} converges in law to a normal random variable with mean zero and variance $(\Sigma_{j=0}^{k}\alpha_j)^2\sigma^2$. As k increases, the variance of the normal distribution converges to $(\Sigma_{j=0}^{\infty}\alpha_j)^2\sigma^2$. The conclusion follows by Lemma 6.3.1. ▲

We now obtain a central limit theorem for a linear function of a realization of a stationary time series. We shall state the assumptions of this theorem in a manner that permits us to use the Lindeberg central limit theorem.

Theorem 6.3.4. Let $\{X_t: t\in T=(0,\pm 1,\pm 2,\ldots)\}$ be a time series defined by

$$X_t=\sum_{j=0}^{\infty}\alpha_j e_{t-j},$$

where $\Sigma_{j=0}^{\infty}|\alpha_j|<\infty$, and the e_t are independent $(0,\sigma^2)$ random variables with distribution functions $F_t(e)$ such that

$$\lim_{\delta\to\infty}\sup_{t\in T}\int_{|e|>\delta} e^2\,dF_t(e)=0.$$

Furthermore, let $\{C_t\}_{t=1}^{\infty}$ be a sequence of fixed real numbers satisfying

$$\lim_{n\to\infty}\sum_{t=1}^{n}C_t^2=\infty,$$

$$\lim_{n\to\infty}\frac{C_n^2}{\sum_{t=1}^{n}C_t^2}=0,$$

$$\lim_{n\to\infty}\frac{\sum_{t=1}^{n-h}C_tC_{t+|h|}}{\sum_{t=1}^{n}C_t^2}=g(h),\qquad h=0,\pm 1,\pm 2,\ldots .$$

Let $V=\Sigma_{h=-\infty}^{\infty}g(h)\gamma_X(h)\neq 0$. Then,

$$\left[\sum_{t=1}^{n}C_t^2\right]^{-1/2}\sum_{t=1}^{n}C_tX_t\xrightarrow{\mathcal{L}}N(0,V).$$

Proof. Following the proof of Theorem 6.3.3, we set

$$Y_{tk} = \sum_{j=0}^{k} \alpha_j e_{t-j}$$

$$W_{tk} = \sum_{j=k+1}^{\infty} \alpha_j e_{t-j}$$

and note that

$$\operatorname{Var}\left\{\left(\sum_{t=1}^{n} C_t^2\right)^{-1/2} \sum_{t=1}^{n} C_t W_{tk}\right\} = \left(\sum_{t=1}^{n} C_t^2\right)^{-1} \sum_{t=1}^{n} \sum_{j=1}^{n} C_t C_j \gamma_W(t-j)$$

$$\leqslant \sum_{h=-(n-1)}^{n-1} |\gamma_W(h)|,$$

which, by the proof of Theorem 6.3.3, can be made arbitrarily small by choosing k sufficiently large.

For fixed k, Y_{tk} is a finite moving average and, following the proof of Theorem 6.3.1, we have

$$\left(\sum_{t=1}^{n} C_t^2\right)^{-1/2} \left(\sum_{t=1}^{n} C_t Y_{tk}\right) = \left(\sum_{t=1}^{n} C_t^2\right)^{-1/2} \sum_{t=1}^{n} \sum_{j=0}^{k} \alpha_j C_{t+j} e_t + o_p(1),$$

where the random variables $(\Sigma_{j=0}^{k} \alpha_j C_{t+j}) e_t$ are independent with mean zero and variance $(\Sigma_{j=0}^{k} \alpha_j C_{t+j})^2 \sigma^2$. Let

$$S_n = V_n^{-1/2} \sum_{t=1}^{n} \left(\sum_{j=0}^{k} \alpha_j C_{t+j}\right) e_t,$$

$$V_n = \sum_{t=1}^{n} \left(\sum_{j=0}^{k} \alpha_j C_{t+j}\right)^2 \sigma^2,$$

$$M_n = \sup_{1 \leqslant t \leqslant n} \left(\sum_{j=0}^{k} \alpha_j C_{t+j}\right)^2,$$

and consider, for $\delta > 0$,

$$\lim_{n\to\infty} V_n^{-1} \sum_{t=1}^{n} \left(\sum_{j=0}^{k} \alpha_j C_{t+j} \right)^2 \int_{R_t} e^2\, dF_t(e)$$

$$\leqslant \lim_{n\to\infty} \sum_{t=1}^{n} V_n^{-1} \left(\sum_{j=0}^{k} \alpha_j C_{t+j} \right)^2 \sup_{1\leqslant t\leqslant n} \int_{R_0} e^2\, dF_t(e),$$

where

$$R_t = \left\{ e\colon |e| > \left[V_n^{-1} \left(\sum_{j=0}^{k} \alpha_j C_{t+j} \right)^2 \right]^{-1/2} \delta \right\}$$

and

$$R_0 = \left\{ e\colon |e| > M_n^{-1/2} V_n^{1/2} \delta \right\}.$$

Clearly, $R_t \subset R_0$ for all $t \leqslant n$. Now the assumption

$$\lim_{n\to\infty} \frac{C_n^2}{\sum_{t=1}^{n} C_t^2} = 0$$

implies that

$$\lim_{n\to\infty} \sup_{1\leqslant t\leqslant n} \frac{C_t^2}{\sum_{j=1}^{n} C_j^2} = 0.$$

Because $V = \lim_{n\to\infty} (\Sigma_{t=1}^{n} C_t^2)^{-1} \Sigma_{t=1}^{n} (\Sigma_{j=0}^{\infty} \alpha_j C_{t+j})^2 \sigma^2 \neq 0$, we have

$$\lim_{n\to\infty} \sup_{1\leqslant t\leqslant n} \frac{\left(\sum_{j=0}^{k} \alpha_j C_{t+j} \right)^2}{\sum_{t=1}^{n} \left(\sum_{j=0}^{k} \alpha_j C_{t+j} \right)^2} = 0,$$

which, in turn, implies that

$$\lim_{n\to\infty} M_n^{-1/2} V_n^{1/2} = \infty.$$

By assumption, the supremum of the integral over R_0 goes to zero. Hence the assumptions of the Lindeberg central limit theorem are met and S_n converges in distribution to a normal random variable with zero mean and unit variance. Since

$$\lim_{n\to\infty} \frac{\sum_{t=1}^{n}\left(\sum_{j=0}^{k}\alpha_j C_{t+j}\right)^2}{\sum_{t=1}^{n} C_t^2} = g(0)\sum_{j=0}^{k}\alpha_j^2 + 2\sum_{h=1}^{k} g(h)\sum_{j=0}^{k-h}\alpha_j\alpha_{j+h}$$

and

$$\lim_{k\to\infty}\sigma^2\left\{g(0)\sum_{j=0}^{k}\alpha_j^2 + 2\sum_{h=1}^{k} g(h)\sum_{j=0}^{k-h}\alpha_j\alpha_{j+h}\right\} = \sum_{h=-\infty}^{\infty} g(h)\gamma_X(h)$$

the result follows by Lemma 6.3.1. ▲

Since a stationary finite order autoregressive time series can be expressed as an infinite moving average with absolutely summable covariance function, the conclusions of Theorems 6.3.3 and 6.3.4 hold for such time series.

We now investigate the large sample properties of the estimated autocovariances and autocorrelations.

Theorem 6.3.5. Let X_t be a finite moving average time series defined by

$$X_t = \sum_{j=0}^{M}\alpha_j e_{t-j},$$

where the e_t are independent $(0,\sigma^2)$ random variables with fourth moment $\eta\sigma^4$ and sixth moment ξ. Let K be fixed. Then the limiting distribution of $n^{1/2}[\hat{\gamma}(0)-\gamma(0),\hat{\gamma}(1)-\gamma(1),\ldots,\hat{\gamma}(K)-\gamma(K)]'$ is multivariate normal with mean zero and covariance matrix $\mathbf{V}$ where the elements of $\mathbf{V}$ are defined by (6.2.4) of Theorem 6.2.1.

Proof. The estimated covariance, for $h=0,1,2,\ldots,K$, is

$$\begin{aligned}\hat{\gamma}(h) &= \frac{1}{n}\sum_{t=1}^{n-h}(X_t-\bar{x}_n)(X_{t+h}-\bar{x}_n)\\ &= \frac{1}{n}\sum_{t=1}^{n-h}X_tX_{t+h} - \frac{1}{n}\bar{x}_n\sum_{t=1}^{n-h}(X_t+X_{t+h}) + \frac{n-h}{n}\bar{x}_n^2 \qquad (6.3.1)\end{aligned}$$

and the last two terms, when multiplied by $n^{1/2}$, converge in probability to zero. Therefore, in investigating the limiting distribution of $n^{1/2}[\hat{\gamma}(h)-\gamma(h)]$ we need only consider the first term on the right of (6.3.1). Let

$$S_n = n^{1/2}\sum_{h=0}^{K}\lambda_h\left[\frac{1}{n}\sum_{t=1}^{n-h}X_tX_{t+h}-\gamma(h)\right]$$

$$= n^{-1/2}\sum_{h=0}^{K}\sum_{t=1}^{n-h}\lambda_h[Z_{th}-E\{Z_{th}\}]-n^{-1/2}\sum_{h=0}^{K}h\lambda_h\gamma(h),$$

where the λ_h are arbitrary real numbers (not all zero) and

$$Z_{th}=X_tX_{t+h}, \qquad h=0,1,2,\ldots,K.$$

Now Z_{th} is an $(M+h)$-dependent covariance stationary time series with mean $\gamma_X(h)$ and covariance function

$$\gamma_{Z_h}(s)=E\{(X_tX_{t+h})(X_{t+s}X_{t+s+h})\}$$

$$=(\eta-3)\sigma^4\sum_{j=-\infty}^{\infty}\alpha_j\alpha_{j+h}\alpha_{j+s}\alpha_{j+s+h}$$

$$+\gamma^2(h)+\gamma^2(s)+\gamma(s+h)\gamma(s-h),$$

where we have used (6.2.5), and it is understood that $\alpha_j=0$ for $j>M$ and $j<0$. Thus, the weighted average of the Z_{th}'s,

$$Y_t=\sum_{h=0}^{K}\lambda_hZ_{th}=\sum_{h=0}^{K}\lambda_hX_tX_{t+h},$$

is a stationary time series. Furthermore, the time series Y_t is $(M+K)$-dependent and has finite third moment, and

$$\lim_{n\to\infty}n^{-1/2}\sum_{h=0}^{K}h\lambda_h\gamma(h)=0.$$

Therefore, by Theorem 6.3.2, S_n converges in distribution to a normal random variable. Since the λ_h were arbitrary, the vector random variable $n^{1/2}[\hat{\gamma}(0)-\gamma(0),\hat{\gamma}(1)-\gamma(1),\ldots,\hat{\gamma}(K)-\gamma(K)]'$ converges in distribution to a multivariate normal by Theorem 5.3.3. ▲

Generally it is the estimated autocorrelations that are subjected to analysis and hence their limiting distribution is of interest.

Corollary 6.3.5.1. Let the assumptions of Theorem 6.3.5 hold. Then the vector $n^{1/2}[\hat{r}(1)-\rho(1),\hat{r}(2)-\rho(2),\ldots,\hat{r}(K)-\rho(K)]'$ converges in distribution to a multivariate normal with mean zero and covariance matrix $\mathbf{G}$, where the hqth element of $\mathbf{G}$ is $\sum_{p=-\infty}^{\infty}[\rho(p)\rho(p-h+q)+\rho(p+q)\rho(p-h)-2\rho(q)\rho(p)\rho(p-h)-2\rho(h)\rho(p)\rho(p-q)+2\rho(h)\rho(q)\rho^2(p)]$.

Proof. Since the $\hat{r}(h)$ are continuous differentiable functions of the $\hat{\gamma}(h)$, the result follows immediately from Theorems 5.1.4, 6.2.3, and 6.3.5. ▲

Observe that if the original time series X_t is a sequence of independent identically distributed random variables with finite moments, the sample correlations will be nearly independent in large samples. Because of the importance of this result in the testing of time series for independence, we state it as a corollary.

Corollary 6.3.5.2. Let the time series e_t be a sequence of independent $(0,\sigma^2)$ random variables with uniformly bounded sixth moments. Let $\hat{\rho}(h)$ be defined by (6.2.12) and let K be a fixed integer. Then $n(n-h)^{-1/2}\hat{\rho}(h)$, $h=1,2,\ldots,K$, converge in distribution to independent normal $(0,1)$ random variables.

Proof. Omitted. ▲

Note that we chose to normalize the $\hat{\rho}(h)$ by $n(n-h)^{-1/2}$ on the basis of the variance results of Corollary 6.2.3. The limiting distribution of the correlation coefficients can be obtained under considerably relaxed conditions. See, for example, Hannan and Heyde (1972).

By arguments analogous to those of Theorem 6.3.3, it is possible to extend the results of Theorem 6.3.5 to the estimated covariances of infinite moving average time series.

Theorem 6.3.6. Let $\{X_t\colon t\in(0,\pm1,\pm2,\ldots)\}$ be a time series defined by

$$X_t=\sum_{j=0}^{\infty}\alpha_j e_{t-j},$$

where

$$\sum_{j=0}^{\infty}|\alpha_j|<\infty$$

and the e_t are independent $(0,\sigma^2)$ random variables with fourth moment $\eta\sigma^4$ and sixth moment ξ. Let K be fixed. Then the limiting distribution of $n^{1/2}[\hat{\gamma}(0)-\gamma(0),\hat{\gamma}(1)-\gamma(1),\ldots,\hat{\gamma}(K)-\gamma(K)]'$ is multivariate normal with mean zero and covariance matrix $\mathbf{V}$ whose elements are defined by (6.2.4) of Theorem 6.2.1.

Proof. Omitted. ▲

6.4. AN EXAMPLE

The quarterly seasonally adjusted United States unemployment rate from 1948 to 1972 is given in Table 6.4.1 and displayed in Figure 6.4.1. The mean unemployment rate is 4.77. The autocorrelation function estimated using (6.2.7) is given in Figure 6.4.2 and Table 6.4.2. This plot is sometimes called a correlogram.

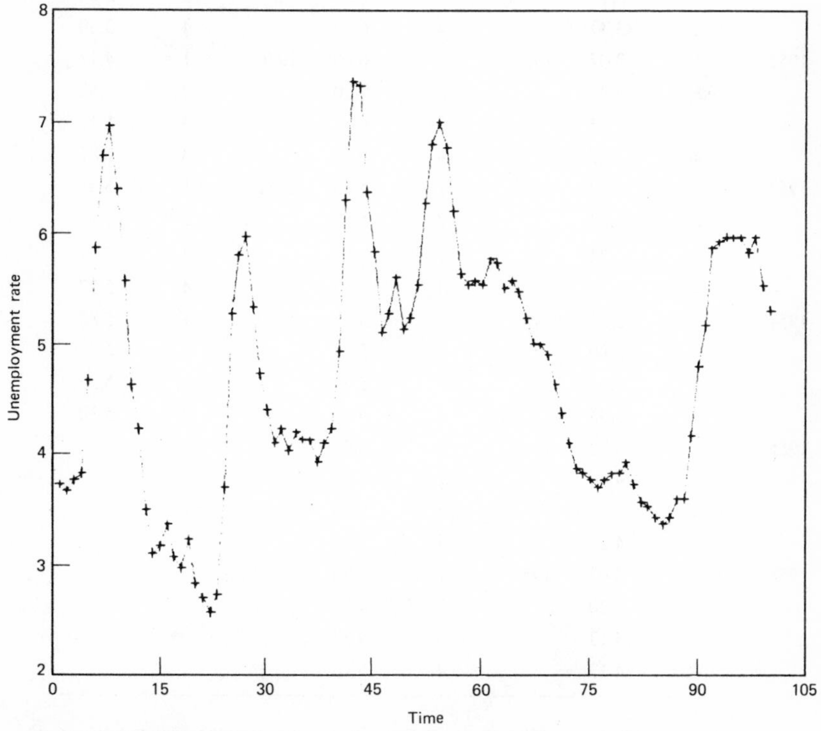

Figure 6.4.1. United States quarterly seasonally adjusted unemployment rate.

Table 6.4.1 U. S. unemployment rate (quarterly seasonally adjusted) 1948–1972

Year	Quarter	Rate	Year	Quarter	Rate	Year	Quarter	Rate
1948	1	3.73	1957	1	3.93	1966	1	3.87
	2	3.67		2	4.10		2	3.80
	3	3.77		3	4.23		3	3.77
	4	3.83		4	4.93		4	3.70
1949	1	4.67	1958	1	6.30	1967	1	3.77
	2	5.87		2	7.37		2	3.83
	3	6.70		3	7.33		3	3.83
	4	6.97		4	6.37		4	3.93
1950	1	6.40	1959	1	5.83	1968	1	3.73
	2	5.57		2	5.10		2	3.57
	3	4.63		3	5.27		3	3.53
	4	4.23		4	5.60		4	3.43
1951	1	3.50	1960	1	5.13	1969	1	3.37
	2	3.10		2	5.23		2	3.43
	3	3.17		3	5.53		3	3.60
	4	3.37		4	6.27		4	3.60
1952	1	3.07	1961	1	6.80	1970	1	4.17
	2	2.97		2	7.00		2	4.80
	3	3.23		3	6.77		3	5.17
	4	2.83		4	6.20		4	5.87
1953	1	2.70	1962	1	5.63	1971	1	5.93
	2	2.57		2	5.53		2	5.97
	3	2.73		3	5.57		3	5.97
	4	3.70		4	5.53		4	5.97
1954	1	5.27	1963	1	5.77	1972	1	5.83
	2	5.80		2	5.73		2	5.77
	3	5.97		3	5.50		3	5.53
	4	5.33		4	5.57		4	5.30
1955	1	4.73	1964	1	5.47			
	2	4.40		2	5.20			
	3	4.10		3	5.00			
	4	4.23		4	5.00			
1956	1	4.03	1965	1	4.90			
	2	4.20		2	4.67			
	3	4.13		3	4.37			
	4	4.13		4	4.10			

SOURCES. *Business Statistics*, 1971 Biennial Edition, pp. 68 and 233 and *Survey of Current Business*, January 1972 and January 1973. Quarterly data are the averages of monthly data.

Table 6.4.2. ***Estimated autocorrelations, quarterly U. S. seasonally adjusted unemployment rate, 1948–72.***

Lag h	Estimated correlations	$\lvert 1.96n^{-1}(n-h)^{1/2}\rvert$	Correlations for Second Order Process
0	1.0000	—	1.0000
1	0.9200	0.1950	0.9200
2	0.7436	0.1940	0.7436
3	0.5348	0.1930	0.5262
4	0.3476	0.1920	0.3105
5	0.2165	0.1910	0.1247
6	0.1394	0.1900	−0.0164
7	0.0963	0.1890	−0.1085
8	0.0740	0.1880	−0.1557
9	0.0664	0.1870	−0.1665
10	0.0556	0.1859	−0.1515
11	0.0352	0.1849	−0.1212
12	0.0109	0.1839	−0.0848
13	−0.0064	0.1828	−0.0490
14	−0.0135	0.1818	−0.0186
15	0.0004	0.1807	0.0043
16	0.0229	0.1796	0.0190
17	0.0223	0.1786	0.0263
18	−0.0126	0.1775	0.0277
19	−0.0762	0.1764	0.0249
20	−0.1557	0.1753	0.0197
21	−0.2351	0.1742	0.0136
22	−0.2975	0.1731	0.0077
23	−0.3412	0.1720	0.0027
24	−0.3599	0.1709	−0.0010
25	−0.3483	0.1697	−0.0033
26	−0.3236	0.1686	−0.0044
27	−0.3090	0.1675	−0.0046
28	−0.3142	0.1663	−0.0041
29	−0.3299	0.1652	−0.0032
30	−0.3396	0.1640	−0.0022
31	−0.3235	0.1628	−0.0012
32	−0.2744	0.1616	−0.0004
33	−0.2058	0.1604	0.0002
34	−0.1378	0.1592	0.0006
35	−0.0922	0.1580	0.0007
36	−0.0816	0.1568	0.0008
37	−0.1027	0.1556	0.0007
38	−0.1340	0.1543	0.0005
39	−0.1590	0.1531	0.0004
40	−0.1669	0.1518	0.0002

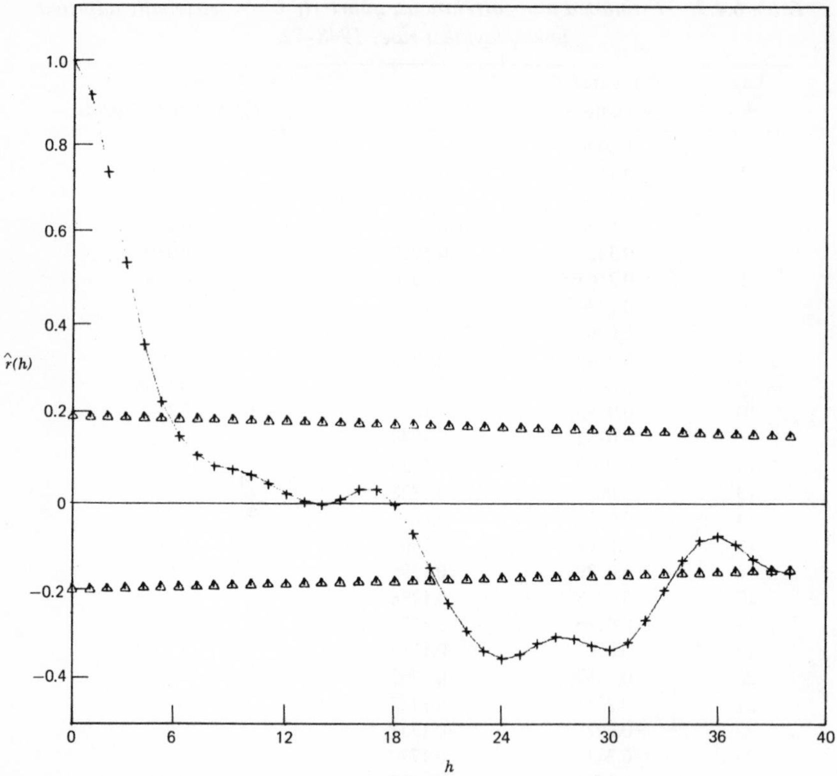

Figure 6.4.2. Correlogram of quarterly seasonally adjusted unemployment rate.

To carry out statistical analyses we assume the time series can be treated as a stationary time series with finite sixth moment. If the original time series was a sequence of uncorrelated random variables, we would expect about 95 percent of the estimated correlations to fall between the lines plus and minus $1.96n^{-1}(n-h)^{1/2}$. Obviously unemployment is not an uncorrelated time series. Casual inspection of the correlogram might lead one to conclude that the time series contains a periodic component with a period of about 54 quarters, since the estimated correlations for h equal to 21 through 33 are negative and below the 1.96 sigma bounds for an uncorrelated time series. However, because the time series is highly correlated at small lags, the variance of the estimated correlations at large lags is much larger than the variance of correlations computed from a white noise sequence.

The first few autocorrelations of the unemployment time series are in

good agreement with those generated by the second order autoregressive process

$$X_t = 1.5356X_{t-1} - 0.6692X_{t-2} + e_t,$$

where the e_t are uncorrelated (0, 0.1155) random variables. We shall discuss the estimation of the parameters of autoregressive time series in Chapter 8. However, the fact that the sample autocorrelations are consistent estimators of the population correlations permits us to obtain consistent estimators of the autoregressive parameters of a second order process from (2.5.7). The general agreement between the correlations for the second order autoregressive process and the sample correlations is clear from Table 6.4.2.

The roots of the second order autoregressive process are $0.768 \pm 0.282\,\iota$. We recall that the correlation function of such a second order process can be written as

$$\rho(h) = b_1 m_1^h + b_2 m_2^h, \qquad h = 0, 1, 2, \ldots,$$

where

$$b_1 = \frac{m_2 - \rho(1)}{m_2 - m_1},$$

$$b_2 = \frac{m_1 - \rho(1)}{m_1 - m_2}.$$

For the unemployment time series the estimated parameters are $\hat{b}_1 = 0.500 - 0.270\iota$ and $\hat{b}_2 = \hat{b}_1^* = 0.500 + 0.270\iota$. Using these values, we can estimate the variance of the estimated autocorrelations for large lags using (6.2.11). We have

$$\begin{aligned} n\hat{\text{V}}\text{ar}\{\hat{r}(h)\} &\doteq \sum_{s=-\infty}^{\infty} \hat{\rho}^2(s) \\ &= \hat{b}_1^2 \frac{1+\hat{m}_1^2}{1-\hat{m}_1^2} + \hat{b}_2^2 \frac{1+\hat{m}_2^2}{1-\hat{m}_2^2} \\ &\quad + 2\hat{b}_1\hat{b}_2 \frac{1+\hat{m}_1\hat{m}_2}{1-\hat{m}_1\hat{m}_2} \\ &= 4.812. \end{aligned}$$

Thus the estimated standard error of the estimated autocorrelations at large lags is about 0.22, and the observed correlations at lags near 27 could arise from such a process.

Given that the time series was generated by the second order autoregressive mechanism, the variance of the sample mean can be estimated by Corollary 6.1.1.2. By that corollary

$$n\hat{\mathrm{V}}\mathrm{ar}\{\bar{x}_n\} = \sum_{h=-\infty}^{\infty} \hat{\gamma}(h) = \hat{\gamma}(0) \sum_{h=-\infty}^{\infty} \hat{\rho}(h)$$

$$= \frac{0.1155}{(1-1.5356+.6692)^2}$$

$$= 6.47.$$

For this highly correlated process the variance of the mean is about five times that of an uncorrelated time series with the same variance.

6.5. ESTIMATION OF THE CROSS COVARIANCES

In our Section 1.7 discussion of vector valued time series we introduced the $k \times k$ covariance matrix

$$\boldsymbol{\Gamma}(h) = E\{(\mathbf{X}_t - \boldsymbol{\mu})(\mathbf{X}_{t+h} - \boldsymbol{\mu})'\},$$

where $\mathbf{X}_t$ is a stationary k-dimensional time series and $\boldsymbol{\mu} = E\{\mathbf{X}_t\}$. The jth diagonal element of the matrix is the autocovariance of X_{jt}, $\gamma_{jj}(h) = E\{(X_{jt} - \mu_j)(X_{j,t+h} - \mu_j)\}$, and the ijth element is the cross covariance between X_{it} and X_{jt},

$$\gamma_{ij}(h) = E\{(X_{it} - \mu_i)(X_{j,t+h} - \mu_j)\}.$$

Expressions for the estimated cross covariance analogous to those of (6.2.1) and (6.2.3) are

$$\tilde{\gamma}_{ij}(h) = \frac{1}{n-h} \sum_{t=1}^{n-h} X_{it} X_{j,t+h}, \qquad h = 0, 1, \ldots, n-1$$

$$= \frac{1}{n+h} \sum_{t=-h}^{n} X_{it} X_{j,t+h}, \qquad h = -1, -2, \ldots, -(n-1) \qquad (6.5.1)$$

for the means known and taken to be zero, and

$$\hat{\gamma}_{ij}(h) = \frac{1}{n} \sum_{t=1}^{n-h} (X_{it} - \bar{x}_{in})(X_{j,t+h} - \bar{x}_{jn}), \qquad h = 0, 1, \ldots, n-1$$

$$= \frac{1}{n} \sum_{t=-h}^{n} (X_{it} - \bar{x}_{in})(X_{j,t+h} - \bar{x}_{jn}), \qquad h = -1, -2, \ldots, -(n-1), \qquad (6.5.2)$$

where the unknown means are estimated by $\bar{x}_{in} = n^{-1}\Sigma_{t=1}^{n} X_{it}$. By (1.7.4), we can also write

$$\hat{\gamma}_{ij}(-h) = \hat{\gamma}_{ji}(h) = \frac{1}{n}\sum_{t=1}^{n-h}(X_{jt} - \bar{x}_{jn})(X_{i,t+h} - \bar{x}_{in}), \qquad h = 0, 1, \ldots, n-1.$$

The corresponding estimators of the cross correlations are

$$\tilde{r}_{ij}(h) = [\tilde{\gamma}_{ii}(0)\tilde{\gamma}_{jj}(0)]^{-1/2}\tilde{\gamma}_{ij}(h) \tag{6.5.3}$$

and

$$\hat{r}_{ij}(h) = [\hat{\gamma}_{ii}(0)\hat{\gamma}_{jj}(0)]^{-1/2}\hat{\gamma}_{ij}(h). \tag{6.5.4}$$

By our earlier results (see Theorem 6.2.2) the estimation of the mean in estimator (6.5.2) introduces a bias that is $O(n^{-1})$ for time series with absolutely summable covariance function. The properties of the estimated cross covariances are analogous to the properties of the estimated autocovariance given in Theorems 6.2.1 and 6.2.3. To simplify the derivation, we only present the results for normal time series.

Theorem 6.5.1. Let the bivariate stationary normal time series $\mathbf{X}_t$ be such that

$$\sum_{h=-\infty}^{\infty} |\gamma_{ii}(h)| < \infty, \qquad i = 1, 2.$$

Then

$$\lim_{n\to\infty} n \operatorname{Cov}\{\hat{\gamma}_{12}(h), \hat{\gamma}_{12}(q)\} = \sum_{p=-\infty}^{\infty} \gamma_{11}(p)\gamma_{22}(p+q-h) + \sum_{p=-\infty}^{\infty} \gamma_{12}(p+q)\gamma_{21}(p-h).$$

Proof. Letting the mean vector be zero and $h, q \geqslant 0$, we have

$$E\{\hat{\gamma}_{12}(h)\hat{\gamma}_{12}(q)\} = E\left\{\frac{1}{n^2}\sum_{t=1}^{n-h} X_{1t}X_{2,t+h}\sum_{s=1}^{n-q} X_{1s}X_{2,s+q}\right\} + O(n^{-1})$$

where the remainder term enters because the mean is estimated. Evaluating

the expectation, we have

$$E\{[\hat{\gamma}_{12}(h)-\gamma_{12}(h)][\hat{\gamma}_{12}(q)-\gamma_{12}(q)]\}$$

$$=\frac{1}{n^2}\sum_{t=1}^{n-h}\sum_{s=1}^{n-q}\gamma_{11}(s-t)\gamma_{22}(s-t+q-h)$$

$$+\frac{1}{n^2}\sum_{t=1}^{n-h}\sum_{s=1}^{n-q}\gamma_{12}(s+q-t)\gamma_{21}(s-t-h)+O(n^{-1}).$$

Using Lemma 6.2.1 to take the limit, we have the stated result. ▲

Since the cross correlations are simple functions of the cross covariances, we can obtain a similar expression for the covariance of the estimated cross correlations.

Corollary 6.5.1.1. Given the bivariate stationary normal time series of Theorem 6.5.1,

$$\lim_{n\to\infty} n\operatorname{Cov}\{\hat{r}_{12}(h),\hat{r}_{12}(q)\}$$

$$=\sum_{p=-\infty}^{\infty}\Big[\rho_{11}(p)\rho_{22}(p+q-h)+\rho_{12}(p+q)\rho_{21}(p-h)$$

$$-\rho_{12}(h)\{\rho_{11}(p)\rho_{21}(p+q)+\rho_{22}(p)\rho_{21}(p-q)\}$$

$$-\rho_{12}(q)\{\rho_{11}(p)\rho_{21}(p+h)+\rho_{22}(p)\rho_{21}(p-q)\}$$

$$+\rho_{12}(h)\rho_{12}(q)\{\tfrac{1}{2}\rho_{11}^2(p)+\rho_{12}^2(p)+\tfrac{1}{2}\rho_{22}^2(p)\}\Big].$$

Proof. By Theorem 5.5.1 we may use the first term in Taylor's series to obtain the leading term in the covariance expression. Evaluating $\operatorname{Cov}\{\hat{\gamma}_{12}(h),\hat{\gamma}_{12}(q)\}$, $\operatorname{Cov}\{\hat{\gamma}_{12}(h),\frac{1}{2}[\hat{\gamma}_{11}(0)+\hat{\gamma}_{22}(0)]\}$, $\operatorname{Cov}\{\hat{\gamma}_{12}(q),\frac{1}{2}[\hat{\gamma}_{11}(0)+\hat{\gamma}_{22}(0)]\}$ and $\operatorname{Var}\{\frac{1}{2}[\hat{\gamma}_{11}(0)+\hat{\gamma}_{22}(0)]\}$ by the methods of Theorem 6.2.1, we obtain the conclusion. ▲

Perhaps the most important aspect of these rather cumbersome results is that the covariances are decreasing at the rate n^{-1}. Also, certain special cases are of interest. One working hypothesis is that the two time series are uncorrelated. If X_{1t} is a sequence of independent normal random variables we obtain a particularly simple result.

Corollary 6.5.1.2. Let $\mathbf{X}_t$ be a bivariate stationary normal time series satisfying

$$\sum_{h=-\infty}^{\infty} |\gamma_{22}(h)| < \infty,$$

$$\gamma_{11}(h) = \sigma_1^2, \qquad h = 0$$
$$= 0, \qquad \text{otherwise,}$$

and

$$\gamma_{12}(h) = 0, \qquad \text{all } h.$$

Then

$$\lim_{n\to\infty} n \operatorname{Cov}\{\hat{r}_{12}(h), \hat{r}_{12}(q)\} = \rho_{22}(q-h).$$

In the null case, the variance of the estimated cross correlation is approximately n^{-1} and the correlation between estimated cross correlations is the autocorrelation of X_{2t} multiplied by n^{-1}. If the two time series are independent and neither time series is autocorrelated, then the estimated cross correlations are uncorrelated with an approximate variance of n^{-1}.

It is possible to demonstrate that the sample covariances are consistent estimators under much weaker conditions.

Lemma 6.5.1. Let

$$X_{1t} = \sum_{j=0}^{\infty} \alpha_j e_{1,t-j},$$

$$X_{2t} = \sum_{j=0}^{\infty} \beta_j e_{2,t-j},$$

where $\{\alpha_j\}$ and $\{\beta_j\}$ are absolutely summable and $\{\mathbf{e}\} = \{(e_{1t}, e_{2t})'\}$ is a sequence of independent $(\mathbf{0}, \mathbf{\Sigma})$ random variables with $E\{|e_{it}|^{2+\delta}\} < L$ for $i = 1, 2$ and some $\delta > 0$. Then

$$\frac{1}{n}\sum_{t=1}^{n-h} X_{1t} X_{2,t+h} \xrightarrow{P} \gamma_{X_1X_2}(h), \qquad h = 0, 1, \ldots .$$

Proof. Define

$$Y_{1t}=\sum_{j=0}^{k}\alpha_j e_{1,t-j},$$

$$Y_{2t}=\sum_{j=0}^{k}\beta_j e_{2,t-j},$$

$$D_{1t}=\sum_{j=k+1}^{\infty}\alpha_j e_{1,t-j},$$

$$D_{2t}=\sum_{j=k+1}^{\infty}\beta_j e_{2,t-j},$$

fix h, and consider

$$\frac{1}{n}\sum_{t=1}^{n-h}Y_{1t}Y_{2,t+h}=\frac{1}{n}\sum_{t=1}^{n-h}\sum_{j=0}^{k}\sum_{i=0}^{k}\alpha_j\beta_i e_{1,t-j}e_{2,t-i+h}.$$

If $j\neq i-h$

$$\operatorname{Var}\left\{\frac{1}{n}\sum_{t=1}^{n-h}\alpha_j\beta_i e_{1,t-j}e_{2,t-i+h}\right\}=\frac{n-h}{n^2}\alpha_j^2\beta_i^2\sigma_1^2\sigma_2^2,$$

where σ_i^2 is the variance of e_{it}. If $j=i-h$ and $\sigma_{12}=E\{e_{1t}e_{2t}\}$, then

$$\frac{1}{n}\sum_{t=1}^{n-h}\alpha_j\beta_{j+h}e_{1,t-j}e_{2,t-j}\xrightarrow{P}\alpha_j\beta_{j+h}\sigma_{12}$$

by the weak law of large numbers. [See, for example, Chung (1968, p. 104).] Now

$$\left|\frac{1}{n}\sum_{t=1}^{n-h}X_{1t}X_{2,t+h}-\frac{1}{n}\sum_{t=1}^{n-h}Y_{1t}Y_{2,t+h}\right|\leqslant\left|\frac{1}{n}\sum_{t=1}^{n-h}Y_{1t}D_{2t}\right|+\left|\frac{1}{n}\sum_{t=1}^{n-h}Y_{2t}D_{1t}\right|$$
$$+\left|\frac{1}{n}\sum_{t=1}^{n-h}D_{1t}D_{2t}\right|.$$

By Chebyshev's inequality

$$P\left\{\left|\frac{1}{n}\sum_{t=1}^{n-h} D_{1t}^2\right| > \epsilon\right\} \leqslant \frac{\sum_{j=k+1}^{\infty} \alpha_j^2 \sigma_1^2}{\epsilon}$$

and it follows that

$$\underset{k\to\infty}{p\lim} \frac{1}{n}\sum_{t=1}^{n-h} Y_{1t}D_{2t} = 0,$$

$$\underset{k\to\infty}{p\lim} \frac{1}{n}\sum_{t=1}^{n-h} D_{1t}D_{2t} = 0,$$

uniformly in n. Convergence in probability implies convergence in distribution and by an application of Lemma 6.3.1 we have that $n^{-1}\sum_{t=1}^{n-h} X_{1t}X_{2,t+h}$ converges in distribution to the constant $\gamma_{X_1X_2}(h)$. The result follows by Lemma 5.2.1. ▲

Theorem 6.5.2. Let $\{e_{1t}\}$ and $\{X_t\}$ be independent time series, where $\{e_{1t}\}$ is a sequence of independent $(0, \sigma_1^2)$ random variables with uniformly bounded third moment and $\{X_t\}$ is defined by

$$X_t = \sum_{j=0}^{\infty} \alpha_j e_{2,t-j},$$

where $\sum_{j=0}^{\infty}|\alpha_j| < \infty$ and $\{e_{2t}\}$ is a sequence of independent $(0, \sigma_2^2)$ random variables with uniformly bounded third moment. Then, for fixed $h > 0$,

$$n^{1/2}\hat{r}_{12}(h) \xrightarrow{\mathcal{L}} N(0,1).$$

Proof. We write

$$n^{1/2}\hat{r}_{12}(h) = \left[\left(n^{-1}\sum_{t=1}^{n} e_{1t}^2\right)^{1/2}\left(n^{-1}\sum_{t=1}^{n} X_t^2\right)^{1/2}\right]^{-1} n^{-1/2}\sum_{t=1}^{n-h} e_{1t}X_{t+h}$$
$$+ O_p(n^{-1/2})$$

and note that, by Lemma 6.5.1,

$$p\lim n^{-1}\sum_{t=1}^{n} e_{1t}^2 = \sigma_1^2,$$

$$p\lim n^{-1}\sum_{t=1}^{n} X_t^2 = \gamma_{XX}(0).$$

For fixed h, the time series

$$Z_{th} = e_{1t}X_{t+h}$$

is weakly stationary with

$$E\{Z_{th}\} = 0$$

$$E\{Z_{th}^2\} = \sigma_1^2 \gamma_{XX}(0)$$

and bounded third moment. Asymptotic normality follows by a modest extension of Theorem 6.3.3. ▲

Example. In Table 6.5.1 we present the sample autocorrelations and cross correlations for the bivariate time series $\mathbf{Y}_t = (Y_{1t}, Y_{2t})'$, where Y_{1t} is the logarithm of suspended sediment in the water of the Des Moines River at Boone, Iowa, and Y_{2t} is the logarithm of suspended sediment in the water at Saylorville, Iowa. Saylorville is approximately 48 miles downstream from Boone. The sample data were 205 daily observations collected from April to October 1973.

There are no large tributaries entering the Des Moines River between Boone and Saylorville, and a correlation between the readings at the two points is expected. Since Saylorville is some distance downstream, the correlation pattern should reflect the time required for water to move between the two points. In fact, the largest sample cross correlation is between the Saylorville reading at time $t+1$ and the Boone reading at time t. Also, estimates of $\gamma_{12}(h)$, $h > 0$, are consistently larger than the estimates of $\gamma_{12}(-h)$.

The Boone time series was discussed in Section 4.5. There it was assumed that the time series Y_{1t} could be represented as the sum of the "true process" and a measurement error. The underlying true value X_{1t} was assumed to be a first order autoregressive process with parameter 0.81. If we define the time series

$$W_{3t} = Y_{1t} - 0.81\, Y_{1,t-1}$$

$$W_{4t} = Y_{2t} - 0.81\, Y_{2,t-1},$$

the transformed true process for Boone is a sequence of uncorrelated random variables, although the observed time series W_{3t} will show a small negative first order autocorrelation.

Table 6.5.2 contains the first few estimated correlations for $\mathbf{W}_t$. Note that the estimated cross correlation at zero is quite small. Under the null hypothesis that the cross correlations are zero and that the autocorrelations

Table 6.5.1 Sample correlation functions for suspended sediment in Des Moines River at Boone, Iowa and Saylorville, Iowa.

h	Autocorrelation Boone $\hat{r}_{11}(h)$	Autocorrelation Saylorville $\hat{r}_{22}(h)$	Cross correlation Boone-Saylorville $\hat{r}_{12}(h)$
−12	0.20	0.10	0.10
−11	0.19	0.13	0.07
−10	0.22	0.16	0.08
− 9	0.25	0.16	0.08
− 8	0.29	0.16	0.13
− 7	0.29	0.17	0.18
− 6	0.29	0.15	0.21
− 5	0.32	0.18	0.21
− 4	0.39	0.27	0.23
− 3	0.48	0.42	0.30
− 2	0.62	0.60	0.40
− 1	0.76	0.81	0.53
0	1.00	1.00	0.64
1	0.76	0.81	0.74
2	0.62	0.60	0.67
3	0.48	0.42	0.53
4	0.39	0.27	0.42
5	0.32	0.18	0.32
6	0.29	0.15	0.26
7	0.29	0.17	0.26
8	0.29	0.16	0.25
9	0.25	0.16	0.29
10	0.22	0.16	0.31
11	0.19	0.13	0.28
12	0.20	0.10	0.33

of W_{3t} and W_{4t} are zero after a lag of two, the estimated variance of the sample cross correlations is

$$\begin{aligned}\hat{\text{V}}\text{ar}\{\hat{r}_{34}(h)\} &= \frac{1}{n}\sum_{p=-2}^{2}\hat{r}_{33}(p)\hat{r}_{44}(p)\\ &= \frac{1}{205}\left[1+2(-0.17)(0.15)+2(0.05)(-0.04)\right]\\ &= 0.0046.\end{aligned}$$

Table 6.5.2. Sample correlation functions for transformed suspended sediment in Des Moines River at Boone, Iowa and Saylorville, Iowa.

h	Autocorrelation W_{3t}	Autocorrelation W_{4t}	Cross correlation W_{3t} with W_{4t}
−12	0.05	0.05	0.04
−11	−0.06	0.02	−0.07
−10	0.03	0.08	0.05
−9	0.01	0.01	0.11
−8	0.11	0.04	0.03
−7	0.05	0.07	0.08
−6	−0.03	−0.06	0.10
−5	0.04	−0.12	0.03
−4	−0.01	−0.07	−0.05
−3	−0.07	0.02	−0.01
−2	0.05	−0.04	−0.04
−1	−0.17	0.15	0.06
0	1.00	1.00	0.06
1	−0.17	0.15	0.41
2	0.05	−0.04	0.24
3	−0.07	0.02	−0.02
4	−0.01	−0.07	0.01
5	0.04	−0.12	0.04
6	−0.03	−0.06	−0.10
7	0.05	0.07	0.08
8	0.11	0.04	−0.08
9	0.01	0.01	0.05
10	0.03	0.08	0.16
11	−0.06	0.02	0.08
12	0.05	0.05	0.01

Under these hypotheses, the estimated standard error of the estimated cross correlation is 0.068.

The hypothesis of zero cross correlation is rejected by the estimates $\hat{r}_{34}(1)$, $\hat{r}_{34}(2)$, since they are several times as large as the estimated standard error. The two nonzero sample cross correlations suggest that the input-output model is more complicated than that of a simple integer delay. It might be a simple delay of over one day, or it is possible that the mixing action of moving water produces a more complicated lag structure.

REFERENCES

Section 6.1. Grenander and Rosenblatt (1957), Hannan (1970), Parzen (1958), (1962).

Section 6.2. Anderson (1971), Bartlett (1946), (1966), Hart (1942), Kendall (1954), Kendall and Stuart (1966), Mariott and Pope (1954), von Neumann (1941), (1942).

Section 6.3. Anderson (1959), (1971), Anderson and Walker (1964), Diananda (1953), Eicker (1963), Hannan and Heyde (1972), Hoeffding and Robbins (1948), Moran (1947).

Section 6.5. Bartlett (1966), Box and Jenkins (1970), Hannan (1970).

EXERCISES

1. Let $Y_t=\mu+X_t$, where $X_t=e_t+0.4e_{t-1}$ and the e_t are uncorrelated $(0,\sigma^2)$ random variables. Compute the variance of $\bar{y}_n=n^{-1}\Sigma_{t=1}^n Y_t$. What is $\lim_{n\to\infty} n\operatorname{Var}\{\bar{y}_n\}$?

2. Let the time series $\{Y_t: t\in(1,2,\ldots)\}$ be defined by

$$Y_t=\alpha+\rho Y_{t-1}+e_t,$$

where Y_0 is fixed, $\{e_t: t\in(1,2,\ldots)\}$ is a sequence of independent $(0,\sigma^2)$ random variables, and $|\rho|<1$. Find $E\{Y_t\}$ and $\operatorname{Var}\{Y_t\}$. Show that Y_t satisfies the conditions of Theorem 6.1.1. What value does the sample mean of Y_t converge to?

3. Let $X_t=e_t+0.5e_{t-1}$, where the e_t are normal independent $(0,\sigma^2)$ random variables. Letting

$$\tilde{\gamma}_X(h)=\frac{1}{n-h}\sum_{t=1}^{n-h} X_t X_{t+h},$$

$$\tilde{r}_X(h)=\tilde{\gamma}(h)/\tilde{\gamma}(0),$$

find $\operatorname{Var}\{\tilde{\gamma}_X(h)\}$ for $h=0,1,2,3$; $\operatorname{Cov}\{\tilde{\gamma}_X(0),\tilde{\gamma}_X(h)\}$ for $h=1,2,3$; $\operatorname{Var}\{\tilde{r}_X(h)\}$ for $h=0,1,2,3$; and $\operatorname{Cov}\{\tilde{r}_X(1),\tilde{r}_X(h)\}$ for $h=2,3,4$.

4. Evaluate (6.2.4) for the first order autoregressive process

$$X_t=\rho X_{t-1}+e_t,$$

where $|\rho|<1$ and the e_t are normal independent $(0,\sigma^2)$ random variables.

5. Given the finite moving average

$$X_t=\sum_{j=1}^{M} a_j e_{t-j},$$

where the e_t are normal independent $(0,\sigma^2)$ random variables, is there a distance $d=h-q$ such that $\tilde{\gamma}(h)$ and $\tilde{\gamma}(q)$ are uncorrelated?

6. Use the realization (10, 1, 10) and equation (6.2.2) to construct the estimated (3×3) covariance matrix for a realization of size 3. Show that the resulting matrix is not positive definite. Use the fact that

$$\frac{1}{n}\sum_{t=1}^{n-h}(X_t-\bar{x}_n)(X_{t+h}-\bar{x}_n)=\frac{1}{n}\sum_{j=1}^{2n-1}Z_{mj}Z_{m+h,j}$$

for $m=0,1,\ldots,n-1$; $h=0,1,\ldots,n-1$, where, for $j=1,2,\ldots,2n-1$,

$$Z_{mj}=\begin{cases}X_{j-m}-\bar{x}_n, & j=m+1,m+2,\ldots,m+n\\ 0, & \text{otherwise,}\end{cases}$$

to prove that (6.2.3) yields an estimated covariance matrix that is always positive semidefinite.

7. Give the variance of $\bar{x}_n$, $n=1,2,\ldots,$ for $\{X_t\colon t\in(1,2,\ldots)\}$ defined by

$$X_t=\mu+e_t-e_{t-1},$$

where $\{e_t\colon t\in(0,1,2,\ldots)\}$ is a sequence of independent identically distributed $(0,\sigma^2)$ random variables. Do you think there is a function, w_n, such that $w_n(\bar{x}_n-\mu)\xrightarrow{\mathcal{L}}N(0,1)$?

8. Prove the following result which is used in Theorem 6.3.4. If the sequence $\{c_t\}$ is such that

$$\lim_{n\to\infty}\left(\sum_{t=1}^{n}c_t^2\right)^{-1}c_n^2=0$$

then

$$\lim_{n\to\infty}\sup_{1\leq t\leq n}\left(\sum_{j=1}^{n}c_j^2\right)^{-1}c_t^2=0.$$

9. Prove Theorem 6.3.5 assuming there exists a $\delta>0$ such that

$$E\{|e_t|^{4+\delta}\}<M<\infty$$

for all t.

10. The data on page 273 are the average weekly gross hours per production worker on the payrolls of manufacturing establishments (seasonally adjusted).

(a) Estimate the covariance function $\gamma(h)$, assuming the mean unknown.
(b) Estimate the correlation function $\rho(h)$, assuming the mean unknown.
(c) Using large sample theory, test the hypothesis, H_0: $\rho(1)=0$, assuming $\rho(h)=0$, $h>1$.

11. Using Hart's (1942) tables for the percentage points of d_v or Anderson's (1971) tables for r_v, obtain the percentage points for $t_v=r_v(n+1)^{1/2}(1-r_v^2)^{-1/2}$ for $n=10$ and 15. Compare these values to the percentage points of Student's t with 13 and 18 degrees of freedom.

12. Using the truncation argument of Theorem 6.3.3, complete the proof of Theorem 6.5.2 by showing that

$$(n-h)^{-1/2}\sum_{t=1}^{n-h}Z_{th}$$

Year	Quarter I	II	III	IV
1948	40.30	40.23	39.97	39.70
1949	39.23	38.77	39.23	39.30
1950	39.70	40.27	40.90	40.97
1951	40.90	40.93	40.43	40.37
1952	40.63	40.33	40.60	41.07
1953	41.00	40.87	40.27	39.80
1954	39.53	39.47	39.60	39.90
1955	40.47	40.73	40.60	40.93
1956	40.60	40.30	40.27	40.47
1957	40.37	39.97	39.80	39.13
1958	38.73	38.80	39.40	39.70
1959	40.23	40.53	40.20	40.03
1960	40.17	39.87	39.63	39.07
1961	39.27	39.70	39.87	40.40
1962	40.27	40.53	40.47	40.27
1963	40.37	40.40	40.50	40.57
1964	40.40	40.77	40.67	40.90
1965	41.27	41.10	41.03	41.30
1966	41.53	41.47	41.33	41.10
1967	40.60	40.43	40.67	40.70
1968	40.60	40.63	40.83	40.77
1969	40.57	40.73	40.63	40.53
1970	40.17	39.87	39.73	39.50
1971	39.80	39.93	39.77	40.07
1972	40.30	40.67	40.67	40.87

SOURCES. *Business Statistics*, 1971 pp. 74 and 237 and *Survey of Current Business*, Jan. 1972, Jan. 1973. The quarterly data are the averages of monthly data.

converges in distribution to a normal random variable.

13. Denoting the data of Exercise 10 by X_{1t} and that of Table 6.4.1 by X_{2t}, compute the cross covariance and cross correlation functions. Define

$$Y_{1t} = X_{1t} - 1.53X_{1,t-1} - 0.66X_{1,t-2},$$

$$Y_{2t} = X_{2t} - 1.53X_{2,t-1} - 0.66X_{2,t-2}.$$

Compute the cross covariance and cross correlation functions for (Y_{1t}, Y_{2t}). Plot the cross correlation function of (Y_{1t}, Y_{2t}). Compute the variance of $\hat{r}_{Y_1Y_2}(h)$ under the assumption that Y_{1t} is a sequence of uncorrelated random variables. Plot the standard error on your figure.

14. Let X_t be a time series with positive continuous spectral density $f_X(\omega)$.

(a) Show that, given $\epsilon > 0$, one may define two moving average time series W_{1t} and W_{2t} with spectral densities

$$f_{W_1}(\omega)=\frac{\sigma^2}{2\pi}\left|\sum_{j=0}^{q_1}\beta_{1j}e^{-i\omega j}\right|^2$$

$$f_{W_2}(\omega)=\frac{\sigma^2}{2\pi}\left|\sum_{j=0}^{q_2}\beta_{2j}e^{-i\omega j}\right|^2$$

such that

$$f_X(\omega)-\epsilon\leqslant f_{W_1}(\omega)\leqslant f_X(\omega)\leqslant f_{W_2}(\omega)\leqslant f_X(\omega)+\epsilon.$$

(b) Let $\{a_t: t=1,2,\ldots\}$ be such that

$$\lim_{n\to\infty}\sum_{t=1}^{n}a_t^2=\infty$$

$$\lim_{n\to\infty}\frac{\sum_{t=1}^{n}a_t a_{t+|h|}}{\sum_{t=1}^{n}a_t^2}=g(h),\qquad h=0,\pm1,\pm2,\ldots,$$

$$\sum_{h=-\infty}^{\infty}|g(h)|<\infty.$$

Define $f_g(\omega)=(2\pi)^{-1}\sum_{h=-\infty}^{\infty}g(h)e^{-i\omega h}$. Show that

$$\lim_{n\to\infty}\operatorname{Var}\left\{\left[\sum_{t=1}^{n}a_t^2\right]^{-1/2}\sum_{t=1}^{n}a_tX_t\right\}=\int_{-\pi}^{\pi}f_X(\omega)f_g(\omega)\,d\omega.$$

See Exercise 4.15 and Grenander and Rosenblatt (1957, Chap. 7).

CHAPTER 7

The Periodogram, Estimated Spectrum

In this chapter we shall investigate estimators of the spectral density of time series with absolutely summable covariance function. The spectral density of a time series was defined in Section 4.1 as the Fourier transform of the covariance function. We also noted in Section 4.2 that the variances of the random variables defined by the Fourier coefficients of the original time series are, approximately, multiples of the spectral density. These two results suggest methods of estimating the spectral density.

The study of the Fourier coefficients was popular among economists in the 1920s and 1930s, and the student of economics may be interested in the discussion of Davis (1941) and the studies cited by Tintner (1952). Granger and Hatanaka (1964) is a more recent study and Nold (1972) is a bibliography of applications of spectral analysis.

7.1. THE PERIODOGRAM

Given a finite realization from a time series, we can represent the n observations by the trigonometric polynomial

$$X_t = \frac{a_0}{2} + \sum_{k=1}^{m} (a_k \cos \omega_k t + b_k \sin \omega_k t), \tag{7.1.1}$$

where

$$\omega_k = \frac{2\pi k}{n}, \qquad k = 0, 1, 2, \ldots, m,$$

$$a_k = \frac{2 \sum_{t=1}^{n} X_t \cos \omega_k t}{n}, \qquad k = 0, 1, 2, \ldots, m,$$

$$b_k = \frac{2 \sum_{t=1}^{n} X_t \sin \omega_k t}{n}, \qquad k = 1, 2, \ldots, m,$$

and we have assumed n odd and equal to $2m+1$. As before, we recognize the Fourier coefficients as regression coefficients. We can use the standard regression analysis to partition the total sum of squares for the n observations. The sum of squares removed by the regression of X_t on $\cos\omega_k t$ is the regression coefficient multiplied by the sum of crossproducts; that is, the sum of squares due to a_k is

$$\frac{n}{2}a_k^2 = \frac{2}{n}\left(\sum_{t=1}^{n} X_t \cos\omega_k t\right)^2, \qquad k=1,2,\ldots,m, \tag{7.1.2}$$

and the sum of squares removed by $\cos\omega_k t$ and $\sin\omega_k t$ is

$$\frac{n}{2}\left(a_k^2+b_k^2\right) = \frac{2}{n}\left\{\left(\sum_{t=1}^{n} X_t \cos\omega_k t\right)^2 + \left(\sum_{t=1}^{n} X_t \sin\omega_k t\right)^2\right\}. \tag{7.1.3}$$

We might call the quantity in (7.1.3) the sum of squares associated with frequency ω_k. Thus the total sum of squares for the $n=2m+1$ observations may be partitioned into $m+1$ components. One component is associated with the mean. Each of the remaining m components is the sum of the two squares associated with the m nonzero frequencies. The partition is displayed in Table 7.1.1.

Table 7.1.1. Analysis of variance table for a sample of size $n=2m+1$

Source	Degrees of freedom	Sum of squares
Mean	1	$n\bar{x}_n^2 = \frac{1}{4}na_0^2$
Frequency $\omega_1 = 2\pi/n$	2	$(n/2)(a_1^2+b_1^2)$
Frequency $\omega_2 = 4\pi/n$	2	$(n/2)(a_2^2+b_2^2)$
$\vdots$		
Frequency $\omega_m = 2\pi m/n$	2	$(n/2)(a_m^2+b_m^2)$
Total	n	$\sum_{t=1}^{n} X_t^2$

Should the number of observations be even and denoted by $2m$, there is only one regression variable associated with the mth frequency: $\cos\pi t$. Then the sum of squares for the mth frequency has one degree of freedom and is given by $n^{-1}(\sum_{t=1}^{n} X_t \cos\pi t)^2 = (1/4)na_m^2$.

One might divide all of the sums of squares of Table 7.1.1 by the degrees of freedom and consider the mean squares. However, it is more common

to investigate the sums of squares, multiplying the sums of squares with one degree of freedom by two. The function of frequency given by these normalized sums of squares is called the *periodogram*. Thus, the periodogram is defined by

$$I_n(\omega_k) = \frac{n}{2}(a_k^2 + b_k^2), \qquad k = 1, 2, \ldots, m \tag{7.1.4}$$

where m is the smallest integer greater than or equal to $(n-1)/2$.

If $\{X_t\}$ is a sequence of normal independent $(0, \sigma^2)$ random variables, then the a_k and b_k, being linear combinations of the X_t, will be normally distributed. Since the sine and cosine functions are orthogonal, the a_k and b_k are independent. In this case those entries in Table 7.1.1 with two degrees of freedom divided by σ^2 are distributed as independent chi-squares with two degrees of freedom.

The periodogram may also be defined in terms of the original observations as

$$I_n(\omega_k) = \frac{2}{n}\left[\left(\sum_{t=1}^{n} X_t \cos \omega_k t\right)^2 + \left(\sum_{t=1}^{n} X_t \sin \omega_k t\right)^2\right], \qquad k = 0, 1, \ldots, m. \tag{7.1.5}$$

Note that if we define the complex coefficients c_k by

$$c_k = \frac{1}{n} \sum_{t=1}^{n} X_t e^{i\omega_k t},$$

then

$$I_n(\omega_k) = 2n c_k c_k^* = \frac{2}{n}\left|\sum_{t=1}^{n} X_t e^{i\omega_k t}\right|^2. \tag{7.1.6}$$

Thus, the periodogram ordinate at ω_k is a multiple of the squared norm of the complex Fourier coefficient of the time series associated with the frequency ω_k.

The periodogram is also expressible as a multiple of the Fourier transform of the estimated covariance function. If $\omega_k \neq 0$, we can write

$$\sum_{t=1}^{n} X_t \cos \omega_k t = \sum_{t=1}^{n} (X_t - \mu) \cos \omega_k t,$$

$$\sum_{t=1}^{n} X_t \sin \omega_k t = \sum_{t=1}^{n} (X_t - \mu) \sin \omega_k t,$$

where $\mu = E\{X_t\}$. Therefore,

$$I_n(\omega_k) = \frac{2}{n}\left[\left\{\sum_{t=1}^{n}(X_t-\mu)\cos\omega_k t\right\}^2 + \left\{\sum_{t=1}^{n}(X_t-\mu)\sin\omega_k t\right\}^2\right]$$

$$= \frac{2}{n}\left[\sum_{t=1}^{n}\sum_{j=1}^{n}(X_t-\mu)(X_j-\mu)\cos\omega_k t\cos\omega_k j\right.$$

$$\left.+\sum_{t=1}^{n}\sum_{j=1}^{n}(X_t-\mu)(X_j-\mu)\sin\omega_k t\sin\omega_k j\right]$$

$$= \frac{2}{n}\left[\sum_{t=1}^{n}\sum_{j=1}^{n}(X_t-\mu)(X_j-\mu)\cos\omega_k(t-j)\right]$$

for $k=1,2,\ldots,m$. In this double sum there are n combinations with $t-j=0$, $n-1$ combinations with $t-j=1$, and so forth. Therefore, by letting $p=t-j$, we obtain the following result.

Result 7.1.1. The kth periodogram ordinate is given by

$$I_n(\omega_k) = \begin{cases} 2n\bar{x}_n^2, & k=0 \\ 4\pi\hat{f}(\omega_k), & k=1,2,\ldots,m, \end{cases} \tag{7.1.7}$$

where

$$\hat{f}(\omega) = \frac{1}{2\pi}\sum_{p=-\infty}^{\infty}\frac{n-|p|}{n}\tilde{\gamma}(p)\cos\omega p, \qquad (-\pi \leqslant \omega \leqslant \pi), \tag{7.1.8}$$

$$\tilde{\gamma}(-p) = \tilde{\gamma}(p) = \begin{cases} \dfrac{1}{n-p}\displaystyle\sum_{j=1}^{n-p}(X_j-\mu)(X_{j+p}-\mu), & 0 \leqslant p \leqslant n-1 \\ 0, & p > n-1. \end{cases} \tag{7.1.9}$$

The coefficients a_k and b_k $(k \neq 0)$ that are computed using $X_t - \bar{x}_n$ are identical to those computed using X_t. Therefore, by substituting $X_t - \bar{x}_n$ for X_t in (7.1.5), we can also write

$$I_n(\omega_k) = 2\sum_{p=-\infty}^{\infty}\hat{\gamma}(p)\cos\omega_k p, \tag{7.1.10}$$

where $k \neq 0$ and

$$\hat{\gamma}(h) = \hat{\gamma}(-h) = \begin{cases} \frac{1}{n} \sum_{t=1}^{n-h} (X_t - \bar{x}_n)(X_{t+h} - \bar{x}_n), & 0 \leqslant h \leqslant n-1 \\ 0, & h > n-1. \end{cases}$$

The function $\hat{f}(\omega)$ of equation (7.1.8) is a continuous periodic function of ω defined for all ω. The periodogram has been defined only for the discrete set of points $\omega_k = 2\pi k/n, k = 0, 1, 2, \ldots, m$. In investigating the limiting properties of the periodogram, it is convenient to have a function defined for all $\omega \in [0, \pi]$. To this end we introduce the function

$$K(n,\omega) = k \quad \text{for} \quad \frac{\pi(2k-1)}{n} < \omega \leqslant \frac{\pi(2k+1)}{n}, \qquad k = 0, \pm 1, \pm 2, \ldots,$$

and take

$$I_n(\omega) = I_n(\omega_{K(n,\omega)}).$$

Thus, $I_n(\omega)$ for $\omega \in [0, \pi]$ is a step function that takes the value $I_n(\omega_k)$ on the interval $(\pi\{2k-1\}/n,\ \pi\{2k+1\}/n]$.

The word periodogram is used with considerable flexibility in the literature. Despite the apparent conflict in terms, our definition of the periodogram as a function of frequency rather than of period is a common one. We have chosen to define the periodogram for the discrete frequencies $\omega_k = 2\pi k/n, k = 0, 1, 2, \ldots, m$, and to extend it to all ω as a step function. An alternative definition of the periodogram is $4\pi\hat{f}(\omega)$, where $\hat{f}(\omega)$ is given in (7.1.8), in which case one automatically has a function for all ω. Our definition will prove convenient in obtaining the limiting properties of estimators of the spectrum.

The distributional properties of the periodogram ordinates are easily obtained when the time series is normal white noise. To establish the properties of the periodogram for other time series, we first obtain the limiting value of the expectation of $I_n(\omega)$ for a time series with absolutely summable covariance function.

Theorem 7.1.1. Let X_t be a stationary time series with $E\{X_t\} = \mu$ and absolutely summable covariance function. Then,

$$\lim_{n \to \infty} E\{I_n(\omega)\} = 4\pi f(\omega), \qquad \omega \neq 0$$

$$\lim_{n \to \infty} E\{I_n(0) - 2n\mu^2\} = 4\pi f(0), \qquad \omega = 0.$$

Proof. Since $E\{\tilde{\gamma}(p)\}=\gamma(p)$, it follows from (7.1.7) that

$$E\{I_n(0)\}=2\sum_{h=-(n-1)}^{n-1}\frac{n-|h|}{n}\gamma(h)+2n\mu^2,$$

$$E\{I_n(\omega_k)\}=2\sum_{h=-(n-1)}^{n-1}\frac{n-|h|}{n}\gamma(h)\cos\omega_k h, \qquad k=1,2,\ldots,m.$$

Now

$$\left|\sum_{h=-n}^{n}\frac{|h|}{n}\gamma(h)\cos\omega_k h\right|\leq\sum_{h=-n}^{n}\frac{|h|}{n}|\gamma(h)|,$$

and the latter sum goes to zero as $n\to\infty$ by Lemma 3.1.4. The sequence $g_n(\omega)=2\sum_{h=-(n-1)}^{n-1}\gamma(h)\cos\omega h$ converges uniformly to $4\pi f(\omega)$ by Corollary 3.1.8, and $\omega_{K(n,\omega)}$ converges to ω by construction. ▲

The normalized coefficients $2^{-1/2}n^{1/2}a_k$ and $2^{-1/2}n^{1/2}b_k$ are the random variables obtained by applying to the observations the transformation discussed in Section 4.2. Therefore, for a time series with absolutely summable covariance function, these random variables are in the limit uncorrelated and have variance given by a multiple of the spectral density evaluated at the associated frequency.

The importance of this result is difficult to overemphasize. For a wide class of time series we are able to transform an observed set of n observations into a set of n statistics that are nearly uncorrelated. All except the first, the sample mean, have zero expected value. The variance of these random variables is, approximately, a simple function of the spectral density.

Theorem 7.1.2. Let X_t be a time series defined by

$$X_t=\sum_{j=0}^{\infty}\alpha_j e_{t-j},$$

where $\{\alpha_j\}$ is absolutely summable and the e_t are independent identically distributed $(0,\sigma^2)$ random variables. Let $f_X(\omega)$ be positive for all ω. Then, for ω and λ in $(0,\pi)$ and $\omega\neq\lambda$, $[2\pi f(\omega)]^{-1}I_n(\omega)$ and $[2\pi f(\lambda)]^{-1}I_n(\lambda)$ converge in distribution to independent chi-square random variables, each with two degrees of freedom.

Proof. Consider

$$2^{-1/2}n^{1/2}a_{K(n,\omega)} = 2^{1/2}n^{-1/2}\sum_{t=1}^{n}\cos\omega_{K(n,\omega)}tX_t$$

$$= 2^{1/2}n^{-1/2}\sum_{t=1}^{n}\cos\omega_{K(n,\omega)}t\left[\sum_{j=0}^{r}\alpha_j e_{t-j} + \sum_{j=r+1}^{\infty}\alpha_j e_{t-j}\right],$$

where

$$\lim_{r\to\infty}\mathrm{Var}\left\{2^{1/2}n^{-1/2}\sum_{t=1}^{n}\cos\omega_{K(n,\omega)}t\sum_{j=r+1}^{\infty}\alpha_j e_{t-j}\right\} = 0$$

uniformly in n. Fixing r, we have

$$2^{1/2}n^{-1/2}\sum_{t=1}^{n}\cos\omega_{K(n,\omega)}t\sum_{j=0}^{r}\alpha_j e_{t-j} = n^{-1/2}\sum_{t=1}^{n}\delta_{n\omega t}e_t + R_n,$$

where

$$\delta_{n\omega t} = 2^{1/2}\sum_{j=0}^{r}\alpha_j\cos\omega_{K(n,\omega)}(t+j),$$

$$R_n = 2^{1/2}n^{-1/2}\left[\sum_{j=0}^{r-1}\sum_{s=j+1}^{r}\alpha_s e_{-j}\cos\omega_{K(n,\omega)}(s-j) - \sum_{j=0}^{r-1}\sum_{s=j+1}^{r}\alpha_s e_{n-j}\cos\omega_{K(n,\omega)}(n+s-j)\right],$$

and R_n converges in probability to zero. As $\delta_{n\omega t}$ is uniformly bounded by, say, M we have, for $\epsilon > 0$,

$$\lim_{n\to\infty} n^{-1}\sum_{t=1}^{n}\int_{|\delta_{n\omega t}e| > \epsilon n^{1/2}}\delta_{n\omega t}^2 e^2\,dF(e) \leq \lim_{n\to\infty} n^{-1}\sum_{t=1}^{n}\int_{|e| > \epsilon M^{-1}n^{1/2}} M^2 e^2\,dF(e) = 0$$

and $n^{-1/2}\sum_{t=1}^{n}\delta_{n\omega t}e_t$ converges in distribution to a normal random variable by the Lindeberg condition. The asymptotic normality of $2^{-1/2}n^{1/2}a_{K(n,\omega)}$

follows by Lemma 6.3.1. The same arguments hold for a linear combination of $n^{1/2}a_{K(n,\omega)}, n^{1/2}b_{K(n,\omega)}, n^{1/2}a_{K(n,\lambda)}, n^{1/2}b_{K(n,\lambda)}$ so that $2^{-1/2}n^{1/2}[a_{K(n,\omega)}, b_{K(n,\omega)}, a_{K(n,\lambda)}, b_{K(n,\lambda)}]$, $\omega \neq \lambda$, converges in distribution to a multivariate normal random variable. The covariance matrix is given by Theorem 4.2.1 and is

$$\operatorname{diag}\{2\pi f(\omega), 2\pi f(\omega), 2\pi f(\lambda), 2\pi f(\lambda)\}.$$

Hence, by Theorem 5.2.4, the limiting distribution of $I_n(\omega)/2\pi f(\omega)$ and $I_n(\lambda)/2\pi f(\lambda)$ is that of two independent chi-square random variables with two degrees of freedom. ▲

Thus, for many nonnormal processes, we may treat the periodogram ordinates as multiples of chi-square random variables. If the original time series is a sequence of independent $(0, \sigma^2)$ random variables, then the periodogram ordinates all have the same expected value. However, for a time series with a nonzero autocorrelation structure, the ordinates will have different expected values. These facts have been used in constructing tests based on the periodogram.

Perhaps it is most natural to use the periodogram to search for "cycles" or "periodicities" in the data. For example, let us hypothesize that a time series is well represented by

$$X_t = \mu + A\cos\omega t + B\sin\omega t + e_t, \tag{7.1.11}$$

where the e_t are normal independent $(0, \sigma^2)$ random variables and A and B are fixed.

First assume that ω is known and of the form $2\pi k/n$, where k is an integer. To test the hypothesis that $A = B = 0$ against the alternative $A \neq 0$ or $B \neq 0$, we can use

$$F_{2m-2}^2 = \frac{(2m-2)(a_k^2 + b_k^2)}{2\sum_{\substack{j=1 \\ j\neq k}}^{m} (a_j^2 + b_j^2)},$$

where F_{2m-2}^2 has the F-distribution with 2 and $2m-2$ degrees of freedom. Note that the sum of squares for the mean is not included in the denominator, since we postulated a general mean, μ. If ω cannot be expressed as $2\pi k/n$, where n is an integer, then the regression associated with (7.1.11) can be computed and the usual regression test constructed.

We sometimes believe a time series contains a periodic component, but are unwilling to postulate the periodic function to be a perfect sine wave. For example, we may feel that a monthly time series contains a seasonal component, a large portion of which is because of a high value for

December. We know that any periodic function defined on the integers, with integral period H, can be represented by

$$\frac{a_0}{2} + \sum_{k=1}^{L[H]} \left\{ a_k \cos\frac{2\pi k}{H} t + b_k \sin\frac{2\pi k}{H} t \right\},$$

where $L[H]$ is the largest integer less than or equal to $H/2$. For the monthly time series one might postulate

$$Y_t = \mu + \sum_{k=1}^{6} \left\{ A_k \cos\frac{2\pi k}{12} t + B_k \sin\frac{2\pi k}{12} t \right\} + e_t.$$

To test the hypothesis of no seasonal effect (i.e., all A_k and B_k equal zero), we form Snedecor's F as the ratio of the mean square for the six seasonal frequencies to the mean square for the remaining frequencies. Note that these tests assume e_t to be a sequence of normal independent $(0,\sigma^2)$ random variables.

The periodogram has also been used to search for "hidden periodicities." In the above examples we postulated the frequency or frequencies of interest and, hence, under the null hypothesis, the ratio of the mean squares has the F-distribution. However, we might postulate the null model

$$X_t = \mu + e_t$$

and the alternative model

$$X_t = \mu + A\cos\omega t + B\sin\omega t + e_t,$$

where ω is unknown.

In such a case one might search out the largest periodogram ordinate and ask if this ordinate can reasonably be considered the largest in a random sample of size m selected from a distribution function that is a multiple of a chi-square with two degrees of freedom. A statistic that can be used to test the hypothesis is

$$\xi = \left[\frac{1}{m} \sum_{k=1}^{m} I_n(\omega_k) \right]^{-1} I_n(L)$$

where $I_n(L)$ is the largest periodogram ordinate in a sample of m periodogram ordinates each with two degrees of freedom. Fisher (1929) demonstrated that, for $g > 0$,

$$P\{m^{-1}\xi > g\} = \sum_{j=1}^{k} (-1)^{j-1} \binom{m}{j} (1 - jg)^{m-1},$$

Table 7.1.2. Percentage points for the ratio of largest periodogram ordinate to the average

Number of ordinates	Probability of a larger value		
	0.10	0.05	0.01
2	1.900	1.950	1.990
3	2.452	2.613	2.827
4	2.830	3.072	3.457
5	3.120	3.419	3.943
6	3.354	3.697	4.331
7	3.552	3.928	4.651
8	3.722	4.125	4.921
9	3.872	4.297	5.154
10	4.005	4.450	5.358
15	4.511	5.019	6.103
20	4.862	5.408	6.594
25	5.130	5.701	6.955
30	5.346	5.935	7.237
40	5.681	6.295	7.663
50	5.937	6.567	7.977
60	6.144	6.785	8.225
70	6.317	6.967	8.428
80	6.465	7.122	8.601
90	6.595	7.258	8.750
100	6.711	7.378	8.882
150	7.151	7.832	9.372
200	7.458	8.147	9.707
250	7.694	8.389	9.960
300	7.886	8.584	10.164
350	8.047	8.748	10.334
400	8.186	8.889	10.480
500	8.418	9.123	10.721
600	8.606	9.313	10.916
700	8.764	9.473	11.079
800	8.901	9.612	11.220
900	9.022	9.733	11.344
1000	9.130	9.842	11.454

where ξ is constructed from the periodogram of a sequence of normal independent (μ, σ^2) random variables and k is the largest integer less than g^{-1}. A table of the distribution of ξ is given by Davis (1941). Wilks (1962, p. 529) contains a derivation of the distribution. In Table 7.1.2 we give the 1, 5, and 10 percentage points for the distribution.

Under the null hypothesis that a time series is normal white noise, the periodogram ordinates are multiples of independent chi-squares, each with two degrees of freedom. Hence, any number of other "goodness of fit" tests are available to test the hypothesis of independence. Bartlett (1966, p. 318) suggested a test based on the normalized cumulative periodogram

$$C_k = \operatorname{cum}\{I_n(\omega_k)\} = \left[\sum_{j=1}^{m} I_n(\omega_j)\right]^{-1} \sum_{j=1}^{k} I_n(\omega_j). \tag{7.1.12}$$

The normalized cumulative periodogram for $k = 1, 2, \ldots, m-1$, has the same distribution function as that of an ordered sample of size $m-1$ selected from the uniform $(0,1)$ distribution. Therefore, if we plot the normalized periodogram as a sample distribution function and apply the Kolmogorov-Smirnov test of the hypothesis that it is a sample distribution function for a sample of $m-1$ selected from a uniform $(0,1)$ distribution, we have a test of the hypothesis that the original time series is white noise. This testing procedure has been discussed by Durbin (1967), (1969).

Example. In Section 9.2 a grafted quadratic trend is fitted to United States wheat yields from 1908 to 1971. The periodogram computed for the deviations[1] from trend is given in Table 7.1.3. To test the hypothesis that the largest ordinate is the largest in a random sample of 31 estimates, we form the ratio of the fourth ordinate to the average of ordinates 1 to 31:

$$\xi = \frac{53.50}{5.391} = 9.92.$$

From Table 7.1.2 we see that the 1% point for this ratio is about 7.28. Thus, the null hypothesis is rejected at this level. While we may be somewhat reluctant to accept the existence of a perfect sine cycle of length 16 years on the basis of 64 observations, it is unlikely that the deviation from trend is a white noise time series.

The cumulative periodogram for the wheat yield data is displayed in Figure 7.1.1. We have followed the common practice of plotting the

[1]As we are working with deviations from regression, the distributions of the test statistics are only approximately those discussed above. Durbin (1969) has demonstrated that the critical points for the Kolmogorov-Smirnov-type test statistic using deviations from regression differ from those for an unaltered time series by a quantity that is $O(n^{-1})$.

Table 7.1.3. Periodogram of deviations of United States wheat yields from trend

k	Period in years	Periodogram ordinate
1	64.0	0.54
2	32.0	5.15
3	21.3	4.74
4	16.0	53.50
5	12.8	9.86
6	10.7	3.02
7	9.1	3.20
8	8.0	2.62
9	7.1	3.56
10	6.4	2.75
11	5.8	7.99
12	5.3	5.89
13	4.9	2.08
14	4.6	2.61
15	4.3	3.44
16	4.0	1.47
17	3.8	1.44
18	3.6	0.12
19	3.4	8.77
20	3.2	1.03
21	3.0	4.00
22	2.9	0.41
23	2.8	0.10
24	2.7	17.03
25	2.6	5.73
26	2.5	0.15
27	2.4	1.90
28	2.3	3.53
29	2.2	1.91
30	2.1	6.64
31	2.1	1.96
32	2.0	12.50

cumulative periodogram against k. The value of the cumulative periodogram for k is plotted on the interval $(k-1,k]$. The upper and lower 5% bounds for the Kolmogorov-Smirnov test have been drawn in the figure. The lines are $k/31+0.245$ and $k/31-0.245$, where 0.245 is the 5% point of the Kolmogorov-Smirnov statistic for sample size 31. The tables of Birnbaum (1952) indicate that for $(m-1)>30$, the 95% point for the Kolmogorov-Smirnov statistic is approximately $1.36\,(m-1)^{-1/2}$ and the 99% point is approximately $1.63\,(m-1)^{-1/2}$. In constructing the cumulative periodogram we included all 32 ordinates, even though the last ordinate has only one degree of freedom. Since the normalized cumulative periodogram passes above the upper 5% line, the data reject the hypothesis of independence, primarily because of the large ordinate at $k=4$.

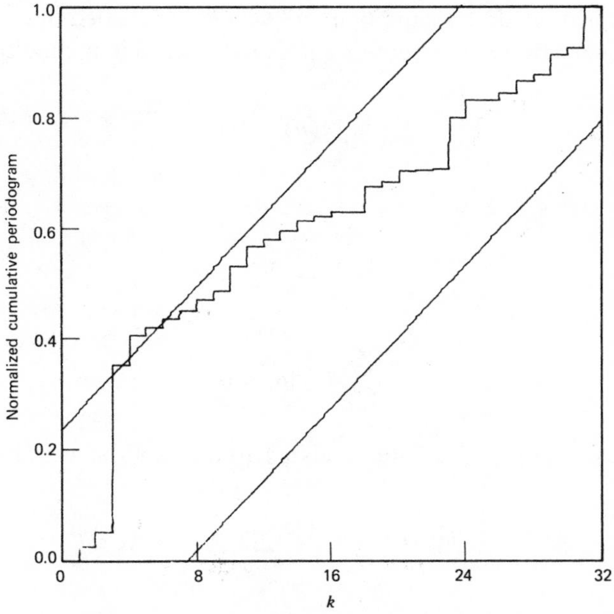

Figure 7.1.1. Normalized cumulative periodogram for deviations of wheat yield from trend.

7.2. SMOOTHING, ESTIMATING THE SPECTRUM

It is clear from the development of the distributional properties of the periodogram that increasing the sample size has little effect on the behavior of the estimated ordinate for a particular frequency. In fact, if the time series is normal (0, 1) white noise, the distribution of the periodogram ordinate is a two-degree-of-freedom chi-square independent of sample size.

The number of periodogram ordinates increases as the sample size increases, but the efficiency of the estimator for a particular frequency remains unchanged.

If the spectral density is a continuous function of ω it is natural to consider an average of local values of the periodogram to obtain a better estimate of the spectral density. Before treating such estimators, we present an additional result on the covariance properties of the periodogram.

In investigating local averages of the periodogram we find it convenient to make a somewhat stronger assumption about the rate at which the covariance function approaches zero. We shall assume that the covariances are such that

$$\sum_{h=-n}^{n} |h||\gamma(h)| = O(n^{1/2}).$$

This is a fairly modest assumption and would be satisfied, for example, by any stationary finite autoregressive process. A sufficient condition for

$$\sum_{h=-n}^{n} |h||\gamma(h)| = O(n^{1/2})$$

for a time series X_t is that

$$X_t = \sum_{j=0}^{\infty} \alpha_j e_{t-j},$$

where

$$\sum_{j=0}^{\infty} j^{1/2}|\alpha_j| < \infty$$

and $\{e_t\}$ is a sequence of uncorrelated $(0,\sigma^2)$ random variables. This is because

$$\begin{aligned} n^{-1/2}\sum_{h=0}^{n} h|\gamma(h)| &\leqslant n^{-1/2}\sum_{h=0}^{n}\sum_{j=0}^{\infty} h|\alpha_j\alpha_{j+h}|\sigma^2 \\ &\leqslant \sum_{h=0}^{n}\sum_{j=0}^{\infty} h^{1/2}|\alpha_j||\alpha_{j+h}|\sigma^2 \\ &\leqslant \sum_{j=0}^{\infty}\sum_{h=0}^{\infty} |\alpha_j|(j+h)^{1/2}|\alpha_{j+h}|\sigma^2 \\ &\leqslant \sigma^2\sum_{j=0}^{\infty}|\alpha_j|\sum_{s=0}^{\infty} s^{1/2}|\alpha_s|. \end{aligned}$$

Before giving a theorem on the covariance properties of the periodogram, we present two lemmas useful in the proof.

Lemma 7.2.1. For integer p, $0 \leqslant p < n$, and real L,

$$\left|\sum_{j=1}^{n-p} \cos\omega_k(L+j)\right| \leqslant \min\{p, n-p\}$$

and

$$\left|\sum_{j=1}^{n-p} \sin\omega_k(L+j)\right| \leqslant \min\{p, n-p\},$$

where $\omega_k = 2\pi k/n$, $k = 1, 2, \ldots, n-1$.

Proof. Now

$$\begin{aligned}\sum_{j=1}^{n} \cos\omega_k(L+j) &= \sum_{j=1}^{n} (\cos\omega_k L \cos\omega_k j - \sin\omega_k L \sin\omega_k j)\\ &= (\cos\omega_k L)\sum_{j=1}^{n} \cos\omega_k j - (\sin\omega_k L)\sum_{j=1}^{n} \sin\omega_k j\\ &= 0 \quad \text{for} \quad \omega_k \neq 0,\end{aligned}$$

where the zero sum follows from the arguments of Theorem 3.1.1. Since $|\cos\theta| \leqslant 1$ for all θ, the result follows. The analogous argument holds for the sum of sines. ▲

Lemma 7.2.2. For real L

$$\left|\sum_{t=1}^{n}\sum_{j=1}^{n} \gamma(t-j)\cos\omega_k(L+j)\right| \leqslant \sum_{p=-(n-1)}^{(n-1)} |p||\gamma(p)|,$$

$$\left|\sum_{t=1}^{n}\sum_{j=1}^{n} \gamma(t-j)\sin\omega_k(L+j)\right| \leqslant \sum_{p=-(n-1)}^{n-1} |p||\gamma(p)|,$$

where $\omega_k = 2\pi k/n$, $k = 1, 2, \ldots, n-1$.

Proof. We have

$$\begin{aligned}\left|\sum_{t=1}^{n}\sum_{j=1}^{n} \gamma(t-j)\cos\omega_k(L+j)\right| &= \left|\sum_{p=0}^{n-1}\sum_{j=1}^{n-p} \gamma(p)\cos\omega_k(L+j)\right.\\ &\qquad \left. + \sum_{p=-(n-1)}^{-1}\ \sum_{j=-p+1}^{n} \gamma(p)\cos\omega_k(L+j)\right|\\ &\leqslant \sum_{p=-(n-1)}^{(n-1)} |p||\gamma(p)|,\end{aligned}$$

where we have used Lemma 7.2.1. The sine result follows in a completely analogous manner. ▲

Theorem 7.2.1. Let the time series X_t be defined by

$$X_t = \sum_{j=0}^{\infty} \alpha_j e_{t-j},$$

where the e_t are independent $(0, \sigma^2)$ random variables with fourth moment $\eta\sigma^4$ and

$$\sum_{j=1}^{\infty} j^{1/2}|\alpha_j| < \infty.$$

Then

$$\operatorname{Cov}\{I_n(\omega_j), I_n(\omega_k)\} = \begin{cases} 2(4\pi)^2 f^2(0) + o(1), & \omega_j = \omega_k = 0 \\ (4\pi)^2 f^2(\omega_k) + o(1), & \omega_j = \omega_k, \omega_k \neq 0, \pi \\ O(n^{-1}), & \omega_j \neq \omega_k. \end{cases}$$

Furthermore, for the sequence composed only of even-sized samples

$$\operatorname{Var}\{I_n(\pi)\} = 2(4\pi)^2 f^2(\pi) + o(1).$$

Proof. By the definition (7.1.6) of the periodogram, we have

$$I_n(\omega_k) = \frac{2}{n} \sum_{t=1}^{n} \sum_{s=1}^{n} X_t X_s e^{i\omega_k(t-s)}$$

and

$$E\{I_n(\omega_k) I_n(\omega_j)\} - E\{I_n(\omega_k)\} E\{I_n(\omega_j)\}$$

$$= E\left\{\frac{4}{n^2} \sum_{t=1}^{n} \sum_{s=1}^{n} \sum_{u=1}^{n} \sum_{v=1}^{n} X_t X_s X_u X_v e^{i\omega_k(t-s)} e^{i\omega_j(u-v)}\right\}$$

$$-4 \sum_{h=-(n-1)}^{n-1} \frac{n-|h|}{n} \gamma(h) e^{-i\omega_k h} \sum_{q=-(n-1)}^{n-1} \frac{n-|q|}{n} \gamma(q) e^{-i\omega_j q}$$

$$= \frac{4(\eta-3)\sigma^4}{n^2} \sum_{t=1}^{n} \sum_{s=1}^{n} \sum_{u=1}^{n} \sum_{v=1}^{n} \sum_{r=-\infty}^{\infty} \alpha_r \alpha_{r+t-s} \alpha_{r+t-u} \alpha_{r+t-v} e^{i\omega_k(t-s)} e^{i\omega_j(u-v)}$$

$$+ \frac{4}{n^2} \sum_{t=1}^{n} \sum_{s=1}^{n} \sum_{u=1}^{n} \sum_{v=1}^{n} \gamma(t-u)\gamma(s-v) e^{i\omega_k(t-u)} e^{i\omega_j(s-v)} e^{i(\omega_j+\omega_k)(u-s)}$$

$$+ \frac{4}{n^2} \sum_{t=1}^{n} \sum_{s=1}^{n} \sum_{u=1}^{n} \sum_{v=1}^{n} \gamma(t-v)\gamma(u-s) e^{i\omega_k(t-v)} e^{i\omega_j(u-s)} e^{i(\omega_j-\omega_k)(s-v)}, \tag{7.2.1}$$

where we have used (6.2.5) and $\alpha_r=0$, $r<0$. Now,

$$\frac{4}{n^2}\left|\sum_{t=1}^{n}\sum_{s=1}^{n}\sum_{u=1}^{n}\sum_{v=1}^{n}\sum_{r=-\infty}^{\infty}\alpha_r\alpha_{r+t-s}\alpha_{r+t-u}\alpha_{r+t-v}e^{i\omega_k(t-s)}e^{i\omega_j(u-v)}\right|$$

$$\leqslant\frac{4}{n}\sum_{j=-\infty}^{\infty}\sum_{p=-\infty}^{\infty}\sum_{q=-\infty}^{\infty}\sum_{h=-\infty}^{\infty}|\alpha_j||\alpha_{j+p}||\alpha_{j+p+q}||\alpha_{j+p+q+h}|$$

$$=O(n^{-1}),$$

by the absolute summability of α_j.

If $\omega_k=\omega_j=0$, or if n is even and $\omega_k=\omega_j=\pi$, the second term of (7.2.1) is

$$\frac{4}{n^2}\left[\sum_{h=-(n-1)}^{n-1}(n-|h|)\gamma(h)\right]^2.$$

For $\omega_k=\omega_j\neq 0,\pi$ the second term of (7.2.1) is

$$\frac{4}{n^2}\sum_{t=1}^{n}\sum_{s=1}^{n}\sum_{u=1}^{n}\sum_{v=1}^{n}\gamma(t-u)\gamma(s-v)e^{i\omega_k[(t-u+2u)-(s-v+2v)]}$$

$$\leqslant\left[2\sum_{p=-(n-1)}^{n-1}\frac{|p|}{n}|\gamma(p)|\right]^2=O(n^{-1}),$$

where the inequality follows from Lemma 7.2.2. For $\omega_k=\omega_j$ the third term reduces to

$$\left[2\sum_{p=-(n-1)}^{n-1}\frac{n-|p|}{n}\gamma(p)e^{i\omega_k p}\right]^2.$$

By Lemma 7.2.2, for $\omega_k\neq\omega_j$,

$$\left|\sum_{t=1}^{n}\sum_{u=1}^{n}\gamma(t-u)e^{i\omega_k(t-u)}e^{i(\omega_j+\omega_k)u}\right|\leqslant\sum_{p=-(n-1)}^{n-1}|p||\gamma(p)|$$

and the absolute value of the second term shown above is less than $[2n^{-1}\Sigma_{p=-(n-1)}^{n-1}|p||\gamma(p)|]^2$ which, by assumption, is $O(n^{-1})$. By a similar argument, the third term of (7.2.1) is $O(n^{-1})$ when $\omega_k\neq\omega_j$. Thus, if $\omega_k=\omega_j=\omega\neq 0,\pi$,

$$\lim_{n\to\infty}\frac{4}{n^2}\sum_{t=1}^{n}\sum_{s=1}^{n}\sum_{u=1}^{n}\sum_{v=1}^{n}\gamma(t-v)\gamma(u-s)e^{i\omega_{K(n,\omega)}(t-v)}e^{i\omega_{K(n,\omega)}(u-s)}$$

$$=\lim_{n\to\infty}\left[2\sum_{h=-(n-1)}^{n-1}\frac{n-|h|}{n}\gamma(h)e^{-i\omega h}\right]^2=(4\pi)^2f^2(\omega).$$

If $\omega_j = \omega_k = 0$, then

$$\lim_{n\to\infty} \frac{4}{n^2} \sum_{t=1}^{n} \sum_{s=1}^{n} \sum_{u=1}^{n} \sum_{v=1}^{n} \gamma(t-u)\gamma(s-v)$$

$$+ \lim_{n\to\infty} \frac{4}{n^2} \sum_{t=1}^{n} \sum_{s=1}^{n} \sum_{u=1}^{n} \sum_{v=1}^{n} \gamma(t-v)\gamma(u-s)$$

$$= 2 \lim_{n\to\infty} \left[2 \sum_{h=-(n-1)}^{n-1} \frac{n-|h|}{n} \gamma(h) \right]^2 = 2(4\pi)^2 f^2(0),$$

and the stated results follow. ▲

Since the covariances between periodogram ordinates are small, the variance of the periodogram estimator of the spectral density at a particular frequency can be reduced by averaging adjacent periodogram ordinates. The simplest such estimator is defined by

$$\bar{f}(\omega_k) = \frac{1}{2d+1} \sum_{j=-d}^{d} \hat{f}(\omega_{k+j}),$$

where

$$\hat{f}(\omega_k) = \frac{1}{4\pi} I_n(\omega_k).$$

In general, we consider the linear function of the periodogram ordinates

$$\bar{f}(\omega_k) = \sum_{j=-d}^{d} W(j) \hat{f}(\omega_{k+j}) \tag{7.2.2}$$

where $\sum_{j=-d}^{d} W(j) = 1$. The weight function $W(j)$ is typically symmetric about zero with a maximum at zero.

We may extend $\bar{f}(\omega_k)$ to all ω by defining $\bar{f}(\omega) = \bar{f}(\omega_{K(n,\omega)})$ where the function $K(n,\omega)$ was introduced in Section 7.1. In practice the values of $\bar{f}(\omega_k)$ are often connected by lines to obtain a continuous function of ω.

Theorem 7.2.2. Let the time series X_t be defined by

$$X_t = \sum_{j=0}^{\infty} \alpha_j e_{t-j},$$

where the e_t are independent $(0, \sigma^2)$ random variables with fourth moment

$\eta\sigma^4$ and

$$\sum_{j=0}^{\infty} j^{1/2}|\alpha_j| < \infty.$$

Let d_n be an increasing sequence of positive integers satisfying

$$\lim_{n\to\infty} d_n = \infty,$$

$$\lim_{n\to\infty} \frac{d_n}{n} = 0.$$

Let the weight function $W_n(j), j=0, \pm 1, \pm 2, \ldots, \pm d_n$ satisfy

$$\sum_{j=-d_n}^{d_n} W_n(j) = 1,$$

$$W_n(j) = W_n(-j),$$

$$\lim_{n\to\infty} \sum_{j=-d_n}^{d_n} W_n^2(j) = 0.$$

Then $\bar{f}(\omega_{K(n,\omega)})$ defined by (7.2.2) satisfies

$$\lim_{n\to\infty} E\left\{\bar{f}(\omega_{K(n,\omega)})\right\} = f(\omega),$$

$$\begin{aligned} \lim_{n\to\infty} \left[\sum_{j=-d_n}^{d_n} W_n^2(j)\right]^{-1} \operatorname{Var}\left\{\bar{f}(\omega_{K(n,\omega)})\right\} &= f^2(\omega), \qquad \omega \neq 0, \pi \\ &= 2f^2(\omega), \qquad \omega = 0, \pi. \end{aligned}$$

Proof. Now

$$E\left\{\bar{f}(\omega_{K(n,\omega)})\right\} = \frac{1}{2\pi} \sum_{j=-d_n}^{d_n} W_n(j) \sum_{h=-(n-1)}^{(n-1)} \frac{n-|h|}{n} \gamma(h) e^{\{-ih\omega_{K(n,\omega+2\pi j/n)}\}}$$

and

$$\lim_{n\to\infty} E\left\{\bar{f}(\omega_{K(n,\omega)})\right\} = \lim_{n\to\infty} \sum_{j=-d_n}^{d_n} W_n(j) f(\omega_{K(n,\omega+2\pi j/n)}).$$

Since $f(\omega)$ is uniformly continuous, given $\epsilon>0$, there exists an N such that for $n>N, |f(\delta)-f(\omega)|<\epsilon$ for δ in the interval $[\omega_{K(n,\omega-2\pi d_n/n)}, \omega_{K(n,\omega+2\pi d_n/n)}]$. Therefore,

$$\lim_{n\to\infty} E\{\bar{f}(\omega)\}=f(\omega).$$

By Theorem 7.2.1, for $2\pi d_n/n \leqslant \omega \leqslant \pi-2\pi d_n/n$,

$$\operatorname{Var}\{\bar{f}(\omega_{K(n,\omega)})\}=\sum_{j=-d_n}^{d_n} W_n^2(j) f^2(\omega_{K(n,\omega+2\pi j/n)})+o\left[\sum_{j=-d_n}^{d_n} W_n^2(j)\right]$$

and, by the argument used for the expectation,

$$\lim_{n\to\infty}\left[\sum_{j=-d_n}^{d_n} W_n^2(j)\right]^{-1} \operatorname{Var}\{\bar{f}(\omega)\}=f^2(\omega).$$

If $\omega=0$ or π only d_n+1 (or d_n) estimates, $\hat{f}(\omega_k)$, are averaged since, for example, $\hat{f}(2\pi/n)=\hat{f}(-2\pi/n)$. Therefore,

$$\operatorname{Var}\{\bar{f}(0)\}=W_n^2(0)f^2(0)$$
$$+4\sum_{j=1}^{d_n} W_n^2(j) f^2(\omega_{K(n,\omega+2\pi j/n)})+o\left[\sum_{j=-d_n}^{d_n} W_n^2(j)\right]$$

and the result follows. ▲

Corollary 7.2.2. Let the assumptions of Theorem 7.2.2 be satisfied with $W_n(j)=(2d_n+1)^{-1}$. Then,

$$\operatorname{Var}\{\bar{f}(\omega_{K(n,\omega)})\}=\frac{1}{2d_n+1} f^2(\omega)+o(d_n^{-1}), \qquad \omega\neq 0,\pi,$$
$$=\frac{2}{2d_n+1} f^2(\omega)+o(d_n^{-1}), \qquad \omega=0,\pi.$$

Perhaps it is worthwhile to pause and summarize our results for the periodogram estimators. First, the periodogram ordinates $I_n(\omega_j)$ are the sums of squares associated with sine and cosine regression variables for the frequency ω_j. For a wide class of time series the $I_n(\omega_j)$ are approximately independently distributed as $[2\pi f(\omega_j)]\chi_2^2$, that is, as a multiple of a chi-square with two degrees of freedom (Theorems 7.1.1 and 7.2.1).

If $f(\omega)$ is a continuous function of ω, then, for large n, adjacent periodogram ordinates have approximately the same mean and variance. Therefore, an average of $(2d+1)$ adjacent ordinates has approximately the same mean and a variance $(2d+1)^{-1}$ times that of the original ordinates. It is possible to construct a sequence of estimators based on realizations of increasing size wherein the number of adjacent ordinates being averaged increases (at a slower rate than n) so that the average, when divided by 4π, is a consistent estimator of $f(\omega)$ (Theorem 7.2.2). The consistency result is less than fully appealing, since it does not tell us how many terms to include in the average for any particular time series. Some general conclusions are possible. For most time series the average of the periodogram ordinates will be a biased estimator of $4\pi f(\omega)$. For the largest portion of the range of most functions, this bias will increase as the number of terms being averaged increases. On the other hand, we can expect the variance of the average to decline as additional terms are added.[2] Therefore, the mean square error of our average as an estimator of the spectral density will decline as long as the increase in the squared bias is less than the decrease in the variance. The white noise time series furnishes the limiting case. Since the spectral density is a constant function, the best procedure is to include all ordinates in the average [i.e., to use $\hat{\gamma}(0)$ to estimate $2\pi f(\omega)$ for all ω]. For a time series of known structure we could determine the optimum number of terms to include in the weight function. However, if we possess that degree of knowledge, the estimation problem is no longer of interest. The practitioner, as he works with data, will develop certain rules of thumb for particular kinds of data. For data with unknown structure it would seem advisable to construct several averages of varying length before reaching conclusions about the nature of the spectral density.

The approximate distributional properties of the smoothed periodogram can be used to construct a confidence interval for the estimated spectral density. Under the conditions of Theorem 7.2.2, the $I_n(\omega_k)$ are approximately distributed as independent chi-squares and, therefore, $\bar{f}(\omega)$ is approximately distributed as a linear combination of chi-square random variables. One common approximation[3] to such a distribution is a chi-square distribution with degrees of freedom determined by the variance of the distribution.

[2]Suppose $X_1, X_2, \ldots, X_{p+1}$ are uncorrelated random variables. Then

$$\operatorname{Var}\left\{(p+1)^{-1}\sum_{i=1}^{p+1} X_i\right\} \leqslant \operatorname{Var}\left\{p^{-1}\sum_{i=1}^{p} X_i\right\} \quad \text{unless}$$

$$\operatorname{Var}\{X_{p+1}\} > (2p+1)\operatorname{Var}\left\{p^{-1}\sum_{i=1}^{p} X_i\right\}.$$

[3]The approximation is discussed by Box (1954).

Result 7.2.1. Let X_t satisfy the assumptions of Theorem 7.2.2 and let $f(\omega)>0$. Then, for $\pi d_n/n<\omega<\pi(1-d_n/n), f^{-1}(\omega)\bar{f}(\omega)$ is approximately distributed as a chi-square random variable divided by its degrees of freedom ν, where

$$\nu=2\left[\sum_{j=-d_n}^{d_n} W_n^2(j)\right]^{-1}.$$

An approximate $1-\alpha$ level confidence interval for $f(\omega)$ can be constructed as

$$\frac{\nu\bar{f}(\omega)}{\chi^2_{\nu,\alpha/2}} \leqslant f(\omega) \leqslant \frac{\nu\bar{f}(\omega)}{\chi^2_{\nu,1-(\alpha/2)}}, \tag{7.2.3}$$

where $\chi^2_{\nu,\alpha/2}$ is the $\alpha/2$ tabular value for the chi-square distribution with ν degrees of freedom. (Point exceeded with probability $\alpha/2$.)

Since the variance of $\bar{f}(\omega)$ is a multiple of $[f(\omega)]^2$, the logarithm of $\bar{f}(\omega)$ for time series with $f(\omega)$ strictly positive will have approximately constant variance,

$$\operatorname{Var}\{\log\bar{f}(\omega)\} \doteq [f(\omega)]^{-2}\operatorname{Var}\{\bar{f}(\omega)\} \doteq \frac{2}{\nu}.$$

Therefore, $\log\bar{f}(\omega)$ is often plotted as a function of ω. Approximate confidence intervals for $\log f(\omega)$ are given by

$$\log\bar{f}(\omega)+\log\left(\frac{\nu}{\chi^2_{\nu,\alpha/2}}\right) \leqslant \log f(\omega) \leqslant \log\bar{f}(\omega)+\log\left(\frac{\nu}{\chi^2_{\nu,1-(\alpha/2)}}\right). \tag{7.2.4}$$

The smoothed periodogram is a weighted average of the Fourier transform of the sample autocovariance. An alternative method of obtaining an estimated spectral density is to apply weights to the estimated covariance function and then transform the "smoothed" covariance function.The impetus for this procedure came from a desire to reduce the computational costs of computing covariances for realizations with very large numbers of observations. Thus, the weight function has traditionally been chosen to be nonzero for the first few autocovariances and zero otherwise. The development of computer routines using the Fast Fourier Transform[4] reduced the

[4]See Cooley, Lewis, and Welch (1967) for references on the Fast Fourier Transform. Singleton (1969) gives a Fortran program for the transform.

cost of computing finite Fourier transforms and reduced the emphasis on the use of the weighted autocovariance estimator of the spectrum.

Let $w(x)$ be a bounded even continuous function satisfying

$$\begin{aligned} &w(0)=1, \\ &w(x)=0, \qquad |x|>1, \\ &|w(x)|\leqslant 1 \quad \text{for all } x. \end{aligned} \tag{7.2.5}$$

Then a weighted estimator of the spectral density is

$$\tilde{f}(\omega)=\frac{1}{2\pi}\sum_{h=-g_n}^{g_n} w\left(\frac{h}{g_n}\right)\hat{\gamma}(h)e^{-\iota\omega h}, \tag{7.2.6}$$

where $g_n \leqslant n$ is the chosen point of truncation and

$$\hat{\gamma}(h)=\hat{\gamma}(-h)=\frac{1}{n}\sum_{t=1}^{n-h}(X_t-\bar{x}_n)(X_{t+h}-\bar{x}_n), \qquad h\geqslant 0.$$

From Section 3.4 we know that the transform of a convolution is the product of the transforms. Conversely, the transform of a product is the convolution of the transforms. In the current context we define the following function (or transform):

$$\begin{aligned} \hat{f}(\omega)&=\frac{1}{2\pi}\sum_{h=-(n-1)}^{(n-1)}\hat{\gamma}(h)e^{-\iota\omega h} \\ &=\frac{1}{2\pi}\left[\hat{\gamma}(0)+2\sum_{h=1}^{(n-1)}\hat{\gamma}(h)\cos\omega h\right]. \end{aligned}$$

The function $\hat{f}(\omega)$ is a continuous function of ω and is an unweighted estimator of the continuous spectral density. By the uniqueness of Fourier transforms, we can write

$$\hat{\gamma}(h)=\int_{-\pi}^{\pi}\hat{f}(\omega)e^{\iota\omega h}\,d\omega, \qquad h=0,\pm 1,\pm 2,\ldots,\pm(n-1).$$

We define the transform of $w(x)$ similarly as

$$W(\omega)=\frac{1}{2\pi}\sum_{h=-(n-1)}^{(n-1)} w\left(\frac{h}{g_n}\right)e^{-\iota\omega h},$$

where $W(\omega)$ is also a continuous function. It follows that

$$w\left(\frac{h}{g_n}\right)=\int_{-\pi}^{\pi} W(\omega)e^{i\omega h}\,d\omega, \qquad h=0,\pm 1,\pm 2,\ldots,\pm(n-1).$$

Note that $w(0)=1$ means that $\int_{-\pi}^{\pi} W(s)\,ds=1$. Then, by Exercise 3.14,

$$\tilde{f}(\omega)=\frac{1}{2\pi}\sum_{h=-(n-1)}^{n-1} w\left(\frac{h}{g_n}\right)\hat{\gamma}(h)e^{-i\omega h}$$

$$=\int_{-\pi}^{\pi} W(s)\hat{f}(\omega-s)\,ds.$$

One should remember that both $W(s)$ and $\hat{f}(s)$ are even periodic functions. Thus, the estimated spectrum obtained from the weighted estimated covariances $w(h/g_n)\hat{\gamma}(h)$ is a weighted average (convolution) of the spectrum estimated from the original covariances, where the weight function is the transform of weights applied to the covariances.

The function $W(\omega)$ is called the *kernel* or *spectral window*. The weight function $w(x)$ is often called the *lag window*.

Theorem 7.2.3. Let g_n be a sequence of positive integers such that

$$\lim_{n\to\infty} g_n=\infty,$$

$$\lim_{n\to\infty} \frac{g_n}{n}=0;$$

and let X_t be a time series defined by

$$X_t=\sum_{j=-\infty}^{\infty} \alpha_j e_{t-j},$$

where the e_t are independent $(0,\sigma^2)$ random variables with fourth moment $\eta\sigma^4$, and $\{\alpha_j\}$ is absolutely summable. Let

$$\tilde{\gamma}(h)=\tilde{\gamma}(-h)=\frac{1}{n-h}\sum_{t=1}^{n-h} X_t X_{t+h}, \qquad h\geqslant 0,$$

and

$$\tilde{f}(\omega)=\frac{1}{2\pi}\sum_{h=-g_n}^{g_n} w\left(\frac{h}{g_n}\right)\left(\frac{n-|h|}{n}\right)\tilde{\gamma}(h)e^{-i\omega h},$$

where $w(x)$ is a bounded even continuous function satisfying conditions (7.2.5). Then

$$E\{\tilde{f}(\omega)\} = \frac{1}{2\pi} \sum_{h=-g_n}^{g_n} \frac{n-|h|}{n} w\left(\frac{h}{g_n}\right)\gamma(h)e^{-i\omega h}$$

and

$$\begin{aligned} \lim_{n\to\infty} \frac{n}{g_n} \operatorname{Cov}\{\tilde{f}(\omega),\tilde{f}(\lambda)\} &= 2f^2(\omega)\int_{-1}^{1} w^2(x)\,dx, & \omega=\lambda=0,\pi \\ &= f^2(\omega)\int_{-1}^{1} w^2(x)\,dx, & \omega=\lambda\neq 0,\pi \\ &= 0, & \omega\neq\lambda. \end{aligned}$$

Proof. See, for example, Anderson (1971, Chap. 9) or Hannan (1970, Chap. 5). ▲

The choice of a truncation point, g_n, for a particular sample of n observations is not determined by the asymptotic theory of Theorem 7.2.3. As in our discussion of the smoothed periodogram, the variance generally increases as g_n increases, but the bias in the estimator will typically decrease as g_n increases. It is possible to determine the order of the bias as a function of the properties of the weight function $w(x)$ and the speed with which $\gamma(h)$ approaches zero [see Parzen (1961)], but this still does not solve the problem for a given sample and unknown covariance structure.

Approximate confidence limits for the spectral density can be constructed in the same manner as that used for the smoothed periodogram estimator.

***Result** 7.2.2.* Let X_t satisfy the assumptions of Theorem 7.2.3, let $w(x)$ satisfy (7.2.5) and let $f(\omega)>0$. Then, for $\pi\nu/2n<\omega<\pi-\pi\nu/2n$, $f^{-1}(\omega)\tilde{f}(\omega)$ is approximately distributed as a chi-square random variable divided by its degrees of freedom ν, where

$$\nu = \frac{2n}{\int_{-1}^{1} w^2(x)\,dx}.$$

Considerable research has been conducted on the weights $w(x)$ to use in estimating the spectrum. One of the simplest windows is obtained by truncating the sequence of autocovariances at g_n. This procedure is

equivalent to applying the window

$$w(x)=\begin{cases}1, & |x|\leqslant 1\\ 0, & \text{otherwise.}\end{cases} \tag{7.2.8}$$

The function $w(x)$ is sometimes called a *truncated* or *rectangular* window. While $w(x)$ does not meet the conditions of Theorem 7.2.3, it can be shown that the conclusion holds. The spectral window for the function (7.2.8),

$$W(s)=\frac{1}{2\pi}\frac{\sin(g_n+1/2)s}{\sin(s/2)}$$

takes on negative values and it is possible that the weighted average $\tilde{f}(\omega)$ will be negative for some ω. This is generally considered an undesirable attribute, since $f(\omega)\geqslant 0$ for all ω.

Bartlett (1950) suggested splitting an observed time series of n observations into p groups of M observations each. The periodogram is then computed for each group, and the estimator for the ordinate associated with a particular frequency is taken to be the average of the p estimators; that is,

$$4\pi\tilde{f}(\omega_k)=\frac{1}{p}\sum_{s=0}^{p-1} I_{M_s}(\omega_k),$$

where $I_{M_s}(\omega_k)$ is the estimator for the ordinate at frequency ω_k obtained from the sth subsample.

Bartlett's estimator is closely related to the estimator with lag window

$$\begin{aligned} w(x)&=1-|x|, \qquad & |x|\leqslant 1\\ &=0, & \text{otherwise.}\end{aligned}$$

This window has been called *modified Bartlett* or *triangular*. Setting $g_n=M$, the spectral window is given by

$$\begin{aligned} W_B(\omega)&=\frac{1}{2\pi}\sum_{h=-M}^{M}\frac{M-|h|}{M}\cos\omega h\\ &=\frac{1}{2\pi M}\sum_{h=0}^{M-1}\sum_{j=-h}^{h}\cos j\omega.\end{aligned}$$

Using Lemma 3.1.2,

$$W_B(\omega) = \frac{\sin^2(M/2)\omega}{2\pi M \sin^2(\omega/2)}.$$

To evaluate the variance of the modified Bartlett estimator, we have

$$\sum_{h=-M}^{M} w^2\left(\frac{h}{M}\right) = \sum_{h=-M}^{M} \left(\frac{M-|h|}{M}\right)^2 \doteq \frac{2}{3}M.$$

Thus, for the modified Bartlett estimator with covariances truncated at M, the variance of the estimated spectrum is approximately

$$\operatorname{Var}\{\tilde{f}(\omega)\} \doteq \frac{2M}{3n} f^2(\omega).$$

Blackman and Tukey (1959) suggested the weight function

$$\begin{aligned} w(x) &= 1 - 2a + 2a\cos\pi x, & |x| \leqslant 1 \\ &= 0, & \text{otherwise.} \end{aligned}$$

The use of the window with $a = 0.23$ they called "hamming" and the use of the window with $a = 0.25$, "hanning." Parzen (1961) suggested the weight function

$$\begin{aligned} w(x) &= 1 - 6x^2 + 6|x|^3, & |x| \leqslant \frac{1}{2} \\ &= 2(1-|x|)^3, & \frac{1}{2} \leqslant |x| \leqslant 1 \\ &= 0, & \text{otherwise.} \end{aligned}$$

This kernel will always produce nonnegative estimators. Brillinger (1975, Chap. 3) contains a discussion of these and other kernels.

7.3. EXAMPLES

As a first illustration of spectral estimation, we consider an artificially generated time series. Table 7.3.1 contains 100 observations for the time series

$$X_t = 0.7X_{t-1} + e_t,$$

Table 7.3.1. One hundred observations from a first order autoregressive time series with $\rho = 0.7$

	First 25	Second 25	Third 25	Fourth 25
1	0.874	−0.613	−0.366	−0.955
2	0.850	0.110	−1.420	−0.948
3	2.345	0.113	−0.183	0.046
4	2.501	−0.308	−0.044	0.091
5	1.657	0.723	−0.391	0.254
6	1.649	−0.257	−0.095	2.750
7	2.498	1.051	−0.971	1.673
8	1.330	0.803	0.371	2.286
9	1.307	0.116	−1.622	1.220
10	3.404	−1.454	−2.941	−0.256
11	2.445	0.296	−2.814	0.252
12	2.805	1.501	−1.784	0.325
13	1.639	0.880	−2.471	−0.338
14	1.240	−0.672	−3.508	0.378
15	1.116	0.436	−2.979	0.127
16	0.448	0.930	−0.779	−2.006
17	0.377	1.168	0.869	−2.380
18	−0.488	1.999	1.786	−2.024
19	−0.960	1.376	0.123	−1.085
20	−0.579	1.613	0.093	1.037
21	−1.674	2.030	−0.731	−0.467
22	−0.366	0.616	−1.253	−0.794
23	−0.922	0.667	−2.213	−0.493
24	−1.174	0.707	−0.252	−0.157
25	−1.685	1.029	0.403	0.659

where the e_t are computer generated normal independent (0, 1) random variables. The periodogram for this sample is given in Figure 7.3.1. We have connected the ordinates with straight lines. The approximate expected value of the periodogram ordinates is $4\pi f(\omega) = 2[1.49 - 1.4\cos\omega]^{-1}$ and has also been plotted in the figure. Both the average value and the variance are much larger for the ordinates associated with the smaller frequencies.

We have labeled the frequency axis in radians. Thus, the fastest frequency we can observe is π, which is 0.50 cycles per time unit. This corresponds to a cycle with a period of two time units. In applications there are natural time units such as hours, months, or years, and one may choose to label the axis in terms of the frequency in these units.

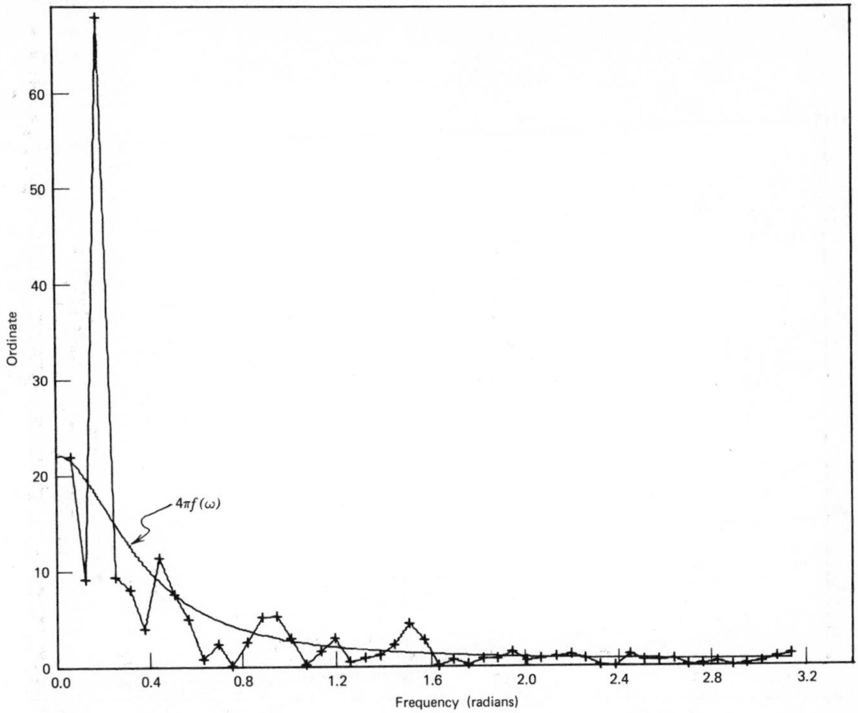

Figure 7.3.1. Periodogram computed from the 100 autoregressive observations of Table 7.3.1 compared with $4\pi f(\omega)$.

In Figure 7.3.2 we display the smoothed periodogram where the smoothed value at ω_k is

$$\bar{I}(\omega_k) = \frac{1}{5} \sum_{j=-2}^{2} I_n(\omega_{k-j}).$$

The smoothed periodogram roughly assumes the shape of the spectral density. Because the standard error of the smoothed estimator for $2 < k < 48$ is about 0.45 of the true value, there is considerable variability in the plot. The smoothing introduces a correlation between observations less than $2d+1$ units apart. In one sense this was the objective of the smoothing, since it produces a plot that more nearly approximates that of the spectral density. On the other hand, if the estimated spectral density is above the true density it is apt to remain so for some distance.

To compute smoothed values near zero and π, the periodic nature of the function $f(\omega)$ is used. In most applications the mean is estimated and,

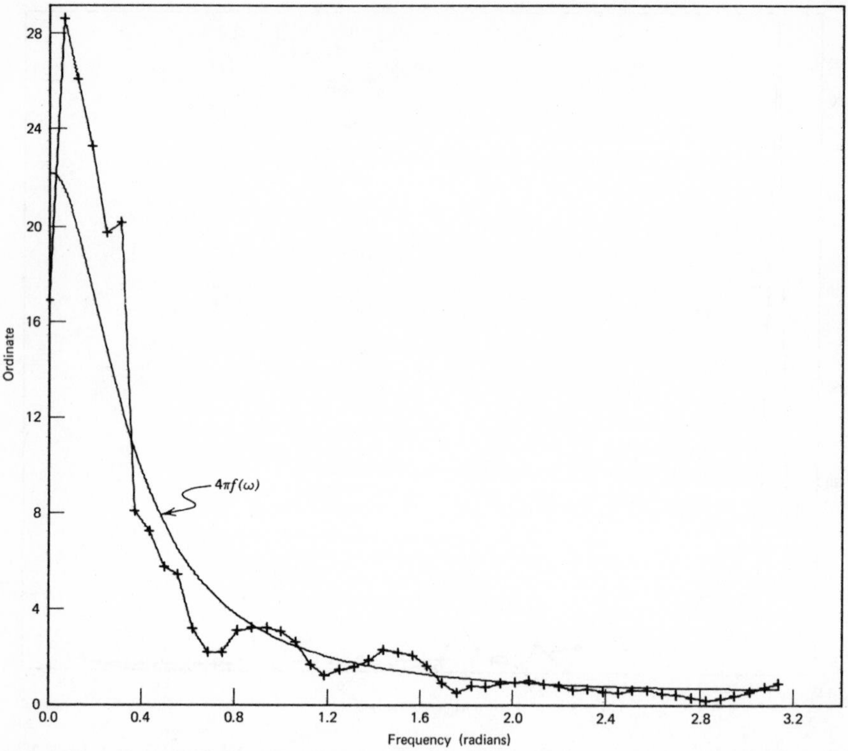

Figure 7.3.2. Smoothed periodogram ($d=2$) computed from 100 autoregressive observations of Table 7.3.1 compared with $4\pi f(\omega)$.

therefore, $I_n(0)$ is not computed. We follow this practice in our example. To compute the smoothed value at zero we set[5] $I_n(\omega_0)=I_n(\omega_1)$ and compute

$$\bar{I}(\omega_0)=\frac{1}{5}\left[I_{100}(\omega_{-2})+I_{100}(\omega_{-1})+I_{100}(\omega_0)+I_{100}(\omega_1)+I_{100}(\omega_2)\right]$$

which, by the even periodic property of $f(\omega)$, is given by

$$\bar{I}(\omega_0)=\frac{1}{5}\left[I_{100}(\omega_0)+2I_{100}(\omega_1)+2I_{100}(\omega_2)\right].$$

In our example, replacing $I_{100}(\omega_0)$ by $I_{100}(\omega_1)$,

$$\bar{I}(\omega_0)=\frac{1}{5}\left[3(22.004)+2(9.230)\right]=16.894.$$

[5] Other methods of estimating the zero ordinate could be used. For example, we could set $I_n(\omega_0)=d^{-1}\sum_{j=1}^{d}I_n(\omega_j)$.

Similarly

$$\bar{I}(\omega_2)=\frac{1}{5}[I_{100}(\omega_0)+I_{100}(\omega_1)+I_{100}(\omega_2)+I_{100}(\omega_3)+I_{100}(\omega_4)]$$

$$=\frac{1}{5}[2(22.004)+(9.230)+(67.776)+(9.360)]=26.075.$$

As the sample size is even, there is a one degree of freedom periodogram ordinate for π. The smoothed estimate at π is

$$\bar{I}(\omega_{50})=\frac{1}{5}[I_{100}(\omega_{50})+2I_{100}(\omega_{49})+2I_{100}(\omega_{48})]$$

$$=\frac{1}{5}[(1.166)+2(0.870)+2(0.440)]=0.757.$$

In Figure 7.3.3 we plot the logarithm of the average of 11 periodogram ordinates ($d=5$). The 95% confidence intervals are also plotted in Figure

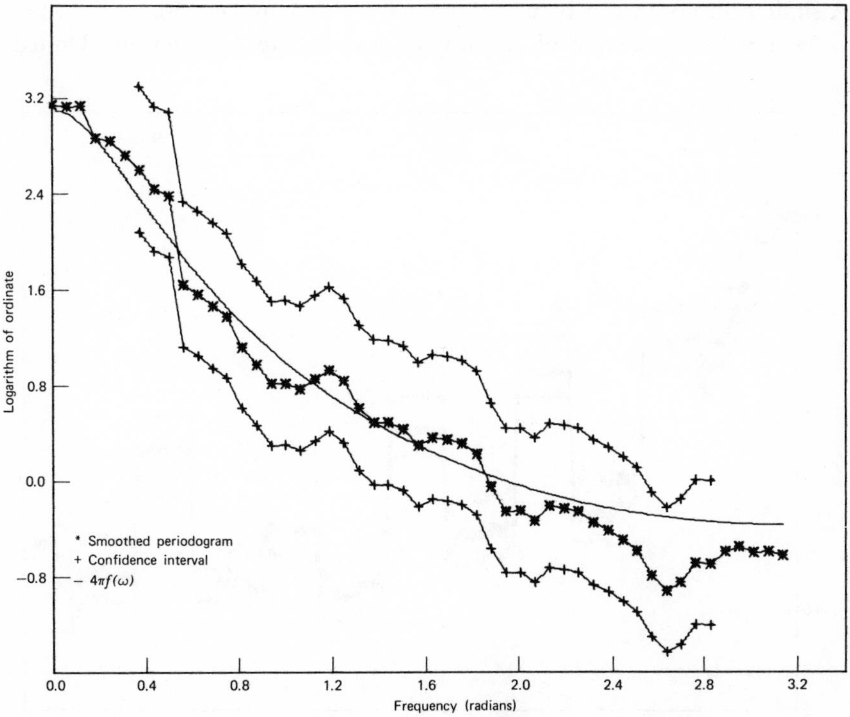

Figure 7.3.3. Logarithm of smoothed periodogram ($d=5$) and confidence interval for logarithm of $4\pi f(\omega)$ computed from the 100 autoregressive observations of Table 7.3.1.

7.3.3. They were constructed using (7.2.4), so that the upper bound is

$$\log \bar{I}(\omega) + \log\left[\frac{22}{10.98}\right]$$

and the lower bound is

$$\log \bar{I}(\omega) + \log\left[\frac{22}{36.78}\right].$$

This interval is appropriate for $6 \leqslant k \leqslant 44$ and adequate for $k=45$. Confidence intervals for other values of k could be constructed using the variance of the estimator. For example, the smoothed value at zero is

$$\bar{I}(0) = \frac{1}{11}\left[3I_{100}(\omega_1) + 2\sum_{j=2}^{5} I_{100}(\omega_j)\right],$$

and the variance is approximately $(25/121)[4\pi f(0)]^2 \doteq [0.21][4\pi f(0)]^2$. Since the variance of a chi-square with 10 degrees of freedom divided by its degrees of freedom is 0.20, we can establish a confidence interval for $4\pi f(0)$ using the critical points for a chi-square with 10 degrees of freedom. A similar approach can be used for $1 \leqslant k \leqslant 5$ and $46 \leqslant k \leqslant 50$.

As a second example of spectral estimation, we consider the United

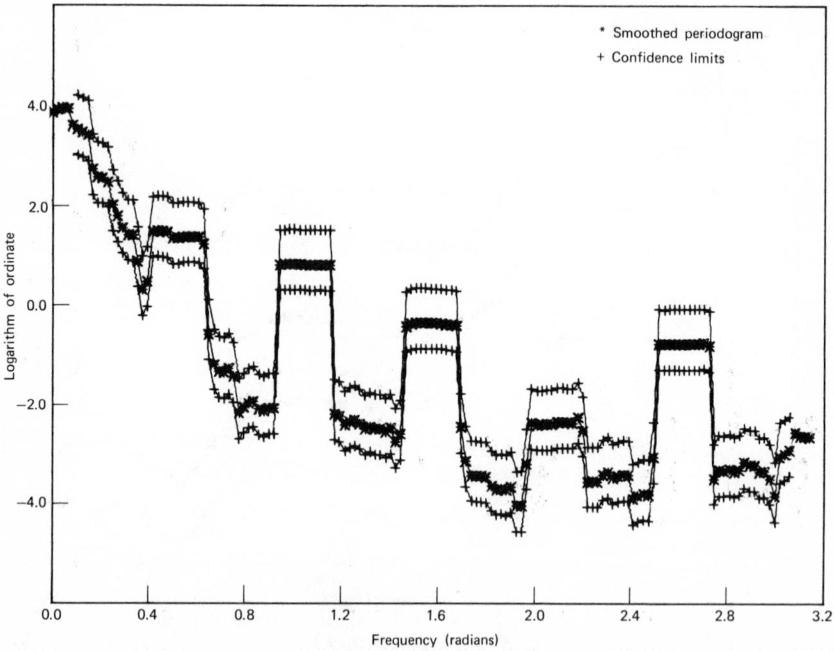

Figure 7.3.4. Logarithm of smoothed periodogram for monthly United States unemployment rate for October 1949 through September 1974 ($n=300$, $d=5$).

States monthly unemployment rate from October 1949, to September 1974. Figure 7.3.4 is a plot of the logarithm of the smoothed periodogram using the rectangular weights and $d=5$. Also included in the figure are lines defining the 95% confidence interval. This plot displays characteristics typical of many economic time series. The smoothed periodogram is high at small frequencies indicating a large positive autocorrelation for observations close together in time. Second, there are peaks at the seasonal frequencies $\pi, 5\pi/6, 2\pi/3, \pi/2, \pi/3, \pi/6$, indicating the presence of a seasonal component in the time series. The periodogram has been smoothed using the simple average, and the flatness of the seasonal peaks indicates that they are being dominated by the center frequencies. That is, the shape of the peak is roughly that of the weights being used in the smoothing. Table 7.3.2 contains the ordinates at and near the seasonal frequencies. With the exception of π, the ordinates at the seasonal frequencies are much larger than the other ordinates. Also, it seems that the ordinates adjacent

Table 7.3.2. Periodogram ordinates near seasonal frequencies

Frequency (Radians)	Ordinate	Frequency (Radians)	Ordinate
0.461	0.127	2.032	0.025
0.482	0.607	2.052	0.021
0.503	5.328	2.073	0.279
0.523	31.584	2.094	0.575
0.545	2.904	2.115	0.012
0.565	0.557	2.136	0.011
0.586	0.069	2.157	0.008
0.984	0.053	2.555	0.028
1.005	0.223	2.576	0.010
1.026	0.092	2.597	0.286
1.047	23.347	2.618	4.556
1.068	0.310	2.639	0.038
1.089	0.253	2.660	0.017
1.110	0.027	2.681	0.022
1.508	0.191	3.079	0.008
1.529	0.041	3.100	0.003
1.550	0.177	3.120	0.294
1.571	6.402	3.142	0.064
1.592	0.478		
1.613	0.142		
1.634	0.012		

to the seasonal frequencies are larger than those farther away. This suggests that the "seasonality" in the time series is not perfectly periodic. That is, more than the six seasonal frequencies are required to completely explain the peaks in the estimated spectral density.

7.4. MULTIVARIATE SPECTRAL ESTIMATES

We now investigate estimators of the spectral parameters for vector time series. Let n observations be available on the bivariate time series $\mathbf{Y}_t = (Y_{1t}, Y_{2t})'$. The Fourier coefficients a_{1k}, b_{1k}, a_{2k}, and b_{2k} can be computed for the two time series by the formulas following (7.1.1). Having studied estimators of $f_{11}(\omega)$ and $f_{22}(\omega)$ in the preceding sections, it remains only to investigate estimators of $f_{12}(\omega)$ and of the associated quantities such as the phase spectrum and the squared coherency.

Recalling the transformation introduced in Section 4.2 which yields the normalized Fourier coefficients, we shall investigate the joint distributional properties of

$$2^{-1}n^{1/2}(a_{1k}+ib_{1k}) = n^{-1/2}\sum_{t=1}^{n} Y_{1t}e^{i\omega_k t},$$

$$2^{-1}n^{1/2}(a_{2k}+ib_{2k}) = n^{-1/2}\sum_{t=1}^{n} Y_{2t}e^{i\omega_k t}, \tag{7.4.1}$$

where $\omega_k = 2\pi k/n$, $k=0,1,\ldots,n-1$. Define a transformation matrix $\mathbf{H}$ by

$$\mathbf{H} = \begin{pmatrix} \mathbf{G} & \mathbf{O} \\ \mathbf{O} & \mathbf{G} \end{pmatrix}, \tag{7.4.2}$$

where $\mathbf{G}$ is an $n\times n$ matrix with rows given by

$$\mathbf{g}_{k.} = n^{-1/2}\left[1, e^{i2\pi k/n}, e^{i2\pi 2k/n}, \ldots, e^{i2\pi(n-1)k/n}\right], \qquad k=0,1,\ldots,n-1.$$

The matrix $\mathbf{G}$ was introduced in (4.2.6), and is the matrix that will diagonalize a circular matrix. Let

$$\mathbf{V} = \begin{pmatrix} \mathbf{V}_{11} & \mathbf{V}_{12} \\ \mathbf{V}_{21} & \mathbf{V}_{22} \end{pmatrix} \tag{7.4.3}$$

be the $2n\times 2n$ covariance matrix of $\mathbf{y} = (\mathbf{y}_1', \mathbf{y}_2')'$, where

$$\mathbf{y}_1' = (Y_{11}, Y_{12}, \ldots, Y_{1n}),$$

$$\mathbf{y}_2' = (Y_{21}, Y_{22}, \ldots, Y_{2n}),$$

and

$$\mathbf{V}_{12}=\begin{bmatrix} \gamma_{12}(0) & \gamma_{12}(1) & \gamma_{12}(2) & \cdots & \gamma_{12}(n-1) \\ \gamma_{12}(-1) & \gamma_{12}(0) & \gamma_{12}(1) & \cdots & \gamma_{12}(n-2) \\ \gamma_{12}(-2) & \gamma_{12}(-1) & \gamma_{12}(0) & \cdots & \gamma_{12}(n-3) \\ \vdots & \vdots & \vdots & & \vdots \\ \gamma_{12}(-n+1) & \gamma_{12}(-n+2) & \gamma_{12}(-n+3) & \cdots & \gamma_{12}(0) \end{bmatrix}. \tag{7.4.4}$$

In Theorem 4.2.1, for time series with absolutely summable covariance function, we demonstrated that the elements of $\mathbf{GV}_{ii}\mathbf{G}^*$ converge to the elements of the diagonal matrix $2\pi\mathbf{D}_{ii}$, where the elements of $\mathbf{D}_{ii}$ are $f_{ii}(\omega)$ evaluated at $\omega_k=2\pi k/n$, $k=0,1,2,\ldots,n-1$.

It remains to investigate the behavior of $\mathbf{GV}_{12}\mathbf{G}^*$. To this end define the circular matrix

$$\mathbf{V}_{12c}=\begin{bmatrix} \gamma_{12}(0) & \gamma_{12}(1) & \gamma_{12}(2) & \cdots & \gamma_{12}(-1) \\ \gamma_{12}(-1) & \gamma_{12}(0) & \gamma_{12}(1) & \cdots & \gamma_{12}(-2) \\ \gamma_{12}(-2) & \gamma_{12}(-1) & \gamma_{12}(0) & \cdots & \gamma_{12}(-3) \\ \vdots & \vdots & \vdots & & \vdots \\ \gamma_{12}(1) & \gamma_{12}(2) & \gamma_{12}(3) & \cdots & \gamma_{12}(0) \end{bmatrix}. \tag{7.4.5}$$

Then, $\mathbf{GV}_{12c}\mathbf{G}^*$ is a diagonal matrix with elements

$$\sum_{h=0}^{M} \gamma_{12}(h)e^{-i2\pi kh/n} + \sum_{h=-M}^{-1} \gamma_{12}(h)e^{-i2\pi k(n+h)/n}$$

$$= \sum_{h=-M}^{M} \gamma_{12}(h)e^{-i2\pi kh/n}, \qquad k=0,1,\ldots,n-1, \tag{7.4.6}$$

where we have assumed n is odd and set $M=(n-1)/2$. If n is even the sum is from $-M+1$ to M, where $M=n/2$. If $\gamma_{12}(h)$ is absolutely summable we obtain the following theorem.

Theorem 7.4.1. Let $\mathbf{Y}_t$ be a stationary bivariate time series with absolutely summable autocovariance function. Let $\mathbf{V}$ of (7.4.3) be the covariance matrix for n observations. Then, given $\epsilon>0$, there exists an N such that for $n>N$, every element of the matrix

$$\mathbf{HVH}^* - 2\pi\mathbf{D}$$

is less than ϵ, where

$$\mathbf{D}=\begin{pmatrix}\mathbf{D}_{11} & \mathbf{D}_{12}\\ \mathbf{D}_{21} & \mathbf{D}_{22}\end{pmatrix},$$

$$\mathbf{D}_{ij}=\operatorname{diag}\{f_{ij}(\omega_0),f_{ij}(\omega_1),\ldots,f_{ij}(\omega_{n-1})\},\qquad i,j=1,2,$$

and $\omega_k=2\pi k/n$, $k=0,1,\ldots,n-1$.

Proof. The result for $\mathbf{D}_{11}$ and $\mathbf{D}_{22}$ follows by Theorem 4.2.1. The result for $\mathbf{D}_{12}$ is obtained by arguments completely analogous to those of Section 4.2, by showing that the elements of $\mathbf{GV}_{12}\mathbf{G}^*-\mathbf{GV}_{12c}\mathbf{G}^*$ converge to zero as n increases. The details are reserved for the reader. ▲

If we make a stronger assumption about the autocovariance function we obtain the stronger result parallel to Corollary 4.2.1.

Corollary 7.4.1.1. Let $\mathbf{Y}_t$ be a stationary bivariate time series with an autocovariance function that satisfies

$$\sum_{h=\infty}^{\infty}|h||\gamma_{ij}(h)|<L<\infty,\qquad i,j=1,2.$$

Let $\mathbf{V}$ be as defined in (7.4.3), $\mathbf{H}$ as defined in (7.4.2), and $\mathbf{D}$ as defined in Theorem 7.4.1. Then every element of the matrix $\mathbf{HVH}^*-2\pi\mathbf{D}$ is less than $3L/n$.

Proof. The proof parallels that of Corollary 4.2.1 and is reserved for the reader. ▲

By Theorem 7.4.1, the complex coefficients $2^{-1}n^{1/2}(a_{1k}+ib_{1k})$ and $2^{-1}n^{1/2}(a_{2j}+ib_{2j})$ are nearly uncorrelated in large samples if $j\neq k$. Since

$$a_{ik}-ib_{ik}=a_{i,n-k}+ib_{i,n-k},\qquad i=1,2,\ k=1,2,\ldots,n-1,$$

it follows that

$$\lim_{n\to\infty}E\{a_{1k}a_{2k}-b_{1k}b_{2k}\}=0,$$

$$\lim_{n\to\infty}E\{b_{1k}a_{2k}+b_{2k}a_{1k}\}=0. \tag{7.4.7}$$

That is, the covariance between a_{1k} and a_{2k} is approximately equal to that between b_{1k} and b_{2k}, while the covariance between b_{1k} and a_{2k} is approximately the negative of the covariance between b_{2k} and a_{1k}.

We define the cross periodogram by

$$I_{12n}(\omega_k)=\begin{cases}\frac{n}{2}\left[a_{1k}a_{2k}+b_{1k}b_{2k}-i(a_{1k}b_{2k}-b_{1k}a_{2k})\right], & \omega_k\neq 0,\pi,\\ 2na_{1k}a_{2k}, & \omega_k=0,\pi.\end{cases} \tag{7.4.8}$$

To obtain a function defined at all ω, we recall the function

$$K(n,\omega)=k \quad \text{for} \quad \frac{\pi(2k-1)}{n}<\omega\leqslant\frac{\pi(2k+1)}{n} \tag{7.4.9}$$

and take $I_{12n}(\omega)=I_{12n}(\omega_{K(n,\omega)})$.

Corollary 7.4.1.2. Let $\mathbf{Y}_t$ be a stationary bivariate time series with absolutely summable covariance function. Then,

$$\lim_{n\to\infty} E\{I_{12n}(\omega)\}=4\pi f_{12}(\omega).$$

Proof. The result is an immediate consequence of Theorem 7.4.1. ▲

We now obtain some distributional results for spectral estimates. To simplify the presentation, we assume that $\mathbf{Y}_t$ is a normal time series.

Theorem 7.4.2. Let $\mathbf{Y}_t$ be a bivariate normal time series with covariance function that satisfies

$$\sum_{h=-\infty}^{\infty}|h||\gamma_{ij}(h)|<L<\infty, \qquad i,j=1,2.$$

Then $\mathbf{r}_k=2^{-1/2}n^{1/2}(a_{1k},b_{1k},a_{2k},b_{2k})'$ is distributed as a multivariate normal random variable with zero mean and covariance matrix

$$E\{\mathbf{r}_k\mathbf{r}_k'\}=2\pi\begin{bmatrix} f_{11}(\omega_k) & 0 & c_{12}(\omega_k) & q_{12}(\omega_k)\\ 0 & f_{11}(\omega_k) & -q_{12}(\omega_k) & c_{12}(\omega_k)\\ c_{12}(\omega_k) & -q_{12}(\omega_k) & f_{22}(\omega_k) & 0\\ q_{12}(\omega_k) & c_{12}(\omega_k) & 0 & f_{22}(\omega_k)\end{bmatrix}+O(n^{-1})$$

for $\omega_k\neq 0,\pi$, where $f_{12}(\omega_k)=c_{12}(\omega_k)-iq_{12}(\omega_k)$. Also,

$$E\{\mathbf{r}_k\mathbf{r}_j'\}=O(n^{-1}), \qquad j\neq k.$$

It follows that

$$E\left\{\frac{n}{2}(a_{1k}a_{2k}+b_{1k}b_{2k})\right\}=4\pi c_{12}(\omega_k)+O(n^{-1}),$$

$$E\left\{\frac{n}{2}(a_{1k}b_{2k}-a_{2k}b_{1k})\right\}=4\pi q_{12}(\omega_k)+O(n^{-1}),$$

$$E\{I_{12n}(\omega_k)\}=4\pi f_{12}(\omega_k)+O(n^{-1}),$$

and, for $\omega_k \neq 0, \pi$,

$$\begin{aligned}\operatorname{Cov}\{I_{12n}(\omega_k), I_{12n}(\omega_j)\} &= (4\pi)^2 f_{11}(\omega_k) f_{22}(\omega_k)+O(n^{-1}), & j=k\\ &= O(n^{-2}), & j\neq k.\end{aligned}$$

Proof. Since the a_{ik} and b_{ik} are linear combinations of normal random variables, the normality is immediate. The moment properties follow from the moments of the normal distribution and from Corollary 7.4.1.1. ▲

We may construct smoothed estimators of the cross spectral density in the same manner that we constructed smoothed estimators in Section 7.2. Let

$$\begin{aligned}\bar{f}_{12}(\omega_k) &= \sum_{j=-d}^{d} W_n(j)\hat{f}_{12}(\omega_{k+j}) \\ &= \frac{1}{4\pi}\bar{I}_{12}(\omega_k),\end{aligned} \tag{7.4.10}$$

where

$$\hat{f}_{12}(\omega_k)=\frac{1}{4\pi}I_{12n}(\omega_k),$$

$$\bar{I}_{12}(\omega_k)=\sum_{j=-d}^{d} W_n(j)I_{12n}(\omega_{k+j})$$

and $W_n(j)$, $j=0, \pm 1, \pm 2, \ldots, \pm d$, is a weight function.

Theorem 7.4.3. Let $\mathbf{Y}_t$ be a bivariate normal time series with covariance function that satisfies $\sum_{h=-\infty}^{\infty}|h||\gamma_{ij}(h)|<L<\infty$, $i,j=1,2$. Let d_n be a sequence of positive integers satisfying

$$\lim_{n\to\infty} d_n=\infty,$$

$$\lim_{n\to\infty}\frac{d_n}{n}=0,$$

and let $W_n(j), j=0, \pm 1, \pm 2, \ldots, \pm d_n$ satisfy

$$\sum_{j=-d_n}^{d_n} W_n(j)=1,$$

$$W_n(j)=W_n(-j),$$

$$\lim_{n\to\infty} \sum_{j=-d_n}^{d_n} W_n^2(j)=0.$$

Then

$$\lim_{n\to\infty} E\left\{\bar{f}_{12}(\omega_{K(n,\omega)})\right\}=f_{12}(\omega)$$

$$\begin{aligned}\lim_{n\to\infty}\left[\sum_{j=-d_n}^{d_n} W_n^2(j)\right]^{-1} \operatorname{Var}\left\{\bar{f}_{12}(\omega_{K(n,\omega)})\right\} &= f_{11}(\omega) f_{22}(\omega), \qquad \omega\neq 0,\pi\\ &= 2f_{11}(\omega) f_{22}(\omega), \qquad \omega=0,\pi,\end{aligned}$$

where $\bar{f}_{12}(\omega_{K(n,\omega)})$ is defined by (7.4.10) and $K(n,\omega)$ is defined in (7.4.9).

Proof. Reserved for the reader. ▲

The properties of the other multivariate spectral estimators follow from Theorems 7.4.2 and 7.4.3. Table 7.4.1 will be used to illustrate the computations and emphasize the relationships to normal regression theory. Notice that the number of entries in a column of Table 7.4.1 is $2(2d+1)$. We set $W_n(j)=(2d+1)^{-1}$ for $j=0, \pm 1, \pm 2, \ldots, \pm d$. Then the cospectrum is

Table 7.4.1. Statistics used in computation of cross spectra at $\omega_k=2\pi k/n$, $d=2$

Fourier Coefficients for Y_2	Fourier Coefficients for Y_1	Fourier Coefficients for Y_1
$a_{2,k-2}$	$a_{1,k-2}$	$-b_{1,k-2}$
$b_{2,k-2}$	$b_{1,k-2}$	$a_{1,k-2}$
$a_{2,k-1}$	$a_{1,k-1}$	$-b_{1,k-1}$
$b_{2,k-1}$	$b_{1,k-1}$	$a_{1,k-1}$
a_{2k}	a_{1k}	$-b_{1k}$
b_{2k}	b_{1k}	a_{1k}
$a_{2,k+1}$	$a_{1,k+1}$	$-b_{1,k+1}$
$b_{2,k+1}$	$b_{1,k+1}$	$a_{1,k+1}$
$a_{2,k+2}$	$a_{1,k+2}$	$-b_{1,k+2}$
$b_{2,k+2}$	$b_{1,k+2}$	$a_{1,k+2}$

estimated by

$$\bar{c}_{12}(\omega_k)=\frac{n}{8\pi(2d+1)}\sum_{j=-d}^{d}(a_{1,k-j}a_{2,k-j}+b_{1,k-j}b_{2,k-j}) \qquad (7.4.11)$$

which is the mean of the cross products of the first two columns of Table 7.4.1 multiplied by $n/4\pi$. The quadrature spectrum is estimated by

$$\bar{q}_{12}(\omega_k)=\frac{n}{8\pi(2d+1)}\sum_{j=-d}^{d}(a_{1,k-j}b_{2,k-j}-a_{2,k-j}b_{1,k-j}) \qquad (7.4.12)$$

which is the mean of the cross products of the first and third column of Table 7.4.1 multiplied by $n/4\pi$.

The estimator of the squared coherency for a bivariate time series computed from the smoothed periodogram estimator of $\boldsymbol{f}(\omega)$,

$$\bar{\boldsymbol{f}}(\omega_k)=\frac{1}{2d+1}\sum_{j=-d}^{d}\begin{pmatrix}\hat{f}_{11}(\omega_{k-j}) & \hat{f}_{12}(\omega_{k-j})\\ \hat{f}_{21}(\omega_{k-j}) & \hat{f}_{22}(\omega_{k-j})\end{pmatrix},$$

is given by

$$\bar{\mathcal{K}}_{12}^2(\omega_k)=\frac{|\bar{f}_{12}(\omega_k)|^2}{\bar{f}_{11}(\omega_k)\bar{f}_{22}(\omega_k)}=\frac{[\bar{c}_{12}(\omega_k)]^2+[\bar{q}_{12}(\omega_k)]^2}{\bar{f}_{11}(\omega_k)\bar{f}_{22}(\omega_k)}. \qquad (7.4.13)$$

This quantity is recognizable as the multiple correlation coefficient of normal regression theory obtained by regressing the first column of Table 7.4.1 on the second and third columns. By construction, the second and third columns are orthogonal.

The estimation of the squared coherency generalizes immediately to higher dimensions. If there is a second explanatory variable, the Fourier coefficients of this variable are added to Table 7.4.1 in the same form as the columns for Y_1. Then the multiple squared coherency is the multiple correlation coefficient associated with the regression of the column for Y_2 on the four columns for the two explanatory variables.

An estimator of the error spectrum or residual spectrum of Y_2 after Y_1 is

$$\bar{f}_{ZZ}(\omega_k)=\bar{f}_{22}(\omega_k)\left[1-\bar{\mathcal{K}}_{12}^2(\omega_k)\right]\left(\frac{2d+1}{2d}\right). \qquad (7.4.14)$$

This is the residual mean square for the regression of the first column of Table 7.4.1 on the second and third multiplied by $n/4\pi$. Many authors define the estimator of the error spectrum without the factor $(2d+1)/2d$. We include the term to make the analogy to multiple regression complete. It also serves to remind us that $\bar{\mathcal{K}}_{12}^2(\omega_k)$ is identically one if computed for $d=0$. A test of the hypothesis that $\mathcal{K}_{12}^2(\omega_k)=0$ is given by the statistic

$$F_{4d}^2 = \frac{4d\,\bar{\mathcal{K}}_{12}^2(\omega_k)}{2\left[1-\bar{\mathcal{K}}_{12}^2(\omega_k)\right]} \tag{7.4.15}$$

which is approximately distributed as Snedecor's F with 2 and $4d\,(d>0)$ degrees of freedom under the null hypothesis. This is the test of the hypothesis that the regression coefficients associated with columns two and three of Table 7.4.1 are zero.

If $\mathcal{K}_{12}^2(\omega_k)\neq 0$, the distribution of $\bar{\mathcal{K}}_{12}^2(\omega_k)$ is approximately that of the multiple correlation coefficient [see, for example, Anderson (1958, p. 93) or Kendall and Stuart, Vol. 2 (1967, p. 337)]. Tables and graphs useful in constructing confidence intervals for $\mathcal{K}_{12}^2(\omega_k)$ have been given by Amos and Koopmans (1963). For large degrees of freedom and $\mathcal{K}_{12}^2(\omega_k)\neq 0$ $\bar{\mathcal{K}}_{12}^2(\omega_k)$ is approximately normally distributed with variance

$$\operatorname{Var}\left\{\bar{\mathcal{K}}_{12}^2(\omega_k)\right\} \doteq 4\mathcal{K}_{12}^2(\omega_k)\left[1-\mathcal{K}_{12}^2(\omega_k)\right]^2/(4d+2). \tag{7.4.16}$$

The estimated phase spectrum is

$$\bar{\varphi}_{12}(\omega_k) = \tan^{-1}\left[-\bar{q}_{12}(\omega_k)/\bar{c}_{12}(\omega_k)\right], \tag{7.4.17}$$

where it is understood that $\bar{\varphi}_{12}(\omega_k)$ is the angle in $(-\pi,\pi]$ between the positive half of the $c_{12}(\omega_k)$ axis and the ray from the origin through $(\bar{c}_{12}(\omega_k), -\bar{q}_{12}(\omega_k))$. The sample distribution of this quantity depends in a critical manner on the true coherency between the two time series. If $\mathcal{K}_{12}^2(\omega_k)=0$, then, conditional on $\bar{f}_{11}(\omega_k)$, $\bar{c}_{12}(\omega_k)/\bar{f}_{11}(\omega_k)$ and $\bar{q}_{12}(\omega_k)/\bar{f}_{11}(\omega_k)$ are approximately distributed as independent normal $(0, f_{ZZ}(\omega_k)[2(2d+1)\bar{f}_{11}(\omega_k)]^{-1})$ random variables. This is because $\bar{c}_{12}(\omega_k)/\bar{f}_{11}(\omega_k)$ and $\bar{q}_{12}(\omega_k)/\bar{f}_{11}(\omega_k)$ are the regression coefficients obtained by regressing column one of Table 7.4.1 on columns two and three of Table 7.4.1. It is well known[6] that the ratio of two independent normal random variables with zero mean and common variance has the Cauchy distribution, and that the arc tangent of the ratio has a uniform distribution. Therefore, if $\mathcal{K}_{12}^2(\omega_k)=0$, the principal

[6]See, for example, Kendall and Stuart, Vol. 1(1969, p. 268).

value of $\bar{\varphi}_{12}(\omega_k)$ will be approximately uniformly distributed on the interval $(-\pi/2, \pi/2)$.

If $\mathcal{K}_{12}^2(\omega_k) \neq 0, \bar{\varphi}_{12}(\omega)$ will converge in distribution to a normal random variable. While approximate confidence limits could be established on the basis of the limiting distribution, it seems preferable to set confidence limits using the normality of $\bar{c}_{12}(\omega_k)/\bar{f}_{11}(\omega_k)$ and $\bar{q}_{12}(\omega_k)/\bar{f}_{11}(\omega_k)$.

Fieller's method [see Fieller (1954)] can be used to construct a confidence interval for $\varphi_{12}(\omega)$. This procedure follows from the fact that the statement $-q_{12}(\omega)/c_{12}(\omega) = R_{12}(\omega)$ is equivalent to the statement $c_{12}(\omega) R_{12}(\omega) + q_{12}(\omega) = 0$. Therefore, the method of setting confidence intervals for the sum $c_{12}(\omega) R_{12}(\omega) + q_{12}(\omega)$ can be used to determine a confidence interval for $R_{12}(\omega) = \tan \varphi_{12}(\omega)$ and hence for $\varphi_{12}(\omega)$. The $(1-\alpha)$-level confidence interval for the principal value of $\varphi_{12}(\omega)$ is the set of $\varphi_{12}(\omega)$ in $[-\pi/2, \pi/2]$ such that

$$\sin^2[\varphi_{12}(\omega) - \bar{\varphi}_{12}(\omega)] \leqslant t_\alpha^2[\bar{c}_{12}^2(\omega) + \bar{q}_{12}^2(\omega)]^{-1} \hat{\text{V}}\text{ar}\{\bar{c}_{12}(\omega)\}, \quad (7.4.18)$$

where

$$\hat{\text{V}}\text{ar}\{\bar{c}_{12}(\omega)\} = (4d+2)^{-1} \bar{f}_{11}(\omega) \bar{f}_{ZZ}(\omega),$$

t_α is such that $P\{|t| > t_\alpha\} = \alpha$ and t is distributed as Student's t with $4d$ degrees of freedom.

To obtain a confidence interval for $\varphi_{12}(\omega)$ in the interval $(-\pi, \pi]$ it is necessary to modify Fieller's method. We suggest the following procedure to establish an approximate $(1-\alpha)$-level confidence interval. Let $F_{4d}^2(\alpha)$ denote the α-percent point of the F-distribution with 2 and $4d$ degrees of freedom. The possibilities for the interval fall into two categories.

1. $\bar{c}_{12}^2(\omega) + \bar{q}_{12}^2(\omega) \leqslant 2F_{4d}^2(\alpha) \hat{\text{V}}\text{ar}\{\bar{c}_{12}(\omega)\}$.
The confidence interval for $\varphi_{12}(\omega)$ is $(-\pi, \pi]$.
2. $\bar{c}_{12}^2(\omega) + \bar{q}_{12}^2(\omega) > 2F_{4d}^2(\alpha) \hat{\text{V}}\text{ar}\{\bar{c}_{12}(\omega)\}$.
The confidence interval for $\varphi_{12}(\omega)$ is $[\bar{\varphi}_{12}(\omega) - \delta, \bar{\varphi}_{12}(\omega) + \delta]$, where

$$\delta = \sin^{-1}\left[t_\alpha^2\{\bar{c}_{12}^2(\omega) + \bar{q}_{12}^2(\omega)\}^{-1} \hat{\text{V}}\text{ar}\{\bar{c}_{12}(\omega)\}\right]^{\frac{1}{2}}$$

$$= \sin^{-1}\left[t_\alpha^2 \frac{1 - \bar{\mathcal{K}}_{12}^2(\omega)}{4d\, \bar{\mathcal{K}}_{12}^2(\omega)}\right]^{\frac{1}{2}}$$

Note that the criterion for category 1 is satisfied when the F-statistic of (7.4.15) is less than $F_{4d}^2(\alpha)$. Assuming $\bar{c}_{12}(\omega)$ and $\bar{q}_{12}(\omega)$ to be normally

distributed, it can be proven that the outlined procedure furnishes a confidence interval with probability at least $(1-\alpha)$ of covering the true $\varphi(\omega)$. If the true coherency is zero the interval will have length 2π with probability $(1-\alpha)$.

Recall that the cross amplitude spectrum is

$$A_{12}(\omega)=\left[c_{12}^2(\omega)+q_{12}^2(\omega)\right]^{1/2}=|f_{12}(\omega)|$$

and the gain of X_{2t} over X_{1t} is

$$\psi_{12}(\omega)=\frac{A_{12}(\omega)}{f_{11}(\omega)}.$$

Estimators of these quantities are

$$\bar{A}_{12}(\omega)=\left[\bar{c}_{12}^2(\omega)+\bar{q}_{12}^2(\omega)\right]^{1/2}, \tag{7.4.19}$$

$$\bar{\psi}_{12}(\omega)=\left[\bar{f}_{11}(\omega)\right]^{-1}\bar{A}_{12}(\omega). \tag{7.4.20}$$

It is possible to establish approximate confidence intervals for these quantities using the approximate normality of $\bar{c}_{12}(\omega)/\bar{f}_{11}(\omega)$ and $\bar{q}_{12}(\omega)/\bar{f}_{11}(\omega)$. As a consequence of this normality,

$$\frac{(2d+1)\left\{\left[\bar{c}_{12}(\omega)-c_{12}(\omega)\right]^2+\left[\bar{q}_{12}(\omega)-q_{12}(\omega)\right]^2\right\}}{\bar{f}_{11}(\omega)\bar{f}_{ZZ}(\omega)} \tag{7.4.21}$$

has, approximately, the F-distribution with 2 and $4d$ degrees of freedom. Therefore, those $c_{12}(\omega)$ and $q_{12}(\omega)$ for which (7.4.21) is less than the α percentage tabular value of the F-distribution form a $(1-\alpha)$-level confidence region. Let

$$A_U(\omega)=\bar{A}_{12}(\omega)+\left[(2d+1)^{-1}\bar{f}_{11}(\omega)\bar{f}_{ZZ}(\omega)F_{4d}^2(\alpha)\right]^{1/2}$$

$$A_L(\omega)=\max\left\{0,\bar{A}_{12}(\omega)-\left[(2d+1)^{-1}\bar{f}_{11}(\omega)\bar{f}_{ZZ}(\omega)F_{4d}^2(\alpha)\right]^{1/2}\right\}. \tag{7.4.22}$$

Assuming normal $\bar{c}_{12}(\omega)$ and $\bar{q}_{12}(\omega)$, $[A_L(\omega), A_U(\omega)]$ is a confidence interval for $A_{12}(\omega)$ of at least level $(1-\alpha)$. The confidence interval for gain is that of $A_{12}(\omega)$ divided by $\bar{f}_{11}(\omega)$.

Example. We use the data on the sediment in the Des Moines River

discussed in Section 6.5 to illustrate some of the cross spectral computations. Table 7.4.2 contains the Fourier coefficients for the first 11 frequencies for the 205 observations. For $d=5$, these are the statistics used to estimate $f(\omega_6)$, where $\omega_6=0.0293$ cycles per day.

Using the rectangular weight function, we have

$$\bar{I}_{22,205}(0.0293)=\frac{1}{11}\sum_{j=-5}^{5}\left(\frac{205}{2}\right)\left(a_{2,6+j}^2+b_{2,6+j}^2\right)=4.0320,$$

$$\bar{I}_{11,205}(0.0293)=\frac{1}{11}\sum_{j=-5}^{5}\left(\frac{205}{2}\right)\left(a_{1,6+j}^2+b_{1,6+j}^2\right)=7.2999,$$

$$\begin{aligned}\bar{I}_{12,205}(0.0293)&=\frac{1}{11}\sum_{j=-5}^{5}\frac{205}{2}\left(a_{1,6+j}a_{2,6+j}+b_{1,6+j}b_{2,6+j}\right)\\&\quad-i\,\frac{205}{22}\sum_{j=-5}^{5}\left(b_{2,6+j}a_{1,6+j}-a_{2,6+j}b_{1,6+j}\right)\\&=4.6768-1.4313i\,.\end{aligned}$$

It follows that

$$\bar{f}_{22}(0.0293)=0.3209,$$

$$\bar{f}_{11}(0.0293)=0.5809,$$

$$\bar{f}_{12}(0.0293)=0.3722-0.1138i\,.$$

The error spectrum for Saylorville is estimated by the residual sum of squares obtained from the regression of the Saylorville column on the Boone columns multiplied by $205[2(10)(4\pi)]^{-1}$. We have

$$\begin{aligned}\bar{f}_{ZZ}(0.0293)&=\frac{205}{80\pi}\left[0.4327-0.6407(0.5019)-(-0.1960)(-0.1536)\right]\\&=0.0661.\end{aligned}$$

Table 7.4.2. Statistics used in computing smoothed estimates of cross spectrum for sediment in the Des Moines River at Boone and Saylorville for frequency of 0.0293 cycles per day, d = 5

Index (k)	Frequency (Cycles per Day)	Period (Days)	Coefficients Saylorville (a_{2k}, b_{2k})	Coefficients Boone (a_{1k}, b_{1k})	Coefficients Boone $(-b_{1k}, a_{1k})$
1	0.0049	205.00	−0.194	−0.103	−0.504
			0.305	0.504	−0.103
2	0.0098	102.50	0.175	0.432	−0.196
			0.161	0.196	0.432
3	0.0146	68.33	0.011	0.047	0.039
			0.013	−0.039	0.047
4	0.0195	51.25	0.058	0.055	−0.088
			0.147	0.088	0.055
5	0.0244	41.00	0.067	0.202	−0.116
			0.042	0.116	0.202
6	0.0293	34.17	0.021	0.133	−0.072
			0.061	0.072	0.133
7	0.0341	29.29	0.322	0.234	0.012
			0.059	−0.012	0.234
8	0.0390	25.63	0.065	−0.017	0.019
			−0.067	−0.019	−0.017
9	0.0439	22.78	−0.053	−0.013	−0.098
			0.019	0.098	−0.013
10	0.0488	20.50	0.281	0.332	0.053
			0.037	−0.053	0.332
11	0.0537	18.64	0.081	0.152	−0.026
			−0.062	0.026	0.152

The squared coherency is

$$\bar{\mathfrak{K}}_{12}^2(0.0293) = \frac{|\bar{f}_{12}(0.0293)|^2}{\bar{f}_{11}(0.0293)\bar{f}_{22}(0.0293)} = 0.8128.$$

The F-statistic to test the hypothesis of zero coherency is

$$F_{20}^2 = \frac{20\,\bar{\mathfrak{K}}_{12}^2(0.0293)}{2\left[1 - \bar{\mathfrak{K}}_{12}^2(0.0293)\right]} = 43.42.$$

This is well beyond the 1% tabular value of 5.85 for 2 and 20 degrees of freedom, and it is clear that the two time series are not independent. The estimate of the phase spectrum is

$$\bar{\varphi}_{12}(0.0293) = \tan^{-1}[-1.4313/4.6768] = -0.2970 \text{ radians.}$$

Let us establish a 95% confidence interval for $\varphi_{12}(0.0293)$. Because the F-test rejects the hypothesis of zero coherency at the 5% level, the criterion of category 2 is satisfied. Consequently the confidence interval for $\varphi_{12}(\omega_6)$ is $[-0.5228, -0.0712]$, where

$$\delta = \sin^{-1}\left[(2.086)^2 \frac{0.1872}{20(0.8128)}\right]^{1/2} = 0.2258.$$

The estimated gain of X_{2t} over X_{1t} is

$$\bar{\psi}_{12}(\omega) = \frac{|\bar{f}_{12}(0.0293)|}{\bar{f}_{11}(0.0293)} = \frac{0.3892}{0.5809} = 0.6700.$$

A 95% confidence interval for $\psi_{12}(\omega_6)$ is given by $[\psi_L, \psi_U]$, where

$$\psi_L = 0.6700 - \left[(11)^{-1}(0.5809)^{-1}(0.0661)(3.49)\right]^{1/2} = 0.4800$$

$$\psi_U = 0.6700 + \left[(11)^{-1}(0.5809)^{-1}(0.0661)(3.49)\right]^{1/2} = 0.8600.$$

Figure 7.4.1 is a plot of $\bar{\varphi}_{12}(\omega)$ and the confidence interval for $\varphi_{12}(\omega)$ ($d=5$ and the rectangular weight function) for the Boone-Saylorville sediment data. Recall that $q_{12}(\omega)$ is an odd function of ω with $q_{12}(0) = q_{12}(\pi) = 0$. Therefore, we define the imaginary part of $I_{12n}(\omega_{-k})$ to be the negative of the imaginary part of $I_{12n}(\omega_k)$. Likewise, the imaginary part of $I_{12n}(\pi+\lambda)$ is set equal to the negative of the imaginary part of $I_{12n}(\pi-\lambda)$. If $I_{12n}(0)$ or $I_{12n}(\pi)$ are computed, the imaginary part is zero. As a result, $\bar{q}_{12}(0)=0$, and $\bar{q}_{12}(\pi)=0$ if n is even. It follows that the estimated phase at zero is zero if $\bar{c}_{12}(0)>0$ and is π if $\bar{c}_{12}(0)<0$. In plotting the estimated phase a continuous function of ω is desirable. Therefore, in creating Figure 7.4.1, that angle in the set

$$\left\{ \bar{\varphi}_{12}(\omega_k) + j\pi : j = 0, \pm 2, \pm 4, \ldots \right\}$$

that differed least from the angle previously chosen for $\omega_{k-1}, k = 2, 3, \ldots, 102$, was plotted.

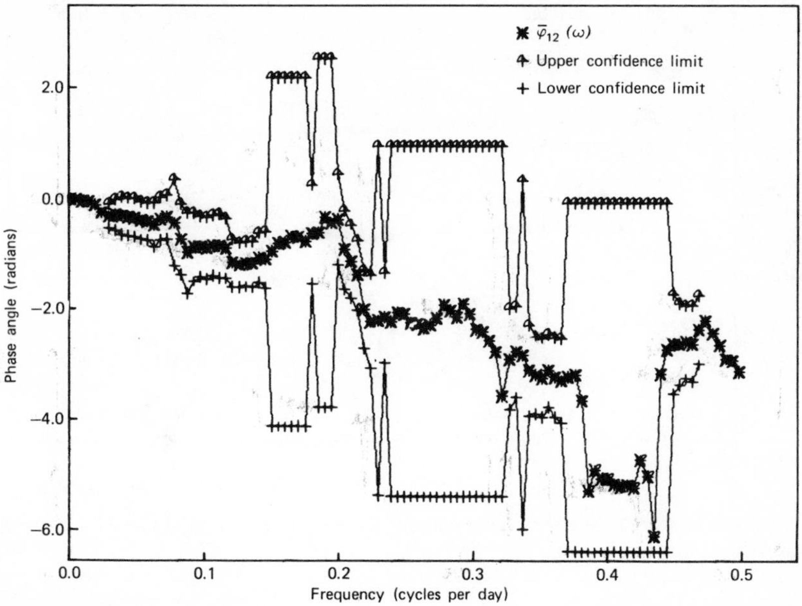

Figure 7.4.1. Estimated phase spectrum for Boone-Saylorville sediment data ($d=5$ and rectangular weight function).

The general downward slope of $\bar{\varphi}_{12}(\omega)$ is associated with the fact that the readings at Saylorville lag behind those at Boone. If the relationship was a perfect one period lag, $\bar{\varphi}_{12}(\omega)$ would be estimating a straight line with a negative slope of one radian per radian. The estimated function seems to differ enough from such a straight line to suggest a more complicated lag relationship.

Figure 7.4.2 contains a plot of squared coherency for the sediment data. The 5% point for the F-distribution with 2 and 20 degrees of freedom is 3.49. On the basis of (7.4.15), any $\overline{\mathcal{K}}_{12}^2(\omega)$ greater than 0.259 would be judged significant at that level. A line has been drawn at this height in the figure.

Similar information is contained in Figure 7.4.3, where the smoothed periodogram for Saylorville and $4\pi\bar{f}_{ZZ}(\omega)$ are plotted on the same graph. The estimated error spectrum lies well below the spectrum of the original time series for low frequencies, but the two nearly coincide for high frequencies. The estimated error spectrum is clearly not that of white noise, since it is considerably higher at low frequencies than at high. One might be led to consider a first or second order autoregressive process as a model for the error.

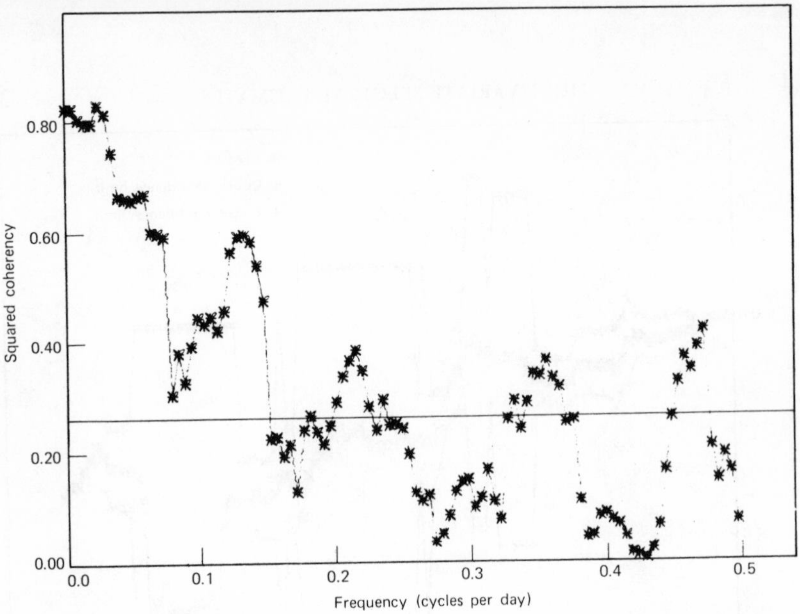

Figure 7.4.2. Squared coherency for the Boone-Saylorville sediment data ($d=5$ and rectangular weight function).

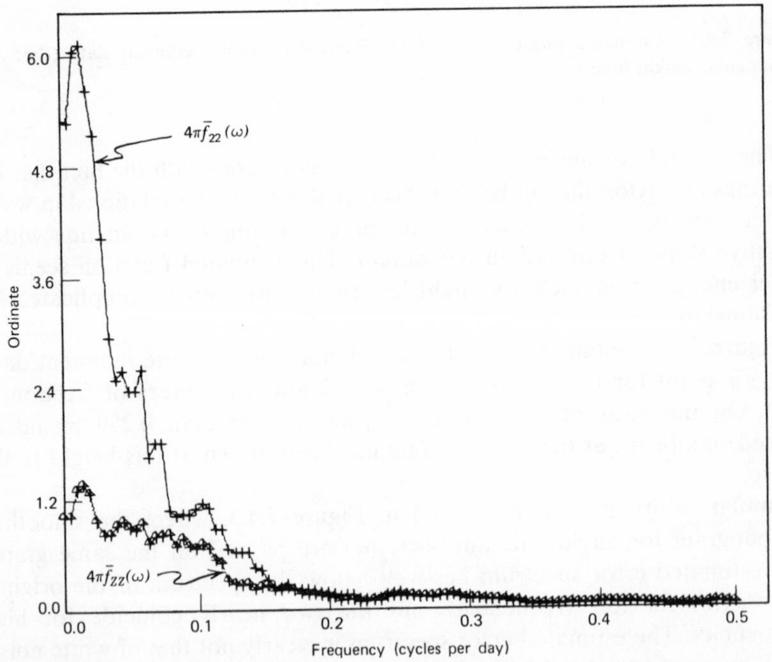

Figure 7.4.3. Plot of $4\pi\bar{f}_{22}(\omega)$ and $4\pi\bar{f}_{ZZ}(\omega)$ for Saylorville sediment ($d=5$ and rectangular weight function).

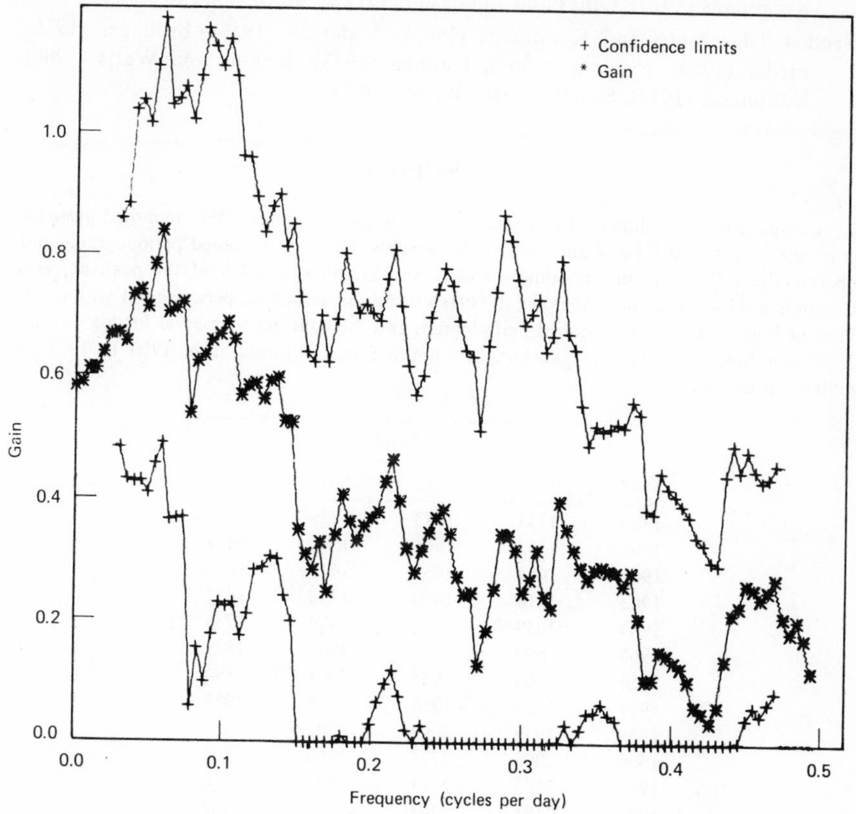

Figure 7.4.4. Gain of Saylorville over Boone ($d=5$ and rectangular weight function).

Figure 7.4.4 is a plot of the estimated gain of Saylorville over Boone. The 95% confidence interval plotted on the graph was computed using the limits (7.4.22) divided by $\bar{f}_{11}(\omega)$. Note that the lower confidence bound for gain is zero whenever the squared coherency falls below 0.259.

REFERENCES

Section 7.1. Bartlett (1950), (1966), Birnbaum (1952), Davis (1941), Durbin (1967), (1968), (1969), Eisenhart, Hastay, and Wallis (1947), Fisher (1929), Grenander and Rosenblatt (1957), Hannan (1958), (1960), Harris (1967), Kendall and Stuart (1966), Koopmans (1974), Tintner (1952), Tukey (1961), Wilks (1962).

Section 7.2. Anderson(1971), Bartlett (1950), Blackman and Tukey (1959), Brillinger (1975), Hannan (1960), (1970), Jenkins (1961), Jenkins and Watts (1968),

Koopmans (1974), Otnes and Enochson (1972), Parzen (1957), (1961).

Section 7.4. Amos and Koopmans (1963), Anderson (1958), Brillinger (1975), Fieller (1954), Fishman (1969), Hannan (1973), Jenkins and Watts (1968), Koopmans (1974), Scheffé (1970), Wahba (1968).

EXERCISES

1. Compute the periodogram for the data of Table 6.4.1. Calculate the smoothed periodogram using 2, 4, and 8 for d and rectangular weights. Plot the smoothed periodograms and observe the differences in smoothness and in the height and width of the peak at zero. Compute a 95% confidence interval for $4\pi f(\omega)$ using the smoothed periodogram with $d=4$. Plot the logarithm of the smoothed periodogram and the confidence interval for $\log 4\pi f(\omega)$.

2. Given below is quarterly gasoline consumption in California from 1960 to 1973 in millions of gallons.

	Quarter			
Year	I	II	III	IV
1960	1335	1443	1529	1447
1961	1363	1501	1576	1495
1962	1464	1450	1611	1612
1963	1516	1660	1738	1652
1964	1639	1754	1839	1736
1965	1699	1812	1901	1821
1966	1763	1937	2001	1894
1967	1829	1966	2068	1983
1968	1939	2099	2201	2081
1969	2008	2232	2299	2204
1970	2152	2313	2393	2278
1971	2191	2402	2450	2387
1972	2391	2549	2602	2529
1973	2454	2647	2689	2549

SOURCE. U.S. Dept. of Transportation (1975), *Review and Analysis of Gasoline Consumption in the United States from 1960 to the Present,* and U.S. Dept. of Transportation, *News*, various issues.

Using these data:

(a) Compute the periodogram.

(b) Obtain the smoothed periodogram by computing the centered moving average

$$\bar{I}_n(\omega_k)=\frac{1}{5}\sum_{j=-2}^{2} I_n(\omega_{k-j}).$$

(c) Fit the regression model

$$Y_t=\alpha+\beta t+Z_t.$$

Repeat parts a and b for the regression residuals $\hat{Z}_t$.

(d) Compute the smoothed periodogram for $\hat{Z}_t$ of part c using the symmetric weight function $W_n(j)$ where

$$\begin{aligned} W_n(j) &= 0.3, & j&=0 \\ &= 0.2, & j&=1 \\ &= 0.1, & j&=2 \\ &= 0.05, & j&=3 \\ &= 0, & |j|&\geq 4. \end{aligned}$$

(e) Fit the regression model

$$Y_t = \sum_{j=1}^{4} \alpha_j \delta_{tj} + \beta t + u_t,$$

where

$$\begin{aligned} \delta_{tj} &= 1, \quad j\text{th quarter}, \\ &= 0, \quad \text{otherwise}. \end{aligned}$$

Repeat parts a and b for the residuals from the fitted regression. Compute and plot a 95% confidence interval for the estimated spectral density.

3. Let the time series, X_t, be defined by

$$X_t = e_t + 0.6 e_{t-1}$$

where $\{e_t\}$ is a sequence of normal independent (0, 1) random variables. Given a sample of 10,000 observations from such a time series, what is the approximate joint distribution of the periodogram ordinates associated with $\omega_{2500} = 2\pi(2500)/10{,}000$ and $\omega_{1250} = 2\pi(1250)/10{,}000$?

4. Prove the sine result of Lemma 7.2.2.

5. Prove that the covariance function of a stationary finite autoregressive process satisfies

$$\sum_{h=-n}^{n} |h||\gamma(h)| = O(1).$$

6. Use the moment properties of the normal distribution to demonstrate the portion of Theorem 7.4.2 that states that

$$\begin{aligned} \operatorname{Cov}\{I_{12n}(\omega_k), I_{12n}(\omega_j)\} &= (4\pi)^2 f_{11}(\omega_k) f_{22}(\omega_k) + O(n^{-1}), & j&=k \\ &= O(n^{-2}), & j&\neq k. \end{aligned}$$

7. Let X_{1t} denote the United States quarterly unemployment rate of Table 6.4.1 and let X_{2t} denote the weekly gross hours per production worker given in Exercise 13 of Chapter 6. Compute the periodogram quantities $I_{11n}(\omega_k)$, $I_{12n}(\omega_k)$, and $I_{22n}(\omega_k)$. Compute the smoothed estimates using $d=5$ and the rectangular weight function. Compute and plot $\bar{\mathfrak{K}}^2_{12}(\omega_k), \bar{\varphi}_{12}(\omega_k)$. Obtain and plot a confidence interval for $\varphi_{12}(\omega_k)$ and for the gain of X_{2t} over X_{1t}. Treating hours per week as the dependent variable, plot $4\pi \bar{f}_{ZZ}(\omega)$.

8. Show that

$$I_{XYn}(\omega_k) = 4\pi \hat{f}_{XY}(\omega_k), \qquad k = 1, 2, \ldots, m$$

where

$$\hat{f}_{XY}(\omega) = \frac{1}{2\pi} \sum_{h=-\infty}^{\infty} \hat{\gamma}_{XY}(h) e^{-i\omega h},$$

$$\hat{\gamma}_{XY}(h) = \begin{cases} \dfrac{1}{n} \displaystyle\sum_{t=1}^{n-h} (X_t - \bar{x}_n)(Y_{t+h} - \bar{y}_n), & h = 0, 1, \ldots, n-1 \\ \dfrac{1}{n} \displaystyle\sum_{t=-h+1}^{n} (X_t - \bar{x}_n)(Y_{t+h} - \bar{y}_n), & h = -1, -2, \ldots, -(n-1) \\ 0, & \text{otherwise.} \end{cases}$$

9. Let X_t be a time series defined by

$$X_t = \sum_{j=0}^{\infty} \alpha_j e_{t-j},$$

where $\{e_t\}$ is a sequence of independent $(0, \sigma^2)$ random variables with fourth moment $\eta\sigma^4$ and

$$\sum_{j=0}^{\infty} j|\alpha_j| < \infty.$$

(a) Show that

$$\sum_{h=-\infty}^{\infty} |h||\gamma(h)| < \infty$$

for such a time series.

(b) Let d_n, $W_n(j)$ and $\bar{f}(\omega)$ be as defined in Theorem 7.2.2. Show that

$$E\{\bar{f}(\omega)\} = f(\omega) + O(n^{-1}d_n).$$

CHAPTER 8

Estimation for Autoregressive and Moving Average Time Series

8.1. FIRST ORDER AUTOREGRESSIVE TIME SERIES

The time series defined by

$$Y_t - \mu = \rho(Y_{t-1} - \mu) + e_t, \tag{8.1.1}$$

where the e_t are normal independent $(0, \sigma^2)$ random variables and $|\rho| < 1$, is one of the simplest and most heavily used models in time series analysis. It is often a satisfactory representation of the error time series in economic models. This model also underlies many tests of the hypothesis that the observed time series is a sequence of independently and identically distributed random variables.

One estimator of ρ is the first order autocorrelation coefficient discussed in Chapter 6,

$$\hat{r}(1) = \frac{\hat{\gamma}(1)}{\hat{\gamma}(0)} = \frac{\sum_{t=1}^{n-1} (Y_t - \bar{y}_n)(Y_{t+1} - \bar{y}_n)}{\sum_{t=1}^{n} (Y_t - \bar{y}_n)^2}, \tag{8.1.2}$$

where

$$\bar{y}_n = \frac{1}{n} \sum_{t=1}^{n} Y_t.$$

To introduce some other estimators of the parameter ρ, let us consider the distribution of the Y_t for the normal time series defined by (8.1.1).

The expected value of Y_t is μ and the expected value of $Y_t - \rho Y_{t-1}$ is λ,

where $\lambda=(1-\rho)\mu$. For a sample of n observations we can write

$$\begin{aligned} Y_1&=\mu+v_1, \\ Y_t-\mu&=\rho(Y_{t-1}-\mu)+e_t, \qquad t=2,3,\ldots,n, \end{aligned} \tag{8.1.3}$$

or

$$\begin{aligned} Y_1&=\mu+v_1 \\ Y_t&=\lambda+\rho Y_{t-1}+e_t, \qquad t=2,3,\ldots,n, \end{aligned}$$

where the vector $(v_1, e_2, e_3, \ldots, e_n)$ is distributed as a multivariate normal with zero mean and covariance matrix

$$\mathbf{\Sigma}=\begin{pmatrix} (1-\rho^2)^{-1}\sigma^2 & 0 & 0 & \cdots & 0 \\ 0 & \sigma^2 & 0 & \cdots & 0 \\ 0 & 0 & \sigma^2 & \cdots & 0 \\ \vdots & \vdots & \vdots & & \vdots \\ 0 & 0 & 0 & \cdots & \sigma^2 \end{pmatrix}. \tag{8.1.4}$$

It follows that the logarithm of the likelihood of a sample of n observations is

$$\log L(y;\mu,\rho,\sigma^2)=-\frac{n}{2}\log 2\pi-\frac{n}{2}\log\sigma^2+\frac{1}{2}\log(1-\rho^2)$$
$$-\frac{1}{2\sigma^2}\left[(Y_1-\mu)^2(1-\rho^2)+\sum_{t=2}^{n}\{(Y_t-\mu)-\rho(Y_{t-1}-\mu)\}^2\right]. \tag{8.1.5}$$

The computation of the maximum likelihood estimators is greatly simplified if we treat Y_1 as fixed and investigate the conditional likelihood. This is also an appropriate model in some experimental situations. For example, if we initiate an experiment at time one with an initial input of Y_1, it is very reasonable to condition on this initial input.

To construct the conditional likelihood, we consider the last $n-1$ equations of (8.1.3). Maximizing the function

$$\log L(y;\lambda,\rho,\sigma^2|Y_1)=-\frac{n-1}{2}\log 2\pi-\frac{n-1}{2}\log\sigma^2$$
$$-\frac{1}{2\sigma^2}\sum_{t=2}^{n}(Y_t-\lambda-\rho Y_{t-1})^2 \tag{8.1.6}$$

leads to the estimators[1]

$$\hat{\rho}=\frac{\sum_{t=2}^{n}(Y_t-\bar{y}_{(0)})(Y_{t-1}-\bar{y}_{(-1)})}{\sum_{t=2}^{n}(Y_{t-1}-\bar{y}_{(-1)})^2},$$

$$\hat{\lambda}=\bar{y}_{(0)}-\hat{\rho}\bar{y}_{(-1)},$$

$$\hat{\sigma}^2=\frac{1}{n-1}\sum_{t=2}^{n}\left[(Y_t-\bar{y}_{(0)})-\hat{\rho}(Y_{t-1}-\bar{y}_{(-1)})\right]^2, \tag{8.1.7}$$

where

$$\bar{y}_{(0)}=\frac{1}{n-1}\sum_{t=2}^{n}Y_t,$$

$$\bar{y}_{(-1)}=\frac{1}{n-1}\sum_{t=2}^{n}Y_{t-1}.$$

The estimators of λ and ρ are those that would be obtained by applying least squares to the last $n-1$ equations. The least squares estimator of σ^2,

$$s^2=\frac{1}{n-3}\sum_{t=2}^{n}\left[(Y_t-\bar{y}_{(0)})-\hat{\rho}(Y_{t-1}-\bar{y}_{(-1)})\right]^2, \tag{8.1.8}$$

is typically used in place of $\hat{\sigma}^2$.

It is clear that the least squares estimator for ρ differs from (8.1.2) by terms whose order in probability is $1/n$. Note that the estimator of ρ defined by (8.1.7) can be greater than one in absolute value while that defined in (8.1.2) cannot.

Let us now return to a consideration of the unconditional likelihood as given by equation (8.1.5). Differentiating the log likelihood with respect to

[1]These estimators are, strictly speaking, not the maximum likelihood estimators for the model stated in (8.1.1). The estimator $\hat{\rho}$ can take on values outside of $(-1,1)$. The maximum likelihood estimator is constrained to the parameter space.

μ, ρ, and σ^2, we have

$$\frac{\partial \log L}{\partial \mu} = \frac{1}{2\sigma^2}\left[2(Y_1-\mu)(1-\rho^2)+2\sum_{t=2}^{n}\{Y_t-\mu-\rho(Y_{t-1}-\mu)\}(1-\rho)\right],$$

$$\frac{\partial \log L}{\partial \rho} = \frac{-\rho}{1-\rho^2}+\frac{1}{\sigma^2}\left[(Y_1-\mu)^2\rho+\sum_{t=2}^{n}\{Y_t-\mu-\rho(Y_{t-1}-\mu)\}(Y_{t-1}-\mu)\right],$$

$$\frac{\partial \log L}{\partial \sigma^2} = -\frac{n}{2}\frac{1}{\sigma^2}+\frac{1}{2\sigma^4}\left[(Y_1-\mu)^2(1-\rho^2)+\sum_{t=2}^{n}\{Y_t-\mu-\rho(Y_{t-1}-\mu)\}^2\right].$$

(8.1.9)

If the maximum of the likelihood function occurs in the parameter space, these derivatives will be zero at the point defining the maximum likelihood estimator. We rearrange the derivatives and equate them to zero to obtain

$$(Y_1-\mu)(1-\rho^2)+\sum_{t=2}^{n}\{Y_t-\mu-\rho(Y_{t-1}-\mu)\}(1-\rho)=0,$$

$$\left[(Y_1-\mu)^2-\frac{\sigma^2}{1-\rho^2}\right]\rho+\sum_{t=2}^{n}\{(Y_t-\mu)-\rho(Y_{t-1}-\mu)\}(Y_{t-1}-\mu)=0,$$

$$\sigma^2=\frac{1}{n}\left[(Y_1-\mu)^2(1-\rho^2)+\sum_{t=2}^{n}\{(Y_t-\mu)-\rho(Y_{t-1}-\mu)\}^2\right].$$

(8.1.10)

To gain further understanding of these equations and to suggest a computational algorithm, we consider the system of equations (8.1.3). If we knew the covariance matrix of $(v_1, e_2, e_3, \ldots, e_n)$ we could transform the observations and estimate ρ and μ by least squares applied to the transformed variables. The appropriate transformation would be the matrix **T**, where

$$\mathbf{T}=\operatorname{diag}\left[(1-\rho^2)^{1/2}, 1, 1, \ldots, 1\right]$$

and the estimators of μ and ρ would be obtained by applying least squares to the system of equations

$$(1-\rho^2)^{1/2}Y_1=(1-\rho^2)^{1/2}\mu+(1-\rho^2)^{1/2}v_1$$

$$Y_t=(1-\rho)\mu+\rho Y_{t-1}+e_t, \qquad t=2,3,\ldots,n. \qquad (8.1.11)$$

However, the true covariance matrix depends on the parameter ρ, which we wish to estimate, and the square of the first error relative to the squares of other errors contains "information on ρ." Specifically, the expectation of v_1^2 is $(1-\rho^2)^{-1}\sigma^2$, while the expected value of the squares of the other errors is σ^2. Therefore, we add to our system the n equations in the expectations of the squares of the errors. (We are arguing heuristically. The justification for this procedure will be the system of equations obtained by maximum likelihood.) We have

$$
\begin{aligned}
Y_1 &= \mu + v_1,\\
Y_t &= (1-\rho)\,\mu + \rho Y_{t-1} + e_t, \quad t=2,3,\ldots,n,\\
(Y_1-\mu)^2 &= (1-\rho^2)^{-1}\sigma^2 + \epsilon_1,\\
\left[Y_t-\mu-\rho(Y_{t-1}-\mu)\right]^2 &= \sigma^2 + \epsilon_t, \qquad t=2,3,\ldots,n.
\end{aligned} \tag{8.1.12}
$$

The $2n$ errors in this system of equations $(v_1, e_2, e_3, \ldots, e_n, \epsilon_1, \epsilon_2, \epsilon_3, \ldots, \epsilon_n)$ are uncorrelated (clearly not independent) with variances $[(1-\rho^2)^{-1}, 1, 1, \ldots, 1, 2\sigma^2(1-\rho^2)^{-2}, 2\sigma^2, 2\sigma^2, \ldots, 2\sigma^2]\sigma^2$. Expanding the right side of the system of equations (8.1.12) in a first order Taylor's expansion about the initial value $(\hat{\mu}, \hat{\rho}, \hat{\sigma}^2)$, we obtain

$$
\begin{aligned}
Y_1-\hat{\mu} &= (\Delta\mu) + v_1 + R_1,\\
Y_t-\hat{\mu}-\hat{\rho}(Y_{t-1}-\hat{\mu}) &= (1-\hat{\rho})(\Delta\mu) + (Y_{t-1}-\hat{\mu})(\Delta\rho) + e_t + R_t,\\
&\qquad t=2,3,\ldots,n,\\
(Y_1-\hat{\mu})^2 - \hat{\sigma}^2(1-\hat{\rho}^2)^{-1} &= (1-\hat{\rho}^2)^{-1}(\Delta\sigma^2) + 2\hat{\rho}\hat{\sigma}^2(1-\hat{\rho}^2)^{-2}(\Delta\rho)\\
&\qquad + \epsilon_1 + R_{n+1},\\
\left[Y_t-\hat{\mu}-\hat{\rho}(Y_{t-1}-\hat{\mu})\right]^2 - \hat{\sigma}^2 &= (\Delta\sigma^2) + \epsilon_t + R_{n+t}, \qquad t=2,3,\ldots,n,
\end{aligned} \tag{8.1.13}
$$

where $(\Delta\mu)=\mu-\hat{\mu}$, $(\Delta\rho)=\rho-\hat{\rho}$, $(\Delta\sigma^2)=\sigma^2-\hat{\sigma}^2$, and the R_i are the Taylor remainders. An initial vector $(\hat{\mu}, \hat{\rho}, \hat{\sigma}^2)$ is required if one is to use (8.1.13) to obtain the maximum likelihood estimator. The estimators $\hat{\rho}$ and $\hat{\sigma}^2$ of (8.1.7) and

$$
\hat{\mu} = \frac{Y_1 + Y_n + (1-\hat{\rho})\sum_{t=2}^{n-1} Y_t}{2+(1-\hat{\rho})(n-2)} \tag{8.1.14}
$$

are suggested unless $|\hat{\rho}| \geqslant 1$. The sample autocorrelation $\hat{r}(1)$ furnishes an alternative initial estimator of ρ that is always less than one in absolute value. Estimating the variances by $[(1-\hat{\rho}^2)^{-1}, 1, 1, \ldots, 1, 2\hat{\sigma}^2(1-\hat{\rho}^2)^{-2}, 2\hat{\sigma}^2, 2\hat{\sigma}^2, \ldots, 2\hat{\sigma}^2]\hat{\sigma}^2$, ignoring the remainder terms, and applying generalized least squares to the system (8.1.13), we obtain the system of equations

$$\begin{bmatrix} a_{11} & a_{12} & a_{13} \\ a_{12} & a_{22} & a_{23} \\ a_{13} & a_{23} & a_{33} \end{bmatrix} \begin{bmatrix} \Delta\tilde{\mu} \\ \Delta\tilde{\rho} \\ \Delta\tilde{\sigma}^2 \end{bmatrix} = \begin{bmatrix} r_1 \\ r_2 \\ r_3 \end{bmatrix}, \tag{8.1.15}$$

where

$$a_{11} = 1 - \hat{\rho}^2 + (n-1)(1-\hat{\rho})^2,$$

$$a_{12} = (1-\hat{\rho}) \sum_{t=2}^{n} (Y_{t-1} - \hat{\mu}),$$

$$a_{13} = 0,$$

$$a_{22} = \sum_{t=2}^{n} (Y_{t-1} - \hat{\mu})^2 + 2(1-\hat{\rho}^2)^{-2}\hat{\rho}^2\hat{\sigma}^2,$$

$$a_{23} = \hat{\rho}(1-\hat{\rho}^2)^{-1},$$

$$a_{33} = (2\hat{\sigma}^2)^{-1} n,$$

$$r_1 = (1-\hat{\rho}^2)(Y_1 - \hat{\mu}) + (1-\hat{\rho}) \sum_{t=2}^{n} [Y_t - \hat{\mu} - \hat{\rho}(Y_{t-1} - \hat{\mu})],$$

$$r_2 = \hat{\rho}\left[(Y_1 - \hat{\mu})^2 - \hat{\sigma}^2(1-\hat{\rho}^2)^{-1}\right] + \sum_{t=2}^{n} [Y_t - \hat{\mu} - \hat{\rho}(Y_{t-1} - \hat{\mu})](Y_{t-1} - \hat{\mu}),$$

$$r_3 = [2\hat{\sigma}^2]^{-1}\left[(1-\hat{\rho}^2)(Y_1 - \hat{\mu})^2 - \hat{\sigma}^2 + \sum_{t=2}^{n} \left(\{Y_t - \hat{\mu} - \hat{\rho}(Y_{t-1} - \hat{\mu})\}^2 - \hat{\sigma}^2\right)\right].$$

Note that r_1, r_2, and r_3 will be zero if $(\hat{\mu}, \hat{\rho}, \hat{\sigma}^2)$ satisfies the maximum likelihood equations (8.1.10). It is clear that the difference between the estimator of ρ defined by (8.1.7) and that defined by (8.1.15) depends primarily on Y_1 and hence is small for large n and $|\rho| < 1$. All of the statistics estimating $\rho, |\rho| < 1$, studied in this section are asymptotically equivalent to $\hat{r}(1)$ of (8.1.2). Therefore, by the results of Section 6.3, they are asymptotically normally distributed.

The alternative estimators of λ and μ vary more in efficiency for moderate sized samples than do the alternative estimators of ρ. If Y_1 is fixed, then the intercept estimator associated with the regression of Y_t on Y_{t-1} is the appropriate estimator of λ. If Y_1 is not fixed, it is suggested that μ be estimated by the estimator (8.1.14) or by $\bar{y}_n$ (see Exercise 8.3).

8.2. HIGHER ORDER AUTOREGRESSIVE TIME SERIES

We consider the estimation of the parameters of the time series

$$Y_t + \sum_{i=1}^{p} \alpha_i Y_{t-i} = \theta_0 + e_t, \tag{8.2.1}$$

where the e_t are independent $(0,\sigma^2)$ random variables with $E\{e_t^4\} = \eta\sigma^4$ and the roots of

$$m^p + \sum_{i=1}^{p} \alpha_i m^{p-i} = 0$$

are all less than one in absolute value. Since the techniques we consider are closely related to those of multiple regression, it is convenient to write (8.2.1) as

$$Y_t = \theta_0 + \theta_1 Y_{t-1} + \theta_2 Y_{t-2} + \dots + \theta_p Y_{t-p} + e_t,$$

where $\theta_1 = -\alpha_1, \theta_2 = -\alpha_2, \dots, \theta_p = -\alpha_p$. We introduce the estimation with $p=2$. Two expressions of the equation are useful:

$$Y_t = \theta_0 + \theta_1 Y_{t-1} + \theta_2 Y_{t-2} + e_t \tag{8.2.2}$$

and

$$(Y_t - \mu) = \theta_1 (Y_{t-1} - \mu) + \theta_2 (Y_{t-2} - \mu) + e_t, \tag{8.2.2a}$$

where $E\{Y_t\} = \mu$ for all t.

Recall from the discussion of autoregressive time series in Chapter 2 that the Y_{t-j} for $j \geqslant 1$ are uncorrelated with e_t. This fact, together with the appearance of (8.2.2), suggests the use of regression to estimate the parameters. Multiplying (8.2.2) by $\mathbf{X}_t' = (1, Y_{t-1}, Y_{t-2})'$ and summing, we have

$$\sum_{t=3}^{n} \mathbf{X}_t' Y_t = \sum_{t=3}^{n} \mathbf{X}_t' \mathbf{X}_t \boldsymbol{\theta} + \sum_{t=3}^{n} \mathbf{X}_t' e_t, \tag{8.2.3}$$

where $\boldsymbol{\theta}=(\theta_0,\theta_1,\theta_2)'$. Although (8.2.3) looks like the classical regression equations, we cannot directly apply the traditional least squares theory to obtain the distributional properties of the estimators because e_t is *not* independent of Y_{t-j} for $j\leqslant 0$. Therefore, we proceed by investigating the properties of the elements of (8.2.3), first dividing by $n-2$. The expectation of the last vector on the right-hand side of (8.2.3) is zero; that is,

$$E\left\{\frac{1}{n-2}\sum_{t=3}^{n}e_t\right\}=E\left\{\frac{1}{n-2}\sum_{t=3}^{n}Y_{t-1}e_t\right\}=E\left\{\frac{1}{n-2}\sum_{t=3}^{n}Y_{t-2}e_t\right\}=0. \tag{8.2.4}$$

The expectation of each of the remaining terms is also easily found. For example,

$$E\left\{\frac{1}{n-2}\sum_{t=3}^{n}Y_{t-1}Y_{t-2}\right\}=\gamma_Y(1)+\mu^2. \tag{8.2.5}$$

From our results on the variances of estimated covariances (Theorems 6.2.1 and 6.5.1) we know that estimators such as those in (8.2.4) and (8.2.5) have variances that are $O(n^{-1})$. We conclude that the estimators defined in (8.2.3) have errors that are $O_p(n^{-1/2})$.

On the basis of our arguments for $p=2$, we define an estimator of the parameter $\boldsymbol{\theta}'=(\theta_0,\theta_1,\theta_2,\ldots,\theta_p)$ of the pth order autoregressive process by

$$\hat{\boldsymbol{\theta}}=\mathbf{A}_n^{-1}\mathbf{v}_n, \tag{8.2.6}$$

where

$$\mathbf{A}_n=\frac{1}{n-p}\sum_{t=p+1}^{n}\mathbf{X}_t'\mathbf{X}_t,$$

$$\mathbf{X}_t=(1,Y_{t-1},Y_{t-2},\ldots,Y_{t-p}),$$

$$\mathbf{v}_n=\frac{1}{n-p}\sum_{t=p+1}^{n}\mathbf{X}_t'Y_t.$$

Note that the estimator of $(\theta_1,\theta_2,\ldots,\theta_p)$ is asymptotically equivalent to the estimator

$$\begin{bmatrix}\hat{\theta}_1^\dagger\\ \hat{\theta}_2^\dagger\\ \vdots\\ \hat{\theta}_p^\dagger\end{bmatrix}=\begin{bmatrix}\hat{\gamma}(0) & \hat{\gamma}(1) & \cdots & \hat{\gamma}(p-1)\\ \hat{\gamma}(1) & \hat{\gamma}(0) & \cdots & \hat{\gamma}(p-2)\\ \vdots & \vdots & & \vdots\\ \hat{\gamma}(p-1) & \hat{\gamma}(p-2) & \cdots & \hat{\gamma}(0)\end{bmatrix}^{-1}\begin{bmatrix}\hat{\gamma}(1)\\ \hat{\gamma}(2)\\ \vdots\\ \hat{\gamma}(p)\end{bmatrix}. \tag{8.2.7}$$

The asymptotic properties of $\hat{\boldsymbol{\theta}}$ are given by the following theorem.

Theorem 8.2.1. Let Y_t be an autoregressive time series with the representation

$$Y_t = \theta_0 + \sum_{i=1}^{p} \theta_i Y_{t-i} + e_t,$$

where the roots of

$$m^p - \sum_{i=1}^{p} \theta_i m^{p-i} = 0$$

are less than one in absolute value and the e_t are independent $(0, \sigma^2)$ random variables with $E\{e_t^4\} = \eta\sigma^4$. Then

$$n^{1/2}(\hat{\boldsymbol{\theta}} - \boldsymbol{\theta}) \xrightarrow{\mathcal{L}} N(\mathbf{0}, \mathbf{A}^{-1}\sigma^2),$$

where $\hat{\boldsymbol{\theta}}$ is defined in (8.2.6) and $\mathbf{A} = p\lim \mathbf{A}_n$.

Proof. Since the time series Y_t satisfies the assumptions of Theorem 6.2.1, it follows that $p\lim \mathbf{A}_n = \mathbf{A}$, where the ijth element of $\mathbf{A}$ is

$$\begin{aligned} a_{ij} &= \gamma_Y(|i-j|) + \mu^2, && i,j = 2,3,\ldots,p+1 \\ &= 1, && i=j=1 \\ &= \mu, && \text{otherwise.} \end{aligned}$$

Now

$$n^{1/2}(\hat{\boldsymbol{\theta}} - \boldsymbol{\theta}) = n^{1/2}\mathbf{A}_n^{-1}\boldsymbol{\epsilon}_n,$$

where

$$\boldsymbol{\epsilon}_n = \frac{1}{n-p} \sum_{t=p+1}^{n} \mathbf{X}_t' e_t$$

and $\mathbf{X}_t$ was defined following (8.2.6). We consider

$$n^{1/2}\boldsymbol{\lambda}'\boldsymbol{\epsilon}_n = n^{-1/2}\left(\lambda_0 \sum_{t=p+1}^{n} e_t + \sum_{i=1}^{p} \lambda_i \sum_{t=p+1}^{n} Y_{t-i} e_t\right) + O_p(n^{-1/2}),$$

where the λ_i are arbitrary real numbers. As in the proof of Theorem 6.3.3, we define

$$Y_{tk} = \sum_{j=0}^{k} w_j e_{t-j},$$

$$W_{tk} = \sum_{j=k+1}^{\infty} w_j e_{t-j},$$

where the w_j are defined in Theorem 2.6.1. Now

$$\begin{aligned} \operatorname{Cov}\{W_{t-i,k}e_t, W_{j-r,k}e_j\} &= \gamma_W(|r-i|)\sigma^2, && t=j \\ &= 0, && \text{otherwise,} \end{aligned}$$

where $\gamma_W(h) = \sigma^2 \sum_{j=k+1}^{\infty} w_j w_{j+h}$, and we have

$$\operatorname{Var}\left\{(n-p)^{-1/2} \sum_{i=1}^{p} \lambda_i \sum_{t=p+1}^{n} W_{t-i,k} e_t\right\} = \sum_{i=1}^{p} \sum_{j=1}^{p} \lambda_i \lambda_j \sigma^2 \gamma_W(|i-j|),$$

which can be made as small as desired by choosing k sufficiently large. Now

$$\lambda_0 e_t + \sum_{i=1}^{p} \lambda_i Y_{t-i,k} e_t \overset{\text{(say)}}{=} Z_{tk}$$

is a stationary $(k+p)$-dependent time series with

$$\begin{aligned} \gamma_{Z_k}(h) &= \left[\lambda_0^2 + \sum_{i=1}^{p} \sum_{j=1}^{p} \lambda_i \lambda_j \gamma_{Y_k}(|i-j|)\right]\sigma^2, && h=0 \\ &= 0, && \text{otherwise,} \end{aligned}$$

and finite fourth moment. Therefore, by Lemma 6.3.1 and Theorem 6.3.2, $n^{1/2}\boldsymbol{\lambda}'\boldsymbol{\epsilon}_n$ converges in distribution to a normal random variable with covariance matrix $\boldsymbol{\lambda}'\mathbf{A}\boldsymbol{\lambda}$. Since $\boldsymbol{\lambda}$ was arbitrary, the result follows. ▲

In order to use the ordinary regression statistics as approximations, it is necessary to show that the residual mean square is a consistent estimator of σ^2.

Theorem 8.2.2. Let the time series Y_t be as defined in Theorem 8.2.1. Then, for $\delta < 1$,

$$p\lim n^{\delta}(\hat{\sigma}^2 - \tilde{\sigma}^2) = 0,$$

where

$$\tilde{\sigma}^2 = \frac{1}{n-p} \sum_{t=p+1}^{n} e_t^2,$$

$$\hat{\sigma}^2 = \frac{1}{n-2p-1} \sum_{t=p+1}^{n} \hat{e}_t^2,$$

$$\hat{e}_t = Y_t - \mathbf{X}_t \hat{\boldsymbol{\theta}},$$

and $\hat{\boldsymbol{\theta}}$ and $\mathbf{X}_t$ are defined in (8.2.6).

Proof. The residual mean square is

$$\frac{1}{n-2p-1} \sum_{t=p+1}^{n} \hat{e}_t^2 = \frac{1}{n-2p-1} \left\{ \sum_{t=p+1}^{n} e_t^2 - 2 \sum_{t=p+1}^{n} e_t \mathbf{X}_t (\hat{\boldsymbol{\theta}} - \boldsymbol{\theta}) + (\hat{\boldsymbol{\theta}} - \boldsymbol{\theta})' \left(\sum_{t=p+1}^{n} \mathbf{X}_t' \mathbf{X}_t \right) (\hat{\boldsymbol{\theta}} - \boldsymbol{\theta}) \right\}$$

$$= \frac{1}{n-2p-1} \sum_{t=p+1}^{n} e_t^2 - (\hat{\boldsymbol{\theta}} - \boldsymbol{\theta})' \mathbf{A}_n (\hat{\boldsymbol{\theta}} - \boldsymbol{\theta}) \frac{n-p}{n-2p-1},$$

$$= \frac{n-p}{n-2p-1} \tilde{\sigma}^2 + O_p(n^{-1}),$$

since, by Theorem 8.2.1, $(\hat{\boldsymbol{\theta}} - \boldsymbol{\theta}) = O_p(n^{-1/2})$ and $\mathbf{A}_n = O_p(1)$. ▲

As an immediate consequence of Theorem 8.2.2 and the fact that

$$\operatorname{Var}\left\{ \frac{1}{n-p} \sum_{t=p+1}^{n} e_t^2 \right\} = O(n^{-1})$$

we have $\hat{\sigma}^2 - \sigma^2 = O_p(n^{-1/2})$.

Because $(n-p)\ (\hat{\boldsymbol{\theta}}-\boldsymbol{\theta})'\mathbf{A}(\hat{\boldsymbol{\theta}}-\boldsymbol{\theta})(\sigma^2)^{-1}$ is in the limit distributed as a chi-square with $p+1$ degrees of freedom, we chose to use the divisor $n-2p-1$ in defining $\hat{\sigma}^2$. This is also in complete analogy to ordinary regression theory. Let the ijth element of $\mathbf{A}_n$ be denoted by a_{nij}. Since $p\lim\hat{\sigma}^2=\sigma^2$ and $p\lim\mathbf{A}_n=\mathbf{A}$, it follows that, for example, $(\hat{\theta}_i-\theta_i)(a_{nii}\hat{\sigma}^2)^{-1/2}$ has a normal distribution in the limit. Therefore, all of the usual regression theory holds, approximately, for the autoregressive problem.

While we have obtained pleasant asymptotic results, the behavior of the estimators in small samples can deviate considerably from that based on asymptotic theory. If the roots of the autoregressive equation are near zero, the approach to normality is quite rapid. For example, if $\rho=0$ in the first order autoregressive process, the normal or t-distribution approximations will perform very well for $n>30$. On the other hand, if the roots are near one in absolute value, very large samples may be required before the distribution is well approximated by the normal.

In Figure 8.2.1 we present the empirical density of the estimator

$$\hat{\rho}=\frac{\sum_{t=2}^{n}(Y_t-\bar{y}_{(0)})(Y_{t-1}-\bar{y}_{(-1)})}{\sum_{t=2}^{n}(Y_{t-1}-\bar{y}_{(-1)})^2},$$

where

$$\bar{y}_{(0)}=(n-1)^{-1}\sum_{t=2}^{n}Y_t,\qquad \bar{y}_{(-1)}=(n-1)^{-1}\sum_{t=2}^{n}Y_{t-1}.$$

The density was estimated from 10,000 samples of size 100. The observations were generated by the autoregressive equation

$$Y_t=0.9Y_{t-1}+e_t,$$

where the e_t are normal independent $(0,1)$ random variables. The empirical distribution displays a skewness similar to that which we would encounter in sampling from the binomial distribution. The mean of the empirical distribution is 0.868 and the variance is 0.0041. The distribution obtained from the normal approximation has a mean of 0.90 and a variance of 0.0019.

The mean of the empirical distribution agrees fairly well with the approximation obtained by the methods of Section 5.4. The bias approximated from a first order Taylor series is $E\{\hat{\rho}-\rho\}\doteq-n^{-1}(1+3\rho)$. This

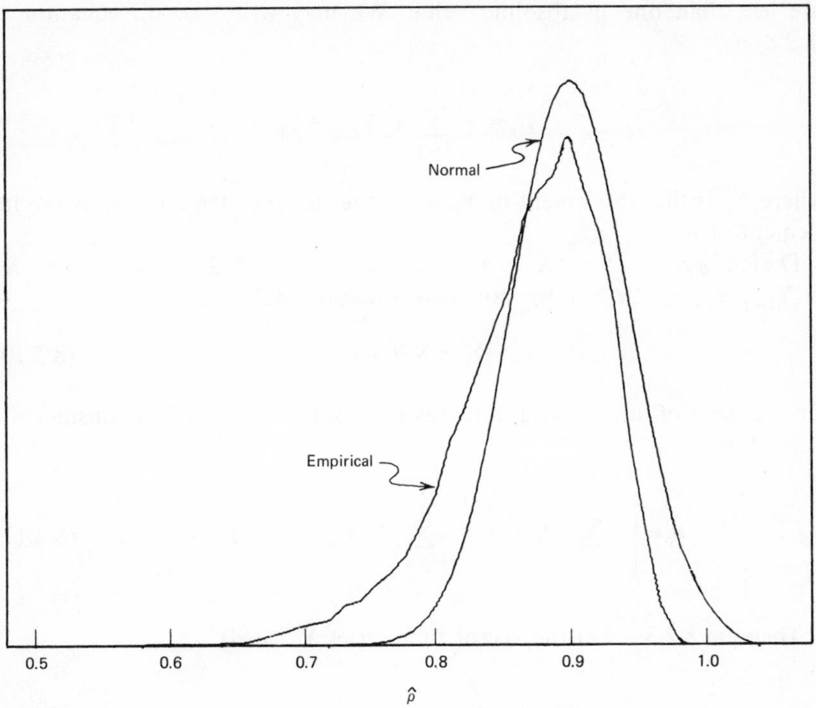

Figure 8.2.1. Estimated density of $\hat{\rho}$ compared with normal approximation for $\rho = 0.9$ and $n = 100$.

approximation to the expectation has been discussed by Mariott and Pope (1954) and Kendall (1954).

Parameter estimation by regression is also applicable to vector autoregressive processes. Let $\mathbf{Y}_t$ be the k-dimensional autoregressive process represented by

$$\sum_{j=0}^{p} \mathbf{A}_j \mathbf{Y}_{t-j} = \mathbf{e}_t, \tag{8.2.8}$$

where the $\mathbf{e}_t$ are independent $(\mathbf{0}, \boldsymbol{\Sigma})$ k-dimensional random variables with bounded fourth moments, and the $\mathbf{A}_j$ are fixed $k \times k$ matrices such that $\mathbf{A}_0 = \mathbf{I}$ and the roots of

$$\left| \sum_{j=0}^{p} \mathbf{A}_j m^{p-j} \right| = 0$$

are less than one in absolute value. We may write the ith equation of (8.2.8) as

$$Y_{it} = -\sum_{j=1}^{p} \mathbf{A}_{ji.}\mathbf{Y}_{t-j} + e_{it}, \tag{8.2.9}$$

where Y_{it} is the ith element of $\mathbf{Y}_t$, $\mathbf{A}_{ji.}$ is the ith row of $\mathbf{A}_j$, and e_{it} is the ith element of $\mathbf{e}_t$.

Defining $-\boldsymbol{\theta}_i = (\mathbf{A}_{1i.}, \mathbf{A}_{2i.}, \ldots, \mathbf{A}_{pi.})'$, $i = 1, 2, \ldots, k$, and $\mathbf{X}_t = (\mathbf{Y}'_{t-1}, \mathbf{Y}'_{t-2}, \ldots, \mathbf{Y}'_{t-p})$, we can write equation (8.2.9) as

$$Y_{it} = \mathbf{X}_t\boldsymbol{\theta}_i + e_{it}. \tag{8.2.10}$$

On the basis of our scalar autoregressive results, we are led to consider the estimators

$$\hat{\boldsymbol{\theta}}_i = \left[\sum_{t=p+1}^{n} \mathbf{X}'_t\mathbf{X}_t\right]^{-1} \sum_{t=p+1}^{n} \mathbf{X}'_t Y_{it}, \qquad i = 1, 2, \ldots, k. \tag{8.2.11}$$

Theorem 8.2.3. Let the vector time series $\mathbf{Y}_t$ satisfy

$$\sum_{j=0}^{p} \mathbf{A}_j\mathbf{Y}_{t-j} = \mathbf{e}_t,$$

where $\{\mathbf{e}_t\}$ is a sequence of independent $(\mathbf{0}, \boldsymbol{\Sigma})$ k-dimensional random variables with bounded fourth moments, and the $\mathbf{A}_j$ are fixed $k \times k$ matrices such that $\mathbf{A}_0 = \mathbf{I}$ and the roots of

$$\left|\sum_{j=0}^{p} \mathbf{A}_j m^{p-j}\right| = 0$$

are less than one in absolute value. Then,

$$n^{1/2}(\hat{\boldsymbol{\theta}} - \boldsymbol{\theta}) \xrightarrow{\mathcal{L}} N(\mathbf{0}, \boldsymbol{\Sigma} \otimes \mathbf{H}^{-1}),$$

where $\hat{\boldsymbol{\theta}}_i$ is defined in (8.2.11),

$$\boldsymbol{\theta}' = (\boldsymbol{\theta}'_1, \boldsymbol{\theta}'_2, \ldots, \boldsymbol{\theta}'_k),$$

$$\mathbf{H} = p\lim \frac{1}{n-p} \sum_{t=p+1}^{n} \mathbf{X}'_t\mathbf{X}_t,$$

and $\boldsymbol{\Sigma}\otimes\mathbf{H}^{-1}$ is the Kronecker product of the matrices $\boldsymbol{\Sigma}$ and $\mathbf{H}^{-1}$.

Proof. We only outline the proof, since it differs in no substantive way from that of Theorem 8.2.1. Since $\mathbf{Y}_t$ is a stationary time series, $p\lim(n-p)^{-1}\Sigma_{t=p+1}^{n}\mathbf{X}_t'\mathbf{X}_t=\mathbf{H}$ by Lemma 6.5.1. Therefore, the limiting distribution of $\hat{\boldsymbol{\theta}}$ follows from that of $n^{1/2}(n-p)^{-1}\Sigma_{t=p+1}^{n}\mathbf{X}_t'e_{it}, i=1,2,\ldots,k$. By arguments parallel to those of Theorem 8.2.1, we obtain the limiting distribution. ▲

Often the practitioner must determine the degree of the autoregressive process as well as estimate the parameters. If it is possible to specify a maximum for the degree of the process, a process of that degree can first be estimated and high order terms discarded using the standard regression statistics. Anderson (1962) gives a procedure for this decision problem. Various other model building methods based on regression theory can be used. Several such procedures are described in Draper and Smith (1966).

Often one inspects the residuals from the fit and perhaps computes the autocorrelations of these residuals. If the model is correct, the sample autocorrelations estimate zero with an error that is $O_p(n^{-1/2})$, but the variances of these estimators are generally smaller than the variances of estimators computed from a time series of independent random variables [Box and Pierce (1970)]. Thus, while it is good practice to inspect the residuals, it is suggested that tests of model adequacy be constructed by adding additional terms to the model and testing the hypothesis that the true value of the added coefficients is zero.

In line with the comments of the previous section, $\bar{y}_n$ or the generalization of $\hat{\mu}$ of (8.1.14) to higher order time series are the preferred estimators for μ with realizations of moderate length.

Example. To illustrate the regression estimation of the autoregressive process, we use the unemployment time series investigated in Section 6.4. The second order process, estimated by regressing $Y_t-\bar{y}_n$ on $Y_{t-1}-\bar{y}_n$ and $Y_{t-2}-\bar{y}_n$, is

$$\hat{Y}_t-4.77=\underset{(0.073)}{1.57}(Y_{t-1}-4.77)-\underset{(0.073)}{0.70}(Y_{t-2}-4.77)$$

where $\bar{y}_{100}=4.77$ and the numbers below the coefficients are the estimated standard errors obtained from the regression analysis. The residual mean square is 0.105.

If we condition the analysis on the first two observations and regress Y_t on Y_{t-1} and Y_{t-2} including an intercept term in the regression, we obtain

$$\hat{Y}_t=\underset{(0.13)}{0.63}+\underset{(0.073)}{1.57}Y_{t-1}-\underset{(0.073)}{0.70}Y_{t-2}.$$

The coefficients are slightly different from those in Section 6.4, since the coefficients in Section 6.4 were obtained from equation (8.2.7).

To check on the adequacy of the second order representation, we fit a fifth order process. The results are summarized in Table 8.2.1. The F-test for the hypothesis that the time series is second order autoregressive against the alternative that it is fifth order is

$$F_{89}^{3}=\frac{0.478}{3(0.101)}=1.578.$$

The tabular 0.05 point for Snedecor's F with 3 and 89 degrees of freedom is 2.71, and so the null hypothesis is accepted at that level.

Table 8.2.1. Analysis of variance for quarterly seasonally adjusted unemployment rate, 1948 to 1972

Source	Degrees of Freedom	Mean Square
Y_{t-1}	1	112.949
Y_{t-2} after Y_{t-1}	1	9.481
Y_{t-3} after Y_{t-1}, Y_{t-2}	1	0.305
Y_{t-4} after $Y_{t-1}, Y_{t-2}, Y_{t-3}$	1	0.159
Y_{t-5} after $Y_{t-1}, Y_{t-2}, Y_{t-3}, Y_{t-4}$	1	0.014
Error	89	0.101

8.3. MOVING AVERAGE TIME SERIES

We have seen that ordinary least squares regression procedures can be used to obtain efficient estimators for the parameters of autoregressive time series. Unfortunately, the estimation for moving average processes is less simple.

By the results of Chapter 2 we know that there is a relationship between the correlation function of a moving average time series and the parameters of the time series. For example, for the first order process, $\rho(1)=\beta(1+\beta^2)^{-1}$, where β is the parameter of the process. On this basis, we might be led to estimate β from an estimator of $\rho(1)$. Since we demonstrated in Chapter 6 that

$$\hat{r}(1)=\frac{\sum_{t=2}^{n}(Y_t-\bar{y}_n)(Y_{t-1}-\bar{y}_n)}{\sum_{t=1}^{n}(Y_t-\bar{y}_n)^2}$$

estimates $\rho(1)$ with an error which is $O_p(n^{-1/2})$, it follows that

$$\begin{aligned} \hat{\beta}_r &= [2\hat{r}(1)]^{-1}\{1-[1-4\hat{r}^2(1)]^{\frac{1}{2}}\}, && 0<|\hat{r}(1)|\leqslant 0.5 \\ &= -1, && \hat{r}(1)<-0.5 \\ &= 1, && \hat{r}(1)>0.5 \\ &= 0, && \hat{r}(1)=0 \end{aligned} \tag{8.3.1}$$

estimates β with an error of the same order. Obviously, if $\hat{r}(1)$ lies outside the range $(-0.5, 0.5)$ by a significant amount, the model of a first order moving average is suspect.

Because the derivative of $\rho(1)$ with respect to β is $(1+\beta^2)^{-2}(1-\beta^2)$, the approximate variance of this estimator of β is related to the approximate variance of $\hat{r}(1)$ by

$$\operatorname{Var}\{\hat{r}(1)\} \doteq \left[\frac{1-\beta^2}{(1+\beta^2)^2}\right]^2 \operatorname{Var}\{\hat{\beta}_r\}.$$

It follows from equation (6.2.8) that

$$\operatorname{Var}\{\hat{r}(1)\} = \frac{1+\beta^2+4\beta^4+\beta^6+\beta^8}{n(1+\beta^2)^4} + O\left(\frac{1}{n^2}\right) \tag{8.3.2}$$

and, therefore, for $|\beta|<1$,

$$\operatorname{Var}\{\hat{\beta}_r\} \doteq \frac{1+\beta^2+4\beta^4+\beta^6+\beta^8}{n(1-\beta^2)^2} \tag{8.3.3}$$

While the estimator (8.3.1) is consistent, we shall see that it is inefficient for $\beta \neq 0$. Roughly, the other sample autocorrelations contain information about β.

To obtain an efficient estimator, we consider an estimation procedure based on the Gauss-Newton method of estimating the parameters of a nonlinear function discussed in Chapter 5. The procedure has computational advantages for higher order processes, and one also obtains estimators of the variances of the estimators, as a direct result of the computations.

To introduce the procedure, consider the first order moving average

$$Y_t = e_t + \beta e_{t-1}, \tag{8.3.4}$$

where $|\beta|<1$ and the e_t are independent $(0,\sigma^2)$ random variables with $E\{e_t^4\}=\eta\sigma^4$. Equation (8.3.4) may also be written

$$e_t = -\beta e_{t-1} + Y_t \tag{8.3.5}$$

and, using our previous difference equation results, we have

$$e_t = \sum_{j=0}^{\infty} (-\beta)^j Y_{t-j}, \tag{8.3.6}$$

$$Y_t = -\sum_{j=1}^{\infty} (-\beta)^j Y_{t-j} + e_t. \tag{8.3.7}$$

The nonlinear nature of the problem of estimating β is clear in expression (8.3.6), which gives e_t as a nonlinear function of β. We write (8.3.6) as

$$\begin{aligned} e_t(Y;\beta) &= \sum_{j=0}^{\infty} (-\beta)^j Y_{t-j} \\ &= \sum_{j=0}^{t-1} (-\beta)^j Y_{t-j} + (-\beta)^t e_0, \end{aligned} \tag{8.3.8}$$

where $e_0 = \sum_{j=0}^{\infty}(-\beta)^j Y_{-j}$ and the notation $e_t(Y;\beta)$ is to emphasize the fact that e_t is being considered a function of the "independent variables" $Y_t, Y_{t-1}, \ldots,$ where the parameter of the function is β. Similarly (8.3.7) can be written

$$Y_t = \beta e_{t-1}(Y;\beta) + e_t. \tag{8.3.9}$$

We assume initial estimators $\tilde{\beta}$ and $\tilde{e}_0$ satisfying $(\tilde{\beta}-\beta)=o_p(n^{-1/4}), \tilde{e}_0 = O_p(1)$. These requirements are satisfied if one uses $\tilde{e}_0=0$ and the estimator $\hat{\beta}_r$. Alternative initial estimators will be discussed later. We then expand the function $e_t(Y;\beta)$ in a first order Taylor series about the point $\tilde{\beta}$, to obtain

$$e_t(Y;\beta) = e_t(Y;\tilde{\beta}) - W_t(Y;\tilde{\beta})[\beta-\tilde{\beta}] + d_t(Y;\tilde{\beta}) \tag{8.3.10}$$

or

$$e_t(Y;\tilde{\beta}) = W_t(Y;\tilde{\beta})[\beta-\tilde{\beta}] - d_t(Y;\tilde{\beta}) + e_t, \tag{8.3.11}$$

where $W_t(Y;\tilde{\beta})$ is the negative of the partial derivative of $e_t(Y;\beta)$ with respect to β evaluated at $\beta=\tilde{\beta}, d_t(Y;\tilde{\beta})$ is one half of $(\beta-\tilde{\beta})^2$ times the

second derivative of $e_t(Y;\beta)$ with respect to β evaluated at $\beta=\beta^\dagger$, and $\beta^\dagger$ is between β and $\tilde{\beta}$. Note that $W_t(Y;\tilde{\beta})$ can be written

$$
\begin{aligned}
W_t(Y;\tilde{\beta}) &= \tilde{e}_0, && t=1\\
&= \sum_{j=1}^{t-1} j(-\tilde{\beta})^{j-1}Y_{t-j} + t(-\tilde{\beta})^{t-1}\tilde{e}_0, && t=2,3,\ldots,n.
\end{aligned}
\tag{8.3.12}
$$

Regressing $e_t(Y;\tilde{\beta})$ on $W_t(Y;\tilde{\beta})$ we obtain an estimator of $(\beta-\tilde{\beta})$. The improved estimator of β is then

$$\hat{\beta}=\tilde{\beta}+\Delta\hat{\beta},$$

where

$$\Delta\hat{\beta}=\frac{\sum_{t=1}^{n} e_t(Y;\tilde{\beta})W_t(Y;\tilde{\beta})}{\sum_{t=1}^{n}\left[W_t(Y;\tilde{\beta})\right]^2}. \tag{8.3.13}$$

The computation of $e_t(Y;\tilde{\beta})$ and $W_t(Y;\tilde{\beta})$ is simplified by noting that both satisfy difference equations:

$$
\begin{aligned}
e_t(Y;\tilde{\beta}) &= Y_1-\tilde{\beta}\tilde{e}_0, && t=1\\
&= Y_t-\tilde{\beta}e_{t-1}(Y;\tilde{\beta}), && t=2,3,\ldots,n,
\end{aligned}
\tag{8.3.14}
$$

and

$$
\begin{aligned}
W_t(Y;\tilde{\beta}) &= \tilde{e}_0, && t=1\\
&= e_{t-1}(Y;\tilde{\beta})-\tilde{\beta}W_{t-1}(Y;\tilde{\beta}), && t=2,3,\ldots,n.
\end{aligned}
\tag{8.3.15}
$$

The difference equation for $e_t(Y;\tilde{\beta})$ follows directly from (8.3.5). Equation (8.3.15) can be obtained by differentiating both sides of

$$e_t(Y;\beta)=Y_t-\beta e_{t-1}(Y;\beta)$$

with respect to β and evaluating the resulting expression at $\beta=\tilde{\beta}$. The asymptotic properties of $\hat{\beta}$ are developed in Theorem 8.3.1.

Theorem 8.3.1. Let Y_t be defined by

$$Y_t = e_t + \beta e_{t-1},$$

where $|\beta| < 1$ and the e_t are independent $(0, \sigma^2)$ random variables with $E\{e_t^4\} = \eta\sigma^4$. Let $\tilde{e}_0$ and $\tilde{\beta}$ be initial estimators satisfying $\tilde{e}_0 = O_p(1)$, $(\tilde{\beta} - \beta) = o_p(n^{-1/4})$, and $|\tilde{\beta}| < 1$. Then,

$$n^{1/2}(\hat{\beta} - \beta) \xrightarrow{\mathcal{L}} N(0, 1 - \beta^2),$$

where $\hat{\beta}$ is defined in (8.3.13).

Proof. From (8.3.11) and (8.3.13),

$$\hat{\beta} - \beta = \left[\sum_{t=1}^{n} W_t^2(Y; \tilde{\beta}) \right]^{-1} \left[\sum_{t=1}^{n} W_t(Y; \tilde{\beta}) e_t + R(Y; \tilde{\beta}) \right],$$

where

$$R(Y; \tilde{\beta}) = \frac{1}{2} \sum_{t=1}^{n} W_t(Y; \tilde{\beta}) H_t(Y; \ddot{\beta}) [\beta - \tilde{\beta}]^2,$$

and $H_t(Y; \ddot{\beta})$ is the derivative of $W_t(Y; \beta)$ with respect to β evaluated at $\beta = \ddot{\beta}$, $\ddot{\beta}$ between $\tilde{\beta}$ and β. We have

$$W_t(Y; \delta) = \sum_{j=1}^{t-1} j(-\delta)^{j-1} Y_{t-j} + t(-\delta)^{t-1} \tilde{e}_0,$$

$$H_t(Y; \delta) = -\sum_{j=2}^{t-1} j(j-1)(-\delta)^{j-2} Y_{t-j} - t(t-1)(-\delta)^{t-2} \tilde{e}_0.$$

Let $\bar{S}$ be a closed interval containing β as an interior point and such that $\max_{\delta \in \bar{S}} |\delta| < \lambda < 1$. Then,

$$W_t(Y; \delta) = \sum_{j=0}^{t-1} a_j e_{t-1-j} + t(-\delta)^{t-1} \tilde{e}_0,$$

$$H_t(Y; \delta) = \sum_{j=0}^{t-2} b_j e_{t-2-j} + t(t-1)(-\delta)^{t-2} \tilde{e}_0,$$

where, by Exercise 2.24 for $\delta \in \bar{S}$, there is an M such that

$$|a_j| < M\lambda^j,$$

$$|b_j| < M\lambda^j.$$

Clearly, $n^{-1}\sum_{t=1}^{n} t^2 \delta^{2(t-1)} \tilde{e}_0^2 = O_p(n^{-1})$ for $\tilde{e}_0 = O_p(1)$. Hence, to simplify the notation for the remainder of the proof, we set $\tilde{e}_0 = 0$. For $\tilde{e}_0 = 0$, and $\delta \in \bar{S}$,

$$E\{W_t(Y;\delta)W_{t+h}(Y;\delta)\} < M^2(1-\lambda^2)^{-1}\lambda^{|h|}\sigma^2,$$

$$E\{H_t(Y;\delta)H_{t+h}(Y;\delta)\} < M^2(1-\lambda^2)^{-1}\lambda^{|h|}\sigma^2,$$

and, by the arguments of Theorem 6.2.1, the variances of

$$n^{-1/2}\sum_{t=1}^{n} W_t^2(Y;\delta) \quad \text{and} \quad n^{-1/2}\sum_{t=1}^{n} H_t^2(Y;\delta)$$

are bounded for $\delta \in \bar{S}$. Therefore,

$$\frac{1}{n}\sum_{t=1}^{n} W_t(Y;\tilde{\beta})H_t(Y;\ddot{\beta}) = O_p(1)$$

and

$$n^{-1}R(Y;\tilde{\beta}) = o_p(n^{-1/2}).$$

For $\delta \in (-1,1)$, $W_t(Y;\delta)$ converges to a stationary time series, the effect of $\tilde{e}_0$ being transient. Therefore,

$$\frac{1}{n}\sum_{t=1}^{n} W_t^2(Y;\delta)$$

converges in probability for $\delta \in (-1,1)$ and the limit is a continuous function of δ, for $\delta \in \bar{S}$. Now

$$n^{-1/2}\sum_{t=1}^{n} W_t(Y;\tilde{\beta})e_t$$

$$= n^{-1/2}\sum_{t=1}^{n} e_t\left[W_t(Y;\beta) + H_t(Y;\beta)(\tilde{\beta}-\beta) + G_t(Y;\beta^\dagger)(\tilde{\beta}-\beta)^2\right]$$

where $G_t(Y;\beta^\dagger)$ is one half of the derivative of $H_t(Y;\beta)$ with respect to β evaluated at $\beta=\beta^\dagger$ and $\beta^\dagger$ is between $\tilde{\beta}$ and β. By the arguments used for $n^{-1}\Sigma_{t=1}^n H_t^2(Y;\delta)$, it follows that $n^{-1}\Sigma_{t=1}^n G_t^2(Y;\beta^\dagger)$ is bounded in probability and

$$n^{-1/2}\sum_{t=1}^n W_t(Y;\tilde{\beta})e_t = n^{-1/2}\sum_{t=1}^n W_t(Y;\beta)e_t + o_p(1).$$

Therefore, the limiting distribution of $n^{1/2}(\hat{\beta}-\beta)$ is the same as the limiting distribution of

$$\left(\frac{\sigma^2}{1-\beta^2}\right)^{-1} n^{-1/2}\sum_{t=1}^n W_t(Y;\beta)e_t.$$

Since $W_t(Y;\beta)$ is a first order autoregressive time series with parameter β, we have

$$\lim_{t\to\infty}\operatorname{Cov}\{W_t(Y;\beta)e_t, W_{t+h}(Y;\beta)e_{t+h}\} = \frac{\sigma^4}{1-\beta^2}, \qquad h=0,$$

$$\operatorname{Cov}\{W_t(Y;\beta)e_t, W_{t+h}(Y;\beta)e_{t+h}\} = 0, \qquad h\neq 0.$$

The proof of asymptotic normality of $n^{-1/2}\Sigma_{t=1}^n W_t(Y;\beta)e_t$ follows the proof of Theorem 8.2.1. ▲

Comparison of the result of Theorem 8.3.1 and equation (8.3.3) establishes the large sample inefficiency of the estimator constructed from the first order autocorrelation.

Theorem 8.3.2. Given the assumptions of Theorem 8.3.1,

$$\hat{\sigma}^2 - \tilde{\sigma}^2 = O_p(n^{-1}),$$

where

$$\tilde{\sigma}^2 = \frac{1}{n}\sum_{t=1}^n e_t^2,$$

$$\hat{\sigma}^2 = \frac{1}{n}\sum_{t=1}^n e_t^2(Y;\hat{\beta}),$$

and $e_t(Y;\hat{\beta})$ is defined by (8.3.14) with $\hat{\beta}$ replacing $\tilde{\beta}$.

Proof. Since $|\beta|<1$,

$$n^{-1}\left\{\sum_{t=1}^{n} e_t^2(Y;\beta)-\sum_{t=1}^{n} e_t^2\right\}=O_p(n^{-1}),$$

where

$$e_t(Y;\beta)=\sum_{j=0}^{t-1}(-\beta)^j Y_{t-j}+(-\beta)^t\tilde{e}_0$$

and $\tilde{e}_0=O_p(1)$. We have, by arguments similar to those of Theorem 8.3.1,

$$\begin{aligned}\frac{1}{n}\sum_{t=1}^{n} e_t^2(Y;\hat{\beta})&=\frac{1}{n}\left\{\sum_{t=1}^{n} e_t^2(Y;\beta)-2(\hat{\beta}-\beta)\sum_{t=1}^{n} e_t(Y;\beta)W_t(Y;\beta)\right.\\&\qquad\left.-(\hat{\beta}-\beta)^2\sum_{t=1}^{n}\left[e_t(Y;\beta^{\dagger})H_t(Y;\beta^{\dagger})-W_t^2(Y;\beta^{\dagger})\right]\right\}\\&=\frac{1}{n}\sum_{t=1}^{n} e_t^2+O_p(n^{-1}),\end{aligned}$$

where $H_t(Y;\beta)$ is defined in the proof of Theorem 8.3.1 and $\beta^{\dagger}$ is between $\hat{\beta}$ and β. ▲

Theorems 8.3.1 and 8.3.2 mean that, for large samples, we can use the regular regression statistics as approximations when drawing inferences about β.

The nonlinear estimation procedure may be generalized to higher order moving average processes. We illustrate with the second order process

$$Y_t=e_t+\beta_1 e_{t-1}+\beta_2 e_{t-2}, \qquad (8.3.16)$$

where the e_t are independent $(0,\sigma^2)$ random variables such that $E\{e_t^4\}=\eta\sigma^4$ and the roots of the characteristic equation

$$m^2+\beta_1 m+\beta_2=0$$

are less than one in absolute value. In our functional notation expression (8.3.16) becomes

$$e_t(Y;\boldsymbol{\beta})=Y_t-\beta_1 e_{t-1}(Y;\boldsymbol{\beta})-\beta_2 e_{t-2}(Y;\boldsymbol{\beta}), \qquad (8.3.17)$$

where $\boldsymbol{\beta}=(\beta_1,\beta_2)'$. Define

$$W_{1t}(Y;\boldsymbol{\beta})=-\frac{\partial e_t(Y;\boldsymbol{\beta})}{\partial\beta_1}$$

$$W_{2t}(Y;\boldsymbol{\beta})=-\frac{\partial e_t(Y;\boldsymbol{\beta})}{\partial\beta_2}.$$

Then, by differentiating both sides of (8.3.17), we establish difference equation relationships for $W_{1t}(Y;\boldsymbol{\beta})$ and $W_{2t}(Y;\boldsymbol{\beta})$:

$$\begin{aligned}W_{1t}(Y;\boldsymbol{\beta})&=e_{t-1}(Y;\boldsymbol{\beta})-\beta_1W_{1,t-1}(Y;\boldsymbol{\beta})-\beta_2W_{1,t-2}(Y;\boldsymbol{\beta}),\\ W_{2t}(Y;\boldsymbol{\beta})&=e_{t-2}(Y;\boldsymbol{\beta})-\beta_1W_{2,t-1}(Y;\boldsymbol{\beta})-\beta_2W_{2,t-2}(Y;\boldsymbol{\beta}).\end{aligned}\quad(8.3.18)$$

We note that the difference equation defining $W_{1t}(Y;\boldsymbol{\beta})$ is identical to that defining $W_{2t}(Y;\boldsymbol{\beta})$ except that the e-value is lagged one more period in the definition of $W_{2t}(Y;\boldsymbol{\beta})$. Hence, if the initial conditions are identical (or are a sufficient distance in the past),

$$W_{2t}(Y;\boldsymbol{\beta})=W_{1,t-1}(Y;\boldsymbol{\beta})\overset{\text{(say)}}{=}W_{t-1}(Y;\boldsymbol{\beta}).$$

The time series $W_t(\text{Y};\boldsymbol{\beta})$ converges to a stationary second order autoregressive time series with the characteristic equation

$$m^2+\beta_1m+\beta_2=0.$$

Consider now the invertible qth order moving average

$$Y_t=\sum_{i=0}^{q}\beta_ie_{t-i},$$

where $\beta_0=1$. Let us assume that we have initial estimators of $\beta_i, i=1,2,\ldots,q$, and initial estimators of e_t for $t=-q+1,-q+2,\ldots,0$. Then estimators of the differences between the true parameter and the initial estimators, $(\beta_i-\tilde{\beta}_i)=\Delta\beta_i$, $i=1,2,\ldots,q$, are obtained by regressing $e_t(Y;\tilde{\boldsymbol{\beta}})$ on $W_{1t}(Y;\tilde{\boldsymbol{\beta}}), W_{2t}(Y;\tilde{\boldsymbol{\beta}}),\ldots,W_{qt}(Y;\tilde{\boldsymbol{\beta}})$, where

$$e_t(Y;\tilde{\boldsymbol{\beta}})=Y_t-\sum_{i=1}^{q}\tilde{\beta}_ie_{t-i}(Y;\tilde{\boldsymbol{\beta}}),\qquad t=1,2,\ldots,n,$$

$$W_{it}(Y;\tilde{\boldsymbol{\beta}})=\begin{cases}0, & t\leqslant0\\ e_{t-i}(Y;\tilde{\boldsymbol{\beta}})-\displaystyle\sum_{j=1}^{q}\tilde{\beta}_jW_{i,t-j}(Y;\tilde{\boldsymbol{\beta}}), & t=1,2,\ldots,n,\end{cases}$$

for $i=1,2,\ldots,q$. Thus, the improved estimator of $\boldsymbol{\beta}$ is defined by

$$\hat{\boldsymbol{\beta}}=\tilde{\boldsymbol{\beta}}+\Delta\hat{\boldsymbol{\beta}}, \tag{8.3.19}$$

where

$$\Delta\hat{\boldsymbol{\beta}}=\mathbf{G}_n^{-1}(n^{-1}\mathbf{W}'\tilde{\mathbf{e}}),$$

$$\mathbf{G}_n=\frac{1}{n}\sum_{t=1}^{n}\mathbf{w}_t'\mathbf{w}_t,$$

$$\mathbf{W}'=[\mathbf{w}_1',\mathbf{w}_2',\ldots,\mathbf{w}_n'],$$

$$\tilde{\mathbf{e}}=[e_1(Y;\tilde{\boldsymbol{\beta}}),e_2(Y;\tilde{\boldsymbol{\beta}}),\ldots,e_n(Y;\tilde{\boldsymbol{\beta}})]',$$

$$\mathbf{w}_t=[W_{1t}(Y;\tilde{\boldsymbol{\beta}}),W_{2t}(Y;\tilde{\boldsymbol{\beta}}),\ldots,W_{qt}(Y;\tilde{\boldsymbol{\beta}})].$$

As with the first order model, the limiting distribution of the properly normalized estimator is normal.

Theorem 8.3.3. Let

$$Y_t=\sum_{j=1}^{q}\beta_j e_{t-j}+e_t,$$

where the roots of

$$m^q+\sum_{j=1}^{q}\beta_j m^{q-j}=0$$

are less than one in absolute value and the e_t are independent $(0,\sigma^2)$ random variables with $E\{e_t^4\}=\eta\sigma^4$. Let $\tilde{\boldsymbol{\beta}}=(\tilde{\beta}_1,\tilde{\beta}_2,\ldots,\tilde{\beta}_q)'$ be an initial estimator for $\boldsymbol{\beta}$ satisfying $(\tilde{\boldsymbol{\beta}}-\boldsymbol{\beta})=o_p(n^{-1/4})$ and such that the roots of $m^q+\sum_{j=1}^{q}\tilde{\beta}_j m^{q-j}=0$ are less than one in absolute value; and let $\tilde{e}_i, i=-q+1,-q+2,\ldots,0$, be bounded in probability. Then

$$n^{1/2}(\hat{\boldsymbol{\beta}}-\boldsymbol{\beta})\xrightarrow{\mathcal{L}}N(\mathbf{0},\mathbf{G}^{-1}\sigma^2),$$

where $\mathbf{G}=p\lim\mathbf{G}_n$, and $\mathbf{G}_n$ and $\hat{\boldsymbol{\beta}}$ are defined in (8.3.19).

Proof. The proof parallels that of Theorem 8.3.1. $W_{it}(Y;\boldsymbol{\beta})$ converges to a stationary qth order autoregressive time series with $W_{it}(Y;\boldsymbol{\beta})$ converging to $W_{i+s,t-s}(Y;\boldsymbol{\beta})$, $s=1,2,\ldots,q-1$. Hence the matrix $\mathbf{G}$ is well defined.

The limiting distribution of $n^{1/2}\mathbf{G}(\hat{\boldsymbol{\beta}}-\boldsymbol{\beta})$ is the same as that of

$$n^{-1/2}\left[\sum_{t=1}^{n} W_{1t}(Y;\boldsymbol{\beta})e_t, \sum_{t=1}^{n} W_{2t}(Y;\boldsymbol{\beta})e_t, \ldots, \sum_{t=1}^{n} W_{qt}(Y;\boldsymbol{\beta})e_t\right]'$$

and the result follows by Theorem 8.2.1. ▲

Theorem 8.3.4. Given the assumptions and estimators of Theorem 8.3.3,

$$\hat{\sigma}^2-\tilde{\sigma}^2=O_p(n^{-1}),$$

where

$$\tilde{\sigma}^2=\frac{1}{n}\sum_{t=1}^{n} e_t^2,$$

$$\hat{\sigma}^2=\frac{1}{n}\sum_{t=1}^{n} e_t^2(Y;\hat{\boldsymbol{\beta}}).$$

Proof. The proof parallels that of Theorem 8.3.2. The details are reserved for the reader. ▲

By Theorems 8.3.3 and 8.3.4, the usual regression statistics obtained in the regression of $e_t(Y;\tilde{\boldsymbol{\beta}})$ on $W_{it}(Y;\tilde{\boldsymbol{\beta}})$ can be used to construct tests and confidence intervals.

In our discussion we have assumed that the mean of the time series was known and taken to be zero. It is easy to demonstrate that all of the asymptotic results of this section hold for Y_t replaced by $Y_t-\bar{y}_n$, where $\bar{y}_n$ is the sample mean.

To obtain initial estimators for higher order processes, we consider an estimation procedure suggested by Durbin (1959). By Theorem 2.6.2, any qth order moving average

$$Y_t=\sum_{s=1}^{q}\beta_s e_{t-s}+e_t \tag{8.3.20}$$

for which the roots of the characteristic equation are less than one can be represented in the form

$$Y_t=-\sum_{j=1}^{\infty}c_j Y_{t-j}+e_t,$$

where the weights satisfy the difference equation

$$\begin{aligned}
c_1 &= -\beta_1 \\
c_2 &= -\beta_2 - \beta_1 c_1 \\
c_3 &= -\beta_3 - \beta_1 c_2 - \beta_2 c_1 \\
&\ \vdots \\
c_q &= -\beta_q - \beta_1 c_{q-1} - \beta_2 c_{q-2} - \cdots - \beta_{q-1} c_1 \\
c_j &= -\sum_{m=1}^{q} \beta_m c_{j-m}, \qquad j = q+1, q+2, \ldots.
\end{aligned} \tag{8.3.21}$$

Since the weights c_j are sums of powers of the roots, they decline in absolute value, and one can terminate the sum at a convenient finite number, say k. Then we can write

$$Y_t \doteq -\sum_{j=1}^{k} c_j Y_{t-j} + e_t. \tag{8.3.22}$$

On the basis of this approximation, we treat Y_t as a finite autoregressive process and estimate c_j, $j=1,2,\ldots,k$, by the regression procedures of Section 8.2. As the true weights satisfy equation (8.3.21), we treat the estimated c_j's as a finite autoregressive process satisfying (8.3.21) and estimate the β's. That is, we treat

$$\hat{c}_j = -\sum_{s=1}^{q} \hat{c}_{j-s}\beta_s \tag{8.3.23}$$

as a regression equation and estimate the β's by regressing $-\hat{c}_j$ on $\hat{c}_{j-1}, \hat{c}_{j-2}, \ldots, \hat{c}_{j-q}$ where the appropriate modifications must be made for $j=1,2,\ldots,q$, as per (8.3.21).

If we let $\{k_n\}$ be a sequence such that

$$\lim_{n\to\infty} k_n = \infty,$$

$$k_n = o(n^{1/3}),$$

it is possible to use the results of Berk (1974) to demonstrate that

$$\sum_{j=1}^{k_n} (\hat{c}_j - c_j)^2 = O_p(k_n n^{-1}).$$

It follows that the preliminary estimators constructed from the $\hat{c}_j$ will have an error that is $o_p(n^{-1/3})$.

In carrying out the Gauss-Newton estimation, it is necessary to estimate the initial values $\tilde{e}_{-q+1}, \tilde{e}_{-q+2}, \ldots, \tilde{e}_0$. The simplest procedure is to set them equal to zero. For small samples alternative estimators may be superior. The autoregressive equation (8.3.22) used to obtain initial estimates of the parameters can be used to estimate the Y-values preceding the sample period.[2] By the discussion of Section 2.9, the extrapolated values of a kth order autoregressive process are given by

$$\hat{Y}_0(Y_1, \ldots, Y_n) = -\sum_{i=1}^{k} \hat{c}_i Y_i,$$

$$\hat{Y}_{-r}(Y_1, \ldots, Y_n) = -\sum_{i=1}^{r} \hat{c}_i \hat{Y}_{-r+i}(Y_1, \ldots, Y_n) - \sum_{i=1}^{k-r} \hat{c}_{r+i} Y_i, \qquad 0 < r \leqslant q-1. \quad (8.3.24)$$

Recall that a stationary autoregressive process can be written in either a forward or backward manner (Corollary 2.6.1.2) and, as a result, the extrapolation formula for Y_0 is of the same form as that for Y_{n+1}. If the true process is a qth order moving average, one uses the autoregressive equation to estimate q values, since the best predictors for $Y_t, t \leqslant -q$, are zero.

The covariance between Y_{-q+1} and $Y_t, t = 1, 2, \ldots,$ is the same as that between e_{-q+1} and $Y_t, t = 1, 2, \ldots$. That is, the best linear estimator of Y_{-q+1} is the best linear estimator of e_{-q+1}. Also, the covariance between Y_{-q+2} and $Y_t, t = 1, 2, \ldots,$ is the same as that between $\beta_1 e_{-q+1} + e_{-q+2}$ and $Y_t, t = 1, 2, \ldots,$ and so forth. Therefore, by setting $\tilde{e}_{-q-j} = 0$ for $j = 0, 1, \ldots, q-1$, one can compute $\tilde{e}_{-q+s}, s = 1, 2, \ldots, q$, from the estimators of $Y_{-q+1}, Y_{-q+2}, \ldots, Y_0$ by $\tilde{e}_{-q+s} = \hat{Y}_{-q+s}(Y_1, \ldots, Y_n) - \sum_{j=1}^{q} \tilde{\beta}_j \tilde{e}_{-q+s-j}$.

Example. We illustrate these procedures by fitting a first order moving average to an artificially generated time series. Table 8.3.1 contains 100 observations on X_t defined by $X_t = 0.7e_{t-1} + e_t$, where the e_t are computer generated normal independent (0, 1) random variables. We assume that we know the mean of the time series is zero. If we did not make this assumption, we would first calculate the mean and then work with $Y_t - \bar{y}_{100}, t = 1, 2, \ldots, 100$.

[2]Box and Jenkins (1970) suggest constructing an autoregressive prediction equation such as (8.3.22) using the estimated moving average parameters. They call the extrapolated values *back forecasts*.

Table 8.3.1. One hundred observations from a first order moving average time series with $\beta = 0.7$

	First 25	Second 25	Third 25	Fourth 25
1	1.432	1.176	−1.311	2.607
2	−0.343	0.846	−0.105	1.572
3	−1.759	0.079	0.313	−0.261
4	−2.537	0.815	−0.890	−0.686
5	−0.295	2.566	−1.778	−2.079
6	0.689	1.675	−0.202	−2.569
7	−0.633	0.933	0.450	−0.524
8	−0.662	0.284	−0.127	0.044
9	−0.229	0.568	−0.463	−0.088
10	−0.851	0.515	0.344	−1.333
11	−3.361	−0.436	−1.412	−1.977
12	−0.912	0.567	−1.525	0.120
13	1.594	1.040	−0.017	1.558
14	1.618	0.064	−0.525	0.904
15	−1.260	−1.051	−2.689	−1.437
16	0.288	−1.845	−0.211	0.427
17	0.858	0.281	2.145	0.061
18	−1.752	−0.136	0.787	0.120
19	−0.960	−0.992	−0.452	1.460
20	1.738	0.321	1.267	−0.493
21	−1.008	2.621	2.316	−0.888
22	−1.589	2.804	0.258	−0.530
23	0.289	2.174	−1.645	−2.757
24	−0.580	1.897	−1.552	−1.452
25	1.213	−0.781	−0.213	0.158

As the first step in the analysis, we fit a seventh order autoregressive model to the data. This yields

$$\hat{Y}_t = \underset{(0.108)}{0.685}\, Y_{t-1} - \underset{(0.131)}{0.584}\, Y_{t-2} + \underset{(0.149)}{0.400}\, Y_{t-3} - \underset{(0.152)}{0.198}\, Y_{t-4}$$

$$+ \underset{(0.149)}{0.020}\, Y_{t-5} + \underset{(0.133)}{0.014}\, Y_{t-6} + \underset{(0.108)}{0.017}\, Y_{t-7},$$

where the numbers in parentheses are the estimated standard errors obtained from a standard regression program. The residual mean square for the regression is 1.22. The first few regression coefficients are declining in absolute magnitude with alternating signs. Since the third coefficient exceeds twice its standard error, a second order autoregressive process would be judged an inadequate representation for this realization. Thus, even if one did not know the nature of the process generating the data, the moving average representation would be suggested as a possibility by the

regression coefficients. The regression coefficients estimate the negatives of the c_j, $j \geqslant 1$, of Theorem 2.6.2. By that theorem, $\beta = -c_1$ and $c_j = -\beta c_{j-1}$, $j = 2, 3, \ldots$. Therefore, we arrange the regression coefficients as in Table 8.3.2. The initial estimator of β is obtained by regressing the first row on the second row of that table. This regression yields $\tilde{\beta} = 0.697$.

Table 8.3.2. Observations for regression estimation of an initial estimate for β

j	1	2	3	4	5	6	7
Regression coefficients $(-c_j)$	0.685	−0.584	0.400	−0.198	0.020	0.014	0.017
Multiplier of β	1	−0.685	0.584	−0.400	0.198	−0.020	−0.014

Our initial estimator for e_0 is

$$\tilde{e}_0 = 0.685\,Y_1 - 0.584\,Y_2 + 0.400\,Y_3 - 0.198\,Y_4 + 0.020\,Y_5 + 0.014\,Y_6 + 0.017\,Y_7$$
$$= 0.974.$$

The values of $e_t(Y; 0.697)$ are calculated using (8.3.14), and the values of $W_t(Y; 0.697)$ are calculated using (8.3.15). We have

$$e_t(Y; 0.697) = \begin{cases} Y_1 - 0.697\tilde{e}_0 = 1.432 - 0.697(0.974) = 0.753, & t = 1 \\ Y_t - 0.697 e_{t-1}(Y; 0.697), & t = 2, 3, \ldots, 100 \end{cases}$$

$$W_t(Y; 0.697) = \begin{cases} \tilde{e}_0 = 0.974, & t = 1 \\ e_{t-1}(Y; 0.697) - 0.697\,W_{t-1}(Y; 0.697), & t = 2, 3, \ldots, 100. \end{cases}$$

The first five observations are displayed in Table 8.3.3. Regressing $e_t(Y; \tilde{\beta})$ on $W_t(Y; \tilde{\beta})$ gives a coefficient of 0.037 and an estimate of $\hat{\beta} = 0.734$. The estimated standard error is 0.076 and the residual mean square is 1.16. If

Table 8.3.3. First five observations used in Gauss-Newton computation

t	$e_t(Y; 0.697)$	$W_t(Y; 0.697)$
1	0.753	0.974
2	−0.868	0.074
3	−1.154	−0.920
4	−1.732	−0.513
5	0.913	−1.375

additional iterations are conducted, one obtains the estimator 0.732 and an estimated standard error of 0.069.

The theory we have presented is all for large samples. There is some evidence that samples must be fairly large before the results are applicable. For example, Macpherson (1975) and Nelson (1974) have conducted Monte Carlo studies in which the empirical variances for the estimated first order parameter are about 1.1 to 1.2 times that based on large sample theory for samples of size 100 and $0 \leqslant \rho \leqslant 0.7$. Unlike the estimator for the autoregressive process, the distribution of the nonlinear estimator of β for β near zero differs considerably from that suggested by asymptotic theory for n as large as 100. In the Macpherson study the variance of $\hat{\beta}$ for $\beta = 0.1$ and $n = 100$ was 1.17 times that suggested by asymptotic theory.

In Figure 8.3.1 we compare an estimate of the density of $\hat{\beta}$ for $\beta = 0.9$ with the normal density suggested by the asymptotic theory. The empirical density is based on 5000 samples of size 100 generated by Macpherson. The start value was constructed by the Durbin procedure. The estimator of β is that obtained on the sixth iteration of the modified Gauss-Newton procedure. In the calculations $\hat{\beta}$ was restricted to be less than or equal to

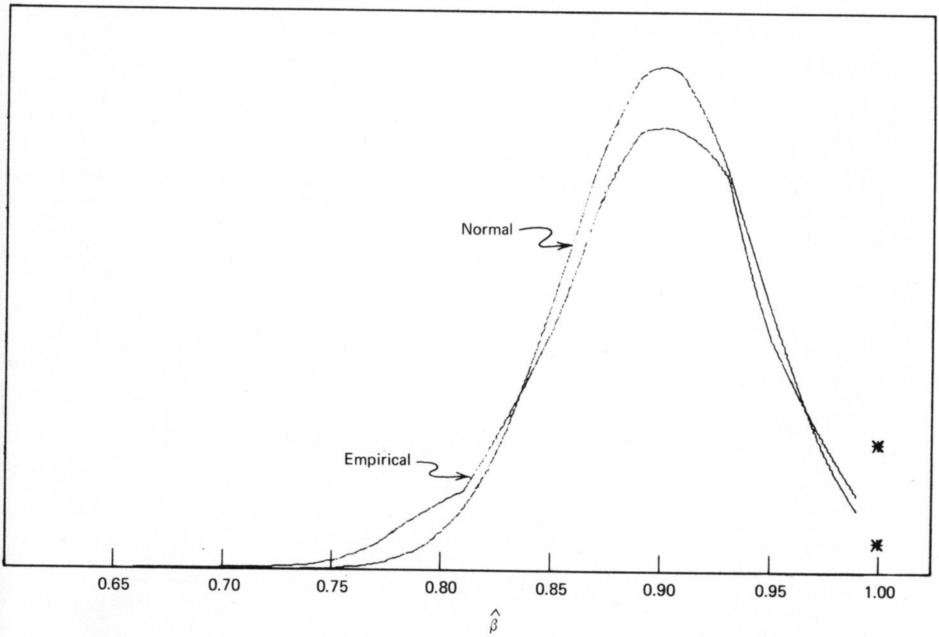

Figure 8.3.1. Estimated density of $\hat{\beta}$ compared with normal approximation for $\beta = = 0.9$ and $n = 100$.

0.999. About 5% of the values achieved this bound. Also, about 1% of the normal distribution with mean 0.9 and variance 0.0019 exceeds one. These two results are responsible for the marks above one in the figure.

The estimated density for $\hat{\beta}$ is fairly symmetric about 0.9 with the exception of the effect of truncating the estimator at 0.999. This is in contrast to the empirical density for the estimator of the autoregressive parameter. However, the empirical density is considerably flatter than the normal approximation. The variance of the empirical distribution is 0.0027 compared to 0.0019 for the normal approximation.

8.4. AUTOREGRESSIVE MOVING AVERAGE TIME SERIES

The procedures of the previous two sections can be generalized to the estimation of the parameters of time series with representation

$$Y_t + \sum_{j=1}^{p} \alpha_j Y_{t-j} = e_t + \sum_{i=1}^{q} \beta_i e_{t-i}, \tag{8.4.1}$$

where the e_t are independent $(0, \sigma^2)$ random variables with $E\{e_t^4\} = \eta\sigma^4$, and the roots of

$$m^p + \sum_{j=1}^{p} \alpha_j m^{p-j} = 0,$$

$$s^q + \sum_{i=1}^{q} \beta_i s^{q-i} = 0$$

are less than one in absolute value.

To obtain the partial derivatives needed in the estimation, we express (8.4.1) as

$$e_t(Y; \boldsymbol{\theta}) = Y_t + \sum_{j=1}^{p} \alpha_j Y_{t-j} - \sum_{i=1}^{q} \beta_i e_{t-i}(Y; \boldsymbol{\theta}), \tag{8.4.2}$$

where $\boldsymbol{\theta}' = (\alpha_1, \alpha_2, \ldots, \alpha_p, \beta_1, \beta_2, \ldots, \beta_q)$. Differentiating both sides of (8.4.2), we have

$$\begin{aligned} -\frac{\partial e_t(Y; \boldsymbol{\theta})}{\partial \alpha_j} &= W_{\alpha_j t}(Y; \boldsymbol{\theta}) \\ &= -Y_{t-j} - \sum_{s=1}^{q} \beta_s W_{\alpha_j, t-s}(Y; \boldsymbol{\theta}), \qquad j = 1, 2, \ldots, p, \\ -\frac{\partial e_t(Y; \boldsymbol{\theta})}{\partial \beta_i} &= W_{\beta_i t}(Y; \boldsymbol{\theta}) \\ &= e_{t-i}(Y; \boldsymbol{\theta}) - \sum_{s=1}^{q} \beta_s W_{\beta_i, t-s}(Y; \boldsymbol{\theta}), \qquad i = 1, 2, \ldots, q. \end{aligned} \tag{8.4.3}$$

Given initial estimators, denoted by $\tilde{\boldsymbol{\theta}}$, improved estimators are given by

$$\hat{\boldsymbol{\theta}} = \tilde{\boldsymbol{\theta}} + \Delta\hat{\boldsymbol{\theta}},$$

where $\Delta\hat{\boldsymbol{\theta}}$ is the vector of estimated regression coefficients for the regression of $e_t(Y;\tilde{\boldsymbol{\theta}})$ on $W_{\alpha_j t}(Y;\tilde{\boldsymbol{\theta}}), j=1,2,\ldots,p$, and $W_{\beta_i t}(Y;\tilde{\boldsymbol{\theta}})$, $i=1,2,\ldots,q$.

Using arguments similar to those of the previous section, it is possible to demonstrate the asymptotic normality of the estimators and that the regression statistics furnish approximations for the parameters of the distribution. The normality result follows because the partial derivatives (8.4.3) evaluated at the true parameters all converge to stationary time series when the original time series is a stationary invertible autoregressive moving average time series.

To obtain initial estimators of the parameters, we use the autoregressive representation technique introduced in Section 8.3. By Theorem 2.7.2, we can represent the invertible autoregressive moving average

$$\sum_{j=0}^{p} \alpha_j Y_{t-j} = \sum_{i=0}^{q} \beta_i e_{t-i}$$

by the infinite autoregressive process

$$Y_t = -\sum_{j=1}^{\infty} d_j Y_{t-j} + e_t,$$

where

$$\begin{aligned} d_1 &= -\beta_1 + \alpha_1, \\ d_2 &= -\beta_1 d_1 - \beta_2 + \alpha_2, \\ &\;\;\vdots \\ d_j &= -\sum_{i=1}^{\min(j,q)} \beta_i d_{j-i} + \alpha_j, \qquad j \leqslant p \\ d_j &= -\sum_{i=1}^{\min(j,q)} \beta_i d_{j-i}, \qquad j > p. \end{aligned} \tag{8.4.4}$$

Hence, by terminating the sum at a convenient finite number, say k, and estimating the autoregressive parameters, $d_1, d_2, \ldots, d_k$, we can use equations (8.4.4) to obtain initial estimates of the parameters.

The estimated parameters can be used to construct initial estimators of e_t

for $t=p-q+1,p-q+2,\dots,p$. Now

$$Z_t = Y_t + \sum_{j=1}^{p} \alpha_j Y_{t-j}$$

is a pure moving average, and we can write

$$Z_t = -\sum_{j=1}^{\infty} c_j Z_{t-j} + e_t,$$

where the c_j are defined in terms of $(\beta_1, \beta_2, \dots, \beta_q)$ by Theorem 2.6.2. Truncating the c_j sequence at a convenient point, say M, we estimate $c_1, c_2, \dots, c_M$ using the initial estimators of $\boldsymbol{\beta}$. Denoting these estimators by $\tilde{c}_j$, we calculate the extrapolated values of Z_t by

$$\tilde{Z}_p = -\sum_{j=1}^{M} \tilde{c}_j \left(Y_{p+j} + \sum_{i=1}^{p} \tilde{\alpha}_i Y_{p+j-i} \right),$$

$$\tilde{Z}_{p-s} = -\sum_{j=1}^{s} \tilde{c}_j \tilde{Z}_{p+j-s} - \sum_{j=s+1}^{M} \tilde{c}_j \left(Y_{p+j} + \sum_{i=1}^{p} \tilde{\alpha}_i Y_{p+j-i} \right), \qquad s=1,2,\dots,q-1,$$

where

$$\tilde{c}_0 = 1$$
$$\tilde{c}_1 = -\tilde{\beta}_1$$
$$\tilde{c}_2 = -\tilde{\beta}_1 \tilde{c}_1 - \tilde{\beta}_2$$
$$\vdots$$
$$\sum_{i=0}^{q} \tilde{\beta}_i \tilde{c}_{j-i} = 0, \qquad j=q,q+1,\dots,$$

$\tilde{\beta}_0 = 1$, and the $\tilde{\alpha}_j$ and $\tilde{\beta}_i$ were obtained from the estimated system (8.4.4). The $\tilde{e}_t$ are then given by

$$\tilde{e}_t = \tilde{Z}_t - \sum_{i=1}^{q} \tilde{\beta}_i \tilde{e}_{t-i}, \qquad t=p-q+1,p-q+2,\dots,p,$$

where $\tilde{e}_t = 0$ for $t \leqslant p-q$. The remaining quantities needed for the calcula-

tion of improved estimators are given by the difference equations

$$\tilde{e}_t(Y;\tilde{\boldsymbol{\theta}}) = \sum_{j=0}^{p} \tilde{\alpha}_j Y_{t-j} - \sum_{i=1}^{q} \tilde{\beta}_i e_{t-i}(Y;\tilde{\boldsymbol{\theta}}), \qquad t=p+1,p+2,\ldots,n,$$

$$W_{\alpha_j t}(Y;\tilde{\boldsymbol{\theta}}) = \begin{cases} 0, & t \leqslant p \\ -Y_{t-j} - \sum_{s=1}^{q} \tilde{\beta}_s W_{\alpha_j,t-s}(Y;\tilde{\boldsymbol{\theta}}), & t=p+1,p+2,\ldots,n, \end{cases}$$

$$W_{\beta_i t}(Y;\tilde{\boldsymbol{\theta}}) = \begin{cases} 0, & t \leqslant p \\ e_{t-i}(Y;\tilde{\boldsymbol{\theta}}) - \sum_{s=1}^{q} \tilde{\beta}_s W_{\beta_i,t-s}(Y;\tilde{\boldsymbol{\theta}}), & t=p+1,p+2,\ldots,n. \end{cases}$$

The estimated changes in the parameter estimates are obtained by regressing $e_t(Y;\tilde{\boldsymbol{\theta}})$ on $W_{\alpha_j t}(Y;\tilde{\boldsymbol{\theta}})$, $j=1,2,\ldots,p$, and $W_{\beta_i t}(Y;\tilde{\boldsymbol{\theta}})$, $i=1,2,\ldots,q$, for $t=p+1,p+2,\ldots,n$.

In using a nonlinear method such as the Gauss-Newton procedure one must beware of certain degeneracies that can occur. For example, consider the autoregressive moving average (1, 1) time series. If we specify zero initial estimates for both parameters, the derivative with respect to α evaluated at $\alpha=\beta=0$ is $-Y_{t-1}$. Likewise, the derivative with respect to β evaluated at $\alpha=\beta=0$ is Y_{t-1} and the matrix of partial derivatives to be inverted is clearly singular. At second glance, this is not a particularly startling result. It means that a first order autoregressive process with small α behaves very much like a first order moving average process with small β and that both behave much like an autoregressive moving average (1, 1) time series where both α and β are small. Therefore, one should consider the autoregressive moving average (1, 1) representation only if at least one of the trial parameter values is well away from zero. Because the autoregressive moving average (1, 1) time series with $\alpha=\beta$ is a sequence of uncorrelated random variables, the singularity occurs whenever the initial values are taken to be equal.

Example. As an example of estimation for a process containing several parameters, we fit an autoregressive moving average to the United States monthly unemployment rate from October 1949 to September 1974 (300 observations). The periodogram of this time series was discussed in Section 7.3. Because of the very large contribution to the total sum of squares from the seasonal frequencies, it seems reasonable to treat the time series as if there is a different mean for each month. Therefore, we analyze the deviations from monthly means, which we denote by Y_t. The fact that the

periodogram ordinates close to the seasonal frequencies are large relative to those separated from the seasonal frequencies leads us to expect a seasonal component in a representation for Y_t.

As the first step in the analysis, we regress Y_t on Y_{t-1}, Y_{t-2}, Y_{t-3}, Y_{t-12}, Y_{t-13}, Y_{t-14}, Y_{t-15}, Y_{t-24}, Y_{t-25}, Y_{t-26}, Y_{t-27}, Y_{t-36}, Y_{t-37}, Y_{t-38}, Y_{t-39}, Y_{t-48}, Y_{t-49}, Y_{t-50}, and Y_{t-51}. Notice that we are anticipating a model of the "component" or "multiplicative" type so that when we include a variable of lag 12, we also include the next three lags. That is, we are anticipating a model of the form

$$\begin{aligned}(1-\theta_1\mathfrak{B}-\theta_2\mathfrak{B}^2-\theta_3\mathfrak{B}^3)(1-\theta_4\mathfrak{B}^{12}-\theta_5\mathfrak{B}^{24}-\theta_6\mathfrak{B}^{36}-\theta_7\mathfrak{B}^{48})Y_t\\ =Y_t-\theta_1Y_{t-1}-\theta_2Y_{t-2}-\theta_3Y_{t-3}-\theta_4Y_{t-12}+\theta_1\theta_4Y_{t-13}+\theta_2\theta_4Y_{t-14}\\ +\cdots+\theta_3\theta_7Y_{t-51}=e_t.\end{aligned} \tag{8.4.5}$$

In calculating the regression equation we added 36 zeros to the beginning of the data set, lagged Y_t the requisite number of times, and used the last 285 observations in the regression. This is a compromise between forms (8.2.6) and (8.2.7) for the estimation of the autoregressive parameters. The regression vectors for the explanatory variables Y_{t-1}, Y_{t-2}, Y_{t-3}, Y_{t-12}, Y_{t-13}, Y_{t-14}, and Y_{t-15} contain all observed values, but the vectors for longer lags contain zeros for some of the initial observations.

The regression coefficients and standard errors are given in Table 8.4.1. The data seem to be consistent with the component type of model. The coefficients for $Y_{t-12i-1}$ are approximately the negative of the coefficients on Y_{t-12i} for $i=1,2,3$. Even more consistently, the sum of the three coefficients for $Y_{t-12i-j}$ for $j=1,2,3$ is approximately the negative of the coefficient for Y_{t-12i}, $i=1,2,3,4$. The individual coefficients show variation about the anticipated relationships, but they give us no reason to reject the component model. The residual mean square for this regression is 0.0634 with 254 degrees of freedom. There are 285 observations in the regression and 19 regression variables. We deduct an additional 12 degrees of freedom for the 12 means previously estimated. We also fit the model with Y_{t-1}, Y_{t-2} and the corresponding lags of 12 as well as the model with Y_{t-1}, Y_{t-2}, Y_{t-3}, Y_{t-4}, Y_{t-5} and the corresponding lags of 12. Since the coefficient on Y_{t-3} is almost twice its standard error, while the coefficients for Y_{t-4} and Y_{t-5} were small, we take the third order autoregressive process as our tentative model for the nonseasonal component.

We next carry out a Gauss-Newton step estimating the multiplicative autoregressive model (8.4.5). We take as our initial estimate $(\tilde{\theta}_1,\tilde{\theta}_2,\tilde{\theta}_3,\tilde{\theta}_4,\tilde{\theta}_5,\tilde{\theta}_6,\tilde{\theta}_7)=(1.08,0.06,-0.16,0.14,0.18,0.08,0.11)$. The deriva-

Table 8.4.1. Regression coefficients obtained in preliminary autoregressive fit to United States monthly unemployment rate

Variable	Coefficient	Standard Error of Coefficient
Y_{t-1}	1.08	0.061
Y_{t-2}	0.06	0.091
Y_{t-3}	−0.16	0.063
Y_{t-12}	0.14	0.060
Y_{t-13}	−0.22	0.086
Y_{t-14}	0.00	0.086
Y_{t-15}	0.06	0.059
Y_{t-24}	0.18	0.053
Y_{t-25}	−0.17	0.075
Y_{t-26}	−0.09	0.075
Y_{t-27}	0.09	0.054
Y_{t-36}	0.08	0.054
Y_{t-37}	−0.13	0.075
Y_{t-38}	0.04	0.075
Y_{t-39}	0.00	0.054
Y_{t-48}	0.11	0.055
Y_{t-49}	−0.01	0.075
Y_{t-50}	−0.01	0.075
Y_{t-51}	−0.09	0.053

tives are

$$\left.\frac{\partial Y_t}{\partial \theta_1}\right|_{\theta=\tilde{\theta}} = Y_{t-1} - \tilde{\theta}_4 Y_{t-13} - \tilde{\theta}_5 Y_{t-25} - \tilde{\theta}_6 Y_{t-37} - \tilde{\theta}_7 Y_{t-49}$$

$$= Y_{t-1} - 0.14\,Y_{t-13} - 0.18\,Y_{t-25} - 0.08\,Y_{t-37} - 0.11\,Y_{t-49}$$

$$\overset{\text{(say)}}{=} M_{t-1}$$

$$\left.\frac{\partial Y_t}{\partial \theta_4}\right|_{\theta=\tilde{\theta}} = Y_{t-12} - \tilde{\theta}_1 Y_{t-13} - \tilde{\theta}_2 Y_{t-14} - \tilde{\theta}_3 Y_{t-15}$$

$$= Y_{t-12} - 1.08\,Y_{t-13} - 0.06\,Y_{t-14} + 0.16\,Y_{t-15}$$

$$\overset{\text{(say)}}{=} L_{t-12}.$$

Regressing

$$\tilde{e}_t = L_t - 0.14L_{t-12} - 0.18L_{t-24} - 0.08L_{t-36} - 0.11L_{t-48}$$

on M_{t-1}, M_{t-2}, M_{t-3}, L_{t-12}, L_{t-24}, L_{t-36}, and L_{t-48}, we obtain the improved estimates $\hat{\boldsymbol{\theta}} = (1.121, 0.056, -0.218, 0.156, 0.168, 0.083, 0.090)'$. The residual mean square for the regression is 0.0644 with 266 degrees of freedom. If we construct the F-statistic as an approximate test for the hypothesis that the restrictions we have imposed are true, we obtain

$$F_{254}^{12} = \frac{0.0862}{0.0634} = 1.36$$

and the hypothesis is accepted at the 0.10 level.

The coefficients for the seasonal component are declining in magnitude, and this suggests the possibility of an autoregressive moving average representation for this component. If the time series has the representation

$$(1 - \theta_1 \mathcal{B} - \theta_2 \mathcal{B}^2 - \theta_3 \mathcal{B}^3)(1 - \delta \mathcal{B}^{12}) Y_t = e_t + \beta e_{t-12}, \qquad (8.4.6)$$

then, using Theorem 2.7.2, the regression coefficients for lags of 12 should satisfy, approximately, the relations of Table 8.4.2. Regressing the first column of that table on the second two columns, we obtain the initial estimates $\tilde{\delta} = 0.957$, $\tilde{\beta} = -0.801$. The estimate of β is of fairly large absolute value to be estimated from only four coefficients, but we proceed without estimating a larger autoregressive model.

Table 8.4.2. Calculation of initial estimates of δ and β

j	Regression Coefficients $(-d_j)$	Multipliers for δ	Multipliers for β
1	0.156	1	1
2	0.168	0	−0.156
3	0.083	0	−0.168
4	0.090	0	−0.083

To obtain initial estimates of e_t, $t = 4, 5, \ldots, 15$, we use the procedure outlined above. That is, we create the time series

$$\begin{aligned} Z_t = {} & Y_t - 1.121 Y_{t-1} - 0.056 Y_{t-2} + 0.218 Y_{t-3} \\ & - 0.957(Y_{t-12} - 1.121 Y_{t-13} - 0.056 Y_{t-14} + 0.218 Y_{t-15}). \end{aligned}$$

On the basis of our initial estimates,

$$Z_t = e_t - 0.801 e_{t-12}$$

and we estimate the e_t by

$$\tilde{e}_t = -\sum_{i=1}^{7} (0.801)^i Z_{t+12i}$$

for $t=4,5,\ldots,15$. The derivatives needed for the first Gauss-Newton iteration are

$$W_{\theta_i t}(Y;\tilde{\lambda}) = Y_{t-i} - 0.957 Y_{t-12-i} + 0.801 W_{\theta_i, t-12}(Y;\tilde{\lambda}), \qquad i=1,2,3,$$

$$W_{\delta t}(Y;\tilde{\lambda}) = Y_{t-12} - 1.121 Y_{t-13} - 0.056 Y_{t-14} + 0.218 Y_{t-15}$$
$$+ 0.801 W_{\delta, t-12}(Y;\tilde{\lambda})$$

$$W_{\beta t}(Y;\tilde{\lambda}) = e_{t-12}(Y;\tilde{\lambda}) + 0.801 W_{\beta, t-12}(Y;\tilde{\lambda}), \quad \text{for} \quad t=16,17,\ldots,300,$$

where $\tilde{\lambda}' = (\tilde{\theta}_1, \tilde{\theta}_2, \tilde{\theta}_3, \tilde{\delta}, \tilde{\beta})$ and it is understood that the W's are zero for $t<16$. Regressing

$$e_t(Y;\tilde{\lambda}) = Y_t - 1.121 Y_{t-1} - 0.056 Y_{t-2} + 0.218 Y_{t-3}$$
$$-0.957(Y_{t-12} - 1.121 Y_{t-13} - 0.056 Y_{t-14} + 0.218 Y_{t-15}) + 0.801 e_{t-12}(Y;\tilde{\lambda})$$

on $W_{\theta_i t}(Y;\tilde{\lambda})$, $i=1,2,3$, $W_{\delta t}(Y;\tilde{\lambda})$, and $W_{\beta t}(Y;\tilde{\lambda})$, we obtain the estimate $\hat{\lambda}' = (1.149, 0.000, -0.193, 0.804, -0.634)$. Additional iteration gives the estimate

$$\hat{\lambda}' = (1.152, -0.002, -0.195, 0.817, -0.651)$$

with estimated standard errors of (0.055, 0.085, 0.053, 0.067, 0.093). The residual mean square error is 0.0626 with 268 degrees of freedom. Since this residual mean square is smaller than that associated with the previous regressions, the hypothesis that the restrictions associated with the autoregressive moving average representation are valid is easily accepted. As a check on model adequacy beyond that given by our initial regressions, we estimated four alternative models with the additional terms, Y_{t-4}, e_{t-1}, Y_{t-24}, e_{t-24}. In no case was the added term significant at the 5% level using the approximate tests based on the regression statistics.

One interpretation of our final model is of some interest. Define $X_t = Y_t - 1.152 Y_{t-1} + 0.002 Y_{t-2} + 0.195 Y_{t-3}$. Then X_t has the autoregressive

moving average representation

$$X_t = 0.817X_{t-12} - 0.651e_{t-12} + e_t,$$

where the e_t are uncorrelated $(0, 0.0626)$ random variables. Now X_t would have this representation if it was the sum of two independent time series $X_t = S_t + v_t$ where v_t is a sequence of uncorrelated $(0, 0.0499)$ random variables,

$$S_t = 0.817S_{t-12} + u_t,$$

and u_t is a sequence of $(0, 0.0059)$ random variables. In such a representation S_t can be viewed as the "seasonal component" and the methods of Section 4.5 could be used to construct a filter to estimate S_t.

8.5. NONSTATIONARY AUTOREGRESSIVE TIME SERIES

In our previous treatment of the estimation of autoregressive parameters we assumed the time series to be stationary with a characteristic equation whose roots were less than one in absolute value. Since not all time series encountered in practice are stationary, we briefly investigate the properties of estimators of the parameters of time series satisfying nonstationary stochastic difference equations. Consider the first order time series

$$Y_t = \rho Y_{t-1} + e_t, \qquad t = 1, 2, \ldots, \tag{8.5.1}$$
$$Y_0 = 0,$$

where the e_t are normal independent $(0, \sigma^2)$ random variables. By Section 8.1 the maximum likelihood estimator of ρ conditional on Y_1, for any real valued ρ, is

$$\hat{\rho} = \frac{\sum_{t=2}^{n} Y_t Y_{t-1}}{\sum_{t=2}^{n} Y_{t-1}^2}, \tag{8.5.2}$$

and

$$\hat{\rho} - \rho = \frac{\sum_{t=2}^{n} Y_{t-1} e_t}{\sum_{t=2}^{n} Y_{t-1}^2}. \tag{8.5.3}$$

If $\rho = 1$, then $Y_t = \Sigma_{j=1}^{t} e_j$ and the mean and variance of the numerator and the denominator of (8.5.3) can be computed. We have

$$E\left\{\sum_{t=2}^{n} Y_{t-1}^2\right\} = E\left\{\sum_{t=1}^{n-1}\left(\sum_{i=1}^{t} e_i\right)^2\right\} = \tfrac{1}{2}n(n-1)\sigma^2 \qquad (8.5.4)$$

$$\operatorname{Var}\left\{\sum_{t=2}^{n} Y_{t-1}^2\right\} = \sum_{t=1}^{n-1} 2(t\sigma^2)^2 + 2\sum_{j=1}^{n-2} 2(n-1-j)(j\sigma^2)^2$$

$$= \tfrac{1}{3}n(n-1)(n^2-n+1)\sigma^4 \qquad (8.5.5)$$

$$\operatorname{Var}\left\{\sum_{t=2}^{n} Y_{t-1}e_t\right\} = \sum_{t=1}^{n-1} t\sigma^4 = \tfrac{1}{2}n(n-1)\sigma^4 \qquad (8.5.6)$$

$$\operatorname{Cov}\left\{\sum_{t=2}^{n} Y_{t-1}e_t, \sum_{t=2}^{n} Y_{t-1}^2\right\} = \tfrac{1}{3}n(n-1)(n-2)\sigma^4. \qquad (8.5.7)$$

These results differ in a notable way from those obtained for $|\rho|<1$. First, with $\rho=1$ the variance of the numerator is order n^2 instead of order n. Second, the mean of the denominator is order n^2 and the variance is order n^4. Thus, the denominator divided by $\frac{1}{2}n(n-1)\sigma^2$ gives a random variable with mean one and variance $4(n^2-n+1)[3(n^2-n)]^{-1}$. Since the variance converges to $\frac{4}{3}$ it is clear that the normalized denominator does not converge to a constant.

Lemma 8.5.1. Let

$$Y_t = \sum_{j=1}^{t} e_j, \qquad t = 1, 2, \ldots,$$

where the e_t are normal independent $(0, \sigma^2)$ random variables, and let

$$\hat{\rho} = \left(\sum_{t=2}^{n} Y_{t-1}^2\right)^{-1} \sum_{t=2}^{n} Y_t Y_{t-1}.$$

Then,

$$\hat{\rho} - 1 = O_p(n^{-1}).$$

Proof. The denominator of $\hat{\rho}$ can be written as a quadratic in the e_t's as

$$\sum_{t=2}^{n} Y_{t-1}^2 = \mathbf{e}'\mathbf{A}\mathbf{e},$$

where $\mathbf{e}' = (e_1, e_2, \ldots, e_{n-1})$ and

$$\mathbf{A} = \begin{bmatrix} n-1 & n-2 & n-3 & \cdots & 1 \\ n-2 & n-2 & n-3 & \cdots & 1 \\ n-3 & n-3 & n-3 & \cdots & 1 \\ \vdots & \vdots & \vdots & & \vdots \\ 1 & 1 & 1 & \cdots & 1 \end{bmatrix}.$$

Therefore, we may write

$$\sum_{t=1}^{n-1} \left(\sum_{j=1}^{t} e_j \right)^2 = \sum_{i=1}^{n-1} \delta_i \xi_i^2,$$

where δ_i are the roots of $\mathbf{A}$ and ξ_i are normal independent $(0, \sigma^2)$ random variables. By moment results (8.5.4) and (8.5.5), we have

$$\sum_{i=1}^{n-1} \delta_i = \frac{n(n-1)}{2}, \tag{8.5.4a}$$

$$\sum_{i=1}^{n-1} \delta_i^2 = \frac{n(n-1)(n^2-n+1)}{6}. \tag{8.5.5a}$$

This means that the largest root δ_M increases at the rate n^2 as n increases. To verify this, note that

$$\begin{aligned} \delta_M &\geq \left(\sum_{i=1}^{n-1} \delta_i^2 \right) \left(\sum_{i=1}^{n-1} \delta_i \right)^{-1} \\ &= \frac{(n^2-n+1)}{3}. \end{aligned}$$

Therefore, for $\epsilon > 0$,

$$\begin{aligned} \text{Prob}\left\{ \sum_{i=1}^{n-1} \delta_i \xi_i^2 < \frac{n(n-1)}{2} \epsilon \right\} &\leq \text{Prob}\left\{ \delta_M \xi_M^2 < \frac{n(n-1)}{2} \epsilon \right\} \\ &\leq \text{Prob}\{ \xi_M^2 < 2\epsilon \}. \end{aligned}$$

Now, given a chi-square random variable, ξ_M^2, and $\Delta>0$, there exists an $\epsilon>0$ such that $P\{\xi_M^2<2\epsilon\}<\Delta$.

It follows that given $\Delta>0$, there exists a $K\epsilon^{-1}$ such that

$$P\left\{\left[\sum_{t=1}^{n}\left(\sum_{j=1}^{t} e_j\right)^2\right]^{-1} n(n-1)>K\epsilon^{-1}\right\}<\Delta$$

for all n and

$$n(n-1)\left[\sum_{t=2}^{n} Y_{t-1}^2\right]^{-1}=O_p(1). \tag{8.5.8}$$

Since the numerator of

$$\hat{\rho}-1=\frac{\sum_{t=2}^{n} Y_{t-1}e_t/n(n-1)}{\sum_{t=2}^{n} Y_{t-1}^2/n(n-1)}$$

is $O_p(n^{-1})$ by (8.5.6), the conclusion follows. ▲

Thus, for $\rho=1$, the estimated value converges in probability to the true value more rapidly than does the estimator for $|\rho|<1$.

The numerator of $\hat{\rho}-1$ can be written in an alternative informative manner as

$$\begin{aligned}\sum_{t=2}^{n} Y_{t-1}e_t &= \sum_{t=2}^{n}\sum_{j=1}^{t-1} e_j e_t\\ &=\frac{1}{2}\left(\sum_{t=1}^{n} e_t\right)^2-\frac{1}{2}\sum_{t=1}^{n} e_t^2\\ &=\frac{1}{2}Y_n^2-\frac{1}{2}\sum_{t=1}^{n} e_t^2.\end{aligned}$$

It follows that, as $n\to\infty$,

$$\frac{2}{n\sigma^2}\sum_{t=2}^{n} Y_{t-1}e_t+1\xrightarrow{\mathcal{L}}\chi_1^2 \tag{8.5.9}$$

where χ_1^2 is the chi-square distribution with one degree of freedom. The probability that a one degree of freedom chi-square random variable is less than one is 0.6826. Therefore, since the denominator is always positive, the probability that $\hat{\rho}<1$ given $\rho=1$ approaches 0.6826 as n gets large. Although a chi-square distribution is skewed to the right, the high correlation between the numerator and denominator of $\hat{\rho}-1$ strongly dampens the skewness. In fact, the distribution of $\hat{\rho}$ displays skewness to the left.

Percentiles for the empirical distribution of $\hat{\rho}$ given $\rho=1$ for $n=25$, 50, 100, 250, and 500 are given in the first part of Table 8.5.1. The total number of observations available is n, while $n-1$ sums of squares and products are used in calculating the regression coefficient. The table may also be used for the distribution of $\hat{\rho}$ for $\rho=-1$, since the distribution for $\rho=-1$ is the mirror image of the distribution for $\rho=1$. To demonstrate that $P\{\hat{\rho}-\rho>a|\rho=1\}=P\{\hat{\rho}-\rho<-a|\rho=-1\}$ for all real a, we show that

$$P\left\{\sum_{t=2}^{n} Y_{t-1}e_t - a\sum_{t=2}^{n} Y_{t-1}^2>0\right\}=P\left\{\sum_{t=2}^{n} X_{t-1}e_t + a\sum_{t=2}^{n} X_{t-1}^2<0\right\},$$

where

$$Y_t=\sum_{j=0}^{t-1} e_{t-j},$$

$$X_t=\sum_{j=0}^{t-1} (-1)^j e_{t-j}.$$

In a manner similar to that used to obtain (8.5.9), we have

$$\begin{aligned}\sum_{t=2}^{n} X_{t-1}e_t &= \sum_{t=2}^{n} e_t \sum_{j=1}^{t-1} (-1)^{j-1} e_{t-j}\\ &= -\sum_{t=2}^{n}\sum_{j=1}^{t-1} (-1)^{t-j} e_j e_t\\ &= -\frac{1}{2}\left[X_n^2-\sum_{t=1}^{n} e_t^2\right].\end{aligned}$$

Therefore,

$$\begin{aligned}&P\left\{\sum_{t=2}^{n} X_{t-1}e_t + a\sum_{t=2}^{n} X_{t-1}^2<0\right\}\\ &\qquad = P\left\{\frac{1}{2}\left\{X_n^2-\sum_{t=1}^{n} e_t^2\right\} - a\sum_{t=1}^{n-1}\left(\sum_{j=0}^{t-1}(-1)^j e_{t-j}\right)^2>0\right\}.\end{aligned}$$

Table 8.5.1. Empirical cumulative distribution of $n(\hat{\rho}-1)$ *for* $\rho=1$

Sample Size	Probability of a Smaller Value							
n	0.01	0.025	0.05	0.10	0.90	0.95	0.975	0.99
				$\hat{\rho}$				
25	−11.9	−9.3	−7.3	−5.3	1.01	1.40	1.79	2.28
50	−12.9	−9.9	−7.7	−5.5	0.97	1.35	1.70	2.16
100	−13.3	−10.2	−7.9	−5.6	0.95	1.31	1.65	2.09
250	−13.6	−10.3	−8.0	−5.7	0.93	1.28	1.62	2.04
500	−13.7	−10.4	−8.0	−5.7	0.93	1.28	1.61	2.04
∞	−13.8	−10.5	−8.1	−5.7	0.93	1.28	1.60	2.03
				$\hat{\rho}_\mu$				
25	−17.2	−14.6	−12.5	−10.2	−0.76	0.01	0.65	1.40
50	−18.9	−15.7	−13.3	−10.7	−0.81	−0.07	0.53	1.22
100	−19.8	−16.3	−13.7	−11.0	−0.83	−0.10	0.47	1.14
250	−20.3	−16.6	−14.0	−11.2	−0.84	−0.12	0.43	1.09
500	−20.5	−16.8	−14.0	−11.2	−0.84	−0.13	0.42	1.06
∞	−20.7	−16.9	−14.1	−11.3	−0.85	−0.13	0.41	1.04
				$\hat{\rho}_\tau$				
25	−22.5	−19.9	−17.9	−15.6	−3.66	−2.51	−1.53	−0.43
50	−25.7	−22.4	−19.8	−16.8	−3.71	−2.60	−1.66	−0.65
100	−27.4	−23.6	−20.7	−17.5	−3.74	−2.62	−1.73	−0.75
250	−28.4	−24.4	−21.3	−18.0	−3.75	−2.64	−1.78	−0.82
500	−28.9	−24.8	−21.5	−18.1	−3.76	−2.65	−1.78	−0.84
∞	−29.5	−25.1	−21.8	−18.3	−3.77	−2.66	−1.79	−0.87

NOTE. This table was constructed by David A. Dickey using the Monte Carlo method. Details are given in Dickey (1975). Standard errors of the estimates vary, but most are less than 0.15 for entries in the left half of the table and less than 0.03 for entries in the right half of the table.

Although the sign of e_{t-j} in the weighted sum $\sum_{j=0}^{t-1}(-1)^j e_{t-j}$ is not the same for all t, the sign is always opposite of that for e_{t-j-1} and e_{t-j+1}, and it follows that

$$\sum_{t=1}^{n-1} X_t^2 = \sum_{t=1}^{n-1}\left(\sum_{j=1}^{t}(-1)^j e_j\right)^2.$$

The distribution of e_t, $t=1,2,\ldots$, is symmetric and hence the distributional properties of the sequence $-e_1, e_2, -e_3, e_4, \ldots$, are precisely the same as the

distributional properties of the sequence $e_1, e_2, e_3, \ldots$, and we conclude that

$$P\left\{\frac{1}{2}\left[\left(\sum_{t=1}^{n} e_t\right)^2 - \sum_{t=1}^{n} e_t^2\right] - a\sum_{t=1}^{n-1}\left(\sum_{j=1}^{t} e_j\right)^2 > 0\right\}$$

$$= P\left\{\frac{1}{2}\left[\left(\sum_{t=1}^{n} (-1)^t e_t\right)^2 - \sum_{t=1}^{n} e_t^2\right] - a\sum_{t=1}^{n-1}\left(\sum_{j=1}^{t} (-1)^j e_j\right)^2 > 0\right\}.$$

A natural statistic to use in testing the hypothesis that $\rho = 1$ is the "t-statistic" one would calculate in ordinary linear regression:

$$\hat{\tau} = \frac{\hat{\rho} - 1}{\left[s^2\left(\sum_{t=2}^{n} Y_{t-1}^2\right)^{-1}\right]^{1/2}}, \tag{8.5.10}$$

where

$$s^2 = \frac{1}{n-2}\sum_{t=2}^{n} \hat{e}_t^2 = \frac{1}{n-2}\sum_{t=2}^{n} (Y_t - \hat{\rho}Y_{t-1})^2.$$

Lemma 8.5.2. Let the model of Lemma 8.5.1 hold. Then $\hat{\tau} = O_p(1)$.

Proof. We have

$$s^2 = \frac{1}{n-2}\sum_{t=2}^{n} \left[e_t - (\hat{\rho} - \rho)Y_{t-1}\right]^2$$

$$= \frac{1}{n-2}\left[\sum_{t=2}^{n} e_t^2 - (\hat{\rho} - \rho)\sum_{t=2}^{n} Y_{t-1}e_t\right]$$

$$= \frac{1}{n-2}\sum_{t=2}^{n} e_t^2 + O_p(n^{-1}).$$

The conclusion follows from (8.5.10), (8.5.8) and Lemma 8.5.1. ▲

Percentage points for the empirical distribution of $\hat{\tau}$ are given in Table 8.5.2.

Table 8.5.2. Empirical cumulative distribution of $\hat{\tau}$ for $\rho = 1$

Sample Size	Probability of a Smaller Value							
n	0.01	0.025	0.05	0.10	0.90	0.95	0.975	0.99
				$\hat{\tau}$				
25	−2.66	−2.26	−1.95	−1.60	0.92	1.33	1.70	2.16
50	−2.62	−2.25	−1.95	−1.61	0.91	1.31	1.66	2.08
100	−2.60	−2.24	−1.95	−1.61	0.90	1.29	1.64	2.03
250	−2.58	−2.23	−1.95	−1.62	0.89	1.29	1.63	2.01
500	−2.58	−2.23	−1.95	−1.62	0.89	1.28	1.62	2.00
∞	−2.58	−2.23	−1.95	−1.62	0.89	1.28	1.62	2.00
				$\hat{\tau}_\mu$				
25	−3.75	−3.33	−3.00	−2.63	−0.37	0.00	0.34	0.72
50	−3.58	−3.22	−2.93	−2.60	−0.40	−0.03	0.29	0.66
100	−3.51	−3.17	−2.89	−2.58	−0.42	−0.05	0.26	0.63
250	−3.46	−3.14	−2.88	−2.57	−0.42	−0.06	0.24	0.62
500	−3.44	−3.13	−2.87	−2.57	−0.43	−0.07	0.24	0.61
∞	−3.43	−3.12	−2.86	−2.57	−0.44	−0.07	0.23	0.60
				$\hat{\tau}_\tau$				
25	−4.38	−3.95	−3.60	−3.24	−1.14	−0.80	−0.50	−0.15
50	−4.15	−3.80	−3.50	−3.18	−1.19	−0.87	−0.58	−0.24
100	−4.04	−3.73	−3.45	−3.15	−1.22	−0.90	−0.62	−0.28
250	−3.99	−3.69	−3.43	−3.13	−1.23	−0.92	−0.64	−0.31
500	−3.98	−3.68	−3.42	−3.13	−1.24	−0.93	−0.65	−0.32
∞	−3.96	−3.66	−3.41	−3.12	−1.25	−0.94	−0.66	−0.33

This table was constructed by David A. Dickey using the Monte Carlo method. Details are given in Dickey (1975). Standard errors of the estimates vary, but most are less than 0.02.

To extend the results for the first order process with $\rho = 1$ to the pth order autoregressive process, we consider the time series

$$Y_t = \sum_{j=1}^{t} Z_j, \qquad t = 1, 2, \ldots, \tag{8.5.11}$$

where $\{Z_t \colon t \in (0, \pm 1, \pm 2, \ldots)\}$ is a $(p-1)$ order autoregressive time series with the representation

$$Z_t + \sum_{i=2}^{p} a_i Z_{t-i+1} = e_t, \tag{8.5.12}$$

the e_t are normal independent $(0,\sigma^2)$ random variables, and the absolute value of the largest root of

$$m^{p-1}+\sum_{i=2}^{p} a_i m^{p-i}=0 \tag{8.5.13}$$

is less than $\lambda<1$. By (8.5.11) and (8.5.12), we may write

$$Y_t+\sum_{j=1}^{p}\alpha_j Y_{t-j}=e_t, \qquad t=p+1,p+2,\ldots, \tag{8.5.14}$$

where $p-1$ of the roots of

$$m^p+\sum_{j=1}^{p}\alpha_j m^{p-j}=0$$

are the $p-1$ roots of (8.5.13) and the remaining root is one. In a problem where a root of one is suspected, it is operationally desirable and theoretically convenient to write the autoregressive equation so that the unit root is isolated as a coefficient. To this end, we write

$$Y_t=\theta_1 Y_{t-1}+\sum_{j=2}^{p}\theta_j(Y_{t-j+1}-Y_{t-j})+e_t, \qquad t=p+1,p+2,\ldots, \tag{8.5.15}$$

where $p\geqslant 2$, $\theta_i=\sum_{j=i}^{p}\alpha_j$, $i=2,3,\ldots,p$, and $\theta_1=-\sum_{j=1}^{p}\alpha_j$. If there is a unit root $\theta_1=1$.

If one knew that $\theta_1=1$, one would regress $Y_t-Y_{t-1}=Z_t$ on $Z_{t-1},Z_{t-2},\ldots,Z_{t-p+1}$ and obtain an estimator of $(\theta_2,\theta_3,\ldots,\theta_p)$, say $(\tilde{\theta}_2,\tilde{\theta}_3,\ldots,\tilde{\theta}_p)$, with the properties described in Section 8.2. It is interesting that given $\theta_1=1$ the regression of Y_t on $Y_{t-1},Z_{t-1},\ldots,Z_{t-p+1}$ yields an estimator of (θ_1-1), say $(\tilde{\theta}_1-1)$, such that the large sample distribution of $nc(\tilde{\theta}_1-1)$ is the same as that of $n(\hat{\rho}-1)$, where $\hat{\rho}$ is the estimator (8.5.2) and c is a constant defined below. The large sample distribution of $(\tilde{\theta}_2,\tilde{\theta}_3,\ldots,\tilde{\theta}_p)$ is the same as that of the regression coefficients in the regression of Z_t on $Z_{t-1},Z_{t-2},\ldots,Z_{t-p+1}$.

Theorem 8.5.1. Let Y_t be defined by (8.5.11). Let $\hat{\boldsymbol{\theta}}=(\hat{\theta}_1,\hat{\theta}_2,\ldots,\hat{\theta}_p)'$ be a vector of regression coefficients obtained by regressing Y_t on $Y_{t-1},Z_{t-1},\ldots,Z_{t-p+1}$, $t=p+1,p+2,\ldots,n$, and define $\mathbf{D}_n$ to be the diagonal matrix

$$\mathbf{D}_n=\operatorname{diag}\{n,n^{1/2},n^{1/2},\ldots,n^{1/2}\}.$$

Define

$$c=\sum_{i=0}^{\infty} w_i,$$

where

$$Z_t=\sum_{i=0}^{\infty} w_i e_{t-i}$$

and the w_i satisfy the homogeneous difference equation with characteristic equation (8.5.13) and initial conditions, $w_0 = 1$, $w_i = 0$, $i < 0$. Let

$$\tilde{\boldsymbol{\theta}} - \boldsymbol{\theta} = \left[\left(c \sum_{t=2}^{n} W_{t-1}^2 \right)^{-1} \sum_{t=2}^{n} W_{t-1} e_t, \tilde{\theta}_2 - \theta_2, \tilde{\theta}_3 - \theta_3, \ldots, \tilde{\theta}_p - \theta_p \right]'$$

where $W_t = \Sigma_{j=1}^{t} e_j$, and $(\tilde{\theta}_2, \tilde{\theta}_3, \ldots, \tilde{\theta}_p)'$ is the vector of regression coefficients obtained by regressing Z_t on $Z_{t-1}, Z_{t-2}, \ldots, Z_{t-p+1}$. Then $\mathbf{D}_n(\hat{\boldsymbol{\theta}} - \tilde{\boldsymbol{\theta}}) = O_p(n^{-1/2})$.

Proof. We can write

$$\begin{aligned} Y_t &= \sum_{j=1}^{t} \sum_{i=0}^{\infty} w_i e_{j-i} \\ &= \sum_{j=1}^{t} \left(\sum_{i=0}^{\infty} w_i \right) e_j - \sum_{j=1}^{t} \sum_{i=j}^{\infty} w_i e_{t-j+1} + \sum_{j=0}^{\infty} e_{-j} \sum_{i=j+1}^{j+t} w_i \\ &= cW_t + U_t + R_t, \end{aligned}$$

where $U_t = \Sigma_{j=1}^{t} g_j e_{t-j+1}$, $g_j = -\Sigma_{i=j}^{\infty} w_i$, and $R_t = \Sigma_{j=0}^{\infty} e_{-j} \Sigma_{i=j+1}^{j+t} w_i$. Since the absolute value of the largest root of (8.5.13) is less than $\lambda < 1$, by Exercise 2.24,

$$|w_i| < M_1 \lambda^i, \qquad i = 0, 1, \ldots,$$

$$|g_j| \leq \sum_{i=j}^{\infty} |w_i| < M_2 \lambda^j, \quad j = 1, 2, \ldots,$$

$$|E\{U_t U_s\}| = \left| \sum_{j=1}^{\min(t,s)} g_j g_{j+|t-s|} \sigma^2 \right| < M_3 \lambda^{|t-s|},$$

$$|E\{Z_t Z_{t+h}\}| < M_4 \lambda^{|h|},$$

$$|E\{W_t U_s\}| \leq \sum_{j=1}^{\infty} |g_j| \sigma^2,$$

$$|E\{Y_t Z_s\}| = \left| \sum_{i=1}^{t} E\{Z_i Z_s\} \right| < M_5,$$

$$|E\{Y_t Y_s\}| = \left| \sum_{i=1}^{t} \sum_{j=1}^{s} E\{Z_j Z_i\} \right| < nM_5 \quad \text{for} \quad t, s, \leq n,$$

$$E\{R_t^2\} < \sum_{j=1}^{\infty} \left[\sum_{i=j}^{\infty} |w_i| \right]^2 \sigma^2 < M_6,$$

where $\{M_i: i=1,2,\ldots,6\}$ are positive constants. If follows that

$$E\left\{\sum_{t=1}^{n} Y_t^2 - c^2 \sum_{t=1}^{n} W_t^2\right\} = E\left\{\sum_{t=1}^{n}\left[2cW_tU_t + U_t^2 + 2cW_tR_t + 2U_tR_t + R_t^2\right]\right\}$$

$$\leqslant \sum_{t=1}^{n}\left[2c\sum_{j=1}^{t} g_j + \sum_{j=1}^{t} g_j^2\right]\sigma^2 + nM_6 = O(n),$$

$$E\left\{\sum_{t=1}^{n} Y_tZ_t\right\} = \sum_{t=1}^{n} E\{Y_tZ_t\} = O(n),$$

$$E\{(Y_{t-1} - cW_{t-1})e_t\} = 0,$$

$$\operatorname{Var}\left\{\sum_{t=1}^{n} Y_t^2 - c^2\sum_{t=1}^{n} W_t^2\right\} = O(n^3),$$

$$\operatorname{Var}\left\{\sum_{t=1}^{n} Y_tZ_t\right\} = \sum_{t=1}^{n}\sum_{s=1}^{n}\left[E\{Y_tY_s\}E\{Z_tZ_s\}\right.$$
$$\left. + E\{Y_tZ_s\}E\{Y_sZ_t\}\right] = O(n^2),$$

$$\operatorname{Var}\left\{\sum_{t=2}^{n} U_{t-1}e_t\right\} = O(n),$$

$$\operatorname{Var}\left\{\sum_{t=2}^{n} R_{t-1}e_t\right\} = O(n).$$

Thus,

$$n^{-2}\left[\sum_{t=p+1}^{n} Y_{t-1}^2 - c^2\sum_{t=p+1}^{n} W_{t-1}^2\right] = O_p(n^{-1/2}),$$

$$n^{-3/2}\left[\sum_{t=p+1}^{n} Y_{t-1}Z_{t-j}\right] = O_p(n^{-1/2}), \qquad j=1,2,\ldots,p-1.$$

Let $\mathbf{A}_n$ be the matrix of sums of squares and cross products of $Y_{t-1}, Z_{t-1}, Z_{t-2}, \ldots, Z_{t-p+1}$, and let $\mathbf{B}_n$ be the matrix defined by

$$\mathbf{B}_n = \begin{bmatrix} c^2\sum_{t=p+1}^{n} W_{t-1}^2 & \mathbf{0} \\ \mathbf{0} & \mathbf{C}_n \end{bmatrix},$$

where $\mathbf{C}_n$ is the $(p-1)\times(p-1)$ matrix of sums of squares and cross products of $Z_{t-1}, Z_{t-2}, \ldots, Z_{t-p+1}$. The lower right $(p-1)\times(p-1)$ submatrix of $\mathbf{A}_n$ is $\mathbf{C}_n$. Furthermore, the ijth element of $n^{-1}\mathbf{C}_n$ is converging in probability to $\gamma_Z(i-j)$. By the moments (8.5.5) and the proof of Lemma 8.5.1, $\mathbf{D}_n^{-1}\mathbf{B}_n\mathbf{D}_n^{-1}=O_p(1)$ and $\mathbf{D}_n\mathbf{B}_n^{-1}\mathbf{D}_n=O_p(1)$. Therefore, by the results above,

$$\mathbf{D}_n(\mathbf{A}_n^{-1}-\mathbf{B}_n^{-1})\mathbf{D}_n=O_p(n^{-1/2}).$$

Similarly,

$$n^{-1}\left[\sum_{t=p+1}^{n} Y_{t-1}e_t - c\sum_{t=p+1}^{n} W_{t-1}e_t\right]=O_p(n^{-1/2})$$

and

$$\frac{n^{-1}\sum_{t=p+1}^{n} Y_{t-1}e_t}{n^{-2}\sum_{t=p+1}^{n} Y_{t-1}^2}=\frac{nc\sum_{t=p+1}^{n} W_{t-1}e_t}{c^2\sum_{t=p+1}^{n} W_{t-1}^2}+O_p(n^{-1/2}).$$

▲

Corollary 8.5.1. Let the assumptions of Theorem 8.5.1 hold. Let

$$K_n^2=\operatorname{diag}\left\{\sum_{t=p+1}^{n} Y_{t-1}^2, \sum_{t=p+1}^{n} Z_{t-1}^2, \sum_{t=p+1}^{n} Z_{t-2}^2, \ldots, \sum_{t=p+1}^{n} Z_{t-p+1}^2\right\},$$

$$\tilde{K}_n^2=\operatorname{diag}\left\{c\sum_{t=p+1}^{n} W_{t-1}^2, \sum_{t=p+1}^{n} Z_{t-1}^2, \sum_{t=p+1}^{n} Z_{t-2}^2, \ldots, \sum_{t=p+1}^{n} Z_{t-p+1}^2\right\}.$$

Then,

$$\mathbf{K}_n(\hat{\boldsymbol{\theta}}-\boldsymbol{\theta})-\tilde{\mathbf{K}}_n(\tilde{\boldsymbol{\theta}}-\boldsymbol{\theta})=O_p(n^{-1/2})$$

where W_t, c, $\hat{\boldsymbol{\theta}}$, and $\tilde{\boldsymbol{\theta}}$ are defined in Theorem 8.5.1.

Proof. By the proof of Theorem 8.5.1,

$$n^{-1}\left(\sum_{t=p+1}^{n} Y_{t-1}^2\right)^{1/2}=n^{-1}\left(c^2\sum_{t=p+1}^{n} W_{t-1}^2\right)^{1/2}+O_p(n^{-1/2}),$$

$$n^{-1}\left(\sum_{t=p+1}^{n} Y_{t-1}^2\right)^{1/2}=O_p(1).$$

Since

$$p\lim \frac{1}{n} \sum_{t=p+1}^{n} Z_{t-1}^2 = \gamma_Z(0),$$

the result follows. ▲

On the basis of Corollary 8.5.1, Table 8.5.2 can be used to investigate the hypothesis that one of the roots in a pth order autoregressive process is unity.

For model (8.5.1) with $|\rho| < 1$, the limiting behavior of the estimator of ρ is the same whether the mean is known or estimated. The result is no longer true when $\rho = 1$. Consider the estimator

$$\hat{\rho}_\mu = \frac{\sum_{t=2}^{n} (Y_t - \bar{y}_{(0)})(Y_{t-1} - \bar{y}_{(-1)})}{\sum_{t=2}^{n} (Y_{t-1} - \bar{y}_{(-1)})^2}, \tag{8.5.16}$$

where

$$\bar{y}_{(0)} = \frac{1}{n-1} \sum_{t=2}^{n} Y_t,$$

$$\bar{y}_{(-1)} = \frac{1}{n-1} \sum_{t=2}^{n} Y_{t-1}.$$

One can establish that

$$E\left\{ \sum_{t=2}^{n} (Y_{t-1} - \bar{y}_{(-1)}) e_t \right\} = -\frac{(n-2)\sigma^2}{2}$$

$$\operatorname{Var}\left\{ \sum_{t=2}^{n} (Y_{t-1} - \bar{y}_{(-1)}) e_t \right\} = \frac{(n+6)(n-2)\sigma^4}{12}$$

$$E\left\{ \sum_{t=2}^{n} (Y_{t-1} - \bar{y}_{(-1)})^2 \right\} = \frac{n(n-2)\sigma^2}{6}$$

$$\operatorname{Var}\left\{ \sum_{t=2}^{n} (Y_{t-1} - \bar{y}_{(-1)})^2 \right\} = \frac{n(n-2)(2n^2 - 4n + 9)\sigma^4}{90}$$

$$\operatorname{Cov}\left\{ \sum_{t=2}^{n} (Y_t - \bar{y}_{(0)})(Y_{t-1} - \bar{y}_{(-1)}), \sum_{t=2}^{n} (Y_{t-1} - \bar{y}_{(-1)})^2 \right\} = -\frac{n(n-2)\sigma^4}{6}.$$

From these moment results we see that the distribution of $\hat{\rho}_\mu$ differs from that of $\hat{\rho}$ when $\rho = 1$. Arguments analogous to those of Lemma 8.5.1 can be used to demonstrate that $\hat{\rho}_\mu - 1 = O_p(n^{-1})$ when $\rho = 1$. The second part of Table 8.5.1 contains empirical percentiles for the distribution of $n(\hat{\rho}_\mu - 1)$, and the second part of Table 8.5.2 contains the empirical percentiles of the corresponding studentized statistic given that $\rho = 1$.

It can be demonstrated that

$$p \lim n(\hat{\rho}_\mu - \hat{\rho}) = 0$$

when $\rho = -1$. Therefore, the first part of Tables 8.5.1 and 8.5.2 can be used to approximate the distributions of $\hat{\rho}_\mu$ and $\hat{\tau}_\mu$ when $\rho = -1$.

The results for the first order autoregressive case with estimated intercept extend to the pth order process in much the same manner as the fixed intercept case.

Theorem 8.5.2. Let Y_t be defined by (8.5.11). Let $\hat{\boldsymbol{\theta}} = (\hat{\theta}_1, \hat{\theta}_2, \ldots, \hat{\theta}_p)'$ be the vector of regression coefficients obtained by regressing Y_t on $Y_{t-1}, Z_{t-1}, \ldots, Z_{t-p+1}$, $t = p+1, p+2, \ldots, n$, with an intercept term in the regression. Define $\mathbf{D}_n$ to be the diagonal matrix

$$\mathbf{D}_n = \operatorname{diag}\{n, n^{1/2}, n^{1/2}, \ldots, n^{1/2}\}$$

and let

$$\tilde{\boldsymbol{\theta}} - \boldsymbol{\theta} = [\tilde{\theta}_1 - \theta_1, \tilde{\theta}_2 - \theta_2, \tilde{\theta}_3 - \theta_3, \ldots, \tilde{\theta}_p - \theta_p]',$$

where

$$\tilde{\theta}_1 - \theta_1 = \left[c \sum_{t=2}^{n} (W_{t-1} - \bar{w}_{(-1)})^2 \right]^{-1} \sum_{t=2}^{n} (W_{t-1} - \bar{w}_{(-1)}) e_t,$$

$$\bar{w}_{(-1)} = \frac{1}{n-1} \sum_{t=1}^{n-1} W_t,$$

W_t and c are defined in Theorem 8.5.1, and $(\tilde{\theta}_2, \tilde{\theta}_3, \ldots, \tilde{\theta}_p)$ is the vector of regression coefficients obtained by regressing Z_t on $Z_{t-1}, Z_{t-2}, \ldots, Z_{t-p+1}$, with an intercept term in the regression. Then,

$$\mathbf{D}_n(\hat{\boldsymbol{\theta}} - \tilde{\boldsymbol{\theta}}) = O_p(n^{-1/2}).$$

Proof. Following Theorem 8.5.1 we write

$$Y_t = cW_t + U_t + R_t,$$

$$Y_t - \bar{y}_{(-1)} = c(W_t - \bar{w}_{(-1)}) + U_t - \bar{u}_{(-1)} + R_t - \bar{r}_{(-1)},$$

where U_t and R_t are defined in the proof of Theorem 8.5.1, and

$$\bar{r}_{(-1)} = \frac{1}{n-1} \sum_{j=0}^{\infty} e_{-j} \sum_{t=1}^{n-1} \sum_{i=j+1}^{j+t} w_i = \sum_{j=0}^{\infty} e_{-j} \sum_{i=j+1}^{j+n-1} \frac{n+j-i}{n-1} w_i,$$

$$R_t - \bar{r}_{(-1)} = \sum_{j=0}^{\infty} e_{-j} \left[\sum_{i=j+1}^{j+t} \frac{i-j-1}{n-1} w_i - \sum_{i=j+t+1}^{j+n-1} \frac{n+j-i}{n-1} w_i \right].$$

We have

$$\sum_{j=0}^{\infty} \left[\sum_{i=j+1}^{j+n-1} \frac{|i-j-1|}{n-1} |w_i| \right]^2 = O(n^{-2})$$

and

$$\begin{aligned} \operatorname{Var}\{R_t - \bar{r}_{(-1)}\} &\leqslant \sum_{j=0}^{\infty} \left(\sum_{i=j+t+1}^{\infty} |w_i| \right)^2 \sigma^2 + O(n^{-2}) \\ &\leqslant M_7 \lambda^{2t} + O(n^{-2}), \end{aligned}$$

where M_7 is a positive constant. Using arguments similar to those of Theorem 8.5.1, one can demonstrate that

$$n^{-2} \left[\sum_{t=p+1}^{n} (Y_{t-1} - \bar{y}_{(-1)})^2 - c^2 \sum_{t=p+1}^{n} (W_{t-1} - \bar{w}_{(-1)})^2 \right] = O_p(n^{-1})$$

$$n^{-3/2} \left[\sum_{t=p+1}^{n} (Y_{t-1} - \bar{y}_{(-1)})(Z_{t-j} - \bar{z}_{(-j)}) \right] = O_p(n^{-1/2}),$$

$$j = 1, 2, \ldots, p-1,$$

$$n^{-1} \left[\sum_{t=p+1}^{n} (Y_{t-1} - \bar{y}_{(-1)}) e_t - c \sum_{t=p+1}^{n} (W_{t-1} - \bar{w}_{(-1)}) e_t \right] = O_p(n^{-1/2}).$$

We illustrate the arguments by considering $\sum_{t=2}^{n} W_{t-1}(R_{t-1} - \bar{r}_{(-1)})$. Since

R_t and W_s are independent for $t,s \geqslant 1$, we have

$$E\left\{\sum_{t=2}^{n} W_{t-1}(R_{t-1}-\bar{r}_{(-1)})\right\}=0,$$

$$\operatorname{Var}\left\{\sum_{t=2}^{n} W_{t-1}(R_{t-1}-\bar{r}_{(-1)})\right\}$$

$$=\sum_{t=2}^{n}\sum_{s=2}^{n} \operatorname{Cov}\{W_{t-1},W_{s-1}\}\operatorname{Cov}\{R_{t-1}-\bar{r}_{(-1)},R_{s-1}-\bar{r}_{(-1)}\}$$

$$\leqslant n\sigma^2 M_7 \sum_{t=2}^{n}\sum_{s=2}^{n}\left[\lambda^{2t}+\lambda^{2s}+O(n^{-2})\right]=O(n^2).$$

The conclusion follows by the arguments of Theorem 8.5.1. ▲

Corollary 8.5.2. Let the assumptions of Theorem 8.5.2 hold. Let

$$\mathbf{M}_n=\operatorname{diag}\left\{\left(\sum_{t=p+1}^{n}(Y_{t-1}-\bar{y}_{(-1)})^2\right)^{1/2},\left(\sum_{t=p+1}^{n}(Z_{t-1}-\bar{z}_{(-1)})^2\right)^{1/2},\ldots,\right.$$
$$\left.\left(\sum_{t=p+1}^{n}(Z_{t-p+1}-\bar{z}_{(-p+1)})^2\right)^{1/2}\right\},$$

$$\tilde{\mathbf{M}}_n=\operatorname{diag}\left\{c\left(\sum_{t=p+1}^{n}(W_{t-1}-\bar{w}_{(-1)})^2\right)^{1/2},\left(\sum_{t=p+1}^{n}(Z_{t-1}-\bar{z}_{(-1)})^2\right)^{1/2},\ldots,\right.$$
$$\left.\left(\sum_{t=p+1}^{n}(Z_{t-p+1}-\bar{z}_{(-p+1)})^2\right)^{1/2}\right\}.$$

Then,

$$\mathbf{M}_n(\hat{\boldsymbol{\theta}}-\boldsymbol{\theta})-\tilde{\mathbf{M}}_n(\tilde{\boldsymbol{\theta}}-\boldsymbol{\theta})=O_p(n^{-1/2}).$$

Proof. See the proof of Corollary 8.5.1. ▲

A competitor for the hypothesis that a time series is nonstationary is sometimes the hypothesis that the mean is a linear function of time. In such a situation one might postulate the regression equation

$$Y_t=\theta_0+\theta_1 t+\rho Y_{t-1}+e_t,\qquad t=1,2,\ldots,$$

where $\{e_t\}$ is a sequence of normal independent $(0, \sigma^2)$ random variables. The third part of Table 8.5.1 gives the empirical percentage points for $\hat{\rho}_\tau$ estimated from such a regression equation given that $\rho = 1$ and $\theta_1 = 0$. The subscript τ is used to denote the fact that ρ is estimated in a regression equation containing time as an explanatory variable. It is worth noting that the probability is greater than 99% that $\hat{\rho}$ is less than one for all of the sample sizes reported in the table. On the other hand, the normalization of $(\hat{\rho}_\tau - 1)$ by n produces a distribution that depends little on the sample size. It is possible to extend Theorem 8.5.2 and Corollary 8.5.2 to the case of an autoregression containing time as an independent variable. This means that the third part of Table 8.5.2 can be used to test the hypothesis that the largest root in a pth order autoregressive equation is one. If the hypothesis is that the root of largest absolute value is negative one, then the first part of Table 8.5.2 furnishes an approximation to the distribution of the τ-statistic.

White (1958) studied the estimator (8.5.2) of ρ in the model (8.5.1) for $\rho \geqslant 1$. He demonstrated that when $|\rho| > 1$ the limiting distribution of $|\rho|^n (\rho^2 - 1)^{-1}(\hat{\rho} - \rho)$ is Cauchy. White also obtained the limit of the moment generating function when $\rho = 1$. Anderson (1959) investigated the estimators and noted that White's results for $\rho > 1$ implied that the limiting distribution of $(\Sigma_{t=1}^n Y_{t-1}^2)^{1/2}(\hat{\rho} - \rho)$ is normal. Rao (1961) discussed the estimation of ρ, $|\rho| > 1$, in higher order autoregressive schemes.

Example. To illustrate an application of the results of this section, we use the unemployment time series studied in Section 8.2. If we regress Y_t on Y_{t-1} and $Y_{t-1} - Y_{t-2}$, including an intercept in the regression, we obtain

$$\hat{Y}_t = \underset{(0.14)}{0.63} + \underset{(0.028)}{0.87}\; Y_{t-1} + \underset{(0.073)}{0.70}\; (Y_{t-1} - Y_{t-2}),$$

where the numbers in parentheses are the standard errors computed by the ordinary regression formulas. To test the hypothesis that one of the roots of the autoregressive process is one, we form

$$\hat{\tau}_\mu = \frac{0.87 - 1.00}{0.028} = -4.64.$$

On the basis of Table 8.5.2 the hypothesis is rejected at the 0.01 level.

8.6. PREDICTION WITH ESTIMATED PARAMETERS

We now investigate the use of the estimated parameters of autoregressive moving average time series in prediction. Prediction was introduced in

Section 2.9 assuming the parameters to be known. The estimators of the parameters of stationary finite order autoregressive invertible moving average time series discussed in this chapter possess errors whose order in probability is $n^{-1/2}$. For such time series, the use of the estimated parameters in prediction increases the prediction error by a quantity of $O_p(n^{-1/2})$.

Let the time series Y_t be defined by

$$Y_t + \sum_{j=1}^{p} \alpha_j Y_{t-j} = \sum_{i=1}^{q} \beta_i e_{t-i} + e_t, \tag{8.6.1}$$

where the roots of

$$m^p + \sum_{j=1}^{p} \alpha_j m^{p-j} = 0$$

and of

$$r^q + \sum_{i=1}^{q} \beta_i r^{q-i} = 0$$

are less than one in absolute value and the e_t are independent $(0,\sigma^2)$ random variables with $E\{e_t^4\} = \eta\sigma^4$. Let

$$\boldsymbol{\theta}' = (-\alpha_1, -\alpha_2, \ldots, -\alpha_p, \beta_1, \beta_2, \ldots, \beta_q)$$

denote the vector of parameters of the process.

When $\boldsymbol{\theta}$ is known, the one period ahead predictor for Y_t is

$$\hat{Y}_{n+1}(Y_1, \ldots, Y_n) = -\sum_{j=1}^{p} \alpha_j Y_{n+1-j} + \sum_{i=1}^{q} \beta_i \tilde{e}_{n+1-i}(Y; \boldsymbol{\theta}), \tag{8.6.2}$$

where

$$\tilde{e}_t(Y; \boldsymbol{\theta}) = \begin{cases} 0, & t = p-q+1, p-q+2, \ldots, p \\ Y_t + \sum_{j=1}^{p} \alpha_j Y_{t-j} - \sum_{i=1}^{q} \beta_i \tilde{e}_{t-i}(Y; \boldsymbol{\theta}), & t = p+1, p+2, \ldots, n. \end{cases}$$

The predictor obtained by replacing α_j and β_i in (8.6.2) by $\hat{\alpha}_j$ and $\hat{\beta}_i$ is denoted by

$$\tilde{Y}_{n+1}(Y_1, \ldots, Y_n) = -\sum_{j=1}^{p} \hat{\alpha}_j Y_{n+1-j} + \sum_{i=1}^{q} \hat{\beta}_i \tilde{e}_{n+1-i}(Y; \hat{\boldsymbol{\theta}}). \tag{8.6.3}$$

Theorem 8.6.1. Let Y_t be the time series defined in (8.6.1). Let $\hat{\boldsymbol{\theta}}$ be an estimator of $\boldsymbol{\theta}=(-\alpha_1,\ldots,-\alpha_p,\beta_1,\ldots,\beta_q)'$ such that $\hat{\boldsymbol{\theta}}-\boldsymbol{\theta}=O_p(n^{-1/2})$. Then

$$\hat{Y}_{n+1}(Y_1,\ldots,Y_n)-\tilde{Y}_{n+1}(Y_1,\ldots,Y_n)=O_p(n^{-1/2}),$$

where $\hat{Y}_{n+1}(Y_1,\ldots,Y_n)$ is defined in (8.6.2) and $\tilde{Y}_{n+1}(Y_1,\ldots,Y_n)$ in (8.6.3).

Proof. We write

$$\begin{aligned}\tilde{Y}_{n+1}(Y_1,\ldots,Y_n)=&-\sum_{j=1}^{p}\alpha_j Y_{n+1-j}-\sum_{j=1}^{p}(\hat{\alpha}_j-\alpha_j)Y_{n+1-j}\\&+\sum_{i=1}^{q}\beta_i\tilde{e}_{n+1-i}(Y;\boldsymbol{\theta})\\&+\sum_{k=1}^{p+q}\frac{\partial\sum_{i=1}^{q}\beta_i^{\dagger}\tilde{e}_{n+1-i}(Y;\boldsymbol{\theta}^{\dagger})}{\partial\theta_k}(\theta_k-\hat{\theta}_k),\end{aligned}$$

where $\boldsymbol{\theta}^{\dagger}$ is between $\hat{\boldsymbol{\theta}}$ and $\boldsymbol{\theta}$ and, for example,

$$\frac{\partial\beta_1^{\dagger}\tilde{e}_n(Y;\boldsymbol{\theta}^{\dagger})}{\partial\beta_1}=\tilde{e}_n(Y;\theta^{\dagger})-\beta_1^{\dagger}\left[\tilde{e}_{n-1}(Y;\boldsymbol{\theta}^{\dagger})+\sum_{i=1}^{q}\beta_i^{\dagger}\frac{\partial\tilde{e}_{n-i}(Y;\boldsymbol{\theta}^{\dagger})}{\partial\beta_1}\right]$$

and

$$\tilde{e}_n(Y;\boldsymbol{\theta}^{\dagger})=\sum_{j=0}^{p}\alpha_j^{\dagger}Y_{n-j}-\sum_{i=1}^{q}\beta_i^{\dagger}\tilde{e}_{n-i}(Y;\boldsymbol{\theta}^{\dagger}).$$

For $\boldsymbol{\theta}$ such that the roots of the characteristic equations are less than one in absolute value, the derivatives multiplying $\theta_k-\hat{\theta}_k$ are, except for the initial effects, stationary time series. Since $(\hat{\theta}_k-\theta_k)=O_p(n^{-1/2})$, the result follows. ▲

Theorem 8.6.1 generalizes immediately to predictions s periods ahead. On the basis of this result, the prediction variance formulas of Section 2.9 can be used as approximations for the estimator $\tilde{Y}_{n+s}(Y_1,\ldots,Y_n)$.

REFERENCES

Section 8.1. Jobson (1972), Koopmans (1942).

Section 8.2. Anderson (1959), (1962), (1971), Box and Pierce (1970), Draper and

Smith (1966), Hannan (1970), Kendall (1954), Mann and Wald (1943a), Marriott and Pope (1954), Salem (1971).

Sections 8.3, 8.4. Berk (1974), Box and Jenkins (1970), Durbin (1959), Kendall and Stuart (1966), Macpherson (1975), Nelson (1974), Pierce (1970a), Walker (1961), Wold (1938).

Section 8.5. Anderson (1959), Dickey (1975), Rao (1961), Reeves (1972), Stigum (1974), White (1958).

EXERCISES

1. Using the first 25 observations of Table 7.3.1, estimate ρ and λ using equations (8.1.7). Compute the estimated standard errors of $\hat{\rho}$ and $\hat{\lambda}$ using the standard regression formulas. Using $\hat{\rho}$ and $\hat{\mu}$ of (8.1.14) as initial values carry out the iteration (8.1.15) to obtain the approximate unconditional maximum likelihood estimates.

2. Let Y_t be a stationary time series. Compare the limiting value of the coefficient obtained in the regression of $Y_t - \hat{r}(1)Y_{t-1}$ on $Y_{t-1} - \hat{r}(1)Y_{t-2}$ with the limiting value of the regression coefficient of Y_{t-2} in the multiple regression of Y_t on Y_{t-1} and Y_{t-2}.

3. Compare the variance of $\bar{y}_n$ for a first order autoregressive process with $\operatorname{Var}\{\tilde{\mu}\}$ and $\operatorname{Var}\{\hat{\mu}\}$, where $\tilde{\mu} = \hat{\lambda}(1-\rho)^{-1}$, and $\hat{\lambda}$ and $\hat{\mu}$ are defined in (8.1.7) and (8.1.14), respectively. In computing $\operatorname{Var}\{\hat{\mu}\}$ and $\operatorname{Var}\{\tilde{\mu}\}$, assume that ρ is known without error. What are the numerical values for $n=10$ and $\rho=0.7$? For $n=10$ and $\rho=0.9$?

4. Assume that 100 observations on a time series gave the following estimates:

$$\hat{\gamma}(0)=200, \qquad \hat{r}(1)=0.8, \qquad \hat{r}(2)=0.7, \qquad \hat{r}(3)=0.5.$$

Test the hypothesis that the time series is first order autoregressive against the alternative that it is second order autoregressive.

5. The estimated autocorrelations for a sample of 100 observations on the time series $\{X_t\}$ are $\hat{r}(1)=0.8$, $\hat{r}(2)=0.5$, and $\hat{r}(3)=0.4$.
   (a) Assuming that the time series $\{X_t\}$ is defined by

$$X_t = \beta_1 X_{t-1} + \beta_2 X_{t-2} + e_t,$$

   where the e_t are normal independent $(0,\sigma^2)$ random variables, estimate β_1 and β_2.
   (b) Test the hypothesis that the order of the autoregressive process is two against the alternative that the order is three.

6. It is asserted in the proof of Theorem 8.3.1 that $W_t(Y;\delta)$ and $H_t(Y;\delta)$ converge to stationary time series. Show that $W_t(Y;\delta)$ converges to an autoregressive moving average of order $(2,1)$ and give the parameters. To what time series does $H_t(Y;\delta)$ converge?

7. Fit a first order moving average to the first fifty observations in Table 8.3.1. Predict the next observation in the realization. Establish an approximate 95% confidence interval for your prediction.

8. Prove Theorem 8.3.4.

9. Prove Theorem 8.3.2 replacing Y_t by $Y_t - \bar{y}_n$ in all defining equations.

10. Fit an autoregressive moving average $(1,1)$ to the data of Table 8.3.1.

11. The sample variance of the Boone sediment time series discussed in Section 6.5 is 0.580 and the sample variance of the Saylorville sediment time series is 0.337. Let X_{1t} and X_{2t} represent the deviations from the sample mean of the Boone and Saylorville sediment,

respectively. Using the correlations of Table 6.5.1, estimate the following models:

$$X_{2t} = \theta_1 X_{1,t-1} + \theta_2 X_{1,t-2}$$

$$X_{2t} = \theta_1 X_{1,t-1} + \theta_2 X_{1,t-2} + \theta_5 X_{2,t-1}$$

$$X_{2t} = \theta_1 X_{1,t-1} + \theta_2 X_{1,t-2} + \theta_3 X_{1,t-3} + \theta_5 X_{2,t-1} + \theta_6 X_{2,t-2}$$

$$X_{2t} = \sum_{i=1}^{4} \theta_i X_{1,t-i} + \sum_{i=5}^{7} \theta_i X_{2,t-i+4}.$$

On the basis of these regressions, suggest a model for predicting Saylorville sediment, given previous observations on Boone and Saylorville sediment.

12. Fit autoregressive equations of order one through seven to the data of Exercise 10 of Chapter 6. Choose a model for these data.

CHAPTER 9

Regression, Trend, and Seasonality

The majority of the theory presented to this point assumes that the time series under investigation is stationary. Many time series encountered in practice are not stationary. They may fail for any of several reasons:

1. The mean is a function of time, other than a constant function.
2. The variance is a function of time, other than the constant function.
3. The time series is generated by a third type of nonstationary stochastic mechanism, for example, the time series might be the sum of independent random variables.

We briefly considered the third type of process in Section 8.5. Examples of hypothesized nonstationarity of the first kind occur most frequently in the applied literature. The traditional model for economic time series is

$$Y_t = T_t + S_t + Z_t,$$

where T_t is the "trend" component, S_t is the "seasonal" component, and Z_t is the "irregular" or "random" component. (In our terminology, Z_t is a stationary time series.) Often T_t is further decomposed into "cyclical" and "long-term" components.

The casual inspection of many economic time series leads one to conclude that the mean is not constant through time, and that monthly or quarterly time series display a type of "periodic" behavior wherein peaks and troughs occur at "nearly the same" time each year. However, these two aspects of the time series typically do not exhaust the variability and, therefore, the random component is included in the representation.

While the model is an old one indeed, a precise definition of the components has not evolved. This is not necessarily to be viewed as a weakness of the representation. In fact, the terms acquire meaning only when a procedure is used to estimate them, and the meaning is determined by the procedure. The reader should not be disturbed by this. An example from another area might serve to clarify the issue. The "Intelligence

Quotient" of a person is his score on an I.Q. test. I.Q. acquires meaning only in the context of the procedure used to determine it. Although the test may be based on a theory of mental behavior, the I.Q. test score should not be taken to be the only estimator of that attribute of humans we commonly call intelligence.

For a particular economic time series and a particular objective, one model and estimation procedure for trend and seasonality may suffice; for a different time series or a different objective, an alternative specification may be preferred.

We shall now study some of the procedures used to estimate trend and seasonality and (or) to reduce a nonstationary time series to stationarity. Since the mean function of a time series may be a function of other time series or of fixed functions of time, we are led to consider the estimation of regression equations wherein the error is a time series.

9.1. GLOBAL LEAST SQUARES

In many situations we are able to specify the mean of a time series to be a simple function of time, often a low order polynomial in t or trigonometric polynomial in t. A sample of n observations can then be represented by by

$$\mathbf{y} = \mathbf{\Phi}\boldsymbol{\beta} + \mathbf{z}, \tag{9.1.1}$$

where

$\boldsymbol{\beta}' = (\beta_1, \beta_2, \ldots, \beta_r)$ is a vector of unknown parameters,

$\mathbf{y}' = (Y_1, Y_2, \ldots, Y_n)$,

$\mathbf{\Phi} = (\boldsymbol{\varphi}_{.1}, \boldsymbol{\varphi}_{.2}, \ldots, \boldsymbol{\varphi}_{.r})$, where $\boldsymbol{\varphi}_{.i}, i = 1, 2, \ldots, r$, are n-dimensional column vectors,

$\mathbf{z}' = (Z_1, Z_2, \ldots, Z_n)$,

Z_t is a stationary time series with absolutely summable covariance function and zero mean.

The elements of $\boldsymbol{\varphi}_{.i}$, say φ_{ti}, are functions of time. For example, we might have $\varphi_{t1} \equiv 1, \varphi_{t2} = t, \varphi_{t3} = t^2$. The elements of $\boldsymbol{\varphi}_{.i}$ might also be random functions of time, for example, a stationary time series. In such a case we shall investigate the behavior of the estimators conditional on a particular realization of φ_{ti}. Thus, in this section, all φ_{ti} shall be treated as fixed functions of time. It is assumed that $\mathbf{z}$ is independent of $\mathbf{\Phi}$. Notice that

$\mathbf{y}, \mathbf{\Phi}, \boldsymbol{\varphi}_{.i}$, and $\mathbf{z}$ might properly be subscripted by n. To simplify the notation, we have omitted this subscript.

The simple least squares estimator of $\boldsymbol{\beta}$ is given by

$$\hat{\boldsymbol{\beta}}_S = (\mathbf{\Phi}'\mathbf{\Phi})^{-1}\mathbf{\Phi}'\mathbf{y}. \tag{9.1.2}$$

Let us assume that the time series is regular (see Section 2.9) so that the matrix

$$\mathbf{V} = E\{\mathbf{z}\mathbf{z}'\}$$

is nonsingular. Then the generalized least squares (best linear unbiased) estimator of $\boldsymbol{\beta}$ is given by

$$\hat{\boldsymbol{\beta}}_G = \left[\mathbf{\Phi}'\mathbf{V}^{-1}\mathbf{\Phi}\right]^{-1}\mathbf{\Phi}'\mathbf{V}^{-1}\mathbf{y}. \tag{9.1.3}$$

The covariance matrix of $\hat{\boldsymbol{\beta}}_S$ is given by

$$E\{(\hat{\boldsymbol{\beta}}_S - \boldsymbol{\beta})(\hat{\boldsymbol{\beta}}_S - \boldsymbol{\beta})'\} = (\mathbf{\Phi}'\mathbf{\Phi})^{-1}\mathbf{\Phi}'\mathbf{V}\mathbf{\Phi}(\mathbf{\Phi}'\mathbf{\Phi})^{-1}, \tag{9.1.4}$$

while that of $\hat{\boldsymbol{\beta}}_G$ is given by

$$E\{(\hat{\boldsymbol{\beta}}_G - \boldsymbol{\beta})(\hat{\boldsymbol{\beta}}_G - \boldsymbol{\beta})'\} = (\mathbf{\Phi}'\mathbf{V}^{-1}\mathbf{\Phi})^{-1}. \tag{9.1.5}$$

It is well known that $\hat{\boldsymbol{\beta}}_G$ is superior to $\hat{\boldsymbol{\beta}}_S$ in that the variance of any linear contrast $\boldsymbol{\lambda}'\hat{\boldsymbol{\beta}}_G$ is no larger than the variance of the corresponding linear contrast $\boldsymbol{\lambda}'\hat{\boldsymbol{\beta}}_S$. However, the construction of $\hat{\boldsymbol{\beta}}_G$ requires knowledge of $\mathbf{V}$ and, generally, $\mathbf{V}$ is not known. In fact, one may wish to estimate the mean function of Y_t prior to investigating the covariance structure of the error time series. Therefore, the properties of the simple least squares estimator are of interest.

To investigate the large sample behavior of least squares estimators, we assume that the real valued functions of t, φ_{ti}, satisfy

$$\lim_{n\to\infty} \sum_{t=1}^{n} \varphi_{ti}^2 = \infty, \qquad i = 1, 2, \ldots, r;$$

$$\lim_{n\to\infty} \frac{\varphi_{ni}^2}{\sum_{t=1}^{n} \varphi_{ti}^2} = 0, \qquad i = 1, 2, \ldots, r; \tag{9.1.6}$$

that

$$\lim_{n\to\infty} \frac{\sum_{t=1}^{n-h} \varphi_{ti}\varphi_{t+h,j}}{\left(\sum_{t=1}^{n}\varphi_{ti}^2 \sum_{t=1}^{n}\varphi_{tj}^2\right)^{1/2}} = a_{hij} \quad = a_{-h,ji} \quad h=0, 1, 2, \ldots, \qquad (9.1.7)$$

and $i,j=1,2,\ldots,r$; that $\mathbf{\Phi}'\mathbf{\Phi}$ is positive definite for all $n>r$; and that $\mathbf{A}_0$ defined by

$$\lim_{n\to\infty} \mathbf{D}_n^{-1}\mathbf{\Phi}'\mathbf{\Phi}\mathbf{D}_n^{-1} = \mathbf{A}_0 \qquad (9.1.8)$$

is a nonsingular matrix, where the diagonal matrix

$$\mathbf{D}_n = \operatorname{diag}\{(\boldsymbol{\varphi}'_{.1}\boldsymbol{\varphi}_{.1})^{1/2}, (\boldsymbol{\varphi}'_{.2}\boldsymbol{\varphi}_{.2})^{1/2}, \ldots, (\boldsymbol{\varphi}'_{.r}\boldsymbol{\varphi}_{.r})^{1/2}\}$$

and $\boldsymbol{\varphi}_{.i}$ is the ith column of the matrix $\mathbf{\Phi}$. Note that the assumptions on the φ_{ti} are quite modest and, for example, are satisfied by polynomial and logarithmic functions of time. Given that $\mathbf{A}_0$ is nonsingular, the assumption that $\mathbf{\Phi}'\mathbf{\Phi}$ is nonsingular for $n>r$ is an assumption of convenience.

The covariance matrix of the least squares estimator is given by (9.1.4). To investigate the limiting behavior of the estimator, we normalize the estimator and consider $\mathbf{D}_n(\hat{\boldsymbol{\beta}}_S - \boldsymbol{\beta})$. It follows that

$$\begin{aligned} \operatorname{Var}\{\mathbf{D}_n(\hat{\boldsymbol{\beta}}_S - \boldsymbol{\beta})\} &= \mathbf{D}_n(\mathbf{\Phi}'\mathbf{\Phi})^{-1}\mathbf{\Phi}'\mathbf{V}\mathbf{\Phi}(\mathbf{\Phi}'\mathbf{\Phi})^{-1}\mathbf{D}_n \\ &= \mathbf{D}_n(\mathbf{\Phi}'\mathbf{\Phi})^{-1}\mathbf{D}_n\mathbf{D}_n^{-1}\mathbf{\Phi}'\mathbf{V}\mathbf{\Phi}\mathbf{D}_n^{-1}\mathbf{D}_n(\mathbf{\Phi}'\mathbf{\Phi})^{-1}\mathbf{D}_n \end{aligned}$$

and the limit is a function of the limit of the normalized quantity

$$\mathbf{D}_n^{-1}\mathbf{\Phi}'\mathbf{V}\mathbf{\Phi}\mathbf{D}_n^{-1}.$$

The ijth element of this matrix is given by

$$\frac{\sum_{s=1}^{n}\sum_{t=1}^{n} \varphi_{ti}\gamma_Z(t-s)\varphi_{sj}}{\left(\sum_{t=1}^{n}\varphi_{ti}^2 \sum_{s=1}^{n}\varphi_{sj}^2\right)^{1/2}}. \qquad (9.1.9)$$

The following theorem demonstrates that the limiting distribution of the standardized estimators is normal for a wide class of time series.

Theorem 9.1.1. Let Z_t be a stationary time series defined by

$$Z_t = \sum_{j=0}^{\infty} \alpha_j e_{t-j}, \qquad t = 0, \pm 1, \pm 2, \ldots,$$

where $\{\alpha_j\}$ is absolutely summable and the e_t are independent $(0, \sigma^2)$ random variables with distribution functions $F_t(e)$ such that

$$\lim_{\delta \to \infty} \sup_{t=1,2,\ldots} \int_{|e| > \delta} e^2 \, dF_t(e) = 0.$$

Let the $\varphi_{ti}, i = 1, 2, \ldots, r,\ t = 1, 2, \ldots,$ be fixed functions of time satisfying assumptions (9.1.6), (9.1.7), and (9.1.8). Let $\mathbf{B}$ be nonsingular, where the ijth element of $\mathbf{B}$ is

$$b_{ij} = \sum_{h=-\infty}^{\infty} a_{hij} \gamma_Z(h).$$

Then,

$$\mathbf{D}_n(\hat{\boldsymbol{\beta}}_S - \boldsymbol{\beta}) \xrightarrow{\mathcal{L}} N(\mathbf{0}, \mathbf{A}_0^{-1} \mathbf{B} \mathbf{A}_0^{-1}),$$

where $\mathbf{A}_0$ is defined in (9.1.8).

Proof. Consider the linear combination

$$\sum_{i=1}^{r} \lambda_i \sum_{t=1}^{n} \frac{\varphi_{ti} Z_t}{\left(\sum_{j=1}^{n} \varphi_{ji}^2 \right)^{1/2}} = \sum_{t=1}^{n} d_t Z_t,$$

where

$$d_t = \sum_{i=1}^{r} \lambda_i \left(\sum_{j=1}^{n} \varphi_{ji}^2 \right)^{-1/2} \varphi_{ti}$$

and the λ_i are arbitrary real numbers. Now, by our assumptions,

$$\lim_{n \to \infty} \sum_{t=1}^{n-h} d_t d_{t+h} = \lim_{n \to \infty} \sum_{t=1}^{n-h} \sum_{i=1}^{r} \sum_{j=1}^{r} \frac{\lambda_i \lambda_j \varphi_{ti} \varphi_{t+h,j}}{\left(\sum_{v=1}^{n} \varphi_{vi}^2 \sum_{s=1}^{n} \varphi_{sj}^2 \right)^{1/2}}$$

$$= \sum_{i=1}^{r} \sum_{j=1}^{r} \lambda_i \lambda_j a_{hij} \overset{\text{(say)}}{=} g_\lambda(h)$$

and d_t is completely analogous to $(\Sigma_{j=1}^{n} C_j^2)^{-1/2} C_t$ of Theorem 6.3.4. By Theorem 6.3.4, $\Sigma_{t=1}^{n} d_t Z_t$ converges to a normal random variable with variance $\Sigma_{h=-\infty}^{\infty} g_\lambda(h)\gamma_Z(h)$. Since $\lim_{n\to\infty} \mathbf{D}_n^{-1}\mathbf{\Phi}'\mathbf{\Phi}\mathbf{D}_n^{-1} = \mathbf{A}_0$ and since $\boldsymbol{\lambda}$ was arbitrary, the result follows from Theorem 5.3.3. ▲

The matrix $\mathbf{A}_h$ with ijth element equal to a_{hij} of (9.1.7) is analogous to the matrix $\mathbf{\Gamma}(h)$ of Section 4.4 and, therefore, the spectral representation

$$\mathbf{A}_h = \int_{-\pi}^{\pi} e^{i\omega h} d\mathbf{M}(\omega) \tag{9.1.10}$$

holds, where $\mathbf{M}(\omega_2) - \mathbf{M}(\omega_1)$ is a positive semidefinite Hermitian matrix for all $-\pi \leqslant \omega_1 < \omega_2 \leqslant \pi$ and $\mathbf{A}_0 = \mathbf{M}(\pi) - \mathbf{M}(-\pi)$.

We state without proof some of the results of Grenander and Rosenblatt (1957), which are based on this representation.

Theorem 9.1.2. Let assumptions (9.1.1), (9.1.6), (9.1.7), and (9.1.8) hold and let the spectral density of Z_t be positive for all ω. Then

$$\lim_{n\to\infty} \mathbf{D}_n^{-1}\mathbf{\Phi}'\mathbf{V}\mathbf{\Phi}\mathbf{D}_n^{-1} = 2\pi \int_{-\pi}^{\pi} f_Z(\omega)\, d\mathbf{M}(\omega)$$

and

$$\lim_{n\to\infty} E\{\mathbf{D}_n(\hat{\boldsymbol{\beta}}_s - \boldsymbol{\beta})(\hat{\boldsymbol{\beta}}_s - \boldsymbol{\beta})'\mathbf{D}_n\} = 2\pi \mathbf{A}_0^{-1}\left[\int_{-\pi}^{\pi} f_Z(\omega) d\mathbf{M}(\omega)\right]\mathbf{A}_0^{-1}, \tag{9.1.11}$$

where $f_Z(\omega)$ is the spectral density function of the process Z_t.

Theorem 9.1.3. Let assumptions (9.1.1), (9.1.6), (9.1.7), and (9.1.8) hold and let the spectral density of Z_t be positive for all ω. Then

$$\lim_{n\to\infty} \mathbf{D}_n^{-1}\mathbf{\Phi}'\mathbf{V}^{-1}\mathbf{\Phi}\mathbf{D}_n^{-1} = \frac{1}{2\pi}\int_{-\pi}^{\pi} f_Z^{-1}(\omega) d\mathbf{M}(\omega)$$

and

$$\lim_{n\to\infty} E\{\mathbf{D}_n(\hat{\boldsymbol{\beta}}_G - \boldsymbol{\beta})(\hat{\boldsymbol{\beta}}_G - \boldsymbol{\beta})'\mathbf{D}_n\} = \left[\frac{1}{2\pi}\int_{-\pi}^{\pi} f_Z^{-1}(\omega) d\mathbf{M}(\omega)\right]^{-1}. \tag{9.1.12}$$

The asymptotic efficiency of the simple least squares and the generalized least squares is the same if the two covariance matrices (9.1.11) and (9.1.12) are equal. Following Grenander and Rosenblatt, we denote the set

of points of increase of $\mathbf{M}(\omega)$ by S. That is, the set S contains all ω such that for any interval (ω_1, ω_2), where $\omega_1 < \omega < \omega_2, \mathbf{M}(\omega_2) - \mathbf{M}(\omega_1)$ is a positive semidefinite matrix and not the null matrix. Grenander and Rosenblatt have proven:

Theorem 9.1.4. Let assumptions (9.1.1), (9.1.6), (9.1.7), and (9.1.8) hold and let the spectral density of Z_t be positive for all ω. Then the simple least squares estimator and the generalized least squares estimator have the same asymptotic efficiency if and only if the set S is composed of q distinct points, $\omega_1, \omega_2, \ldots, \omega_q, q \leqslant r$.

It can be shown that polynomials and trigonometric polynomials satisfy the conditions of Theorem 9.1.4. For example, we established the result for the special case of a constant mean function and autoregressive Y_t in Section 6.1. One may easily establish for the linear trend that

$$\lim_{n\to\infty} \frac{1}{2\pi n}\left(\frac{6}{n(n+1)(2n+1)}\right)\left(\sum_{s=1}^{n}\sum_{t=1}^{n} te^{-i\omega t}se^{-i\omega s}\right) = \begin{cases} 0, & \omega \neq 0 \\ \dfrac{3}{8\pi}, & \omega = 0 \end{cases}$$

from which it follows that the set S of points of increase is $\omega_1 = 0$.

The reader should not forget that these are asymptotic results. If the sample is of moderate size it may be desirable to estimate $\mathbf{V}$ and transform the data to obtain final estimates of the trend function. Also the simple least squares estimators may be asymptotically efficient, but it does not follow that the simple formulas for the estimated variances of the coefficients are consistent. In fact, the estimated variances may be badly biased. See Section 9.7.

9.2. GRAFTED POLYNOMIALS

In many applications the mean function is believed to be a "smooth" function of time but the functional form is not known. While it is difficult to define the term "smooth" in this context, several aspects of functional behavior can be identified. For a function defined on the real line the function would be judged to be continuous and, in most situations, to have a continuous first derivative. This specification is incomplete in that one also often expects few changes in the sign of the first derivative.

Obviously low order polynomials satisfy the stated requirements. Also, by the Weierstrass approximation theorem, we know that any continuous function defined on a compact interval of the real line can be uniformly approximated by a polynomial. Consequently, polynomials have been heavily used to approximate the mean function. However, if the mean

function is such that higher order polynomials are required, the approximating function may be judged unsatisfactory in that it contains a large number of changes in sign of the derivative. An alternative approximation that generally overcomes this problem is to approximate segments of the function by low order polynomials and then join the segments to form a continuous function. The segments may be joined together so that the derivatives of a desired order are continuous.

Our presentation is based on the fact that, on the real line, the functions

$$\varphi_{ti} = \begin{cases} (t-A_i)^k, & t > A_i \\ 0, & \text{otherwise,} \end{cases}$$

where $i = 1, 2, \ldots, M$, k and M are positive integers, and $A_i > A_{i-1}$, are continuous with continuous $(k-1)$st derivatives. It follows that

$$g(t) = b_0 + b_1 t + \ldots + b_k t^k + \sum_{i=1}^{M} b_{k+i} \varphi_{ti}$$

is also continuous with continuous $(k-1)$st derivative. We call the function $g(t)$ a *grafted polynomial of degree* k.

To illustrate the use of grafted polynomials in the estimation of the mean function of a time series let us make the assumption: "the time series may be divided into periods of length A such that the mean function in each period is adequately approximated by a quadratic in time. Furthermore the mean function possesses a continuous first derivative." Let n observations indexed by t, $t = 1, 2, \ldots, n$, be available. We construct the functions φ_{ti}, $i = 1, 2, \ldots, M$, for $k = 2$ and $A_i = Ai$, where M is an integer such that $|A(M+1) - n| < A$. In this formulation we assume that, if $n \neq A(M+1)$, the last interval is the only one whose length is not A. At the formal level all we need do is regress our time series upon $t, t^2, \varphi_{t1}, \varphi_{t2}, \ldots, \varphi_{tM}$ to obtain estimates of b_j, $j = 0, 1, \ldots, M+2$. However, if M is at all large, we can expect to encounter numerical problems in obtaining the inverse and regression coefficients. To reduce the possibility of numerical problems, we suggest that the φ's be replaced by the linear combinations

$$w_{ti} = \varphi_{ti} - 3\varphi_{t,i+1} + 3\varphi_{t,i+2} - \varphi_{t,i+3}, \qquad i = 1, 2, \ldots, M,$$

where, for convenience, we define

$$\varphi_{t,M+1} = \varphi_{t,M+2} = \varphi_{t,M+3} = 0$$

for all t. Note that

$$w_{ti}=\begin{cases}[t-Ai]^2, & Ai<t<(i+1)A\\ [t-Ai]^2-3[t-(i+1)A]^2, & (i+1)A\leqslant t<(i+2)A\\ [t-(i+3)A]^2, & (i+2)A\leqslant t<(i+3)A\\ 0, & \text{otherwise.}\end{cases}$$

Since the function w_{ti} is symmetric about $(i+1.5)A$, the w-variables can be written down immediately. The w_{ti} remain correlated, but there should be little trouble in obtaining the inverse.

This procedure is felt to have merit when one is called on to extrapolate a time series. Since polynomials tend to plus or minus infinity as t increases, practitioners typically hesitate to use high order polynomials in extrapolation. Although a time series may display a nonlinear trend, we might wish to extrapolate on the basis of a linear trend. To accomplish this, we approximate the trend of the last K periods of a time series of n observations by a straight line tangent to a higher degree trend for the earlier portion of the time series. If the mean function for the first part of the sample is to be approximated by a grafted quadratic one could construct the regression variables

$$\varphi_{t1}=t,$$

$$\varphi_{t,1+i}=\begin{cases}[n-K-(i-1)A-t]^2, & t<n-K-(i-1)A\\ 0, & \text{otherwise,}\end{cases}$$

for $i=1,2,\ldots,M$, where $|K+AM-n|<A$. Using these variables, the estimated regression equation can be written as

$$\hat{b}_0+\sum_{j=1}^{M+1}\hat{b}_j\varphi_{tj},$$

where the $\hat{b}$'s are the estimated regression coefficients. The forecast equation for the mean function is $\hat{b}_0+\hat{b}_1 t, t=n+1,n+2,\ldots$.

We can also estimate a mean function that is linear for both the first and last portions of the observational period. Using a grafted quadratic in periods of length A for the remainder of the function the required regression variables would be, for example,

$$\varphi_{t1} = t,$$

$$\varphi_{t,1+i} = \begin{cases} [t-K-A(i-1)]^2, & K+A(i-1)<t\leqslant K+Ai \\ A^2+2A(t-K-Ai), & t>K+Ai \\ 0, & \text{otherwise} \end{cases}$$

for $i=1,2,\ldots,M$, where $n>K+MA$ and the mean function is linear for $t<K$ and $t>K+MA$. If M is large a transformation similar to that discussed above can be used to reduce computational problems.

Example. As an example, we investigate the trend in United States wheat yields from 1908 to 1971. The data are given in Table 9.2.1. There has clearly been an increase in yields occurring over the last 30 to 40 years. The data for the first 20 to 30 years display very little trend. Therefore, as an approximation to the trend line, we fit a function that is constant for the first 25 years, increases at a quadratic rate until 1961 and is linear for the last 10 years. A continuous function that satisfies these requirements is

$$\varphi_{t1} = \begin{cases} 0, & 1\leqslant t\leqslant 25 \\ (t-25)^2, & 25\leqslant t\leqslant 54 \\ 841+58(t-54), & 54\leqslant t\leqslant 64, \end{cases}$$

where $t=1$ for 1908.

The trend line obtained by regressing yield on the constant vector and $\varphi_{.1}$ is displayed in Figure 9.2.1. The equation for the estimated trend line is

$$\hat{Y}_t = 13.97 + 0.0123\varphi_{t1}$$

and the residual mean square is 2.80.

In the example and in our presentation we assumed that one was willing to specify the points at which the different polynomials are joined. The problem of using the data to estimate the join points has been considered by Gallant and Fuller (1973).

Grafted polynomials passing through a set of given points and joined in such a way that they satisfy certain restrictions on the derivatives are called *spline functions* in approximation theory. A discussion of such functions and their properties is contained in Ahlberg *et al.* (1967) and Greville (1969).

Table 9.2.1 United States wheat yields (1908 to 1971)

Year	Bushels per Acre	Year	Bushels per Acre
1908	14.3	1940	15.3
1909	15.5	1941	16.8
1910	13.7	1942	19.5
1911	12.4	1943	16.4
1912	15.1	1944	17.7
1913	14.4	1945	17.0
1914	16.1	1946	17.2
1915	16.7	1947	18.2
1916	11.9	1948	17.9
1917	13.2	1949	14.5
1918	14.8	1950	16.5
1919	12.9	1951	16.0
1920	13.5	1952	18.4
1921	12.7	1953	17.3
1922	13.8	1954	18.1
1923	13.3	1955	19.8
1924	16.0	1956	20.2
1925	12.8	1957	21.8
1926	14.7	1958	27.5
1927	14.7	1959	21.6
1928	15.4	1960	26.1
1929	13.0	1961	23.9
1930	14.2	1962	25.0
1931	16.3	1963	25.2
1932	13.1	1964	25.8
1933	11.2	1965	26.5
1934	12.1	1966	26.3
1935	12.2	1967	25.9
1936	12.8	1968	28.4
1937	13.6	1969	30.6
1938	13.3	1970	31.0
1939	14.1	1971	33.9

SOURCES. U.S. Department of Agriculture, *Agricultural Statistics*, various issues.

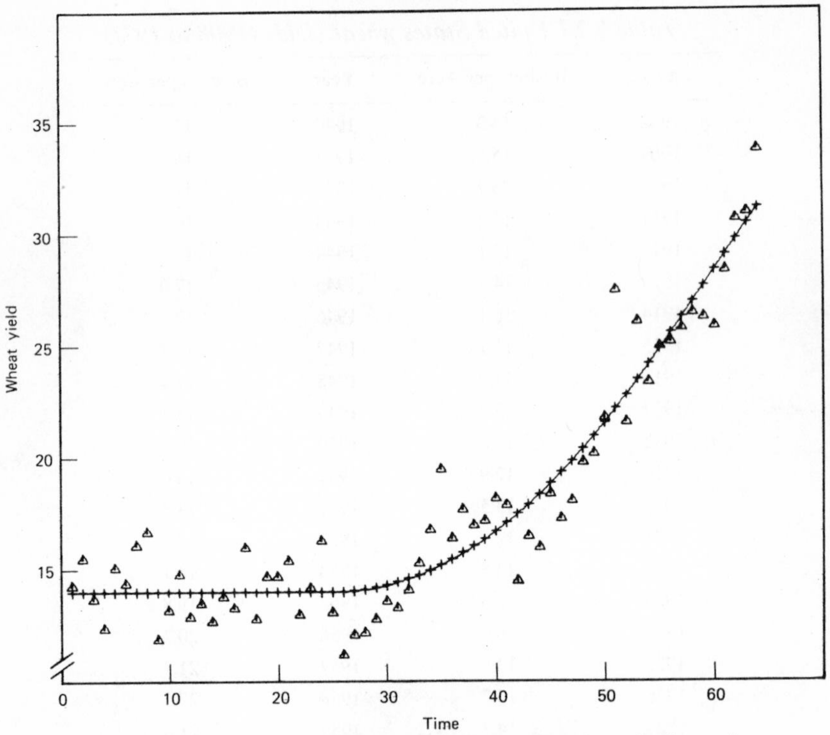

Figure 9.2.1. U.S. wheat yields 1908–1971 and grafted quadratic trend.

9.3. AUTOCORRELATIONS ESTIMATED FROM THE LEAST SQUARES RESIDUALS

As we have pointed out, one of the reasons for estimating the trend function is to enable one to investigate the properties of the time series Z_t. Let our regression model be that defined in (9.1.1):

$$\mathbf{y} = \mathbf{\Phi}\boldsymbol{\beta} + \mathbf{z}. \tag{9.3.1}$$

The calculated residuals using the simple least squares estimator are given by

$$\hat{Z}_t = Y_t - \varphi_{t.}\hat{\beta}_S = Z_t - \varphi_{t.}(\hat{\beta}_S - \beta), \tag{9.3.2}$$

where

$$\hat{\boldsymbol{\beta}}_S = (\mathbf{\Phi}'\mathbf{\Phi})^{-1}\mathbf{\Phi}'\mathbf{y}$$

and $\varphi_{t.}$ is the tth row of $\mathbf{\Phi}$. The estimated autocovariance of Z_t computed using the estimated Z_t is

$$\hat{\gamma}_{\hat{Z}}(h) = \frac{1}{n} \sum_{t=1}^{n-h} \hat{Z}_t \hat{Z}_{t+h}, \qquad h = 0, 1, 2, \ldots, n-1. \tag{9.3.3}$$

Theorem 9.3.1. Let Y_t be defined by model (9.3.1) where $Z_t = \sum_{j=0}^{\infty} \alpha_j e_{t-j}$, $\{\alpha_j\}$ is absolutely summable, $\{e_t\}$ is a sequence of independent $(0, \sigma^2)$ random variables with bounded $2+\delta$ $(\delta > 0)$ moments, and the φ_{ti}, $i = 1, 2, \ldots, r$, $t = 1, 2, \ldots$, are fixed functions of time satisfying assumptions (9.1.6), (9.1.7), and (9.1.8). Then,

$$\hat{\gamma}_{\hat{Z}}(h) - \hat{\gamma}_Z(h) = O_p(n^{-1}),$$

where $\hat{\gamma}_{\hat{Z}}(h)$ is defined in (9.3.3) and

$$\hat{\gamma}_Z(h) = \frac{1}{n} \sum_{t=1}^{n-h} Z_t Z_{t+h}, \qquad h = 0, 1, 2, \ldots, n-1.$$

Proof. We have, for $h = 0, 1, \ldots, n-1$,

$$\begin{aligned}
\sum_{t=1}^{n-h} \hat{Z}_t \hat{Z}_{t+h} &= \sum_{t=1}^{n-h} Z_t Z_{t+h} - \sum_{t=1}^{n-h} Z_t \varphi_{t+h,.}(\hat{\boldsymbol{\beta}}_S - \boldsymbol{\beta}) - \sum_{t=1}^{n-h} Z_{t+h} \varphi_{t.}(\hat{\boldsymbol{\beta}}_S - \boldsymbol{\beta}) \\
&\quad + \sum_{t=1}^{n-h} (\hat{\boldsymbol{\beta}}_S - \boldsymbol{\beta})' \varphi_{t.}' \varphi_{t+h,.}(\hat{\boldsymbol{\beta}}_S - \boldsymbol{\beta}) \\
&= \sum_{t=1}^{n-h} Z_t Z_{t+h} - \sum_{t=1}^{n-h} Z_t \varphi_{t+h,.} \mathbf{D}_n^{-1} \mathbf{D}_n (\hat{\boldsymbol{\beta}}_S - \boldsymbol{\beta}) \\
&\quad - \sum_{t=1}^{n-h} Z_{t+h} \varphi_{t.} \mathbf{D}_n^{-1} \mathbf{D}_n (\hat{\boldsymbol{\beta}}_S - \boldsymbol{\beta}) \\
&\quad + \sum_{t=1}^{n-h} (\hat{\boldsymbol{\beta}}_S - \boldsymbol{\beta})' \mathbf{D}_n \mathbf{D}_n^{-1} \varphi_{t.}' \varphi_{t+h,.} \mathbf{D}_n^{-1} \mathbf{D}_n (\hat{\boldsymbol{\beta}}_S - \boldsymbol{\beta}),
\end{aligned}$$

where $\mathbf{D}_n$ is the diagonal matrix whose elements are the square roots of the elements of the principal diagonal of $\mathbf{\Phi}'\mathbf{\Phi}$. By Theorems 6.3.4 and 9.1.1,

$$\sum_{t=1}^{n-h} Z_t \varphi_{t+h,.} \mathbf{D}_n^{-1} = O_p(1),$$

$$\mathbf{D}_n (\hat{\boldsymbol{\beta}}_S - \boldsymbol{\beta}) = O_p(1).$$

Since each element of the $r\times r$ matrix

$$\sum_{t=1}^{n-h} \mathbf{D}_n^{-1}\boldsymbol{\varphi}_{t.}'\boldsymbol{\varphi}_{t+h,.}\mathbf{D}_n^{-1}$$

is less than one in absolute value, the result follows. ▲

It is an immediate consequence of Theorem 9.3.1 that the limiting behavior of the estimated autocovariances and autocorrelations defined by Theorem 6.3.5 and Corollary 6.3.5 also holds for autocovariances and autocorrelations computed using $\hat{Z}_t$ in place of Z_t. The bias in the estimated autocorrelations is $O(n^{-1})$ and, therefore, can be ignored in large samples. However, if the sample is small and if several φ_{ti} are included in the regression, the bias in $\hat{\gamma}_{\hat{Z}}(h)$ may be sizeable.

To briefly investigate the nature of the bias in $\hat{\gamma}_{\hat{Z}}(h)$ we use the statistic studied by Durbin and Watson (1950, 1951). They suggested the von Neumann ratio (see Section 6.2.) computed from the calculated residuals as a test of the hypothesis of independent errors. The statistic is

$$d=\frac{\sum_{t=2}^{n}(\hat{Z}_t-\hat{Z}_{t-1})^2}{\sum_{t=1}^{n}\hat{Z}_t^2}=\frac{\hat{\mathbf{z}}'\mathbf{A}\hat{\mathbf{z}}}{\hat{\mathbf{z}}'\hat{\mathbf{z}}}, \tag{9.3.4}$$

where

$$\mathbf{A}=\begin{bmatrix} 1 & -1 & 0 & \cdots & 0 & 0 \\ -1 & 2 & -1 & \cdots & 0 & 0 \\ \vdots & \vdots & \vdots & & & \vdots \\ 0 & 0 & 0 & \cdots & -1 & 1 \end{bmatrix},$$

$\hat{\mathbf{z}}=[\mathbf{I}-\boldsymbol{\Phi}(\boldsymbol{\Phi}'\boldsymbol{\Phi})^{-1}\boldsymbol{\Phi}']\mathbf{z}$ is the vector of calculated residuals, and $\mathbf{z}$ is the vector of observations on the original time series. Recall that

$$d=2-2\hat{r}_{\hat{Z}}(1)-\frac{\hat{Z}_1^2+\hat{Z}_n^2}{\sum_{t=1}^{n}\hat{Z}_t^2},$$

where

$$\hat{r}_{\hat{Z}}(1)=\sum_{t=2}^{n}\hat{Z}_t\hat{Z}_{t-1}\left(\sum_{t=1}^{n}\hat{Z}_t^2\right)^{-1}.$$

The expected value of the numerator and denominator of d are given by

$$E\{\hat{\mathbf{z}}'\mathbf{A}\hat{\mathbf{z}}\} = E\left\{\mathbf{z}'\left(\mathbf{I}-\boldsymbol{\Phi}(\boldsymbol{\Phi}'\boldsymbol{\Phi})^{-1}\boldsymbol{\Phi}'\right)\mathbf{A}\left(\mathbf{I}-\boldsymbol{\Phi}(\boldsymbol{\Phi}'\boldsymbol{\Phi})^{-1}\boldsymbol{\Phi}'\right)\mathbf{z}\right\}$$
$$= \operatorname{tr}\mathbf{VA} - \operatorname{tr}\mathbf{V}\boldsymbol{\Phi}(\boldsymbol{\Phi}'\boldsymbol{\Phi})^{-1}\boldsymbol{\Phi}'\mathbf{A} - \operatorname{tr}\mathbf{VA}\boldsymbol{\Phi}(\boldsymbol{\Phi}'\boldsymbol{\Phi})^{-1}\boldsymbol{\Phi}'$$
$$+ \operatorname{tr}\mathbf{V}\boldsymbol{\Phi}(\boldsymbol{\Phi}'\boldsymbol{\Phi})^{-1}\boldsymbol{\Phi}'\mathbf{A}\boldsymbol{\Phi}(\boldsymbol{\Phi}'\boldsymbol{\Phi})^{-1}\boldsymbol{\Phi}' \tag{9.3.5}$$

and

$$E\{\hat{\mathbf{z}}'\hat{\mathbf{z}}\} = \operatorname{tr}\left[\mathbf{V} - \mathbf{V}\boldsymbol{\Phi}(\boldsymbol{\Phi}'\boldsymbol{\Phi})^{-1}\boldsymbol{\Phi}'\right], \tag{9.3.6}$$

where

$$\mathbf{V} = E\{\mathbf{z}\mathbf{z}'\}.$$

It is easy to see that the expectation (9.3.5) may differ considerably from $2(n-1)[\gamma_Z(0) - \gamma_Z(1)]$. The expectation of the ratio is not necessarily equal to the ratio of the expectations, and this may further increase the bias in some cases (e.g., see Exercise 9.3).

We now turn to a consideration of d as a test of independence. For normal independent errors Durbin and Watson have shown that the ratio can be reduced to a cannonical form by a simultaneous diagonalization of the numerator and denominator quadratic forms. Therefore, in this case, we can write

$$d = \frac{\sum_{j=1}^{n} \lambda_j \epsilon_j^2}{\sum_{j=1}^{n} \delta_j \epsilon_j^2} = \frac{\sum_{j=1}^{n} \lambda_j \epsilon_j^2}{\sum_{j=1}^{n-k'-1} \epsilon_j^2},$$

where it is assumed that the model contains an intercept and k' other independent variables, the λ_j are the roots of $(\mathbf{I}-\boldsymbol{\Phi}(\boldsymbol{\Phi}'\boldsymbol{\Phi})^{-1}\boldsymbol{\Phi}')\mathbf{A}(\mathbf{I}-\boldsymbol{\Phi}(\boldsymbol{\Phi}'\boldsymbol{\Phi})^{-1}\boldsymbol{\Phi}')$, the δ_j are the roots of $(\mathbf{I}-\boldsymbol{\Phi}(\boldsymbol{\Phi}'\boldsymbol{\Phi})^{-1}\boldsymbol{\Phi}')$, and the ϵ_j are normal independent random variables with mean zero and variance $\sigma_\epsilon^2 = \sigma_Z^2$. Furthermore, the distribution of the ratio is independent of the distribution of the denominator. The exact distribution of d, depending on the roots λ and δ, can be obtained. For normal independent errors, Z_t, Durbin and Watson were able to obtain upper and lower bounds for the distribution, and they tabled the percentage points of these bounding distributions.

For small sample sizes and (or) a large number of independent variables the bounding distributions may differ considerably. Durbin and Watson

suggested that the distribution of $\frac{1}{4}d$ for a particular $\mathbf{\Phi}$ matrix could be approximated by a beta distribution with the same first two moments as $\frac{1}{4}d$.

As pointed out in Section 6.2, in the null case the t-statistic associated with the first order autocorrelation computed from the von Neumann ratio has percentage points approximately equal to those of Student's t with $n+3$ degrees of freedom. For the d-statistic computed from the regression residuals it is possible to use the moments of the d-statistic to develop a similar approximation. The expected value of d under the assumption of independent errors is

$$E\{d\} = \left[\operatorname{tr}\mathbf{A} - \operatorname{tr}\left\{\mathbf{\Phi}'\mathbf{A}\mathbf{\Phi}(\mathbf{\Phi}'\mathbf{\Phi})^{-1}\right\}\right](n-k'-1)^{-1}$$

$$= \left[2(n-1) - \operatorname{tr}\left\{(\Delta\mathbf{\Phi})'(\Delta\mathbf{\Phi})(\mathbf{\Phi}'\mathbf{\Phi})^{-1}\right\}\right](n-k'-1)^{-1}, \qquad (9.3.7)$$

where $(\Delta\mathbf{\Phi})'(\Delta\mathbf{\Phi})$ is the matrix of sums of squares and products of the first differences. Consider the statistics

$$\hat{\rho} = r_d + \frac{1}{2}(E\{d\} - 2)\frac{n-k'+1}{n-k'}(1-r_d^2) \qquad (9.3.8)$$

and

$$t_d = (n-k'+1)^{1/2}\hat{\rho}(1-\hat{\rho}^2)^{-1/2}, \qquad (9.3.9)$$

where $r_d = \frac{1}{2}(2-d)$. The term $(1-r_d^2)$ keeps $\hat{\rho}$ in the interval $[-1,1]$ for most samples. For normal independent $(0,\sigma^2)$ errors the expected value of $\hat{\rho}$ is $O(n^{-2})$ and the variance of $\hat{\rho}$ differs from that of an estimated autocorrelation computed from a sample of size $n-k'$ by terms of order n^{-2}. This suggests that the calculated t_d be compared with the tabular value of Student's t for $(n-k'+3)$ degrees of freedom. By using the second moment associated with the particular $\mathbf{\Phi}$ matrix the t-statistic could be further modified to improve the approximation, though there is little reason to believe that the resulting approximation would be as accurate as the beta approximation suggested by Durbin and Watson.

9.4. MOVING AVERAGES—LINEAR FILTERING

In the previous sections we considered methods of estimating the mean function for the entire period of observation. One may be interested in a simple approximation to the mean as a part of a preliminary investigation where one is not willing to specify the mean function for the entire period,

or one may desire a relatively simple method of removing the mean function that will permit simple extrapolation of the time series. In such cases the method of moving averages may be appropriate.

One basis for the method of moving averages is the presumption that for a period of M observations, the mean is adequately approximated by a specified function. The function is typically linear in the parameters and most commonly is a polynomial in t. Thus, the specification is

$$Y_{t+j} = g(j;\boldsymbol{\beta}_t) + Z_{t+j}, \qquad j = -M_1, -M_1+1, \ldots, M_2-1, M_2, \tag{9.4.1}$$

where Z_t is a stationary time series with zero expectation, $\boldsymbol{\beta}_t$ is a vector of parameters, and $M = M_1 + M_2 + 1$. The form of the approximating function g is assumed to hold for all t, but the parameters are permitted to be a function of t. Both the "local" and "approximate" nature of the specification should now be clear. If the functional form for the expectation of Y_t held exactly in the interval, then it would hold for the entire realization, and the constants M_1 and M_2 would become $t-1$ and $n-t$, respectively.

Given specification (9.4.1), a set of weights, $w_j(s)$, are constructed that, when applied to the Y_{t+j}, $j = -M_1, -M_1+1, \ldots, M_2-1, M_2$, furnish an estimator of $g(s;\boldsymbol{\beta}_t)$ for the specified s. It follows that an estimator of Z_{t+s} is given by

$$Y_{t+s} - \hat{g}(s;\boldsymbol{\beta}_t),$$

where

$$\hat{g}(s;\boldsymbol{\beta}_t) = \sum_{j=-M_1}^{M_2} Y_{t+j} w_j(s).$$

In the terminology of Section 4.3 the set of weights $\{w_j(s)\}$ is a linear filter.

Let us consider an example. Assume that for a period of five observations the time series is adequately represented by

$$Y_{t+j} = \beta_{0t} + \beta_{1t} j + \beta_{2t} j^2 + Z_{t+j}, \qquad j = -2, -1, 0, 1, 2, \tag{9.4.2}$$

where Z_t is a stationary time series with zero expectation. Using this specification we calculate the least squares estimator for the trend value of the center observation, $s=0$, as a linear function of the five observations. The model may be written in matrix form as

$$\mathbf{y} = \boldsymbol{\Phi}\boldsymbol{\beta} + \mathbf{z},$$

where

$$\boldsymbol{\beta}' = (\beta_{0t}, \beta_{1t}, \beta_{2t}),$$

$\mathbf{y}=(Y_{t-2}, Y_{t-1}, Y_t, Y_{t+1}, Y_{t+2})'$ is the vector in the first column of Table 9.4.1,

$\mathbf{\Phi}$ is the matrix defined by the second, third, and fourth columns of Table 9.4.1 (those columns headed by β_0, β_1, and β_2),

$\mathbf{z}=(Z_{t-2}, Z_{t-1}, Z_t, Z_{t+1}, Z_{t+2})'$ is the vector of (unobservable) elements of the stationary time series.

Table 9.4.1. Calculation of weights for a five period quadratic moving average.

				Weights for Trend			Weights for Trend Adjusted Series
$\mathbf{y}$	β_0	β_1	β_2	$g(0;\boldsymbol{\beta}_t)$	$g(1;\boldsymbol{\beta}_t)$	$g(2;\boldsymbol{\beta}_t)$	$(s=0)$
Y_{t-2}	1	−2	4	−6/70	−10/70	6/70	6/70
Y_{t-1}	1	−1	1	24/70	12/70	−10/70	−24/70
Y_t	1	0	0	34/70	24/70	−6/70	36/70
Y_{t+1}	1	1	1	24/70	26/70	18/70	−24/70
Y_{t+2}	1	2	4	−6/70	18/70	62/70	6/70

The least squares estimator of the mean at $j=0, g(0;\boldsymbol{\beta}_t)$, is given by

$$\hat{\beta}_{0t}=(1,0,0)(\mathbf{\Phi}'\mathbf{\Phi})^{-1}\mathbf{\Phi}'\mathbf{y}$$

$$=\sum_{j=-2}^{2} w_j(0)Y_{t+j}=\mathbf{w}'\mathbf{y},$$

where the vector $(1,0,0)$ is the value of the vector of independent variables associated with the third observation. That is, $(1,0,0)$ is the third row of the $\mathbf{\Phi}$ matrix and is associated with $j=0$. The vector of weights

$$\mathbf{w}'=(1,0,0)(\mathbf{\Phi}'\mathbf{\Phi})^{-1}\mathbf{\Phi}'$$

once computed can be applied to any vector $\mathbf{y}$ of five observations. These weights are given in the fifth column of Table 9.4.1 under the heading "$g(0;\boldsymbol{\beta}_t)$." The least squares estimator of Z_t for the third observation $(s=0)$ is given by

$$Y_t-\hat{\beta}_{0t}=Y_t-(1,0,0)(\mathbf{\Phi}'\mathbf{\Phi})^{-1}\mathbf{\Phi}'\mathbf{y}.$$

The vector of weights giving the least squares estimator of Z_t is presented in the last column of Table 9.4.1. One may readily verify that the

inner product of this vector of weights with each of the columns of the original $\mathbf{\Phi}$ matrix is zero. Thus, if the original time series satisfies specification (9.4.2), the time series of estimated residuals created by applying the weights in the last column will be a stationary time series with zero mean. The created time series X_t is given by

$$\begin{aligned} X_t &= \sum_{j=-2}^{2} r_j Y_{t+j} \\ &= \sum_{j=-2}^{2} r_j \{ \beta_{0t} + \beta_{1t}(t+j) + \beta_{2t}(t+j)^2 + Z_{t+j} \} \\ &= \sum_{j=-2}^{2} r_j Z_{t+j} \\ &= \frac{1}{70} \{ 6Z_{t-2} - 24Z_{t-1} + 36Z_t - 24Z_{t+1} + 6Z_{t+2} \}, \end{aligned}$$

where r_j denotes the weights in the last column of Table 9.4.1. Although X_t can be thought of as an estimator of Z_t, X_t is a linear combination of the five original Z_t included in the moving average.

If one has available a machine regression program that computes the estimated value and deviation from fit for each observation on the dependent variable, the following procedure provides a convenient method of computation. Form a regression problem with the independent variables given by the trend specification (e.g., constant, linear, and quadratic) and the dependent variable defined by

$$\begin{aligned} v_{t+j} &= 1, \qquad j = s \\ &= 0, \qquad \text{otherwise}, \end{aligned}$$

where the estimated mean is to be computed for $j = s$. The vector of weights for the estimated mean is then given by

$$\mathbf{w} = \hat{\mathbf{v}} = \mathbf{\Phi}(\mathbf{\Phi}'\mathbf{\Phi})^{-1}\mathbf{\Phi}'\mathbf{v} = \mathbf{\Phi}(\mathbf{\Phi}'\mathbf{\Phi})^{-1}\boldsymbol{\varphi}_{s.}', \tag{9.4.3}$$

where $\boldsymbol{\varphi}_{s.}$ is the vector of observations on the independent variables associated with $j = s$. Similarly, the vector of weights for computing the trend adjusted time series at $j = s$ is given by

$$\mathbf{v} - \hat{\mathbf{v}} = \mathbf{v} - \mathbf{\Phi}(\mathbf{\Phi}'\mathbf{\Phi})^{-1}\mathbf{\Phi}'\mathbf{v} = \mathbf{v} - \mathbf{\Phi}(\mathbf{\Phi}'\mathbf{\Phi})^{-1}\boldsymbol{\varphi}_{s.}'. \tag{9.4.4}$$

The computation of (9.4.3) is recognized as the computation of **y**-hat associated with the regression of the vector **v** on $\mathbf{\Phi}$. The weights in (9.4.4) are then the deviations from fit of the same regression.

The moving average estimator of the trend and of the trend adjusted time series are frequently calculated for $j=0$ and $M_1 = M_2$. This configuration is called a *centered moving average*. The popularity of the centered moving average is, perhaps, due to the following theorem.

Theorem 9.4.1. The least squares weights constructed for the estimator of trend for a centered moving average of $2M+1$ observations, M a positive integer, under the assumption of a pth degree polynomial, p nonnegative and even, are the same as those computed for the centered moving average of $2M+1$ observations under the assumption of a $(p+1)$st degree polynomial.

Proof. We construct our $\mathbf{\Phi}$ matrix in a manner analogous to that of Table 9.4.1, that is, with columns given by $j^k, k=0,1,2,\ldots,p, j=0,\pm 1, \pm 2,\ldots,\pm M$.

Since we are predicting trend for $j=0$, the regression coefficients of all odd powers are multiplied by zero. That is, the elements of $\boldsymbol{\varphi}'_{s.}$ in (9.4.3) associated with the odd powers are all zero. It remains only to show that the regression coefficients of the even powers remain unchanged by the presence or absence of the $(p+1)$st polynomial in the regression. But

$$\sum_{j=-M}^{M} j^{2k+1} = 0$$

for k a positive integer. Therefore, the presence or absence of odd powers of j in the matrix $\mathbf{\Phi}$ leaves the coefficients of the even powers unchanged and the result follows. ▲

The disadvantage of a centered moving average is the loss of observations at the beginning and at the end of the realization. The loss of observations at the end of the observed time series is particularly critical if the objective of the study is to forecast future observations. If the trend and trend adjusted time series are computed for the end observations using the same model, the variance-covariance structure of these estimates differs from those computed for the center of the time series. We illustrate with our example.

Assume that the moving average associated with equation (9.4.2) and Table 9.4.1 is applied to a sequence of independent identically distributed random variables. That is, the trend value for the first observation is given by applying weights $g(-2,\beta_t)$ to the first five observations and the trend adjusted value for the first observation is obtained by subtracting the trend

value from Y_1. The weights $g(-2,\boldsymbol{\beta}_t)$ are the weights $g(2,\boldsymbol{\beta}_t)$ arranged in reverse order. For the second observation the weights $g(-1,\boldsymbol{\beta}_t)$ are used. For $t=3,4,\ldots,n-2$, the weights $g(0,\boldsymbol{\beta}_t)$ are used. Denote the trend adjusted time series by $X_t, t=1,2,\ldots,n$. If $3 \leqslant t \leqslant n-2$ and $3 \leqslant t+h \leqslant n-2$, then

$$\operatorname{Cov}\{X_t, X_{t+h}\} = \begin{cases} 2520/4900, & h=0 \\ -2016/4900, & h=\pm 1 \\ 1008/4900, & h=\pm 2 \\ -288/4900, & h=\pm 3 \\ 36/4900, & h=\pm 4 \\ 0, & \text{otherwise.} \end{cases}$$

For the first observation,

$$\operatorname{Cov}\{X_1, X_t\} = \begin{cases} 560/4900, & t=1 \\ -1260/4900, & t=2 \\ 420/4900, & t=3 \\ 252/4900, & t=4 \\ -420/4900, & t=5 \\ 204/4900, & t=6 \\ -36/4900, & t=7 \\ 0, & \text{otherwise.} \end{cases}$$

In Section 8.5 we discussed the nonstationary time series that could be represented as an autoregressive process with root of unit absolute value. We have been constructing moving average weights to remove a mean that is polynomial in time. We shall see that weights constructed to remove such a mean will also eliminate the nonstationarity arising from an autoregressive component with unit root. The time series $\{W_t, t \in (0,1,2,\ldots)\}$ is called an *integrated time series of order s* if it is defined by

$$W_t = \sum_{j_s=0}^{t} \cdots \sum_{j_2=0}^{j_3} \sum_{j_1=0}^{j_2} Z_{j_1},$$

where $\{Z_t, t \in (0,1,2,\ldots)\}$ is a stationary time series with zero mean.

Theorem 9.4.2. A moving average constructed to remove the mean will reduce a first order integrated time series to stationarity and a moving average constructed to remove a linear trend will reduce a second order integrated time series to stationarity.

Proof. We first prove that a moving average constructed to remove the mean (zero degree polynomial) will reduce a first order integrated time series $W_t = \sum_{s=0}^{t} Z_s$ to stationarity. Define the time series created by applying a moving average to remove the mean by

$$X_t = \sum_{j=1}^{M} c_j W_{t+j}, \qquad t = 0, 1, 2, \ldots,$$

where the weights satisfy

$$\sum_{j=1}^{M} c_j = 0.$$

Then,

$$\begin{aligned} X_t &= \sum_{j=1}^{M} c_j \sum_{s=0}^{t+j} Z_s \\ &= \sum_{j=1}^{M} c_j \left[W_t + \sum_{r=1}^{j} Z_{t+r} \right] \\ &= \sum_{j=1}^{M} c_j \sum_{r=1}^{j} Z_{t+r}, \end{aligned}$$

which is a finite moving average of a stationary time series and, therefore, is stationary.

Consider next the second order integrated time series

$$U_t = \sum_{r=0}^{t} \sum_{s=0}^{r} Z_s = \sum_{r=0}^{t} W_r = \sum_{j=1}^{t+1} j Z_{t-j+1},$$

where W_r is a first order integrated time series. The weights d_j constructed to remove a linear trend satisfy $\sum_{j=1}^{M} d_j = 0$ and $\sum_{j=1}^{M} j d_j = 0$. Therefore, the time series created by applying such weights is given by

$$\begin{aligned} X_t = \sum_{j=1}^{M} d_j U_{t+j} &= \sum_{j=1}^{M} d_j \left[U_t + \sum_{r=1}^{j} W_{t+r} \right] \\ &= \sum_{j=1}^{M} d_j \sum_{r=1}^{j} \left[W_t + \sum_{s=1}^{r} Z_{t+s} \right] \\ &= \sum_{j=1}^{M} d_j j W_t + \sum_{j=1}^{M} d_j \sum_{s=1}^{j} s Z_{t+j-s+1} \\ &= \sum_{j=1}^{M} \sum_{s=1}^{j} d_j s Z_{t+j-s+1} \end{aligned}$$

which, once again, is a finite moving average of a stationary time series. ▲

The reader may extend this theorem to higher orders.

Moving averages are sometimes repeatedly applied to the same time series. Theorem 9.4.3 can be used to show that the repeated application of a moving average constructed to remove a low order polynomial trend will remove a high order polynomial trend.

Theorem 9.4.3. Let p and q be integers, $p \geqslant 0, q \geqslant 1$. A moving average constructed to remove a pth degree polynomial trend will reduce a $(p+q)$ th degree polynomial trend to degree $q-1$.

Proof. We write the trend function as

$$T_t = \sum_{r=0}^{p+q} b_r t^r .$$

If a moving average with weights $\{w_j : j = -M, -M+1, \ldots, M\}$ is applied to this function, we obtain

$$\sum_{j=-M}^{M} w_j T_{t+j} = \sum_{j=-M}^{M} w_j \sum_{r=0}^{p+q} b_r (t+j)^r$$

$$= \sum_{j=-M}^{M} w_j \sum_{r=0}^{p+q} b_r \sum_{k=0}^{r} \binom{r}{k} t^k j^{r-k} .$$

Interchanging the order of summation and using the fact that a filter constructed to remove a pth degree polynomial trend satisfies

$$\sum_{j=-M}^{M} w_j j^r = 0, \qquad r = 0, 1, 2, \ldots, p,$$

we obtain the conclusion. ▲

Moving averages have been heavily used in the analysis of time series displaying seasonal variation. Assume that Y_t is a monthly time series that can be represented locally by

$$Y_{t+j} = \alpha_t + \beta_t j + \sum_{m=1}^{12} \delta_{tm} D_{t+j,m} + Z_{t+j}, \tag{9.4.5}$$

where $\alpha_t, \beta_t, \delta_{tm}$ are parameters,

$$D_{tm} = \begin{cases} 1, & \text{if } \; Y_t \text{ is observed in month } m \\ 0, & \text{otherwise,} \end{cases}$$

$$\sum_{m=1}^{12} \delta_{tm} = 0,$$

and Z_t is a stationary time series with zero mean. In this representation $\alpha_t+\beta_t j$ is the trend component and $\Sigma_{m=1}^{12}\delta_{tm}D_{t+j,m}$ is the seasonal component. Since the sum of the seasonal effects is zero over a period of 12 observations, it follows that a moving average such as

$$\hat{\tau}_t=\frac{1}{12}\left[\frac{1}{2}Y_{t-6}+Y_{t-5}+Y_{t-4}+\ldots+Y_{t+4}+Y_{t+5}+\frac{1}{2}Y_{t+6}\right] \quad (9.4.6)$$

or

$$\frac{1}{36}\left[\frac{1}{2}Y_{t-18}+Y_{t-17}+Y_{t-16}+\ldots+Y_{t+16}+Y_{t+17}+\frac{1}{2}Y_{t+18}\right]$$

will not contain the seasonal component. Letting

$$d_0=\frac{11}{12},$$

$$d_j=-\frac{1}{12}, \qquad j=\pm 1, \pm 2, \ldots, \pm 5,$$

$$d_{-6}=d_6=-\frac{1}{24},$$

the difference

$$Y_t-\hat{\tau}_t=\sum_{m=1}^{12}\delta_{tm}D_{tm}+\sum_{j=-6}^{6}d_jZ_{t+j}, \quad (9.4.7)$$

constructed for the time series (9.4.5) contains the original seasonal component but no trend. A moving average of the time series $Y_t-\hat{\tau}_t$ can then be used to estimate the seasonal component of the time series. For example,

$$\frac{1}{5}\sum_{i=-2}^{2}(Y_{t+12i}-\hat{\tau}_{t+12i})=\hat{S}_t \quad (9.4.8)$$

furnishes an estimator of the seasonal component for time t based on five years of data. The 12 values of $\hat{S}_t$ computed for a year by this formula do not necessarily sum to zero. Therefore, one may modify the estimators to achieve a zero sum for the year by defining

$$\tilde{S}_{t(k)}=\hat{S}_t-\frac{1}{12}\sum_{j=-k+1}^{12-k}\hat{S}_{t+j}, \quad (9.4.9)$$

where $\tilde{S}_{t(k)}$ is the seasonal component at time t that is associated with the kth month. The seasonally adjusted time series is then given by the difference $Y_t - \tilde{S}_{t(k)}$.

The seasonal adjustment procedure described above was developed from an additive model and the adjustment was accomplished with a difference. It is quite common to specify a multiplicative model and use ratios in the construction.

It is also possible to construct directly a set of weights using regression procedures and a model such as (9.4.5). To illustrate the procedure of seasonal adjustment based directly on a regression model, we assume that a quarterly time series can be represented for a period of 21 quarters by

$$Y_{t+j} = \sum_{r=0}^{3} \beta_r j^r + \sum_{k=1}^{4} \alpha_k D_{t+j,k} + Z_{t+j}, \qquad j=0, \pm 1, \ldots, \pm 10,$$

where

$$D_{tk} = \begin{cases} 1, & \text{if } \quad Y_t \text{ is observed in quarter } k \\ 0, & \text{otherwise,} \end{cases}$$

and

$$\sum_{k=1}^{4} \alpha_k = 0.$$

We call $\sum_{k=1}^{4} \alpha_k D_{tk}$ the quarter (or seasonal) effect and $\sum_{r=0}^{3} \beta_r j^r$ the trend effect. The first seven columns of Table 9.4.2 give a $\mathbf{\Phi}$ matrix for this problem. We have incorporated the restriction $\sum_{k=1}^{4} \alpha_k = 0$ by setting $\alpha_4 = -\alpha_1 - \alpha_2 - \alpha_3$.

With our coding of the variables, the trend value for t, where t is the center observation, is given by β_0 and the seasonal value for t is given by α_1. To compute the weights needed for the trend value at t, we regress the β_0-column on the remaining six columns, compute the deviations, and divide each deviation by the sum of squares of the deviations. It is readily verified that this is equivalent to computing the vector of weights by

$$\mathbf{w}'_{\beta_0} = \mathbf{J}_{\beta_0} (\mathbf{\Phi}'\mathbf{\Phi})^{-1} \mathbf{\Phi}',$$

where

$$\mathbf{J}_{\beta_0} = (1, 0, 0, 0, 0, 0, 0).$$

Table 9.4.2. ***Calculation of weights for the trend and seasonal components of a quarterly time series***

	Φ							Weights		
Index j	β_0	β_1	β_2	β_3	α_1	α_2	α_3	w_{β_0}	w_{α_1}	w_A
−10	1	−10	100	−1000	0	0	1	−0.0477	−0.0314	0.0791
−9	1	−9	81	−729	−1	−1	−1	−0.0304	−0.0407	0.0711
−8	1	−8	64	−512	1	0	0	−0.0036	0.1562	−0.1526
−7	1	−7	49	−343	0	1	0	0.0232	−0.0469	0.0237
−6	1	−6	36	−216	0	0	1	0.0595	−0.0437	−0.0158
−5	1	−5	25	−125	−1	−1	−1	0.0634	−0.0515	−0.0119
−4	1	−4	16	−64	1	0	0	0.0768	0.1469	−0.2237
−3	1	−3	9	−27	0	1	0	0.0902	−0.0546	−0.0356
−2	1	−2	4	−8	0	0	1	0.1131	−0.0499	−0.0632
−1	1	−1	1	−1	−1	−1	−1	0.1036	−0.0562	−0.0474
0	1	0	0	0	1	0	0	0.1036	0.1438	0.7526
1	1	1	1	1	0	1	0	0.1036	−0.0562	−0.0474
2	1	2	4	8	0	0	1	0.1131	−0.0499	−0.0632
3	1	3	9	27	−1	−1	−1	0.0902	−0.0546	−0.0356
4	1	4	16	64	1	0	0	0.0768	0.1469	−0.2237
5	1	5	25	125	0	1	0	0.0634	−0.0515	−0.0119
6	1	6	36	216	0	0	1	0.0595	−0.0437	−0.0158
7	1	7	49	343	−1	−1	−1	0.0232	−0.0469	0.0237
8	1	8	64	512	1	0	0	−0.0036	0.1562	−0.1526
9	1	9	81	729	0	1	0	−0.0304	−0.0407	0.0711
10	1	10	100	1000	0	0	1	−0.0477	−0.0314	0.0791

The weights for the seasonal component at time t can be calculated by regressing the column associated with α_1 on the remaining columns, computing the deviations, and dividing each deviation by the sum of squares. These operations are equivalent to computing

$$\mathbf{w}'_{\alpha_1} = \mathbf{J}_{\alpha_1}(\mathbf{\Phi}'\mathbf{\Phi})^{-1}\mathbf{\Phi}',$$

where

$$\mathbf{J}_{\alpha_1} = (0,0,0,0,1,0,0).$$

Let $\mathbf{v}$ be a vector with 1 in the 11th (tth) position and zeros elsewhere. Then the weights for the trend and seasonally adjusted time series are given by

$$\mathbf{w}_A = \mathbf{v} - \mathbf{w}_{\beta_0} - \mathbf{w}_{\alpha_1}.$$

The weights $\mathbf{w}_A$ can also be obtained by regressing $\mathbf{v}$ on $\mathbf{\Phi}$ and computing the deviations from regression.

9.5. DIFFERENCES

Differences have been used heavily when the objective is to reduce a time series to stationarity, and there is little interest in estimating the mean function of the time series. Differences of the appropriate order will remove nonstationarity associated with locally polynomial trends in the mean and will reduce to stationarity integrated time series. For example, if Y_t is defined by

$$Y_t = \alpha + \beta t + W_t, \tag{9.5.1}$$

where

$$W_t = \sum_{j=0}^{t} Z_{t-j}$$

and Z_t is a stationary time series with zero mean, then

$$\Delta Y_t = Y_t - Y_{t-1} = \beta + Z_t$$

is a stationary time series.

The first difference operator can be viewed as a multiple of the moving average constructed to remove a zero degree polynomial trend from the second of two observations. The second difference is a multiple of the moving average constructed to remove a linear trend from the third of three observations[1], and so on. Therefore, the results of the previous section and of Section 2.4 may be combined in the following lemma.

Lemma 9.5.1. Let Y_t be a time series that is the sum of a polynomial trend of degree r and an integrated time series of order p. Then the qth difference of Y_t, where $r \leqslant q$ and $p \leqslant q$, is a stationary time series. The mean of $\Delta^q Y_t$ is zero for $r \leqslant q-1$.

Proof. See Corollary 2.4.1 and Theorems 9.4.2 and 9.4.3. ▲

Differences of lag other than one are important in transforming time series. For the function $f(t)$ defined on the integers, we define the *difference of lag H* by

$$\Delta_{(H)} f(t) = f(t) - f(t-H), \tag{9.5.2}$$

where H is a positive integer.

[1] The second difference can also be viewed as a multiple of the moving average constructed to remove a linear trend from the second of three observations, and so forth.

In Section 1.6 we defined a periodic function of period H with domain T to be a function satisfying

$$f(t+H)=f(t) \qquad \forall t, t+H \in T.$$

For T the set of integers and H a positive integer, the difference of lag H of a periodic function of period H is identically zero. Therefore differences of lag H have been used to remove seasonal and other periodic components from time series. For example, if the expected value of a monthly time series is written as

$$E\{Y_t\}=\mu+\beta t+\sum_{i=1}^{12}\alpha_i D_{ti}, \tag{9.5.3}$$

where

$$D_{ti}=\begin{cases}1, & \text{if } \; Y_t \text{ is observed in month } i\\ 0, & \text{otherwise,}\end{cases}$$

$$\sum_{i=1}^{12}\alpha_i=0,$$

then the expected value of $\Delta_{(12)}Y_t$ is $E\{Y_t-Y_{t-12}\}=12\beta$. The difference of lag 12 removed the periodic component and reduced the linear trend to a constant as well. Also, a difference of any finite lag will reduce a first order integrated time series to stationarity.

Mixtures of differences of different lags can be used. For example, if we take the first difference of the difference of lag 12 (or the difference of lag 12 of the first difference) of the time series Y_t of (9.5.3), the expected value of the resultant time series is zero; that is, $E\{\Delta_{(12)}\Delta Y_t\}=0$.

The effect of repeated application of differences of different lags is summarized in Lemma 9.5.2.

Lemma 9.5.2. Let the time series Y_t be the sum of (i) a polynomial trend of degree r, (ii) an integrated time series of order p, and (iii) a sum of periodic functions of order $H_1, H_2, \ldots, H_q$, where $r \leqslant q$ and $p \leqslant q$. Then the difference $X_t=\Delta_{(H_1)}\Delta_{(H_2)}\cdots\Delta_{(H_q)}Y_t$, where $1 \leqslant H_i < \infty$ for all i, is a stationary time series. If $r \leqslant q-1$, the mean of X_t is zero.

Proof. Reserved for the reader. ▲

9.6. SOME EFFECTS OF MOVING AVERAGE OPERATORS

Linear moving averages have well-defined effects on the correlation and spectral properties of stationary time series. These were discussed earlier

(see Theorems 2.1.1 and 4.3.1), but the effects of trend-removal filters are of sufficient importance to merit special investigation.

Proposition 9.6.1. Let X_t be a stationary time series with absolutely summable covariance function and let $Y_t = \sum_{j=-L}^{M} a_j X_{t-j}$, where L and M are nonnegative integers and the weights a_j satisfy $\sum_{j=-L}^{M} a_j = 0$. Then the spectral density of Y_t evaluated at zero is zero; that is, $f_Y(0) = 0$.

Proof. By Theorem 4.3.1, the spectral density of Y_t is given by

$$f_Y(\omega) = (2\pi)^2 f_a(\omega) f_a^*(\omega) f_X(\omega),$$

where

$$2\pi f_a(\omega) = \sum_{j=-L}^{M} a_j e^{-\iota\omega j}.$$

Since $\sum_{j=-L}^{M} a_j = 0$, it follows that $f_a(0) = 0$. ▲

Since $f_Y(0) = (2\pi)^{-1} \sum_{h=-\infty}^{\infty} \gamma_Y(h)$, it follows that the sum of the autocovariances is zero for a stationary time series that has been filtered to remove the mean. For example, one may check that the covariances of X_t and X_{t+h}, $h = -4, -3, \ldots, 4$, for the example of Table 9.4.1 sum to zero.

The weights in the last column of Table 9.4.1 were constructed to remove a quadratic trend. The transfer function of that filter is

$$2\pi f_a(\omega) = (70)^{-1}(36 - 48\cos\omega + 12\cos 2\omega)$$

and the squared gain is

$$|2\pi f_a(\omega)|^2 = (70)^{-2}(36 - 48\cos\omega + 12\cos 2\omega)^2.$$

The squared gain is plotted in Figure 9.6.1. The squared gain is zero at zero and rises very slowly. Since the weights remove a quadratic trend, the filter removes much of the power from the spectral density at low frequencies. Since the spectral density of white noise is constant, the squared gain is a multiple of the spectral density of a moving average of uncorrelated random variables where the coefficients are given in Table 9.4.1.

The squared gain of the first difference operator is displayed in Figure 9.6.2. While it is of the same general appearance as the function of Figure 9.6.1, the squared gain of the first difference operator rises from zero more rapidly.

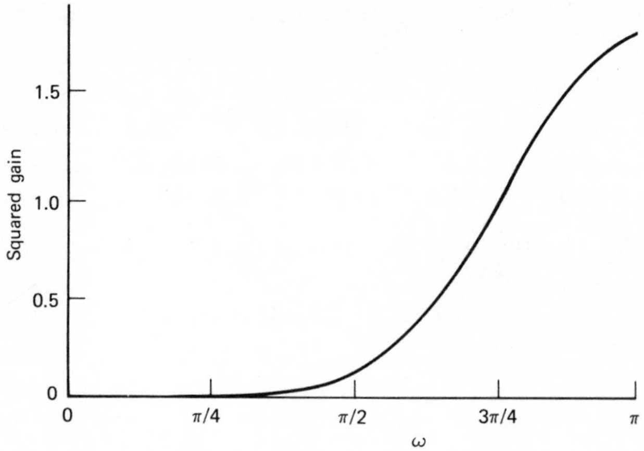

Figure 9.6.1. Squared gain of filter in Table 9.4.1.

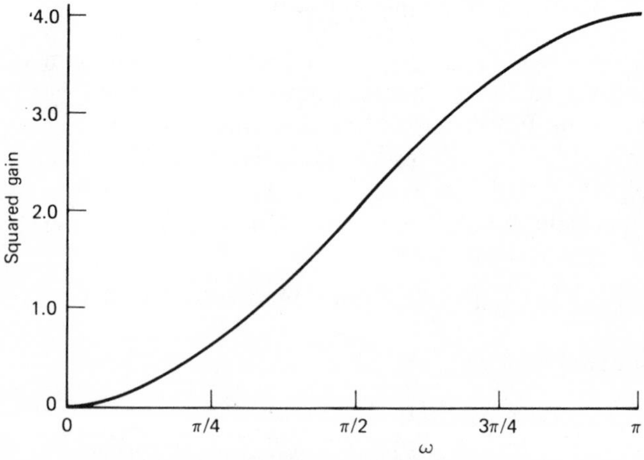

Figure 9.6.2. Squared gain of first difference operator.

If a time series contains a perfect sine component the difference of the time series will also contain a perfect sine component of the same period, but of different amplitude and phase. That is, if $Y_t = \sin \omega t$, the difference is

$$\begin{aligned}\Delta Y_t &= \sin \omega t - \sin \omega(t-1) \\ &= 2 \sin \frac{1}{2}\omega \cos \omega\left(t - \frac{1}{2}\right).\end{aligned}$$

We note that the amplitude of the sine wave is changed. If ω is such that $|\sin \frac{1}{2}\omega| < \frac{1}{2}$, the amplitude will be reduced. This will be true for long period waves. On the other hand, for short period waves, $\pi/3 < \omega < \pi$, the amplitude will be increased. Note that this agrees with Figure 9.6.2, which shows the transfer function to be greater than one for $\omega > \pi/3$.

Filtering a stationary time series with least squares weights to remove the seasonal effects will reduce the power of the spectral density to zero at the seasonal frequencies. For a time series with p observations per period of interest, the seasonal frequencies are defined to be $2mp^{-1}\pi, m = 1, 2, \ldots, L[p]$, where $L[p]$ is the largest integer less than or equal $p/2$. For example, with a monthly time series, the seasonal frequencies are $\pi/6$, $\pi/3$, $\pi/2$, $2\pi/3$, $5\pi/6$, and π. We have not included the zero frequency because most seasonal adjustment schemes are not constructed to remove the mean.

Proposition 9.6.2. Let a_j be a least squares linear filter of length R constructed to remove seasonal variation from a time series with p $(R \geqslant p)$ observations per period of interest. Let the time series have an absolutely summable covariance function. Then the spectral density of the filtered time series is zero at the seasonal frequencies.

Proof. A linear least squares filter constructed to remove the seasonal effects satisfies

$$\sum_{j=1}^{R} a_j \sin \frac{2\pi m}{p} j = 0,$$

$$\sum_{j=1}^{R} a_j \cos \frac{2\pi m}{p} j = 0, \qquad m = 1, 2, \ldots, L[p].$$

This is so because any periodic function of period p defined on the integers can be represented as a sum of p sines and cosines. We have

$$2\pi f_a(\omega) = \sum_{j=1}^{R} a_j e^{-\iota\omega j}$$

and, setting $\omega = 2\pi m/p$,

$$2\pi f_a\left(\frac{2\pi m}{p}\right) = \sum_{j=1}^{R} a_j \left(\cos \frac{2\pi m}{p} j - \iota \sin \frac{2\pi m}{p} j\right) = 0$$

for $m = 1, 2, \ldots, L[p]$. ▲

The first difference is a filter satisfying the conditions of Proposition 9.6.1, and the difference of lag p satisfies the conditions of Proposition 9.6.2. We now consider the effects of difference operators on autoregressive moving average time series.

Proposition 9.6.3. Let X_t be an autoregressive moving average time series of order (p,q) expressible as

$$\sum_{j=0}^{p} \alpha_j X_{t-j} = \sum_{r=0}^{q} \beta_r e_{t-r},$$

where $\{e_t\}$ is a sequence of uncorrelated $(0,\sigma^2)$ random variables. Let the roots of $m^p + \alpha_1 m^{p-1} + \cdots + \alpha_p = 0$ be less than one in absolute value and let the roots of $s^q + \beta_1 s^{q-1} + \cdots + \beta_q = 0$ be $s_1, s_2, \ldots, s_q$. Then the first difference $Y_t = X_t - X_{t-1}$ is an autoregressive moving average $(p,q+1)$ with the autoregressive portion unchanged and the roots of the moving average portion given by $s_1, s_2, \ldots, s_q, 1$.

Proof. The spectral density of X_t is given by

$$f_X(\omega) = \frac{\sigma^2}{2\pi} \frac{\left(\sum_{j=0}^{q} \beta_j e^{ij\omega}\right)\left(\sum_{j=0}^{q} \beta_j e^{-ij\omega}\right)}{\left(\sum_{j=0}^{p} \alpha_j e^{ij\omega}\right)\left(\sum_{j=0}^{p} \alpha_j e^{-ij\omega}\right)}$$

$$= \frac{\sigma^2}{2\pi} \frac{\prod_{j=1}^{q} \left(e^{i\omega} - s_j\right) \prod_{j=1}^{q} \left(e^{-i\omega} - s_j\right)}{\prod_{j=1}^{p} \left(e^{i\omega} - m_j\right) \prod_{j=1}^{p} \left(e^{-i\omega} - m_j\right)}.$$

The spectral density of Y_t is given by

$$f_Y(\omega) = (1 - e^{i\omega})(1 - e^{-i\omega}) f_X(\omega)$$

$$= (e^{i\omega} - 1)(e^{-i\omega} - 1) f_X(\omega)$$

$$= \frac{\sigma^2}{2\pi} \frac{\prod_{j=1}^{q+1} \left(e^{i\omega} - s_j\right) \prod_{j=1}^{q+1} \left(e^{-i\omega} - s_j\right)}{\prod_{j=1}^{p} \left(e^{i\omega} - m_j\right) \prod_{j=1}^{p} \left(e^{-i\omega} - m_j\right)},$$

where $s_{q+1} = 1$. ▲

It follows immediately from Proposition 9.6.3 that the kth difference of the stationary autoregressive moving average time series of order (p,q) is an autoregressive moving average time series of order $(p,q+k)$, where at least k of the roots of the auxiliary equation associated with the moving average are one. Differences of lag r have similar effects on a time series.

Proposition 9.6.4. Given a time series X_t satisfying the assumptions of Proposition 9.6.3, the difference of lag r of X_t is an autoregressive moving average time series of order $(p,q+r)$ where the autoregressive portion is unchanged and the roots of the moving average portion are $s_1,s_2,\ldots,s_q$ plus the r roots of unity.

Proof. The spectral density of $Y_t = X_t - X_{t-r}$ is given by

$$f_Y(\omega) = (1-e^{-i\omega r})(1-e^{i\omega r})f_X(\omega)$$

and, for example, the factor $(e^{-i\omega r}-1)$ can be written as

$$\prod_{j=q+1}^{q+r} \left(e^{-i\omega} - s_j\right),$$

where the s_j, $j=q+1,\ q+2,\ldots,q+r$, are the r roots of the equation

$$s^r = 1. \qquad \blacktriangle$$

These results have important practical implications. If one is attempting to identify an autoregressive moving average model for a time series that has been differenced, it is wise to consider a moving average of order at least as large as the order of differences applied to the original time series. If the time series was stationary before differencing, the characteristic polynomial of the moving average portion of the differenced time series contains at least one unit root and the estimation theory of Section 8.3 is not applicable.

9.7. REGRESSION WITH TIME SERIES ERRORS

In this section we treat the problem of obtaining estimates of the parameters of regression equations with time series errors. We investigated the properties of the ordinary least squares estimators in Section 9.1 and found that for some special types of independent variables the ordinary least squares estimators are asymptotically fully efficient. For other independent variables, including most stationary time series, the ordinary least squares estimators are not efficient. Moreover, in most situations, the variance estimators associated with ordinary least squares are biased.

To illustrate some of these ideas, consider the simple model

$$\begin{aligned} Y_t &= \beta X_t + Z_t, \\ Z_t &= \rho Z_{t-1} + e_t, \\ X_t &= \lambda X_{t-1} + u_t, \end{aligned} \tag{9.7.0}$$

where the e_t are normal independent $(0, \sigma_e^2)$ random variables, the u_t are normal independent $(0, \sigma_u^2)$ random variables, e_t is independent of u_j for all $t, j, |\rho| < 1$, and $|\lambda| < 1$.

Under model (9.7.0) the expected value of

$$\hat{\beta}_S = \frac{\sum_{t=1}^{n} X_t Y_t}{\sum_{t=1}^{n} X_t^2} = \beta + \frac{\sum_{t=1}^{n} X_t Z_t}{\sum_{t=1}^{n} X_t^2}$$

conditional on the observed $X_t, t = 1, 2, \ldots, n$, is β and, hence, $\hat{\beta}_S$ is unbiased.

The variance of the least squares estimator conditional on $X_1, X_2, \ldots, X_n$ is given by equation (9.1.4),

$$\operatorname{Var}\{\hat{\beta}_S | X_1, X_2, \ldots, X_n\} = \frac{(1/n^2) \sum_{t=1}^{n} \sum_{j=1}^{n} X_t X_j \rho^{|t-j|} \sigma_z^2}{\left((1/n) \sum_{t=1}^{n} X_t^2 \right)^2}.$$

Now

$$p\lim \frac{1}{n} \sum_{t=1}^{n} X_t^2 = \sigma_x^2$$

and

$$\begin{aligned} p\lim \frac{1}{n} \sum_{t=1}^{n} \sum_{j=1}^{n} X_t X_j \rho^{|t-j|} \sigma_z^2 &= \lim_{n \to \infty} \sum_{h=-(n-1)}^{n-1} \frac{n - |h|}{n} \lambda^{|h|} \rho^{|h|} \sigma_x^2 \sigma_z^2 \\ &= \frac{(1 + \rho\lambda) \sigma_x^2 \sigma_z^2}{1 - \rho\lambda}, \end{aligned}$$

where $\sigma_x^2 = \gamma_X(0)$ and $\sigma_z^2 = \gamma_Z(0)$. Hence,

$$\underset{n\to\infty}{p\lim}\, n \operatorname{Var}\{\hat{\beta}_S | X_1, X_2, \dots, X_n\} = \frac{(1+\rho\lambda)\sigma_z^2}{(1-\rho\lambda)\sigma_x^2}.$$

If ρ is known, the generalized least squares estimator is given by

$$\hat{\beta}_G = \frac{(1-\rho^2)X_1Y_1 + \sum_{t=2}^{n}(X_t - \rho X_{t-1})(Y_t - \rho Y_{t-1})}{(1-\rho^2)X_1^2 + \sum_{t=2}^{n}(X_t - \rho X_{t-1})^2}$$

with variance, conditional on $X_1, X_2, \dots, X_n$,

$$\operatorname{Var}\{\hat{\beta}_G | X_1, X_2, \dots, X_n\} = \frac{\sigma_e^2}{(1-\rho^2)X_1^2 + \sum_{t=2}^{n}(X_t - \rho X_{t-1})^2}.$$

It follows that the large sample relative efficiency of generalized least squares to ordinary least squares is

$$\underset{n\to\infty}{p\lim} \frac{\operatorname{Var}\{\hat{\beta}_S | X_1, X_2, \dots, X_n\}}{\operatorname{Var}\{\hat{\beta}_G | X_1, X_2, \dots, X_n\}} = \frac{(1+\rho\lambda)(1-2\rho\lambda+\rho^2)}{(1-\rho\lambda)(1-\rho^2)}.$$

This expression is greater than one for all nonzero (ρ, λ) less than one in absolute value. For $\rho = \lambda = (0.5)^{1/2}$, the relative efficiency is 300%!

The estimated variance of the ordinary least squares estimator obtained by the usual formulas is

$$\hat{V}_S\{\hat{\beta}_S\} = \left(\sum_{t=1}^{n} X_t^2\right)^{-1} s^2,$$

where

$$s^2 = \frac{1}{n-1}\sum_{t=1}^{n}(Y_t - \hat{\beta}_S X_t)^2.$$

Since $\operatorname{Var}\{\hat{\beta}_S\} = O(n^{-1})$, we have $\hat{\beta}_S - \beta = O_p(n^{-1/2})$ and

$$s^2 = \frac{1}{n-1}\sum_{t=1}^{n}\left[Z_t^2 - 2(\hat{\beta}_S - \beta)X_tZ_t + (\hat{\beta}_S - \beta)^2X_t^2\right]$$

$$= \frac{1}{n-1}\sum_{t=1}^{n} Z_t^2 + O_p(n^{-1}).$$

Therefore,

$$p\lim n\hat{V}_S\{\hat{\beta}_S\} = p\lim\left\{\frac{ns^2}{\sum_{t=1}^{n} X_t^2}\right\} = \frac{\sigma_z^2}{\sigma_x^2}$$

and

$$p\lim\left\{\frac{\hat{V}_S\{\hat{\beta}_S\} - \operatorname{Var}\{\hat{\beta}_S|X_1, X_2, \ldots, X_n\}}{\operatorname{Var}\{\hat{\beta}_S|X_1, X_2, \ldots, X_n\}}\right\} = \frac{-2\rho\lambda}{1+\rho\lambda}.$$

The common least squares estimator of variance will be an over- or under-estimate, depending on the signs of ρ and λ. If $\rho = \lambda = (0.5)^{1/2}$, the ordinary least squares estimator will be estimating a quantity approximately one third of the true variance! Given these results on the efficiency of ordinary least squares and on the bias in the ordinary least squares estimator of variance, it is natural to consider an estimator such as (9.1.3), where the estimated autocorrelations are used in place of the unknown parameters.

In Section 9.3 we established that the autocorrelations of the error time series could be estimated from the calculated regression residuals. Therefore, we are now in a position to estimate all of the parameters of a model such as

$$Y_t = \sum_{i=1}^{k} \varphi_{ti}\beta_i + Z_t,$$

$$Z_t = \sum_{j=1}^{p} \theta_j Z_{t-j} + e_t, \tag{9.7.1}$$

where the roots of

$$m^p - \sum_{j=1}^{p} \theta_j m^{p-j} = 0$$

are less than one in absolute value, the φ_{ti} satisfy (9.1.6), (9.1.7), and (9.1.8), and $\{e_t\}$ is a sequence of independent $(0,\sigma^2)$ random variables with $E\{e_t^4\}=\eta\sigma^4$. We denote the simple least squares estimator by $\hat{\boldsymbol{\beta}}=(\hat{\beta}_1,\hat{\beta}_2,\ldots,\hat{\beta}_k)'$ and the simple least squares residuals by $\hat{Z}_t$.

Proposition 9.7.1. Let Y_t be defined by model (9.7.1). If $\tilde{\boldsymbol{\theta}}$ is the estimator of $\boldsymbol{\theta}=(\theta_1,\theta_2,\ldots,\theta_p)'$ obtained by regressing $\hat{Z}_t$ on $\hat{Z}_{t-1},\hat{Z}_{t-2},\ldots,\hat{Z}_{t-p}$, $t=p+1,p+2,\ldots,n$, and $\hat{\boldsymbol{\theta}}$ is the estimator obtained by regressing Z_t on $Z_{t-1},Z_{t-2},\ldots,Z_{t-p}$, then

$$\tilde{\boldsymbol{\theta}}-\hat{\boldsymbol{\theta}}=O_p(n^{-1}).$$

Proof. The result is an immediate consequence of Theorem 9.3.1. ▲

Given an estimator of the correlational structure of the error time series, we wish to construct the estimated generalized least squares estimator. Often the most expeditious way to obtain this estimator is to transform the data. In the present case the Gram-Schmidt orthogonalization leads one to the transformed variables.

$$\begin{aligned}
\epsilon_1&=\delta_{11}Z_1,\\
\epsilon_2&=\delta_{22}Z_2-\delta_{21}Z_1,\\
&\;\vdots\\
\epsilon_p&=\delta_{pp}Z_p-\sum_{j=1}^{p-1}\delta_{p,p-j}Z_{p-j},\\
\epsilon_t&=e_t=Z_t-\sum_{j=1}^{p}\theta_jZ_{t-j},\qquad t=p+1,p+2,\ldots,n,
\end{aligned}\tag{9.7.2}$$

where $\delta_{11}=\gamma_Z^{-1/2}(0)\sigma$, $\delta_{22}=[\{1-\rho_Z^2(1)\}\gamma_Z(0)]^{-1/2}\sigma$,

$$\delta_{21}=\rho_Z(1)\left[\{1-\rho_Z^2(1)\}\gamma_Z(0)\right]^{-1/2}\sigma, \text{ etc.}$$

The ϵ_t are uncorrelated with constant variance σ^2. For the first order autoregressive time series the transformed variables are

$$\begin{aligned}
\epsilon_1&=(1-\rho^2)^{1/2}Z_1,\\
\epsilon_t&=Z_t-\rho Z_{t-1},\qquad t=2,3,\ldots,n.
\end{aligned}$$

As an explicit example of the transformation, consider the second order autoregressive process

$$Z_t=1.53Z_{t-1}-0.66Z_{t-2}+e_t.$$

If the variance of e_t is σ^2, then by (2.5.8) the variance of Z_t is $11.77\sigma^2$. The correlations for the time series are given in Table 6.4.2. The transformation is

$$\epsilon_1=(11.77)^{-1/2}Z_1,$$

$$\epsilon_2=(1.7718)^{-1/2}(Z_2-0.9217Z_1),$$

$$\epsilon_t=e_t=Z_t-1.53Z_{t-1}+0.66Z_{t-2}, \qquad t=3,4,\ldots,n.$$

To define the estimated generalized least squares estimator, we can express our original model (9.7.1) in the matrix notation of Section 9.1 as

$$\mathbf{y}=\mathbf{\Phi}\boldsymbol{\beta}+\mathbf{z}.$$

We let $\mathbf{V}=E\{\mathbf{zz}'\}$ and $\boldsymbol{\epsilon}=\mathbf{Tz}$, where $\mathbf{T}$ is the $n\times n$ transformation matrix defined in (9.7.2). Then the estimated generalized least squares estimator is obtained by regressing $\hat{\mathbf{T}}\mathbf{y}$ on $\hat{\mathbf{T}}\mathbf{\Phi}$,

$$\tilde{\boldsymbol{\beta}}=\left[\mathbf{\Phi}'\hat{\mathbf{T}}'\hat{\mathbf{T}}\mathbf{\Phi}\right]^{-1}\mathbf{\Phi}'\hat{\mathbf{T}}'\hat{\mathbf{T}}\mathbf{y}=\left[\mathbf{\Phi}'\hat{\mathbf{V}}^{-1}\mathbf{\Phi}\right]^{-1}\mathbf{\Phi}'\hat{\mathbf{V}}^{-1}\mathbf{y}, \tag{9.7.3}$$

where $\hat{\mathbf{V}}^{-1}=\hat{\mathbf{T}}'\hat{\mathbf{T}}\hat{\sigma}^{-2}$, $\hat{\sigma}^2=(n-2p)^{-1}\sum_{t=p+1}^{n}(\hat{Z}_t-\sum_{j=1}^{p}\tilde{\theta}_j\hat{Z}_{t-j})^2$, and $\hat{\mathbf{T}}$ is obtained from $\mathbf{T}$ by replacing θ_j by $\tilde{\theta}_j, \sigma^2$ by $\hat{\sigma}^2$, and so forth. The following theorem demonstrates that the loss in efficiency associated with the estimation of the autocorrelation structure is modest in large samples.

Theorem 9.7.1. Let Y_t satisfy model (9.7.1), where the φ_{ti} satisfy assumptions (9.1.6), (9.1.7), and (9.1.8). Then

$$\mathbf{D}_n(\hat{\boldsymbol{\beta}}_G-\tilde{\boldsymbol{\beta}})=O_p(n^{-1/2}),$$

where $\hat{\boldsymbol{\beta}}_G=[\mathbf{\Phi}'\mathbf{V}^{-1}\mathbf{\Phi}]^{-1}\mathbf{\Phi}'\mathbf{V}^{-1}\mathbf{y}, \tilde{\boldsymbol{\beta}}$ is defined in (9.7.3), and

$$\mathbf{D}_n=\operatorname{diag}\left\{\left(\sum_{t=1}^{n}\varphi_{t1}^2\right)^{1/2},\left(\sum_{t=1}^{n}\varphi_{t2}^2\right)^{1/2},\ldots,\left(\sum_{t=1}^{n}\varphi_{tk}^2\right)^{1/2}\right\}.$$

Proof. We have, for $\hat{\mathbf{T}}_{t.}$ the tth row of $\hat{\mathbf{T}}$ and $\boldsymbol{\varphi}_{.i}$ the ith column of $\mathbf{\Phi}$,

$$\hat{\epsilon}_t=\hat{\mathbf{T}}_{t.}\mathbf{z}=\epsilon_t-\sum_{j=1}^{p}(\tilde{\theta}_j-\theta_j)Z_{t-j},$$

$$\hat{w}_{ti}=\hat{\mathbf{T}}_{t.}\boldsymbol{\varphi}_{.i}=\varphi_{ti}-\sum_{j=1}^{p}\theta_j\varphi_{t-j,i}-\sum_{j=1}^{p}(\tilde{\theta}_j-\theta_j)\varphi_{t-j,i}$$

$$=w_{ti}-\sum_{j=1}^{p}(\tilde{\theta}_j-\theta_j)\varphi_{t-j,i},$$

for $t=p+1,p+2,\ldots,n$ with similar expressions holding for $t=1,2,\ldots,p$. By Proposition 9.7.1 and Theorem 8.2.1, $\tilde{\boldsymbol{\theta}}-\boldsymbol{\theta}=O_p(n^{-1/2})$, and it follows that

$$\frac{\sum_{t=1}^{n} \hat{\epsilon}_t \hat{w}_{ti}}{\left(\sum_{t=1}^{n} \varphi_{ti}^2\right)^{1/2}} = \frac{\sum_{t=1}^{n} \epsilon_t w_{ti}}{\left(\sum_{t=1}^{n} \varphi_{ti}^2\right)^{1/2}} + O_p(n^{-1/2}),$$

$$\frac{\sum_{t=1}^{n} \hat{w}_{ti} \hat{w}_{tj}}{\left(\sum_{t=1}^{n} \varphi_{ti}^2 \sum_{t=1}^{n} \varphi_{tj}^2\right)^{1/2}} = \frac{\sum_{t=1}^{n} w_{ti} w_{tj}}{\left(\sum_{t=1}^{n} \varphi_{ti}^2 \sum_{t=1}^{n} \varphi_{tj}^2\right)^{1/2}} + O_p(n^{-1/2}).$$

Therefore,

$$\begin{aligned}
\mathbf{D}_n(\tilde{\boldsymbol{\beta}}-\boldsymbol{\beta}) &= \left[\mathbf{D}_n^{-1}\boldsymbol{\Phi}'\hat{\mathbf{T}}'\hat{\mathbf{T}}\boldsymbol{\Phi}\mathbf{D}_n^{-1}\right]^{-1}\mathbf{D}_n^{-1}\boldsymbol{\Phi}'\hat{\mathbf{T}}'\hat{\mathbf{T}}\mathbf{z} \\
&= \left[\mathbf{D}_n^{-1}\boldsymbol{\Phi}'\mathbf{V}^{-1}\boldsymbol{\Phi}\mathbf{D}_n^{-1}\right]^{-1}\mathbf{D}_n^{-1}\boldsymbol{\Phi}'\mathbf{V}^{-1}\mathbf{z}+O_p(n^{-1/2}) \\
&= \mathbf{D}_n(\hat{\boldsymbol{\beta}}_G-\boldsymbol{\beta})+O_p(n^{-1/2}).
\end{aligned}$$

▲

As an immediate consequence of Theorem 9.7.1, the limiting distribution of $\mathbf{D}_n(\tilde{\boldsymbol{\beta}}-\boldsymbol{\beta})$ is the same as that of $\mathbf{D}_n(\hat{\boldsymbol{\beta}}_G-\boldsymbol{\beta})$. Since the residual mean square for the regression of $\hat{\mathbf{T}}\mathbf{y}$ on $\hat{\mathbf{T}}\boldsymbol{\Phi}$ converges in probability to σ^2, one can use the ordinary regression statistics for approximate tests and confidence intervals. We proved Theorem 9.7.1 for an autoregressive error structure, but the analogous result holds for moving average and autoregressive moving average Z_t.

Because of the infinite autoregressive nature of invertible moving average time series, an exact transformation of the form (9.7.2) is cumbersome. However, the approximate difference equation transformation will be adequate for many purposes. For example, if

$$Z_t = e_t + \beta e_{t-1}, \qquad |\beta|<1,$$

the transformation

$$\begin{aligned}
\epsilon_1 &= (1+\beta^2)^{-1/2} Z_1, \\
\epsilon_2 &= (1+\beta^2)^{1/2}(1+\beta^2+\beta^4)^{-1/2}\left[Z_2-(1+\beta^2)^{-1}\beta Z_1\right], \qquad (9.7.4)\\
\epsilon_t &= Z_t - \beta\epsilon_{t-1}, \qquad t=3,4,\ldots,n,
\end{aligned}$$

will generally be satisfactory when β is not too close to one in absolute value.

Example. To illustrate the fitting of a regression model with time series errors, we use the data studied by A. R. Prest (1949). The data were discussed by Durbin and Watson (1951), and we use the data as they appear in that article. The data, displayed in Table 9.7.1, pertain to the consumption of spirits in the United Kingdom from 1870 to 1938. The dependent variable Y_t is the annual per capita consumption of spirits in the United Kingdom. The explanatory variables φ_{t1} and φ_{t2} are per capita income and price of spirits, respectively, both deflated by a general price index. All data are in logarithms. The model suggested by Prest can be written

$$Y_t = \beta_0 + \beta_1 \varphi_{t1} + \beta_2 \varphi_{t2} + \beta_3 t + \beta_4 (t-35)^2 + Z_t,$$

where 1869 is the origin for t and we assume Z_t is a stationary time series.

The simple least squares regression gives the following equation:

$$\hat{Y}_t = \underset{(0.28)}{2.14} + \underset{(0.14)}{0.69}\,\varphi_{t1} - \underset{(0.05)}{0.63}\,\varphi_{t2} - \underset{(0.00087)}{0.0095}\,t - \underset{(0.000016)}{0.00011}\,(t-35)^2.$$

The residual mean square is 0.00098. The numbers in parentheses are the standard errors computed by the ordinary regression formulas and would be a part of the output in any standard regression computer program. They are *not* proper estimators of the standard errors when the error in the equation is autocorrelated. We shall return to this point.

The Durbin-Watson d is 0.5265. Under the null hypothesis of normal independent errors, the expected value of d is

$$E\{d\} = [2(68) - 0.67](64)^{-1} = 2.1145,$$

where

$$\operatorname{tr}\{(\Delta\Phi)'(\Delta\Phi)(\Phi'\Phi)^{-1}\} = 0.67.$$

Therefore, from equations (9.3.8) and (9.3.9),

$$\begin{aligned}\hat{\rho} &= 0.7367 + (0.5)(0.1145)(66)(65)^{-1}(1 - 0.5427)\\ &= 0.7633\end{aligned}$$

and

$$t_d = (66)^{1/2}(0.7633)(0.4174)^{-1/2} = 9.60.$$

Table 9.7.1. Annual consumption of spirits in the United Kindom from 1870 to 1938

Year	Consumption Y_t	Income φ_{t1}	Price φ_{t2}	Year	Consumption Y_t	Income φ_{t1}	Price φ_{t2}
1870	1.9565	1.7669	1.9176	1905	1.9139	1.9924	1.9952
1871	1.9794	1.7766	1.9059	1906	1.9091	2.0117	1.9905
1872	2.0120	1.7764	1.8798	1907	1.9139	2.0204	1.9813
1873	2.0449	1.7942	1.8727	1908	1.8886	2.0018	1.9905
1874	2.0561	1.8156	1.8984	1909	1.7945	2.0038	1.9859
1875	2.0678	1.8083	1.9137	1910	1.7644	2.0099	2.0518
1876	2.0561	1.8083	1.9176	1911	1.7817	2.0174	2.0474
1877	2.0428	1.8067	1.9176	1912	1.7784	2.0279	2.0341
1878	2.0290	1.8166	1.9420	1913	1.7945	2.0359	2.0255
1879	1.9980	1.8041	1.9547	1914	1.7888	2.0216	2.0341
1880	1.9884	1.8053	1.9379	1915	1.8751	1.9896	1.9445
1881	1.9835	1.8242	1.9462	1916	1.7853	1.9843	1.9939
1882	1.9773	1.8395	1.9504	1917	1.6075	1.9764	2.2082
1883	1.9748	1.8464	1.9504	1918	1.5185	1.9965	2.2700
1884	1.9629	1.8492	1.9723	1919	1.6513	2.0652	2.2430
1885	1.9396	1.8668	2.0000	1920	1.6247	2.0369	2.2567
1886	1.9309	1.8783	2.0097	1921	1.5391	1.9723	2.2988
1887	1.9271	1.8914	2.0146	1922	1.4922	1.9797	2.3723
1888	1.9239	1.9166	2.0146	1923	1.4606	2.0136	2.4105
1889	1.9414	1.9363	2.0097	1924	1.4551	2.0165	2.4081
1890	1.9685	1.9548	2.0097	1925	1.4425	2.0213	2.4081
1891	1.9727	1.9453	2.0097	1926	1.4023	2.0206	2.4367
1892	1.9736	1.9292	2.0048	1927	1.3991	2.0563	2.4284
1893	1.9499	1.9209	2.0097	1928	1.3798	2.0579	2.4310
1894	1.9432	1.9510	2.0296	1929	1.3782	2.0649	2.4363
1895	1.9569	1.9776	2.0399	1930	1.3366	2.0582	2.4552
1896	1.9647	1.9814	2.0399	1931	1.3026	2.0517	2.4838
1897	1.9710	1.9819	2.0296	1932	1.2592	2.0491	2.4958
1898	1.9719	1.9828	2.0146	1933	1.2635	2.0766	2.5048
1899	1.9956	2.0076	2.0245	1934	1.2549	2.0890	2.5017
1900	2.0000	2.0000	2.0000	1935	1.2527	2.1059	2.4958
1901	1.9904	1.9939	2.0048	1936	1.2763	2.1205	2.4838
1902	1.9752	1.9933	2.0048	1937	1.2906	2.1205	2.4636
1903	1.9494	1.9797	2.0000	1938	1.2721	2.1182	2.4580
1904	1.9332	1.9772	1.9952				

SOURCE: Durbin and Watson (1951). Reproduced with permission of the Trustees of Biometrika and of the authors.

Obviously, the hypothesis of independent errors is rejected. To investigate the nature of the autocorrelation, we fit several autoregressive models to the regression residuals. The results are summarized in Table 9.7.2. The statistics of that table are consistent with the hypothesis that the error time series is a first order autoregressive process. Therefore, we construct the transformation of (9.7.2) for a first order autoregressive process with $\hat{\rho}=0.7633$. The transformation matrix is

$$\hat{\mathbf{T}}=\begin{bmatrix} 0.6460 & 0 & 0 & \cdots & 0 \\ -0.7633 & 1 & 0 & \cdots & 0 \\ 0 & -0.7633 & 1 & \cdots & 0 \\ \vdots & \vdots & \vdots & & \vdots \\ 0 & 0 & 0 & \cdots & 1 \end{bmatrix},$$

where $\hat{\mathbf{T}}$ is a 69×69 matrix and $0.6460=[1-\hat{\rho}^2]^{1/2}=[1-(0.7633)^2]^{1/2}$.

Regressing the transformed Y_t on the transformed $\varphi_{ti}, i=0,1,\ldots,4$, we obtain the estimated generalized least squares equation

$$\hat{Y}=\underset{(0.30)}{2.36}+\underset{(0.15)}{0.72}\,\varphi_{t1}-\underset{(0.07)}{0.80}\,\varphi_{t2}-\underset{(0.0011)}{0.0081}\,t-\underset{(0.000026)}{0.000092}\,(t-35)^2.$$

The numbers in parentheses are consistent estimates of the standard errors of the estimators. They are computed from the transformed data by the usual regression formulas. The error mean square for the transformed regression is 0.000417.

Every one of the estimated standard errors for the generalized least squares estimator based on $\hat{\rho}=0.7633$ is greater than the standard error calculated by the usual formulas for the simple least squares estimates. Of course, we know that the standard formulas give biased estimators for simple least squares when the errors are autocorrelated. To demonstrate

Table 9.7.2. Analysis of variance for residuals from spirit consumption regression

Source	Degrees of Freedom	Mean Square
$\hat{Z}_{t-1}$	1	0.03174
$\hat{Z}_{t-2}$ after $\hat{Z}_{t-1}$	1	0.00036
$\hat{Z}_{t-3}$ after $\hat{Z}_{t-1}, \hat{Z}_{t-2}$	1	0.00013
$\hat{Z}_{t-4}$ after $\hat{Z}_{t-1}, \hat{Z}_{t-2}, \hat{Z}_{t-3}$	1	0.00004
Error	56	0.00049

the magnitude of this bias, we estimate the variance of the simple least squares estimators, assuming that the error time series is

$$Z_t = 0.7633 Z_{t-1} + e_t,$$

where e_t is a sequence of uncorrelated (0, 0.000417) random variables. The covariance matrix of the simple least squares estimator is estimated by $(\mathbf{\Phi}'\mathbf{\Phi})^{-1}\mathbf{\Phi}'\hat{\mathbf{V}}\mathbf{\Phi}(\mathbf{\Phi}'\mathbf{\Phi})^{-1}$, where $\mathbf{\Phi}$ is the 69×5 matrix of observations on the independent variables and $\hat{\mathbf{V}}$ is the estimated covariance matrix for 69 observations from a first order autoregressive time series with $\hat{\rho} = 0.7633$.

Alternative estimates of the standard errors of the coefficients are compared in Table 9.7.3. The consistent estimates of the standard errors of simple least squares obtained from the matrix $(\mathbf{\Phi}'\mathbf{\Phi})^{-1}\mathbf{\Phi}'\hat{\mathbf{V}}\mathbf{\Phi}(\mathbf{\Phi}'\mathbf{\Phi})^{-1}$ are approximately twice those obtained from the simple formulas. This means that the variance estimates computed under the assumption of zero correlation are approximately one fourth of the consistent estimates. The estimated standard errors for the generalized least squares estimators range from about one half (0.56 for income) to nine tenths (0.85 for time squared) of those for simple least squares. This means that the estimated efficiency of simple least squares relative to generalized least squares ranges from about three tenths to about seven tenths.

Table 9.7.3. Comparison of alternative estimators of standard errors of estimators for spirits example

Coefficients	Estimator for Simple Least Squares Assuming $\rho = 0$	Consistent Estimator for Simple Least Squares	Consistent Estimator for Generalized Least Squares
β_0	0.2821	0.5449	0.3045
β_1	0.1375	0.2618	0.1456
β_2	0.0545	0.1123	0.0720
β_3	0.000875	0.001791	0.001080
β_4	0.0000161	0.0000313	0.0000266

9.8. REGRESSION EQUATIONS WITH LAGGED DEPENDENT VARIABLES AND TIME SERIES ERRORS

In the classical regression problem the error in the equation is assumed to be distributed independently of the independent variables in the equation. Independent variables with this property we call *ordinary*. In our investigation of the estimation of the autoregressive parameters in Chapter

8 we noted that the autoregressive representation had the appearance of a regression problem, but that the vector of errors was not independent of the vector of lagged values of the time series. A generalization of the autoregressive model is one containing ordinary independent variables and also lagged values of the dependent variable. One of the simpler such models is

$$Y_t = \beta_0 + \beta_1 \varphi_t + \lambda Y_{t-1} + e_t, \qquad t = 1, 2, \ldots, \tag{9.8.1}$$

where $|\lambda| < 1$. It is assumed that the e_t are independently distributed with zero mean and variance σ^2, and that there is a positive real L and some $\delta > 0$ such that $E\{|e_t|^{2+\delta}\} < L^{2+\delta}$ for all t. It is assumed that $\{\varphi_t\}$ is a sequence of constants satisfying:

$$\lim_{n\to\infty} \sum_{t=1}^{n} \varphi_t^2 = \infty;$$

$$\lim_{n\to\infty} \left(\sum_{t=1}^{n} \varphi_t^2 \right)^{-1} \sum_{t=1}^{n-h} \varphi_t \varphi_{t+h} = a_h, \qquad h = 0, 1, 2, \ldots; \tag{9.8.2}$$

$$\lim_{n\to\infty} \sup_{1 \le t \le n} \frac{n^{1-\epsilon} \varphi_t^2}{\sum_{j=1}^{n} \varphi_j^2} = 0, \text{ for } \epsilon = (5 + 5\delta)^{-1}\delta.$$

The third condition is stronger than the analogous condition of (9.1.6) used in Section 9.1. However, a wide variety of functions of time, including polynomials, satisfies the conditions (9.8.2), for all $\epsilon > 0$.

Given model (9.8.1), we investigate the least squares estimator of $\boldsymbol{\theta} = (\beta_0, \beta_1, \lambda)'$,

$$\hat{\boldsymbol{\theta}} = (\hat{\beta}_0, \hat{\beta}_1, \hat{\lambda})' = \mathbf{H}_n^{-1} \sum_{t=1}^{n} (1, \varphi_t, Y_{t-1})' Y_t, \tag{9.8.3}$$

where

$$\mathbf{H}_n = \begin{bmatrix} n & \sum_{t=1}^{n} \varphi_t & \sum_{t=1}^{n} Y_{t-1} \\ \sum_{t=1}^{n} \varphi_t & \sum_{t=1}^{n} \varphi_t^2 & \sum_{t=1}^{n} \varphi_t Y_{t-1} \\ \sum_{t=1}^{n} Y_{t-1} & \sum_{t=1}^{n} \varphi_t Y_{t-1} & \sum_{t=1}^{n} Y_{t-1}^2 \end{bmatrix}.$$

By substituting the definition of Y_t from (9.8.1) into (9.8.3), we have $(\hat{\boldsymbol{\theta}}-\boldsymbol{\theta})=\mathbf{H}_n^{-1}\mathbf{R}_n$, where

$$\mathbf{R}_n'=\left(\sum_{t=1}^{n} e_t,\ \sum_{t=1}^{n}\varphi_t e_t,\ \sum_{t=1}^{n} Y_{t-1}e_t\right).$$

Theorem 9.8.1. Let Y_t satisfy

$$Y_t=\beta_0+\beta_1\varphi_t+\lambda Y_{t-1}+e_t,\qquad t=1,2,\ldots,$$

where $|\lambda|<1$ and $\{e_t\}$ is a sequence of independent $(0,\sigma^2)$ random variables with $E\{|e_t|^{2+\delta}\}<L^{2+\delta}$ for some $L>0,\delta>0$, and all t. Let assumptions (9.8.2) hold. Let

$$\mathbf{G}=p\lim \mathbf{D}_n^{-1}\mathbf{H}_n\mathbf{D}_n^{-1}$$

be nonsingular, where

$$\mathbf{D}_n=\operatorname{diag}\left\{n^{1/2},\left(\sum_{t=1}^{n}\varphi_t^2\right)^{1/2},\left(\sum_{t=1}^{n}Y_{t-1}^2\right)^{1/2}\right\}.$$

Then,

$$\mathbf{D}_n(\hat{\boldsymbol{\theta}}-\boldsymbol{\theta})\xrightarrow{\mathcal{L}}N(\mathbf{0},\mathbf{G}^{-1}\sigma^2).$$

Proof. The model (9.8.1) is a difference equation in Y_t and, by (2.4.10), can be written as

$$\begin{aligned}Y_t&=\beta_0\frac{1-\lambda^t}{1-\lambda}+\beta_1\sum_{j=0}^{t-1}\lambda^j\varphi_{t-j}+\lambda^tY_0+\sum_{j=0}^{t-1}\lambda^je_{t-j}\\&=\beta_0(1-\lambda)^{-1}(1-\lambda^t)+\beta_1\xi_t+\lambda^tY_0+u_t,\qquad t=1,2,\ldots,\end{aligned}\tag{9.8.4}$$

where

$$\xi_t=\sum_{j=0}^{t-1}\lambda^j\varphi_{t-j},$$

$$u_t=\sum_{j=0}^{t-1}\lambda^je_{t-j}.$$

We first show that $\{\xi_t\}$ satisfies assumptions (9.8.2). Now

$$\sum_{t=1}^{n}\left(\sum_{j=0}^{t-1}\lambda^j\varphi_{t-j}\right)^2 \leqslant \frac{1}{2}\sum_{t=1}^{n}\sum_{j=0}^{t-1}\sum_{i=0}^{t-1}|\lambda|^j|\lambda|^i\left[\varphi_{t-j}^2+\varphi_{t-i}^2\right]$$

$$\leqslant \sum_{t=1}^{n}\sum_{j=0}^{t-1}\frac{|\lambda|^j}{1-|\lambda|}\varphi_{t-j}^2$$

$$=(1-|\lambda|)^{-1}\sum_{t=1}^{n}\sum_{i=0}^{n-t}|\lambda|^i\varphi_t^2$$

$$\leqslant(1-|\lambda|)^{-2}\sum_{t=1}^{n}\varphi_t^2.$$

Consider

$$\frac{\sum_{t=1}^{n}\xi_t^2}{\sum_{t=1}^{n}\varphi_t^2}=\frac{\sum_{t=1}^{n}\left(\lambda^M\xi_{t-M}+\lambda^{M-1}\varphi_{t-M+1}+\dots+\varphi_t\right)^2}{\sum_{t=1}^{n}\varphi_t^2}$$

where $\varphi_t, \xi_t=0$ for $t\leqslant 0$ and M is a positive integer. Since

$$\left(\sum_{t=1}^{n}\varphi_t^2\right)^{-1}\lambda^{2M}\sum_{t=1}^{n}\xi_{t-M}^2\leqslant\lambda^{2M}(1-|\lambda|)^{-2},$$

$$\left(\sum_{t=1}^{n}\varphi_t^2\right)^{-1}\lambda^{M}\sum_{t=1}^{n}\xi_{t-M}\varphi_t\leqslant(1-|\lambda|)^{-1}\lambda^{M},$$

and

$$\left(\sum_{t=1}^{n}\varphi_t^2\right)^{-1}\sum_{t=1}^{n}\left(\sum_{j=0}^{M-1}\lambda^j\varphi_{t-j}\right)^2$$

converges for fixed M, the first two assumptions of (9.8.2) are satisfied by $\{\xi_t\}$. Because

$$\sup_{1\leqslant t\leqslant n}\xi_t^2\leqslant(1-|\lambda|)^{-2}\sup_{1\leqslant t\leqslant n}\varphi_t^2$$

the third condition is also satisfied.

By the properties of ξ_t, the fact that u_t converges to a first order autoregressive time series, and Lemma 6.5.1, the probability limit $\mathbf{G}$ is well defined. To obtain the limiting distribution of $\mathbf{D}_n^{-1}\mathbf{R}_n$, we follow the proofs of Theorems 6.3.3 and 6.3.2. By (9.8.4), Y_t is a sum of fixed and random parts, and we define a truncated version of the random portion

$$u_{tm} = \sum_{j=0}^{\min(m-1,t-1)} \lambda^j e_{t-j}$$

in the same manner as in Theorem 6.3.3. Let

$$W_t = (g_{t-1} + u_{t-1,m})e_t,$$

where $g_{t-1} = \beta_0(1-\lambda)^{-1}(1-\lambda^{t-1}) + \beta_1\xi_{t-1} + \lambda^{t-1}Y_0$. It follows that W_t is a sequence of m-dependent random variables with zero mean and covariance, for $t \geqslant m+1$,

$$\begin{aligned} E\{W_t W_{t+h}\} &= g_{t-1}^2\sigma^2 + \frac{1-\lambda^{2m}}{1-\lambda^2}\sigma^4, && h=0 \\ &= 0, && \text{otherwise.} \end{aligned}$$

Let

$$V_n = \sigma^2 \sum_{t=1}^{n} \left\{ g_{t-1}^2 + (1-\lambda^2)^{-1}(1-\lambda^{2m})\sigma^2 \right\}$$

and consider

$$V_n^{-1/2} \sum_{t=1}^{n} Y_{t-1}e_t = V_n^{-1/2} \left\{ \sum_{t=1}^{n} W_t + D_{mn} \right\},$$

where

$$\begin{aligned} D_{mn} &= 0, && n \leqslant m+1 \\ &= \sum_{t=m+2}^{n} \sum_{j=m}^{t-2} \lambda^j e_{t-j-1} e_t, && n > m+1. \end{aligned}$$

Since $\lim_{m\to\infty} \operatorname{Var}\{V_n^{-1/2}D_{mn}\} = 0$ uniformly in n, by Lemma 6.3.1 we need only establish the asymptotic normality of $V_n^{-1/2}\Sigma_{t=1}^n W_t$ for a fixed m. We

follow the proof of Theorem 6.3.2 and define

$$X_i = \sum_{j=1}^{k-m} W_{(i-1)k+j}, \qquad i=1,2,\ldots,p,$$

$$S_p = V_n^{-1/2} \sum_{i=1}^{p} X_i,$$

where k is the largest integer less than n^α, $(5+5\delta)^{-1}\delta < \alpha < (4+4\delta)^{-1}\delta$, and p is the largest integer less than $n^{1-\alpha}$. Now

$$\operatorname{Var}\left\{ V_n^{-1/2} \sum_{t=1}^{n} W_t - S_p \right\} \leq V_n^{-1}\sigma^2 \left\{ \sup_{1 \leq t \leq n} g_t^2 + (1-\lambda^2)^{-1}\sigma^2 \right\}(m+1)p$$

and the difference converges to zero in mean square. We have

$$\left[E\left\{ \left| \sum_{j=1}^{k-m} W_{(i-1)k+j} \right|^{2+\delta} \right\} \right]^{1/(2+\delta)} \leq \sum_{j=1}^{k-m} \left[E\left\{ |W_{(i-1)k+j}|^{2+\delta} \right\} \right]^{1/(2+\delta)}$$

$$\leq kL\left[\sup_{1 \leq t \leq n} |g_{t-1}| + L(1-|\lambda|)^{-1} \right], \qquad i=1,2,\ldots,p.$$

Hence,

$$\lim_{n\to\infty} V_n^{-1} \left[\sum_{i=1}^{p} E\left\{ |X_i|^{2+\delta} \right\} \right]^{2/(2+\delta)}$$

$$\leq \lim_{n\to\infty} V_n^{-1} p^{2/(2+\delta)} k^2 L^2 \left[\sup_{1 \leq t \leq n} |g_t| + L(1-|\lambda|)^{-1} \right]^2 = 0,$$

since $p^{2/(2+\delta)}k^2 < n^{1-\epsilon}$. Therefore, S_p satisfies the conditions of Liapounov's central limit theorem and, by Lemma 6.3.1,

$$\left[\sum_{t=1}^{n} \left\{ g_{t-1}^2 + (1-\lambda^2)^{-1}\sigma^2 \right\} \sigma^2 \right]^{-1/2} \sum_{t=1}^{n} Y_{t-1} e_t$$

converges to a normal (0, 1) random variable. As

$$p\lim \left\{ \sum_{t=1}^{n} \left[g_{t-1}^2 + (1-\lambda^2)^{-1}\sigma^2 \right] \right\}^{-1} \sum_{t=1}^{n} Y_{t-1}^2 = 1,$$

$(\Sigma_{t=1}^{n} Y_{t-1}^{2})^{-1/2}\Sigma_{t=1}^{n} Y_{t-1}e_t$ converges to a normal $(0,\sigma^2)$ random variable.

If we consider a linear combination of the elements of $\mathbf{D}_n^{-1}\mathbf{R}_n$, say

$$\sum_{t=1}^{n}\left[\frac{c_0}{n^{1/2}}+c_1\frac{\varphi_t}{\left(\sum_{j=1}^{n}\varphi_j^2\right)^{1/2}}+c_2\frac{g_{t-1}+u_{t-1}}{\left(\sum_{j=1}^{n}Y_{j-1}^2\right)^{1/2}}\right]e_t,$$

we see that each element of the sum is the sum of a fixed and random part multiplied by e_t. This is of the same general form as

$$\left(\sum_{t=1}^{n}Y_{t-1}^2\right)^{-1/2}\sum_{t=1}^{n}Y_{t-1}e_t.$$

Therefore, for a linear combination with $c_2\neq 0$ the arguments developed for $\Sigma Y_{t-1}e_t$ apply. For a linear combination with $c_2=0$ asymptotic normality follows from Theorem 6.3.4. ▲

There was nothing in the proof of Theorem 9.8.1 that precluded the equation from containing additional independent variables, φ_{ti}, as long as they satisfied the assumptions. On the other hand, the independence of the e_t was critical to the proof. If the e_t are autocorrelated, $E\{Y_{t-1}e_t\}$ may not be zero, in which case the least squares estimator is no longer consistent.

To obtain consistent estimators for the parameters in the presence of autocorrelated errors, we use the method of instrumental variables introduced in Section 5.6. The lagged values of φ_{ti} are the natural instrumental variables.

To simplify the presentation we consider a model with two independent variables and two lagged values of Y_t,

$$Y_t=\beta_1\varphi_{t1}+\beta_2\varphi_{t2}+\lambda_1Y_{t-1}+\lambda_2Y_{t-2}+Z_t, \tag{9.8.5}$$

where $Z_t=\Sigma_{j=0}^{\infty}b_je_{t-j}$, the b_j are absolutely summable, the roots of $m^2-\lambda_1m-\lambda_2=0$ are less than one in absolute value, and the e_t are independent $(0,\sigma^2)$ random variables with $E\{|e_t|^{2+\delta}\}=L^{2+\delta}$ for some $\delta>0$. The conclusions will generalize immediately to any number of φ_{ti} and any number of lagged Y's.

The first step in the instrumental variable estimation is the regression of Y_{t-1} and Y_{t-2} on φ_{t1}, $\varphi_{t-1,1}$, $\varphi_{t-2,1}$, φ_{t2}, $\varphi_{t-1,2}$, and $\varphi_{t-2,2}$. The predicted values $\hat{Y}_{t-1}$ and $\hat{Y}_{t-2}$ are obtained from these regressions and the instrumental variable estimates are obtained by regressing Y_t on φ_{t1}, φ_{t2}, $\hat{Y}_{t-1}$, $\hat{Y}_{t-2}$.

We assume the matrix of sums of squares and products of $\varphi_{t1}, \varphi_{t2}, \varphi_{t-1,1}, \varphi_{t-2,1}, \varphi_{t-1,2}, \varphi_{t-2,2}$ to be nonsingular and to satisfy assumption 1 of Theorem 5.6.1. We require only two instrumental variables in order to estimate the parameters λ_1 and λ_2, and some of the variables may be omitted if singularities appear in practice. For example, if $\mathbf{\Phi}=(\boldsymbol{\varphi}_{.0}, \boldsymbol{\varphi}_{.1}, \boldsymbol{\varphi}_{.2})$, where $\varphi_{t0}\equiv 1$ and $\varphi_{t1}=t$, then the matrix with rows $(1, t, \varphi_{t2}, t-1, \varphi_{t-1,2}, t-2, \varphi_{t-2,2})$ will be singular. One may either omit variables from the regression, in the example case omitting $\varphi_{t-1,1}$ and $\varphi_{t-2,1}$, or use a generalized inverse in the computation of $\hat{Y}_{t-1}$ and $\hat{Y}_{t-2}$.

Theorem 9.8.2. Let $\{Y_t\colon t\in(1,2,\ldots)\}$ satisfy

$$Y_t=\beta_1\varphi_{t1}+\beta_2\varphi_{t2}+\lambda_1 Y_{t-1}+\lambda_2 Y_{t-2}+Z_t,$$

where

$$Z_t=\sum_{j=0}^{\infty} b_j e_{t-j},$$

$$\sum_{j=0}^{\infty} |b_j|<\infty,$$

the roots of $m^2-\lambda_1 m-\lambda_2=0$ are less than one in absolute value, $\beta_1\neq 0$ or $\beta_2\neq 0$, and the e_t are independent $(0,\sigma^2)$ random variables with $E\{|e_t|^{2+\delta}\}=\mathrm{L}^{2+\delta}$ for some $\delta>0$. Let φ_{ti}, $i=1, 2$, satisfy assumptions (9.8.2) and (9.1.7). Let

$$\mathbf{Q}=\lim_{n\to\infty}\mathbf{Q}_n$$

be nonsingular where

$$\mathbf{Q}_n=\mathbf{D}_{1n}^{-1}(\mathbf{\Phi};\boldsymbol{\psi})'(\mathbf{\Phi};\boldsymbol{\psi})\mathbf{D}_{1n}^{-1},$$

$(\mathbf{\Phi};\boldsymbol{\psi})$ is the matrix whose tth row is $(\varphi_{t1}, \varphi_{t2}, \varphi_{t-1,1}, \varphi_{t-2,1}, \varphi_{t-1,2}, \varphi_{t-2,2})$ for $t=3,4,\ldots,n$, and

$$\mathbf{D}_{1n}^2=\operatorname{diag}\left[\sum_{t=3}^{n}\varphi_{t1}^2, \sum_{t=3}^{n}\varphi_{t2}^2, \sum_{t=2}^{n-1}\varphi_{t1}^2, \sum_{t=1}^{n-2}\varphi_{t1}^2, \sum_{t=2}^{n-1}\varphi_{t2}^2, \sum_{t=1}^{n-2}\varphi_{t2}^2\right].$$

Let $\hat{\boldsymbol{\theta}}=(\hat{\beta}_1,\hat{\beta}_2,\hat{\lambda}_1,\hat{\lambda}_2)$ be the instrumental variable estimator, and let

$$\mathbf{D}_{2n}=\operatorname{diag}\left\{\left(\sum_{t=1}^{n}\varphi_{t1}^2\right)^{1/2}, \left(\sum_{t=1}^{n}\varphi_{t2}^2\right)^{1/2}, S_n, S_n\right\},$$

where $S_n^2 = \sum_{t=1}^{n}(\beta_1\varphi_{t1} + \beta_2\varphi_{t2})^2$. Then,

$$\mathbf{D}_{2n}(\hat{\boldsymbol{\theta}} - \boldsymbol{\theta}) = O_p(1).$$

Proof. By the difference equation properties of our model, we may write

$$Y_t = \beta_1 \sum_{j=0}^{t-1} w_j \varphi_{t-j,1} + \beta_2 \sum_{j=0}^{t-1} w_j \varphi_{t-j,2} + U_t + c_{1t}Y_0 + c_{2t}Y_{-1}, \quad (9.8.6)$$

where $U_t = \sum_{j=0}^{t-1} w_j Z_{t-j}$, the w_j are defined in terms of λ_1 and λ_2 by Theorem 2.6.1, and c_{1t} and c_{2t} go to zero as t goes to infinity. Equation (9.8.6) gives Y_{t-1} and Y_{t-2} as a function of lagged values of φ_{t1} and φ_{t2} and of a random term U_t. By our assumptions on the φ_{ti} and since β_1 and β_2 are not both zero, the current and lagged values of φ_{ti} satisfy the conditions placed on $\boldsymbol{\Phi}$ and $\boldsymbol{\psi}$ in Theorem 5.6.1. Likewise, Z_t satisfies the conditions on Z_t of Theorem 5.6.1.

The $\mathbf{B}_n$ of assumption 4 of Theorem 5.6.1 is

$$\mathbf{B}_n = \mathbf{D}_{1n}^{-1}(\boldsymbol{\Phi};\boldsymbol{\psi})'\boldsymbol{\Gamma}(\boldsymbol{\Phi};\boldsymbol{\psi})\mathbf{D}_{1n}^{-1},$$

where the ijth element of $\boldsymbol{\Gamma}$ is $\gamma_Z(|i-j|)$, and the matrix $\mathbf{B} = \lim_{n\to\infty}\mathbf{B}_n$ is of the same form as the matrix $\mathbf{B}$ of Theorem 9.1.1. The matrices $\mathbf{G}_{nij}$ of assumption 4 of Theorem 5.6.1 are of a similar form with the covariance matrix of U_{t-1}, of U_{t-2}, or of U_{t-1} with U_{t-2} replacing that of Z_t.

Therefore, the conditions of Theorem 5.6.1 are satisfied and the order in probability of the error in the estimators is given by that theorem. ▲

By Theorem 9.8.2 the instrumental variable estimator is consistent when the error in the equation is a stationary moving average time series. In most applications we will estimate a parametric model for the error time series.

Given the instrumental variable estimates $\hat{\beta}_1$, $\hat{\beta}_2$, $\hat{\lambda}_1$, $\hat{\lambda}_2$, we can calculate the estimated residuals

$$\hat{Z}_t = Y_t - \hat{\beta}_1\varphi_{t1} - \hat{\beta}_2\varphi_{t2} - \hat{\lambda}_1 Y_{t-1} - \hat{\lambda}_2 Y_{t-2}. \quad (9.8.7)$$

The presence of the lagged Y-values in the equation means that the results of Theorem 9.3.1 do not hold for autocorrelations computed from these residuals. However, we are able to obtain a somewhat weaker result.

Theorem 9.8.3. Given the assumptions of Theorem 9.8.2, the sample

covariances calculated from the instrumental variable residuals satisfy

$$\hat{\gamma}_{\hat{Z}}(h) - \hat{\gamma}_Z(h) = O_p(r_n),$$

where

$$\hat{\gamma}_{\hat{Z}}(h) = \frac{1}{n}\sum_{t=3}^{n-h} \hat{Z}_t\hat{Z}_{t+h}, \quad \hat{\gamma}_Z(h) = \frac{1}{n}\sum_{t=3}^{n-h} Z_tZ_{t+h}, \qquad h=0,1,2,\ldots,n-1,$$

$\hat{Z}_t$ is defined in (9.8.7), $r_n = \max\{n^{-1/2}, S_n^{-1}\}$, and S_n is defined in Theorem 9.8.2.

Proof. Using the definition of $\hat{Z}_t$, we have, for $h=0,1,\ldots,n-1$,

$$\begin{aligned}\hat{\gamma}_{\hat{Z}}(h) &= \frac{1}{n}\sum_{t=3}^{n-h} \hat{Z}_t\hat{Z}_{t+h} \\ &= \frac{1}{n}\Bigg[\sum_{t=3}^{n-h} Z_tZ_{t+h} - (\hat{\boldsymbol{\theta}}-\boldsymbol{\theta})'\sum_{t=3}^{n-h}\mathbf{A}'_{t.}Z_{t+h} \\ &\quad - \sum_{t=3}^{n-h} Z_t\mathbf{A}_{t+h,.}(\hat{\boldsymbol{\theta}}-\boldsymbol{\theta}) \\ &\quad + (\hat{\boldsymbol{\theta}}-\boldsymbol{\theta})'\sum_{t=3}^{n-h}\mathbf{A}'_{t.}\mathbf{A}_{t+h,.}(\hat{\boldsymbol{\theta}}-\boldsymbol{\theta})\Bigg],\end{aligned}$$

where $\mathbf{A}_{t.} = (\varphi_{t1}, \varphi_{t2}, Y_{t-1}, Y_{t-2})$. Now, for example,

$$\begin{aligned}\frac{1}{n}\sum_{t=3}^{n-h} Y_{t-1}Z_{t+h} &\leqslant n^{-1/2}\left(\sum_{t=3}^{n-h} Y_{t-1}^2\right)^{1/2}\left(\frac{1}{n}\sum_{t=3}^{n-h} Z_{t+h}^2\right)^{1/2} \\ &= O_p\left(\max\{1, n^{-1/2}S_n\}\right).\end{aligned}$$

Therefore,

$$\frac{1}{n}(\hat{\boldsymbol{\theta}}-\boldsymbol{\theta})'\mathbf{D}_{2n}\mathbf{D}_{2n}^{-1}\sum_{t=3}^{n-h}\mathbf{A}'_{t.}Z_{t+h} = O_p\left(\max\{n^{-1/2}, S_n^{-1}\}\right),$$

$$\begin{aligned}\frac{1}{n}(\hat{\boldsymbol{\theta}}-\boldsymbol{\theta})'\mathbf{D}_{2n}\mathbf{D}_{2n}^{-1}&\sum_{t=3}^{n-h}\mathbf{A}'_{t.}\mathbf{A}_{t+h,.}\mathbf{D}_{2n}^{-1}\mathbf{D}_{2n}(\hat{\boldsymbol{\theta}}-\boldsymbol{\theta}) \\ &= O_p\left(\max\{n^{-1/2}, S_n^{-1}\}\right),\end{aligned}$$

where $\mathbf{D}_{2n}$ is defined in Theorem 9.8.2. ▲

Since the estimated autocorrelations are consistent, the estimators of a parametric model for the time series based on a continuous function of a finite number of these autocorrelations will have an error of the same order as the estimated autocorrelations. To introduce such models, let Z_t be the first order autoregressive process

$$Z_t = \rho Z_{t-1} + e_t, \tag{9.8.8}$$

where $|\rho|<1$ and the e_t are independent $(0,\sigma^2)$ random variables. We may then write (9.8.5) as a nonlinear model with lagged values of Y_t among the explanatory variables and whose error term, e_t, is a sequence of independent random variables. That is, substituting (9.8.5) into (9.8.8), we obtain

$$\begin{aligned} Y_t &= \beta_1\varphi_{t1} + \beta_2\varphi_{t2} + (\lambda_1+\rho)Y_{t-1} + (\lambda_2-\rho\lambda_1)Y_{t-2} \\ &\quad - \rho\beta_1\varphi_{t-1,1} - \rho\beta_2\varphi_{t-1,2} - \rho\lambda_2 Y_{t-3} + e_t \\ &\overset{\text{(say)}}{=} f(\mathbf{W}_t;\boldsymbol{\theta}) + e_t, \qquad t=4,5,\ldots,n, \end{aligned} \tag{9.8.9}$$

where $\mathbf{W}_t = (\varphi_{t1}, \varphi_{t2}, \varphi_{t-1,1}, \varphi_{t-1,2}, Y_{t-1}, Y_{t-2}, Y_{t-3})$ and $\boldsymbol{\theta}' = (\theta_1, \theta_2, \theta_3, \theta_4, \theta_5) = (\beta_1, \beta_2, \lambda_1, \lambda_2, \rho)$. Note that we have expanded the vector of parameters to include ρ.

The instrumental variables estimators are consistent estimators of $(\beta_1, \beta_2, \lambda_1, \lambda_2)$ and, by Theorem 9.8.3,

$$\hat{\rho} = [\hat{\gamma}_{\hat{Z}}(0)]^{-1}\hat{\gamma}_{\hat{Z}}(1) \tag{9.8.10}$$

is a consistent estimator of ρ. Therefore, we are in a position to carry out an iteration of the Gauss-Newton procedure discussed in Section 5.5. Our present situation violates some of the assumptions of that section, but we shall be able to obtain the conclusion of Theorem 5.7.1 under our current conditions.

The estimation procedure consists of the following operations.

1. Use the method of instrumental variables to estimate the parameters of (9.8.5).
2. Calculate the estimated residuals (9.8.7) and estimate the parameter of the autoregressive error structure by (9.8.10).
3. On the basis of a first order Taylor's expansion of the model (9.8.9) about the initial estimates, $\hat{\boldsymbol{\theta}} = (\hat{\beta}_1, \hat{\beta}_2, \hat{\lambda}_1, \hat{\lambda}_2 \hat{\rho})$, estimate the differences between the initial estimators and the parameters by using the regression

equation

$$\begin{aligned}\hat{e}_t \doteq{}& (\varphi_{t1}-\hat{\rho}\varphi_{t-1,1})(\beta_1-\hat{\beta}_1)+(\varphi_{t2}-\hat{\rho}\varphi_{t-1,2})(\beta_2-\hat{\beta}_2)\\&+(Y_{t-1}-\hat{\rho}Y_{t-2})(\lambda_1-\hat{\lambda}_1)+(Y_{t-2}-\hat{\rho}Y_{t-3})(\lambda_2-\hat{\lambda}_2)\\&+\hat{Z}_{t-1}(\rho-\hat{\rho})+e_t, \qquad t=4,5,\ldots,n,\end{aligned} \tag{9.8.11}$$

where

$$\begin{aligned}\hat{e}_t={}& Y_t-\hat{\rho}Y_{t-1}-(\varphi_{t1}-\hat{\rho}\varphi_{t-1,1})\hat{\beta}_1-(\varphi_{t2}-\hat{\rho}\varphi_{t-1,2})\hat{\beta}_2\\&-(Y_{t-1}-\hat{\rho}Y_{t-2})\hat{\lambda}_1-(Y_{t-2}-\hat{\rho}Y_{t-3})\hat{\lambda}_2,\end{aligned}$$

$\hat{Z}_{t-1}$ is defined in (9.8.7), and $\hat{\rho}$ is defined in (9.8.10). On the basis of transformation (9.7.2), the first observation in the regression is given by

$$\begin{aligned}(1-\hat{\rho}^2)^{1/2}\hat{Z}_3 \doteq{}& (1-\hat{\rho}^2)^{1/2}\varphi_{31}(\beta_1-\hat{\beta}_1)+(1-\hat{\rho}^2)^{1/2}\varphi_{32}(\beta_2-\hat{\beta}_2)\\&+(1-\hat{\rho}^2)^{1/2}Y_2(\lambda_1-\hat{\lambda}_1)+(1-\hat{\rho}^2)^{1/2}Y_1(\lambda_2-\hat{\lambda}_2)\\&+(1-\hat{\rho}^2)^{1/2}Z_3.\end{aligned}$$

We note that by rearranging terms, the regression equations can be written as

$$\begin{aligned}(1-\hat{\rho}^2)^{1/2}Y_3 \doteq{}& (1-\hat{\rho}^2)^{1/2}\varphi_{31}\beta_1+(1-\hat{\rho}^2)^{1/2}\varphi_{32}\beta_2+(1-\hat{\rho}^2)^{1/2}Y_2\lambda_1\\&+(1-\hat{\rho}^2)^{1/2}Y_1\lambda_2+(1-\hat{\rho}^2)^{1/2}Z_3, \qquad t=3,\end{aligned}$$

$$\begin{aligned}Y_t-\hat{\rho}Y_{t-1} \doteq{}& (\varphi_{t1}-\hat{\rho}\varphi_{t-1,1})\beta_1+(\varphi_{t2}-\hat{\rho}\varphi_{t-1,2})\beta_2\\&+(Y_{t-1}-\hat{\rho}Y_{t-2})\lambda_1+(Y_{t-2}-\hat{\rho}Y_{t-3})\lambda_2+\hat{Z}_{t-1}(\rho-\hat{\rho})\\&+e_t, \qquad t=4,5,\ldots,n.\end{aligned} \tag{9.8.12}$$

Thus, if the regression equations of (9.8.12) are used, the improved estimators $\tilde{\beta}_1$, $\tilde{\beta}_2$, $\tilde{\lambda}_1$, and $\tilde{\lambda}_2$ are the regression coefficients of $\varphi_{t1}-\hat{\rho}\varphi_{t-1,1}$, $\varphi_{t2}-\hat{\rho}\varphi_{t-1,2}$, $Y_{t-1}-\hat{\rho}Y_{t-2}$, and $Y_{t-2}-\hat{\rho}Y_{t-3}$, respectively (with appropriate modifications for the first observation). The improved estimator of ρ is

$$\tilde{\rho}=\hat{\rho}+\Delta\tilde{\rho},$$

where $\Delta\tilde{\rho}$ is the regression coefficient of $\hat{Z}_{t-1}$.

We now obtain the asymptotic properties of these estimators.

Theorem 9.8.4. Let the conditions of Theorem 9.8.2 hold with the added assumption that Z_t is the first order autoregressive process of (9.8.8). Let the r_n of Theorem 9.8.3 be $o(n^{-1/4})$. Then one iteration of the Gauss-Newton estimator using the instrumental variable estimator of $(\beta_1, \beta_2, \lambda_1, \lambda_2)$ and the estimator (9.8.10) of ρ as the start values is such that

$$\mathbf{D}_{3n}(\tilde{\boldsymbol{\theta}} - \boldsymbol{\theta}) \xrightarrow{\mathcal{L}} N(\mathbf{0}, \mathbf{H}^{-1}\sigma^2),$$

where $\tilde{\boldsymbol{\theta}}$ is the one step Gauss-Newton estimator of $\boldsymbol{\theta} = (\beta_1, \beta_2, \lambda_1, \lambda_2, \rho)'$,

$$\mathbf{H} = p\lim \mathbf{D}_{3n}^{-1}[\mathbf{F}'(\boldsymbol{\theta})\mathbf{F}(\boldsymbol{\theta})]\mathbf{D}_{3n}^{-1},$$

$\mathbf{D}_{3n}$ is a diagonal matrix whose elements are the square roots of the diagonal elements of $\mathbf{F}'(\boldsymbol{\theta})\mathbf{F}(\boldsymbol{\theta})$ and $\mathbf{F}(\boldsymbol{\theta})$ is the $(n-2)\times 5$ matrix with rows

$$\begin{aligned}
&(1-\rho^2)^{-1/2}(\varphi_{31}, \varphi_{32}, Y_2, Y_1, 0), && t=3\\
&(\varphi_{t1} - \rho\varphi_{t-1,1}, \varphi_{t2} - \rho\varphi_{t-1,2}, Y_{t-1} - \rho Y_{t-2}, Y_{t-2} - \rho Y_{t-3}, Z_{t-1}), && t \geq 4.
\end{aligned}$$

Proof. Using equation (9.8.12) and the approach of Section 5.5, we can write

$$\begin{aligned}
(\tilde{\boldsymbol{\theta}} - \boldsymbol{\theta}) = {} & [\mathbf{F}'(\hat{\boldsymbol{\theta}})\mathbf{F}(\hat{\boldsymbol{\theta}})]^{-1}\mathbf{F}'(\boldsymbol{\theta})\mathbf{e}\\
& + [\mathbf{F}'(\hat{\boldsymbol{\theta}})\mathbf{F}(\hat{\boldsymbol{\theta}})]^{-1}\mathbf{R}(\hat{\boldsymbol{\theta}}),
\end{aligned}$$

where $\hat{\boldsymbol{\theta}}$ is the vector of initial estimators, $\mathbf{e}' = [(1-\rho^2)^{-1/2} Z_3, e_4, e_5, \ldots, e_n]$, and the jth row of $\mathbf{R}(\hat{\boldsymbol{\theta}})$ is

$$\sum_{t=4}^{n} \sum_{s=1}^{5} f^{(js)}(\mathbf{W}_t; \boldsymbol{\theta})[\hat{\theta}_s - \theta_s]e_t + O_p(1),$$

with, for $t > 3$,

$$\begin{aligned}
f^{(ij)}(\mathbf{W}_t; \boldsymbol{\theta}) &= 0, && i=1,2,3,4, j=1,2,3,4,\\
&= -\varphi_{t-1,1}, && (i,j)=(1,5) \text{ or } (5,1)\\
&= -\varphi_{t-1,2}, && (i,j)=(2,5) \text{ or } (5,2)\\
&= -Y_{t-2}, && (i,j)=(3,5) \text{ or } (5,3),\\
&= -Y_{t-3}, && (i,j)=(4,5) \text{ or } (5,4),\\
&= 0, && (i,j)=(5,5).
\end{aligned}$$

The contribution to $\mathbf{R}(\hat{\boldsymbol{\theta}})$ from $t=3$ is $O_p(1)$. Therefore,

$$\mathbf{D}_{3n}^{-1}\mathbf{R}(\hat{\boldsymbol{\theta}})=o_p(1)$$

and, by similar arguments,

$$p\lim \mathbf{D}_{3n}^{-1}\mathbf{F}'(\hat{\boldsymbol{\theta}})\mathbf{F}(\hat{\boldsymbol{\theta}})\mathbf{D}_{3n}^{-1}=\mathbf{H}.$$

The elements of the vector $\mathbf{D}_{3n}^{-1}\mathbf{F}'(\boldsymbol{\theta})\mathbf{e}$ are of the same form as the elements of $\mathbf{D}_n^{-1}\mathbf{R}_n$ investigated in Theorem 9.8.1, and asymptotic normality follows by the arguments of that theorem. ▲

Clearly, the results of Theorem 9.8.4 generalize immediately to an equation with any number of ordinary independent variables, any number of lagged values of the dependent variable and a stationary pth order autoregressive error.

As in our previous applications of the Gauss-Newton procedure, the estimator of the covariance matrix of the estimator is given at the last step of the calculations by the usual regression estimator.

Theorem 9.8.5. Let the assumptions of Theorem 9.8.4 be satisfied. Then

$$p\lim \mathbf{D}_{3n}^{-1}\left[\mathbf{F}'(\tilde{\boldsymbol{\theta}})\mathbf{F}(\tilde{\boldsymbol{\theta}})\right]\mathbf{D}_{3n}^{-1}s^2=\mathbf{H}^{-1}\sigma^2,$$

where

$$s^2=\frac{1}{n-7}\sum_{t=3}^{n}\tilde{e}_t^2$$

and

$$\begin{aligned}
\tilde{e}_3&=(1-\tilde{\rho}^2)^{1/2}(Y_3-\varphi_{31}\tilde{\beta}_1-\varphi_{32}\tilde{\beta}_2-Y_2\tilde{\lambda}_1-Y_1\tilde{\lambda}_2),\\
\tilde{e}_t&=Y_t-\tilde{\rho}Y_{t-1}-(\varphi_{t1}-\tilde{\rho}\varphi_{t-1,1})\tilde{\beta}_1-(\varphi_{t2}-\tilde{\rho}\varphi_{t-1,2})\tilde{\beta}_2\\
&\quad-(Y_{t-1}-\tilde{\rho}Y_{t-2})\tilde{\lambda}_1-(Y_{t-2}-\tilde{\rho}Y_{t-3})\tilde{\lambda}_2, \qquad t=4,5,\ldots,n.
\end{aligned}$$

Proof. Reserved for the reader. ▲

The model of Section 9.7 contained only ordinary independent variables. In that situation we obtained a consistent estimator of the variance from the sums of squares and cross products of the transformed explanatory variables. For a model with lagged values of the dependent variable

among the explanatory variables and autocorrelated errors, consistent estimation of the covariance matrix requires that the appropriate lagged values of $\hat{Z}_t$ be included in the matrix.

Example. To illustrate the estimation of a model containing lagged values of the dependent variable, we use a model studied by Ball and St. Cyr (1966). The data, displayed in Table 9.8.1, are for production and employment in the bricks, pottery, glass, and cement industry of the United Kingdom. The quarterly data cover the period 1955–1 to 1964–2.

Table 9.8.1. Production index and employment of the bricks, pottery, glass, and cement industry of the United Kingdom

Year-Quarter	Logarithm of Production Index	Logarithm of Employment (000)
1955–1	4.615	5.844
2	4.682	5.846
3	4.644	5.852
4	4.727	5.864
1956–1	4.625	5.855
2	4.700	5.844
3	4.635	5.841
4	4.663	5.838
1957–1	4.625	5.823
2	4.644	5.817
3	4.575	5.823
4	4.635	5.817
1958–1	4.595	5.802
2	4.625	5.790
3	4.564	5.790
4	4.654	5.784
1959–1	4.595	5.778
2	4.700	5.781
3	4.654	5.802
4	4.727	5.814
1960–1	4.719	5.820
2	4.787	5.826
3	4.727	5.838
4	4.787	5.841

Table 9.8.1. (Continued)

1961–1	4.771	5.838
2	4.812	5.844
3	4.796	5.846
4	4.828	5.855
1962–1	4.762	5.852
2	4.868	5.864
3	4.804	5.864
4	4.844	5.861
1963–1	4.754	5.844
2	4.875	5.835
3	4.860	5.841
4	4.963	5.852
1964–1	4.956	5.849
2	5.004	5.858

SOURCE. Ball and St. Cyr (1966). Reproduced from *Review of Economic Studies* Vol. 33, No. 3 with the permission of the authors and editor.

The model studied by Ball and St. Cyr can be written as

$$Y_t = \beta_0 + \beta_1 \varphi_t + \beta_2 t + \beta_3 D_{t1} + \beta_4 D_{t2} + \beta_5 D_{t3} + \lambda Y_{t-1} + Z_t,$$

where

$Y_t =$ the logarithm of employment in thousands of the bricks, pottery, glass, and cement industry of the United Kingdom,

$\varphi_t =$ the logarithm of the production index of the industry,

$t =$ time with 1954–4 as the origin,

$$D_{tj} = \begin{cases} 1, & \text{if observation } t \text{ occurs in quarter } j \\ -1, & \text{if observation } t \text{ occurs in quarter four} \\ 0, & \text{otherwise.} \end{cases}$$

We assume Z_t is a zero mean stationary time series, $\beta_1 \neq 0$, and $|\lambda| < 1$.

To construct the instrumental variable estimators, we first regress Y_{t-1} on φ_t, φ_{t-1}, t, D_{t1}, D_{t2}, D_{t3}, and a constant term. Lagged values of time and the quarter dummies are not included in the regression, since they are a linear function of the current values and the constant function. This regression gives the estimate

$$\hat{Y}_{t-1} = 4.20 + 0.375\varphi_{t-1} - 0.020\varphi_t - 0.0023t$$
$$- 0.0064D_{t1} + 0.0082D_{t2} - 0.0146D_{t3}.$$

Regressing Y_t on $\hat{Y}_{t-1}$ and the other explanatory variables, we obtain

$$\hat{Y}_t = 0.27 + 0.077\varphi_t - 0.00042t - 0.00492D_{t1}$$
$$- 0.00336D_{t2} + 0.00728D_{t3} + 0.894Y_{t-1}$$

as the instrumental variable estimate. The next step in the estimation is to calculate

$$\hat{Z}_t = Y_t - 0.27 - 0.077\varphi_t + 0.00042t + 0.00492D_{t1}$$
$$+ 0.00336D_{t2} - 0.00728D_{t3} - 0.894Y_{t-1}.$$

Note that the original Y_{t-1}, not its predicted value, is used in the calculation of $\hat{Z}_t$. An analysis of the $\hat{Z}_t$ such as that in Table 9.7.2 indicated that the data are consistent with the hypothesis that Z_t is a first order autoregressive process. The estimated parameter is 0.42.

The data are transformed using the transformation (9.7.2) for a first order autoregressive process with $\rho = 0.42$. The lagged residuals $\hat{Z}_{t-1}$ are added to the set of explanatory variables, after the other explanatory variables have been transformed. The regression on the transformed explanatory variables and $\hat{Z}_{t-1}$ gives

$$\hat{Y}_t = \underset{(0.41)}{0.88} + \underset{(0.032)}{0.097\varphi_t} - \underset{(0.00030)}{0.00055t} - \underset{(0.00186)}{0.00368D_{t1}}$$
$$- \underset{(0.00181)}{0.00432D_{t2}} + \underset{(0.00162)}{0.00734D_{t3}} + \underset{(0.081)}{0.773Y_{t-1}}.$$

The first five observations used in this regression are displayed in Table 9.8.2. The column headed φ_{t0} is for the constant term in the regression equation. The estimator for ρ obtained at this step is $0.420 + 0.048 = 0.468$, where 0.048 is the coefficient of $\hat{Z}_{t-1}$. The residual mean square is

0.0000334. If the procedure is iterated the estimates converge to

$$\begin{aligned}\hat{Y}_t = \underset{(0.43)}{0.84} + \underset{(0.031)}{0.098}\varphi_t - \underset{(0.00029)}{0.00057}t - \underset{(0.00192)}{0.00365}D_{t1}\\ - \underset{(0.00182)}{0.00437}D_{t2} + \underset{(0.00163)}{0.00742}D_{t3} + \underset{(0.086)}{0.777}Y_{t-1},\end{aligned}$$

$$\hat{\rho} = \underset{(0.194)}{0.386},$$

where the numbers in parentheses are the estimated standard errors. Although this is a relatively small sample with a fairly large number of explanatory variables, iteration did not change the estimates to any substantial extent. Because of the small sample and the presence of time among the explanatory variables, it is a reasonable conjecture that the estimator $\hat{\rho}$ possesses a negative bias.

Table 9.8.2. First five observations for the Gauss-Newton regression for the bricks, pottery, glass, and cement industry, $\hat{\rho}=0.42$

Year-Quarter	Transformed Observations								
	Y_t	φ_{t0}	φ_t	t	D_{t1}	D_{t2}	D_{t3}	Y_{t-1}	$\hat{Z}_{t-1}$
1955–2	5.306	0.908	4.249	1.82	0.00	0.908	0.00	5.303	0
1955–3	3.397	0.580	2.678	2.16	0.00	−0.420	1.00	3.392	0.0035
1955–4	3.406	0.580	2.777	2.74	−1.00	−1.000	−1.42	3.397	−0.0007
1956–1	3.392	0.580	2.639	3.32	1.42	0.420	0.42	3.406	0.0059
1956–2	3.384	0.580	2.758	3.90	−0.42	1.000	0.00	3.392	0.0014

There are two important reasons for recognizing the potential for autocorrelated errors in an analysis such as this one. First, the simple least squares estimators are no longer consistent when the errors are autocorrelated. Second, the estimators of the variances of the estimators may be seriously biased. A hypothesis of some interest in the study of employment was the hypothesis that $\beta_1 + \lambda = 1$. This corresponds to the hypothesis of constant returns to scale for labor input. In the simple least squares analysis the sum of these coefficients is 0.89 with an estimated standard error of 0.045. The resulting "t-statistic" of -2.5 suggests that there are increasing returns to scale to labor in this industry. On the other hand, the analysis recognizing the possibility of autocorrelated errors estimates the sum as 0.88, but with a standard error of 0.072. The associated "t-statistic" of -1.7 and the potential for small sample bias in the estimators could lead one to accept the hypothesis that the sum is one.

REFERENCES

Section 9.1. Anderson (1971), Grenander (1954), Grenander and Rosenblatt (1957), Hannan (1970).

Section 9.2. Ahlberg, Nilson, and Walsh (1967), Fuller (1969), Gallant and Fuller (1973), Greville (1969), Poirier (1973).

Section 9.3. Durbin and Watson (1950), (1951), Hannan (1969), von Neumann (1941), (1942).

Section 9.4. Davis (1941), Grether and Nerlove (1970), Jorgenson (1964), Kendall and Stuart (1966), Lovell (1963), Malinvaud (1970a), Nerlove (1964), Stephenson and Farr (1972), Tintner (1942), (1952).

Sections 9.5 and 9.6. Davis (1941), Durbin (1962), (1963), Hannan (1958), Tintner (1940), (1952), Rao and Tintner (1963).

Section 9.7. Amemiya (1973), Cochran and Orcutt (1949), Durbin (1960), Hildreth (1969), Malinvaud (1970a), Prest (1949).

Section 9.8. Aigner (1971), Amemiya and Fuller (1967), Ball and St. Cyr (1966), Dhyrmes (1971), Fuller and Martin (1961), Griliches (1967), Hatanaka (1974), Malinvaud (1970a), Wallis (1967).

EXERCISES

1. Let the time series X_t be defined by

$$X_t = \beta_0 + \beta_1 t + Z_t,$$

where

$$Z_t = \rho Z_{t-1} + e_t, \qquad |\rho| < 1,\ t = 0, \pm 1, \pm 2, \ldots,$$

and $\{e_t\}$ is a sequence of independent $(0, \sigma^2)$ random variables. Compute the covariance matrices (9.1.4) and (9.1.5) for $\rho = 0.9$ and $n = 5$ and 10.

2. The data below are the forecasts of the yield of corn in Illinois released by the United States Department of Agriculture in August of the production year.
   - (a) For this set of data estimate a mean function meeting the following restrictions.
     - (i) The mean function is linear from 1944 through 1956, quadratic from 1956 through 1960, linear from 1960 through 1965, quadratic from 1965 through 1969 and linear from 1969 through 1974.
     - (ii) The mean function, considered as a continuous function of time, has a continuous first derivative.
   - (b) Reestimate the mean function subject to the restriction that the derivative is zero for the last five years. Test the hypothesis that the slope for the last five years is zero.
   - (c) Plot the data and the estimated mean function.

August forecast of corn yield for Illinois

Year	Yield	Year	Yield	Year	Yield
1944	45.5	1955	58.0	1965	88.0
1945	43.0	1956	59.0	1966	82.0
1946	55.0	1957	52.0	1967	96.0
1947	45.0	1958	64.0	1968	103.0
1948	58.0	1959	63.0	1969	95.0
1949	61.0	1960	63.0	1970	93.0
1950	58.0	1961	73.0	1971	92.0
1951	56.0	1962	79.0	1972	97.0
1952	56.0	1963	81.0	1973	105.0
1953	58.0	1964	85.0	1974	86.0
1954	45.0				

SOURCE. U. S. Department of Agriculture, *Crop Production*, various issues.

3. Evaluate (9.3.5) and (9.3.6) for $n=10$ and the model of Exercise 9.1. The covariance between $\Sigma_{t=2}^{n} Z_t Z_{t-1}$ and $\Sigma_{t=1}^{n} Z_t^2$ for the first order autoregressive process can be evaluated using Theorem 6.2.1. Using these approximations and the approximation

$$E\left\{\frac{A}{B}\right\} \doteq \frac{E\{A\}}{E\{B\}} + [E\{B\}]^{-2}\left[\operatorname{Cov}\{A,B\} - \left(\frac{E\{A\}}{E\{B\}}\right)\operatorname{Var}\{B\}\right]$$

find the approximate expected value of $\hat{r}_{\hat{Z}}(1)$ for the model and sample sizes of Exercise 9.1.

4. Using a period of three observations, construct a moving average to remove a linear trend from the third observation and, hence, show that the second difference operator is a multiple of the moving average constructed to remove a linear trend from the third observation in a group of three observations.

5. Define X_t by

$$X_t = \rho X_{t-1} + e_t, \qquad |\rho| < 1,$$

where the e_t are independent $(0, \sigma^2)$ random variables. Find the autocorrelation functions and spectral densities of

$$Y_t = X_t - X_{t-1} \quad \text{and} \quad W_t = X_t - 2X_{t-1} + X_{t-2}.$$

Plot the spectral densities for $\rho = 0.5$.

6. Consider the model

$$Y_t = \beta_0 + \beta_1 t + \beta_2 t^2 + \sum_{i=1}^{4} \alpha_i \delta_{ti} + Z_t,$$

where

$$\delta_{ti} = \begin{cases} 1, & \text{for quarter } i, \quad (i=1,2,3,4) \\ 0, & \text{otherwise,} \end{cases}$$

$$\sum_{i=1}^{4} \alpha_i = 0,$$

and Z_t is a stationary time series. Assuming that the model holds for a period of 20 observations (five years), construct the weights needed to decompose the most recent observation in a set of 20 observations into "trend," "seasonal," and "remainder." Define trend to be $\beta_0+\beta_1 t+\beta_2 t^2$ for the tth observation and seasonal for quarter i to be the α_i associated with that quarter.

7. Obtain the conclusion of Proposition 9.6.3 directly by differencing the autoregressive moving average time series.

8. Prove Lemma 9.5.1.

9. Prove Lemma 9.5.2.

10. Using the data of Exercise 2 and the model of part b of that exercise construct the statistics (9.3.8) and (9.3.9).

11. Assume that the data of Exercise 2 of Chapter 7 satisfy the model

$$Y_t=\sum_{j=1}^{4}\alpha_j\,\delta_{tj}+\beta t+Z_t,$$

where

$$\begin{aligned}\delta_{tj}&=1, &&\text{for quarter } j\\ &=0, &&\text{otherwise,}\end{aligned}$$

Z_t is the stationary autoregressive process

$$Z_t=\theta_1 Z_{t-1}+\theta_2 Z_{t-2}+e_t,$$

and $\{e_t\}$ is a sequence of normal independent $(0,\sigma^2)$ random variables.

(a) Estimate the parameters of the model. Estimate the standard errors of your estimators. Do the data support the model for the error time series?

(b) Using the model of part (a) estimate gasoline consumption for the four quarters of 1974. Construct 95% confidence intervals for your predictions using normal theory. Actual consumption during 1974 was [2223, 2540, 2596, 2529]. What happened?

Computational hint: After estimating the model for the Z_t, but before making the transformation for the final estimation of α_j and β, add four "observations" and four independent variables to the data set. The four added independent variables are defined by

$$\begin{aligned}\varphi_{t,5+i}&=-1, &&\text{for 1974–}i\\ &=\ 0, &&\text{otherwise,}\end{aligned}$$

for $i=1,2,3,4$. The four observations on the original data for 1973 and the four added 'observations' for 1974 are displayed in the table below. The values of the original independent variables for the prediction period are the true values of these independent variables while the values of Y_t for the prediction period are zero.

The augmented data set should then be transformed as per (9.7.2). The regression coefficients for the added independent variables give the predictions for the four quarters of 1974. Furthermore, the regression estimated standard errors of these coefficients are the estimated standard errors of the predictions. The estimated standard errors ignore the error in estimating the coefficients of the model for Z_t but do include a term arising from the error in estimating α_j and β.

Year	Quarter	Dependent variable	δ_{t1}	δ_{t2}	δ_{t3}	δ_{t4}	t	Independent variables for predictions			
1973	1	2454	1	0	0	0	53	0	0	0	0
	2	2647	0	1	0	0	54	0	0	0	0
	3	2689	0	0	1	0	55	0	0	0	0
	4	2549	0	0	0	1	56	0	0	0	0
1974	1	0	1	0	0	0	57	−1	0	0	0
	2	0	0	1	0	0	58	0	−1	0	0
	3	0	0	0	1	0	59	0	0	−1	0
	4	0	0	0	0	1	60	0	0	0	−1

12. Using the Prest data of Table 9.7.1, estimate the parameters of the model

$$Y_t = \beta_0 + \beta_1 t + \beta_2 (t-35)^2 + \beta_3 \varphi_{t1} + \beta_4 \varphi_{t2} + \lambda Y_{t-1} + Z_t,$$

$$Z_t = \rho Z_{t-1} + e_t,$$

where $\{e_t\}$ is a sequence of independent $(0, \sigma^2)$ random variables with $E\{e_t^4\} = \eta\sigma^4$ and $|\rho| < 1$, $|\lambda| < 1$.

13. Let Y_t satisfy

$$Y_t = \alpha\varphi_t + Z_t,$$

$$Z_t = e_t + \beta e_{t-1}, \qquad t = 1, 2, \ldots,$$

where $|\beta| < 1$, $\{e_t\}$ is a sequence of independent $(0, \sigma^2)$ random variables with $E\{e_t^4\} = \eta\sigma^4$, and $\{\varphi_t\}$ satisfies assumptions (9.1.6), (9.1.7), and (9.1.8). Let

$$\hat{\alpha}_S = \left(\sum_{t=1}^{n} \varphi_t^2 \right)^{-1} \sum_{t=1}^{n} \varphi_t Y_t,$$

$$\hat{Z}_t = Y_t - \hat{\alpha}_S \varphi_t,$$

$$\begin{aligned} e_t(\hat{Z}; \tilde{\tilde{\beta}}) &= 0, && t = 1 \\ &= \hat{Z}_t - \tilde{\tilde{\beta}} e_{t-1}(\hat{Z}; \tilde{\tilde{\beta}}), && t = 2, 3, \ldots, n, \end{aligned}$$

$$\begin{aligned} W_t(Y; \tilde{\tilde{\beta}}) &= 0, && t = 1 \\ &= e_{t-1}(\hat{Z}; \tilde{\tilde{\beta}}) - \tilde{\tilde{\beta}} W_{t-1}(Y; \tilde{\tilde{\beta}}), && t = 2, 3, \ldots, n, \end{aligned}$$

and $\tilde{\tilde{\beta}}$ is constructed from the first order autocorrelation of the $\hat{Z}_t$ by rule (8.3.1) and restricted to $(-1, 1)$.

(a) Show that

$$(\hat{\hat{\beta}} - \hat{\beta}) = O_p(n^{-1}),$$

where

$$\hat{\hat{\beta}} = \tilde{\tilde{\beta}} + \Delta\hat{\hat{\beta}},$$

$$\Delta\hat{\hat{\beta}} = \frac{\sum_{t=1}^{n} e_t(\hat{Z};\tilde{\tilde{\beta}})W_t(\hat{Z};\tilde{\tilde{\beta}})}{\sum_{t=1}^{n}\left[W_t(\hat{Z};\tilde{\tilde{\beta}})\right]^2},$$

and $\hat{\beta}$ is the analogous estimator constructed by replacing $\hat{Z}$ by Z.

(b) Let $\hat{\boldsymbol{\epsilon}} = \hat{\mathbf{T}}\mathbf{z}$, where

$$\begin{aligned} \hat{\epsilon}_t = \hat{\mathbf{T}}_{t.}\mathbf{z} &= \left(1+\hat{\hat{\beta}}^2\right)^{-1} Z_1, & t&=1 \\ &= Z_t - \hat{\hat{\beta}}\hat{\epsilon}_{t-1}, & t&=2,3,\ldots,n, \end{aligned}$$

and $\hat{\mathbf{T}}_{t.}$ is the tth row of $\hat{\mathbf{T}}$. Show that

$$\left[\sum_{t=1}^{n}\varphi_t^2\right]^{1/2}(\hat{\alpha}_G - \tilde{\alpha}) = O_p(n^{-1/2}),$$

where

$$\tilde{\alpha} = [\boldsymbol{\Phi}'\hat{\mathbf{T}}'\hat{\mathbf{T}}\boldsymbol{\Phi}]^{-1}\boldsymbol{\Phi}'\hat{\mathbf{T}}'\hat{\mathbf{T}}\mathbf{y},$$

$$\hat{\alpha}_G = [\boldsymbol{\Phi}'\mathbf{V}^{-1}\boldsymbol{\Phi}]^{-1}\boldsymbol{\Phi}'\mathbf{V}^{-1}\mathbf{y},$$

and

$$\mathbf{V} = E\{\mathbf{z}\mathbf{z}'\}.$$

14. Let $\{Y_t\}$ satisfy

$$Y_t = \sum_{i=0}^{r}\beta_i\varphi_{ti} + \sum_{j=1}^{s}\lambda_j Y_{t-j} + e_t, \qquad t=1,2,\ldots,$$

where $\{e_t\}$ is a sequence of independent $(0,\sigma^2)$ random variables with $E\{e_t^4\} = \eta\sigma^4$ and the roots of $m^s - \sum_{j=1}^{s}\lambda_j m^{s-j} = 0$ are less than one in absolute value. State and prove Theorem 9.8.1 for this model.

Bibliography

Ahlberg, J. H., Nilson, E. N., and Walsh, J. L. (1967), *The Theory of Splines and Their Applications.* Academic Press, New York.

Aigner, D. J. (1971), A compendium on estimation of the autoregressive moving average model from time series data. *International Economic Review* **12**, 348–371.

Amemiya, T. (1973), Generalized least squares with an estimated autocovariance matrix. *Econometrica* **41**, 723–732.

Amemiya, T., and Fuller, W. A. (1967), A comparative study of alternative estimators in a distributed lag model. *Econometrica* **35**, 509–529.

Amos, D. E., and Koopmans, L. H. (1963), Tables of the distribution of the coefficient of coherence for stationary bivariate Gaussian processes. Sandia Corporation Monograph SCR–483, Albuquerque, New Mexico.

Anderson, R. L. (1942), Distribution of the serial correlation coefficient. *Ann. Math. Statist.* **13**, 1–13.

Anderson, R. L., and Anderson, T. W. (1950), Distribution of the circular serial correlation coefficient for residuals from a fitted Fourier series. *Ann. Math. Statist.* **21**, 59–81.

Anderson, T. W. (1958), *Introduction to Multivariate Statistical Analysis.* Wiley, New York.

Anderson, T. W. (1959), On asymptotic distributions of estimates of parameters of stochastic difference equations. *Ann. Math. Statist.* **30**, 676–687.

Anderson, T. W. (1962), The choice of the degree of a polynomial regression as a multiple decision problem. *Ann. Math. Statist.* **33**, 255–265.

Anderson, T. W. (1971), *The Statistical Analysis of Time Series.* Wiley, New York.

Anderson, T. W., and Walker, A. M. (1964), On the asymptotic distribution of the autocorrelations of a sample from a linear stochastic process. *Ann. Math. Statist.* **35**, 1296–1303.

Ball, R. J., and St.Cyr, E. B. A. (1966), Short term employment functions in British manufacturing industry. *Review of Economic Studies* **33**, 179–207.

Bartle, R. G. (1964), *The Elements of Real Analysis.* Wiley, New York.

Bartlett, M. S. (1946), On the theoretical specification and sampling properties of autocorrelated time series. *Suppl. J. Roy. Statist. Soc.* **8**, 27–41.

Bartlett, M. S. (1950), Periodogram analysis and continuous spectra. *Biometrika* **37**, 1–16.

Bartlett, M. S. (1955), *An Introduction to Stochastic Processes with Special Reference to Methods and Applications.* (Second Edition, 1966), Cambridge University Press, Cambridge.

Basmann, R. L. (1957), A generalized classical method of linear estimation of coefficients in a structural equation. *Econometrica* **25**, 77–83.

Bellman, R. (1960), *Introduction to Matrix Analysis.* McGraw-Hill, New York.

Berk, K. N. (1974), Consistent autoregressive spectral estimates. *Ann. Statist.* **2**, 489–502.

Bernstein, S. (1927), Sur l'extension du théorème limite du calcul des probabilités aux sommes de quantitiés dépendantes. *Mathematische Annalen* **97**, 1–59.

Birnbaum, Z. W. (1952), Numerical tabulation of the distribution of Kolomogorov's statistic for finite sample size. *J. Amer. Statist. Assoc.* **47**, 425–441.

Blackman, R. B., and Tukey, J. W. (1959), *The Measurement of Power Spectra from the Point of View of Communications Engineering.* Dover Publications, Inc., New York.

Box, G. E. P. (1954), Some theorems on quadratic forms applied in the study of analysis of variance problems, I. Effect of inequality of variance in the one-way classification. *Ann. Math. Statist.* **25**, 290–316.

Box, G. E. P., and Jenkins, G. M. (1970), *Time Series Analysis Forecasting and Control.* Holden-Day, San Francisco.

Box, G. E. P., and Pierce, D. A. (1970), Distribution of residual autocorrelations in autoregressive-integrated moving average time series models. *J. Amer. Statist. Assoc.* **65**, 1509–1526.

Brillinger, D. R. (1975), *Time Series: Data Analysis and Theory.* Holt, Rinehart and Winston, New York.

Brown, R. G. (1962), *Smoothing, Forecasting and Prediction of Discrete Time Series.* Prentice-Hall, Englewood Cliffs, N. J.

Burman, J. P. (1965), Moving seasonal adjustment of economic time series. *J. Roy. Statist. Soc. Ser. A* **128**, 534–558.

Chernoff, H. (1956), Large sample theory: parametric case. *Ann. Math. Statist.* **27**, 1–22.

Chung, K. L. (1968), *A Course in Probability Theory.* Harcourt, Brace and World, New York.

Cleveland, W. S. (1971), Fitting time series models for prediction. *Technometrics* **13**, 713–723.

Cochran, W. T. et al. (1967), What is the fast Fourier transform? *IEEE Transactions on Audio and Electroacoustics*, **AU–15**, No. 2: 45–55.

Cochrane, D., and Orcutt, G. H. (1949), Application of least squares regression to relationships containing autocorrelated error terms. *J. Amer. Statist. Assoc.* **44**, 32–61.

Cooley, J. W., Lewis, P. A. W., and Welch, P. D. (1967), Historical notes on the Fast Fourier Transform. *IEEE Transactions on Audio and Electroacoustics*, **AU–15**, No. 2: 76–79.

Cox, D. R., and Miller, H. D. (1965), *The Theory of Stochastic Processes.* Wiley, New York.

Cramér, H. (1940), On the theory of stationary random processes. *Annals of Mathematics* **41**, 215–230.

Cramér, H. (1946), *Mathematical Methods of Statistics.* Princeton University Press, Princeton., N. J.

Davis, H. T. (1941), *The Analysis of Economic Time Series.* Principia Press, Bloomington, Ind.

DeGracie, J. S., and Fuller, W. A. (1972), Estimation of the slope and analysis of covariance when the concomitant variable is measured with error. *J. Amer. Statist. Assoc.* **67**, 930–937.

Dhrymes, P. J. (1971), *Distributed Lags: Problems of Estimation and Formulation.* Holden-Day, San Francisco.

Diananda, P. H. (1953), Some probability limit theorems with statistical applications. *Proc. Cambridge Philos. Soc.* **49**, 239–246.

Dickey, D. A. (1975), Hypothesis testing for nonstationary time series. Unpublished manuscript. Iowa State University, Ames, Iowa.

Draper, N. R., and Smith, H. (1966), *Applied Regression Analysis.* Wiley, New York.

Durbin, J. (1959), Efficient estimation of parameters in moving-average models. *Biometrika* **46**, 306–316.

Durbin, J. (1960), Estimation of parameters in time series regression models. *J. Roy. Statist. Soc. Ser. B* **22**, 139–153.

Durbin, J. (1961), Efficient fitting of linear models for continuous stationary time series from discrete data. *Bull. Int. Statist. Inst.* **38**, 273–281.

Durbin, J. (1962), Trend elimination by moving-average and variate-difference filters. *Bull. Int. Statist. Inst.* **39**, 131–141.

Durbin, J. (1963), Trend elimination for the purpose of estimating seasonal and periodic components of time series. In *Proc. Symp. Time Series Anal. Brown University.* (Rosenblatt, M., Ed.), Wiley, New York.

Durbin, J. (1967), Tests of serial independence based on the cumulated periodogram. *Bull. Int. Statist. Inst.* **42**, 1039–1049.

Durbin, J. (1968), The probability that the sample distribution function lies between two parallel straight lines. *Ann. Math. Statist.* **39**, 398–411.

Durbin, J. (1969), Tests for serial correlation in regression analysis based on the periodogram of least-squares residuals. *Biometrika* **56**, 1–15.

Durbin, J., and Watson, G. S. (1950), Testing for serial correlation in least squares regression, I. *Biometrika* **37**, 409–428.

Durbin, J., and Watson, G. S. (1951), Testing for serial correlation in least squares regression, II. *Biometrika* **38**, 159–178.

Eicker, F. (1963), Asymptotic normality and consistency of the least squares estimators for families of linear regressions. *Ann. Math. Statist.* **34**, 447–456.

Eicker, F. (1967), Limit theorems for regressions with unequal and dependent errors. *Proc. Fifth Berkeley Sympos. Math. Śtatist. and Probability* **1** Statistics, 59–82. University of California Press, Berkeley.

Eisenhart, C., Hastay, M. W., and Wallis, W. A. (1947), *Selected Techniques of Statistical Analysis.* McGraw-Hill, New York.

Ezekiel, M., and Fox, K. A. (1959), *Methods of Correlation and Regression Analysis.* Wiley, New York.

Fieller, E. C. (1932), The distribution of the index in a normal bivariate population. *Biometrika* **24**, 428–440.

Fieller, E. C. (1954), Some problems in interval estimation. *J. Roy. Statist. Soc. Ser. B* **26**, 175–185.

Finkbeiner, D. T. (1960), *Introduction to Matrices and Linear Transformations.* W. H. Freeman, San Francisco.

Fisher, R. A. (1929), Tests of significance in harmonic analysis. *Proc. Roy. Soc. London Ser. A* **125**, 54–59.

Fishman, G. S. (1969), *Spectral Methods in Econometrics.* Harvard University Press, Cambridge, Mass.

Fuller, W. A. (1969), Grafted polynomials as approximating functions. *Australian J. of Agr. Econ.* **13**, 35–46.

Fuller, W. A., and Martin, J. E. (1961), The effects of autocorrelated errors on the statistical estimation of distributed lag models. *J. Farm Econ.* **43**, 71–82.

Gallant, A. R. (1971), *Statistical Inference for Nonlinear Regression Models.* Ph.D. thesis, Iowa State University, Ames, Iowa.

Gallant, A. R. (1975), Nonlinear regression. *The American Statistician* **29**, 73–81.

Gallant, A. R., and Fuller, W. A. (1973), Fitting segmented polynomial regression models whose join points have to be estimated. *J. Amer. Statist. Assoc.* **68**, 144–147.

Gnedenko, B. V. (1967), *The Theory of Probability.* (Translated by B. D. Seckler), Chelsea, New York.

Goldberg, S. (1958), *Introduction to Difference Equations.* Wiley, New York.

Granger, C. W., and Hatanaka, M. (1964), *Spectral Analysis of Economic Time Series.* Princeton University Press, Princeton, N. J.

Grenander, U. (1954), On the estimation of regression coefficients in the case of an autocorrelated disturbance. *Ann. Math. Statist.* **25**, 252–272.

Grenander, U. (1957), Modern trends in time series analysis. *Sankhyā* **18**, 149–158.

Grenander, U., and Rosenblatt, M. (1957), *Statistical Analysis of Stationary Time Series.* Wiley, New York.

Grether, D. M., and Nerlove, M. (1970), Some properties of "Optimal" seasonal adjustment. *Econometrica* **38**, 682–703.

Greville, T. N. E. (1969), *Theory and Applications of Spline Functions.* Academic Press, New York.

Griliches, Z. (1957), Distributed lags: a survey. *Econometrica* **35**, 16–49.

Grizzle, J. E., and Allen, D. M. (1969), Analysis of growth and dose response curves. *Biometrics* **25**, 357–381.

Hannan, E. J. (1958), The estimation of spectral density after trend removal. *J. Roy. Statist. Soc. Ser. B* **20**, 323–333.

Hannan, E. J. (1960), *Time Series Analysis.* Methuen, London.

Hannan, E. J. (1961), A central limit theorem for systems of regressions. *Proc. Cambridge Phil. Soc.* **57**, 583–588.

Hannan, E. J. (1963), The estimation of seasonal variation in economic time series. *J. Amer. Statist. Assoc.* **58**, 31–44.

Hannan, E. J. (1969), A note on an exact test for trend and serial correlation. *Econometrica* **37**, 485–489.

Hannan, E. J. (1970), *Multiple Time Series.* Wiley, New York.

Hannan, E. J., and Heyde, C. C. (1972), On limit theorems for quadratic functions of discrete time series. *Ann. Math. Statist.* **43**, 2058–2066.

Hansen, M. H., Hurwitz, W. N., and Madow, W. G. (1953), *Sample Survey Methods and Theory, II.* Wiley, New York.

Harris, B. (1967), *Spectral Analysis of Time Series.* Wiley, New York.,

Hart, B. I. (1942), Significance levels for the ratio of the mean square successive difference to the variance. *Ann. Math. Statist.* **13**, 445–447.

Hartley, H. O. (1961) The modified Gauss-Newton method for the fitting of non-linear regression functions by least squares. *Technometrics* **3**, 269–280.

Hartley, H. O., and Booker, A. (1965), Nonlinear least squares estimation. *Ann. Math. Statist.* **36**, 638–650.

Hatanaka, M. (1973), On the existence and the approximation formulae for the moments of the k-class estimators. *The Economic Studies Quarterly,* **24**, 1–15.

Hatanaka, M. (1974), An efficient two-step estimator for the dynamic adjustment model with autoregressive errors. *J. Econometrics* **2**, 199–220.

Hildebrand, F. B. (1968), *Finite-Difference Equations and Simulations.* Prentice-Hall, Englewood Cliffs, N. J.

Hildreth, C. (1969), Asymptotic distribution of maximum likelihood estimators in a linear model with autoregressive disturbances. *Ann. Math. Statist.* **40**, 583–594.

Hoeffding, W., and Robbins, H. (1948), The central limit theorem for dependent random variables. *Duke Math. J.* **15**, 773–780.

Jenkins, G. M. (1961), General considerations in the analysis of spectra. *Technometrics* **3**, 133–166.

Jenkins, G. M., and Watts, D. G. (1968), *Spectral Analysis and its Application.* Holden-Day, San Francisco.

Jennrich, R. I. (1969), Asymptotic properties of non-linear least squares estimators. *Ann. Math. Statist.* **40**, 633–643.

Jobson, J. D. (1972), *Estimation for Linear Models with Unknown Diagonal Covariance Matrix.* Unpublished Ph.D. thesis, Iowa State University, Ames, Iowa.

Jorgenson, D. W. (1964), Minimum variance, linear, unbiased seasonal adjustment of economic time series. *J. Amer. Statist. Assoc.* **59**, 681–724.

Kempthorne, O., and Folks, L. (1971), *Probability, Statistics and Data Analysis.* Iowa State University Press, Ames, Iowa.

Kendall, M. G. (1954), Note on bias in the estimation of autocorrelation. *Biometrika* **41**, 403–404.

Kendall, M. G., and Stuart, A. (1966), *The Advanced Theory of Statistics* **3**, Hafner, New York.

Kendall, M. G., and Stuart, A. (1967), *The Advanced Theory of Statistics* **2**, Hafner, New York.

Kendall, M. G., and Stuart, A. (1969), *The Advanced Theory of Statistics* **1**, Hafner, New York.

Koopmans, L. H. (1974), *The Spectral Analysis of Time Series.* Academic Press, New York.

Koopmans, T. (1942), Serial correlation and quadratic forms in normal variables. *Ann. Math. Statist.* **13**, 14–33.

Kramer, K. H. (1963), Tables for constructing confidence limits on the multiple correlation coefficient. *J. Amer. Statist. Assoc.* **58**, 1082–1085.

Lighthill, M. J. (1970), *Introduction to Fourier Analysis and Generalised Functions.* Cambridge University Press, Cambridge.

Liviatan, N. (1963), Consistent estimation of distributed lags. *International Economic Review* **4**, 44–52.

Loève, M. (1963), *Probability Theory* (Third Edition), Van Nostrand, Princeton, N. J.

Lomnicki, Z. A., and Zaremba, S. K. (1957), On estimating the spectral density function of a stochastic process. *J. Roy. Statist. Soc. Ser. B* **19**, 13–37.

Lovell, M. C. (1963), Seasonal adjustment of economic time series and multiple regression analysis. *J. Amer. Statist. Assoc.* **58**, 993–1010.

Macpherson, B. D. (1975), Empirical distribution of alternative estimators of moving average parameters. Unpublished research. University of Manitoba, Winnipeg, Canada.

Malinvaud, E. (1970a), *Statistical Methods in Econometrics.* North-Holland, Amsterdam.

Malinvaud, E. (1970b), The consistency of non-linear regressions. *Ann. Math. Statist.* **41**, 956–969.

Mann, H. B., and Wald, A. (1943a), On the statistical treatment of linear stochastic difference equations. *Econometrica* **11**, 173–220.

Mann, H. B., and Wald, A. (1943b), On stochastic limit and order relationships. *Ann. Math. Statist.* **14**, 217–226.

Mariott, F. H. C., and Pope, J. A. (1954), Bias in the estimation of autocorrelations. *Biometrika* **41**, 390–402.

McClave, J. T. (1974), A comparison of moving average estimation procedures. *Comm. Statist.* **3**, 865–883.

Miller, K. S. (1963), *Linear Differential Equations in the Real Domain.* W. W. Norton, New York.

Miller, K. S. (1968), *Linear Difference Equations.* W. A. Benjamin, New York.

Moran, P. A. P. (1947), Some theorems on time series I. *Biometrika* **34**, 281–291.

Nelson, C. R. (1974), The first-order moving average process: identification, estimation and prediction. *J. Econometrics* **2**, 121–141.

Nerlove, M. (1964), Notes on Fourier series and analysis for economists. Multilith prepared under Grant NSF-GS-142 to Stanford University. Palo Alto, Calif.

Nerlove, M. (1964), Spectral analysis of seasonal adjustment procedures. *Econometrica* **32**, 241–286.

Nicholls, D. F., Pagan, A. R., and Terrell, R. D. (1975), The estimation and use of models with moving average disturbance terms: a survey. *International Economic Review* **16**, 113–134.

Nold, F. C. (1972), A bibliography of applications of techniques of spectral analysis to economic time series. Technical Report No. 66. Institute for Mathematical Studies in the Social Sciences, Stanford University, Stanford, Calif.

Olshen, R. A. (1967), Asymptotic properties of the periodogram of a discrete stationary process. *J. Appl. Prob.* **4**, 508–528.

Orcutt, G. H., and Winokur, H. S. (1969), First order autoregression: inference, estimation and prediction. *Econometrica* **37**, 1–14.

Otnes, R. K., and Enochson, L. (1972), *Digital Time Series Analysis.* Wiley, New York.

Parzen, E. (1957), On consistent estimates of the spectrum of a stationary time series. *Ann. Math. Statist.* **28**, 329–348.

Parzen, E. (1958), Conditions that a stochastic process be ergodic. *Ann. Math. Statist.* **29**, 299–301.

Parzen, E. (1961), Mathematical considerations in the estimation of spectra. *Technometrics* **3**, 167–190.

Parzen, E. (1962), *Stochastic Processes.* Holden-Day, San Francisco.

Parzen, E. (1969), Multiple Time Series Modeling. In *Multivariate Analysis II* (Kreshnaiah, P. R., Ed.), Academic Press, New York.

Pierce, D. A. (1970a), A duality between autoregressive and moving average processes concerning their least squares parameter estimates. *Ann. Math. Statist.* **41**, 422–426.

Pierce, D. A. (1970b), Rational distributed lag models with autoregressive-moving average errors. *Proceedings of the Business and Economics Statistics Section of the American Statistical Association.* 591–596.

Pierce, D. A. (1971), Least squares estimation in the regression model with autoregressive-moving average errors. *Biometrika* **58**, 299–312.

Pierce, D. A. (1972), Least squares estimation in dynamic-disturbance time series models. *Biometrika* **59**, 73–78.

Poirier, D. J. (1973), Piecewise regression using cubic splines. *J. Amer. Statist. Assoc.* **68**, 515–524.

Pratt, J. W. (1959), On a general concept of "in probability." *Ann. Math. Statist.* **30**, 549–558.

Prest, A. R. (1949), Some experiments in demand analysis. *Review of Economics and Statistics* **31**, 33–49.

Quenouille, M. H. (1957), *The Analysis of Multiple Time Series*. Hafner, New York.

Rao, C. R. (1959), Some problems involving linear hypothesis in multivariate analysis. *Biometrika* **46**, 49–58.

Rao, C. R. (1965a), *Linear Statistical Inference and its Applications.* Wiley, New York.

Rao, C. R. (1965b), The theory of least squares when the parameters are stochastic and its application to the analysis of growth curves. *Biometrika* **52**, 447–458.

Rao, C. R. (1967), Least squares theory using an estimated dispersion matrix and its application to measurement of signals. *Fifth Berkeley Symposium* **1**, 355–372.

Rao. J. N. K., and Tintner, G. (1963). On the variate difference method. *Australian J. of Statist.* **5**, 106–116.

Rao, M. M. (1961), Consistency and limit distributions of estimators of parameters in explosive stochastic difference equations. *Ann. Math. Statist.* **32**, 195–218.

Reeves, J. E. (1972), The distribution of the maximum likelihood estimator of the parameter in the first-order autoregressive series. *Biometrika* **59**, 387–394.

Rogosinski, W. (1959), *Fourier Series.* (Translated by H. Cohn and F. Steinhardt), Chelsea, New York.

Rosenblatt, M. (1963), Proceedings of the Symposium on Time Series Analysis held at Brown University, June 11–14, 1962, Wiley, New York.

Royden, H. L. (1963), *Real Analysis.* Macmillan, New York.

Rubin, H. (1950), Consistency of maximum-likelihood estimates in the explosive case. In *Statistical Inference in Dynamic Economic Models* (Koopmans, T. C., Ed.), Wiley, New York.

Salem, A. S. (1971), *Investigation of Alternative Estimators of the Parameters of Autoregressive Processes.* Unpublished MS. thesis, Iowa State University, Ames, Iowa.

Sargan, J. D. (1958), The estimation of economic relationships using instrumental variables. *Econometrica* **26**, 393–415.

Sargan, J. D. (1959), The estimation of relationships with autocorrelated residuals by the use of instrumental variables. *J. Roy. Statist. Soc. Ser. B* **21**, 91–105.

Scheffé, H. (1970), Multiple testing versus multiple estimation. Improper confidence sets. Estimation of directions and ratios. *Ann. Math. Statist.* **41**, 1–29.

Scott, D. J. (1973), Central limit theorems for martingales and for processes with stationary increments using a Skorokhod representation approach. *Advances in Applied Probability* **5**, 119–137.

Singleton, R. C. (1969), An algorithm for computing the mixed radix fast Fourier thransform. *IEEE Transactions on Audio and Electroacoustics* **AU–17**, No. 2, 93–103.

Stephenson, J. A., and Farr, H. T. (1972), Seasonal adjustment of economic data by application of the general linear statistical model. *J. Amer. Statist. Assoc.* **67**, 37–45.

Stigum, B. P. (1974), Asymptotic properties of dynamic stochastic parameter estimates (III). *J. Multivariate Analysis* **4**, 351–381.

Theil, H. (1961), *Economic Forecasts and Policy*. North-Holland, Amsterdam.

Thompson, L. M. (1969), Weather and technology in the production of wheat in the United States. *J. of Soil and Water Conservation* **24**, No. 6: 219–224.

Thompson, L. M. (1973), Cyclical weather patterns in the middle latitudes. *J. of Soil and Water Conservation* **28**, No. 2: 87–89.

Tintner, G. (1940), *The Variate Difference Method*. Principia Press, Bloomington, Ind.

Tintner, G. (1952), *Econometrics*. Wiley, New York.

Tolstov, G. P. (1962), *Fourier Series*. (Translated by R. A. Silverman), Prentice-Hall, Englewood Cliffs, N.J.

Tucker, H. G. (1967), *A Graduate Course in Probability*. Academic Press, New York.

Tukey, J. W. (1961), Discussion, emphasizing the connection between analysis of variance and spectrum analysis. *Technometrics* **3**, 1–29.

Varadarajan, V. S. (1958), A useful convergence theorem. *Sankhyā* **20**, 221–222.

Von Neumann, J. (1941), Distribution of ratio of the mean square successive difference to the variance. *Ann. Math. Statist.* **12**, 367–395.

Von Neumann, J. (1942), A further remark concerning the distribution of the ratio of the mean square successive difference to the variance. *Ann. Math. Statist.* **13**, 86–88.

Wahba, G. (1968), On the distribution of some statistics useful in the analysis of jointly stationary time series. *Ann. Math. Statist.* **39**, 1849–1862.

Walker, A. M. (1961), Large-sample estimation of parameters for moving -average models. *Biometrika* **48**, 343–357.

Wallis, K. F. (1967), Lagged dependent variables and serially correlated errors: a reappraisal of three pass least squares. *Rev. Econ. Statist.* **49**, 555–567.

Wallis, K. F. (1972), Testing for fourth order autocorrelation in quarterly regression equations. *Econometrica* **40**, 617–638.

Watson, G. S. (1955), Serial correlation in regression analysis I. *Biometrika* **42**, 327–341.

Watson, G. S. and Hannan, E. J. (1956), Serial correlation in regression analysis II. *Biometrika* **43**, 436–445.

White, J. S. (1958), The limiting distribution of the serial correlation coefficient in the explosive case. *Ann. Math. Statist.* **29**, 1188–1197.

Whittle, P. (1951), *Hypothesis Testing in Time Series Analysis.* Almqvist and Wiksell, Uppsala.

Whittle, P. (1953), The analysis of multiple stationary time series. *J. Roy. Statist. Soc. Ser. B* **15**, 125–139.

Whittle, P. (1963), *Prediction and Regulation by Linear Least-Square Methods.* Van Nostrand, Princeton, N. J.

Wilks, S. S. (1962), *Mathematical Statistics.* Wiley, New York.

Williams, E. J. (1959), *Regression Analysis.* Wiley, New York.

Wold, H. O. A. (1938), *A Study in the Analysis of Stationary Time Series* (Second Edition, 1954), Almqvist and Wiksell, Uppsala.

Wold, H. O. A. (1965), *Bibliography on Time Series and Stochastic Processes.* Oliver and Boyd Ltd., Edinburgh.

Yaglom, A. M. (1962), *An Introduction to the Theory of Stationary Random Functions.* (Translated by R. A. Silverman), Prentice-Hall, Englewood Cliffs, N. J.

Yule, G. U. (1927), On a method of investigating periodicities in disturbed series with special reference to Wolfer's sunspot numbers. *Philos. Trans. Roy. Soc. London Ser. A* **226**, 267–298.

Zygmund, A. (1959), *Trigonometric Series.* Cambridge University Press, Cambridge.

Index

Applied Probability and Statistics (Continued)

GROSS and CLARK · Survival Distributions: Reliability Applications in the Biomedical Sciences

GROSS and HARRIS · Fundamentals of Queueing Theory

GUTTMAN, WILKS, and HUNTER · Introductory Engineering Statistics, *Second Edition*

HAHN and SHAPIRO · Statistical Models in Engineering

HALD · Statistical Tables and Formulas

HALD · Statistical Theory with Engineering Applications

HARTIGAN · Clustering Algorithms

HILDEBRAND, LAING, and ROSENTHAL · Prediction Analysis of Cross Classifications

HOEL · Elementary Statistics, *Fourth Edition*

HOLLANDER and WOLFE · Nonparametric Statistical Methods

HUANG · Regression and Econometric Methods

JAGERS · Branching Processes with Biological Applications

JESSEN · Statistical Survey Techniques

JOHNSON and KOTZ · Distributions in Statistics
Discrete Distributions
Continuous Univariate Distributions-1
Continuous Univariate Distributions-2
Continuous Multivariate Distributions

JOHNSON and KOTZ · Urn Models and Their Application: An Approach to Modern Discrete Probability Theory

JOHNSON and LEONE · Statistics and Experimental Design in Engineering and the Physical Sciences, Volumes I and II, *Second Edition*

KEENEY and RAIFFA · Decisions with Multiple Objectives

LANCASTER · An Introduction to Medical Statistics

LEAMER · Specification Searches: Ad Hoc Inference with Nonexperimental Data

McNEIL · Interactive Data Analysis

MANN, SCHAFER, and SINGPURWALLA · Methods for Statistical Analysis of Reliability and Life Data

MEYER · Data Analysis for Scientists and Engineers

OTNES and ENOCHSON · Digital Time Series Analysis

PRENTER · Splines and Variational Methods

RAO and MITRA · Generalized Inverse of Matrices and Its Applications

SARD and WEINTRAUB · A Book of Splines

SEARLE · Linear Models

THOMAS · An Introduction to Applied Probability and Random Processes

WHITTLE · Optimization under Constraints

WILLIAMS · A Sampler on Sampling

WONNACOTT and WONNACOTT · Econometrics

WONNACOTT and WONNACOTT · Introductory Statistics, *Third Edition*

WONNACOTT and WONNACOTT · Introductory Statistics for Business and Economics, *Second Edition*

YOUDEN · Statistical Methods for Chemists

ZELLNER · An Introduction to Bayesian Inference in Econometrics